Aquatic Oligochaetes

Developments in Hydrobiology 139

Series editor
H. J. Dumont

Aquatic Oligochaetes

Proceedings of the 7th International Symposium
on Aquatic Oligochaetes
held in Presque Isle, Maine, USA, 18–22 August 1997

Edited by

B.M. Healy, T.B. Reynoldson and K.A. Coates

Reprinted from Hydrobiologia, volume 406 (1999)

Springer Science+Business Media, B.V.

Library of Congress Cataloging-in-Publication Data

A C.I.P. Catalogue record for this book is available from the Library of Congress.

ISBN 978-94-010-5829-2 ISBN 978-94-011-4207-6 (eBook)
DOI 10.1007/978-94-011-4207-6

Printed on acid-free paper

Originally published by Kluwer Academic Publishers in 1999
Softcover reprint of the hardcover 1st edition 1999

Hydrobiologia **406:** v–vii, 1999.
B. M. Healy, T. B. Reynoldson & K. A. Coates (eds), Aquatic Oligochaetes.

Contents

Distribution, abundance and habitats

Population dynamics

Histology and ultrastructure

Development, physiology and behaviour

Pollution studies

Progress Report

Preface

The Seventh International Symposium on Aquatic Oligochaetes (SISAO) was organized by Stuart R. Gelder of the University of Maine at Presque Isle with assistance from Mary Lawrence, Susan Peterson, Joy Rayburn and Keith Madore, and support from the local organizing committee of John Badger, James Dwyer, Dave Lavway and Michael Corey. Additional help came from colleagues and friends around the world. As a result of this team's effort, everyone agreed the Symposium was a success.

Over 400 invitations were mailed and 78 scientists from 30 countries expressed interest in attending. However, illness and financial restrictions reduced these numbers so that only 50 scientists plus 9 guests from 20 countries were finally able to attend. A more realistic representation in terms of science can be gathered from the 53 papers presented involving 90 authors.

Following, and developing, the themes used at the seventh symposium in Tjärnö, Sweden, three guest presentations, and four workshops, in addition to the traditional research papers were arranged. The presentation by Ralph O. Brinkhurst entitled, 'Retrospective and Perspective', summarized the history and contributions of workers on aquatic oligochaetes with comments on possible future trends. This was a timely work to present at the SISAO meeting for two reasons. Firstly, 20 years have passed since the need for such a forum on aquatic oligochaete research was recognized, and secondly, it marks Ralph's retirement after some 40 years involvement with aquatic oligochaete research. Clive A. Edwards was invited to give an over-view on the biology of earthworms, particularly with respect to agriculture. This aspect of clitellate study is often ignored by workers in the aquatic field, and the topic was particularly appropriate given the importance of farming in Aroostook County. The third speaker, Kristian Fauchald, reviewed current developments in morphological and molecular research, particularly on polychaetes, as a means of characterizing a hypothetical ancestral annelid. Both of these presenters felt that readers interested in their speeches would be best served by obtaining copies of their recent major review papers.

The workshop on enchytreaid anatomical nomenclature chaired by Brenda Healy and Rüdiger M. Schmelz, soon expanded into a discussion on homologous and analogous structures in clitellates. One result of the discussion by the members was to adopt the term 'chaeta/ chaetae' for the bristle-like appendages in annelids. This enables the use of 'seta/ setae' to be used for arthropods etc., and thus these analogous characters become clearly separated in a data matrix. A seminar on the phylogenetic relationships within the Clitellata was held at the sixth symposium in Tjärnö, Sweden, and although an enthusiastic discussion ensued no new consensus of opinion emerged. Given the importance of this topic and the amount of new information, a continuation seminar was included in the current symposium entitled, 'Clitellate Phylogeny – round 2', chaired by Christer Erséus. The theme of the discussion rapidly focused on the relationships between the oligochaetes and leeches, and in particular the lumbriculids and branchiobdellidans. In order for the background and current research results to be made available to a broader readership, Brinkhurst was asked by the organizer to submit the text of his presentation to the Proceedings for publication, "Lumbriculids, branchiobdellidans and leeches: an overview of recent progress in phylogenetic research on clitellates". The evening ended with a default acceptance, for the time being, of the current taxonomic arrangement of the Clitellata (paraphyletic 'oligochaetes', and the debatable position of branchiobdellidans alone, with the latter or the leeches). This acceptance was based on the fact that no significant body of new evidence had been found since the last meeting three years ago, and that important information is anticipated from a number of research groups currently working on genetic sequencing that involves a large numbers of clitellate species. The microscope workshop enabled members to examine specimens, and particularly type material brought from China and Europe to the meeting. The workshop on 'whirling disease in salmonid and other fishes' chaired by Deedee Kathman ended the Symposium with the same full attendance and vigorous enthusiasm as the opening sessions.

The Business Meeting was chaired by Ralph O. Brinkhurst and the members received an invitation and slide presentation from Pilar Rodriguez, Universidad del Pais Vasco, Bilbao, Spain, for the Eighth International Symposium on Aquatic Oligochaetes to be held in Bilbao in July 2000. This was unanimously adopted by the membership. Two tentative proposals were presented for the meeting to be held in 2003. Piet F. M. Verdonschot proposed the symposium be held in Wageningen, The Netherlands, and Marco Ferraguti conveyed an invitation from Giulio Lanzavecchia and his colleagues at the University of Varese, Italy. The majority of the membership voted to accept the proposal for The Netherlands. The location of the 2003 meeting will be confirmed at the Eighth Symposium in Bilbao.

Financial support for the symposium came primarily from contributions made by the University of Maine at Presque Isle, Maine Potato Board, Maine Public Service, Presque Isle Rotary Club and MFX, along with other donations. The field trip to Sinclair and Long Lake, northern Maine, was completely supported and coordinated through the St. John Aroostook Resource Conservation and Development organization and the Fish River Lakes Water Quality Association. Attendees would like to thank all of these organizations and individuals for their generous gifts and time.

All of the papers in the Proceedings have been peer reviewed. I would like to thank the editors for their time and efforts, and all of the reviewers for their constructive criticisms.

STUART R. GELDER
University of Maine at Presque Isle
June, 1998

Hydrobiologia 406: xi–xii, 1999.
B. M. Healy, T. B. Reynoldson & K. A. Coates (eds), Aquatic Oligochaetes.

List of attendees

Michael AUBELE, Texas A & M Marine Laboratory, 5007 Avenue U, Galveston, Texas 77551, U.S.A.

Kathy A. BEAUCHAMP, Department of Medicine and Epidemiology, School of Veterinary Medicine, University of California, Davis, CA 95616, U.S.A.

Alexandra E. BELY, Department of Ecology and Evolution, SUNY at Stony Brook, Stony Brook, NY 11794-5245 U.S.A.

Carla BONACINA, CNR-Istituto Italiano di Idrobiologia, I-28048 Pallanza VB, Italy

Giuliano BONOMI, Dipartimento di Biologia Evoluzionistica Sperimentale, Università degli studi di Bologna, Via Selmi 3, I-40126 Bologna, Italy

Ralph O. BRINKHURST, 205 Cameron Court, Hermitage, TN, 37076, U.S.A.

Sandra CASELLATO, Dipartimento di Bologna, Università degli Studi di Padova, Via Trieste, 75, I-35121 Padova, Italy

Charles DREWES, Department of Zoology and Genetics, Iowa State University, Ames, IA 50011, U.S.A.

Clive A. EDWARDS, Ohio State University, Department of Entomology, 103 Botany and Zoology Building, 1735 Neil Avenue, Columbus, OH 43210-1220, U.S.A.

Philipp EGELER, ECT Oekotoxikologie GmbH. Böttgerstr. 2-14, D-65439 Flörsheim, Germany

Christer ERSÉUS, Department of Invertebrate Zoology, Swedish Museum of Natural History, Box 50007, S-104 05 Stockholm, Sweden

Kristian FAUCHALD, Deptartment of Invertebrate Zoology, MRL 163, National Museum Natural History, Washington D.C. 20560, U.S.A.

Steve FEND, U.S. Geological Survey, 345 Middlefield Road, Menlo Park, CA 94025, U.S.A.

Marco FERRAGUTI, Dipartimento di Biologia, sez. Zoologia e Citologia, Università degli Studi di Milano, 26, Via Celoria, I-20133 Milano, Italy

Stuart R. GELDER, Department of Biology, University of Maine at Presque Isle, 181 Main Street, Presque Isle, ME 04769-2888, U.S.A.

Olav GIERE, Universiata Hamburg, Zoologisches Institut und Zoologisches Museum, Martin-Luther-King-Platz 3, D-20146 Hamburg, Germany

Patti GILLIS, National Water Research Institute, Environment Canada, CCIW, 867 Lakeshore Road, Burlington, Ont. L7R 4A6, Canada

Sascha L. HALLETT, Department of Parasitology, The University of Queensland, Brisbane, Queensland, 4072 Australia

Brenda HEALY, Zoological Department, University College, Belfield, Stillorgan Road, Dublin 4, Ireland

René HESSLING, Universität Osnabrück, FB5 Biologie/Chemie, Abt. Spez. Zool., Postfach 4469, D-49069 Osnabrück, Germany

Deedee KATHMAN, Aquatic Resource Center, P.O. Box 680818, Franklin, TN 37068 - 0818, U.S.A.

Billie KERANS, Department of Biology, Montana State University, Bozeman, Montana 59717-0346, U.S.A.

Jan LOCKE, Bermuda Biological Station for Research, 17 Biological Lane, Ferry Reach, St. Georges, GE 01, Bermuda

Boris LÖHLEIN, Ecosystem Research Centre, Schauenburgerstr. 112, D-24114 Keil, Germany

Patrick MARTIN, Freshwater Biology, Royal Belgian Institute of Natural Sciences, 29 Rue Vautier, B-1000 Bruxelles, Belgium

Damhnait McHUGH, Department of Organismic and Evolutionary Biology, Museum of Comparative Zoology, Harvard University, 26 Oxford St., Cambridge, MA 02138, U.S.A.

Göran MILBRINK, Department of Zoology, Uppsala University, Villavägen 9, S-752 36 Uppsala, Sweden

Michael R. MILLIGAN, Center for Systematics and Taxonomy, P.O. Box 37534, Sarasota, Florida 34237, U.S.A.

William MOSER, Deptartment of Invertebrate Zoology, MRL 163, National Museum Natural History, Washington D.C. 20560, U.S.A.

Akifumi OHTAKA, Department of Natural Science, Faculty of Education, Hirosaki University, Hirosaki 036 Japan

Manuela PAGANI, Dip. di Biologia sez. Ecologia, Università degli Studi di Milano, Via Celoria 26, I-20133 Milano, Italy

Andrea PASTERIS, Dipartimento di Biologia Evoluzionistica Sperimentale, Università degli studi di Bologna, Via Selmi 3, I-40126 Bologna, Italy

Susan PETERSON, Department of Biology, University of Maine at Presque Isle, 181 Main Street, Presque Isle, Maine 04769-2888 U.S.A.

Günter PURSCHKE, Spezielle Zoologie, Dept. of Biology/Chemistry, University of Osnabrück, D-49069 Osnabrück, Germany

Charlotte RASMUSSEN, Biological Resources Division, United States Geological Survey, Northwest Biological Research Center, 6505 NE 65th Street, Seattle WA 98115 U.S.A.

Joy RAYBURN, Department of Biology, University of Maine at Presque Isle, 181 Main Street, Presque Isle, Maine 04769-2888 U.S.A.

Trefor B. REYNOLDSON, National Water Research Institute, 867 Lakeshore Rd., Burlington Ontario L7R 4A6, Canada

Geta RÎSNOVEANU, University of Bucharest, Faculty of Biology, Dept. of Ecology, Spl. Independentel 91-95, 76201, Bucharest, Romania

Pilar RODRIGUEZ, Departamento de Biologia animal y genetica, Facultad de Ciencias, Universidad del Pais Vasco, Apdo 644, E-48080 Bilbao, Spain

Rüdiger M. SCHMELZ, Universität Osnabrück, FB5 Biologie/Chemie, Abt. Spez. Zool., Postfach 4469, D-49069 Osnabrück, Germany

Jan SEYS, Institute of Nature Conservation, Kliniekstraat 25, 1070 Brussel, Belgium

Michael SMITH, Valdosta State College, Department of Biology, Valdosta, Georgia 31698, U.S.A.

Ivana TEODOROVIC, University of Novi Sad, Tempus Center, 21000 Novi Sad, Yugoslavia

Tarmo TIMM, Võrtsjärv Limnological Station of the Institute of Zoology and Botany, EE 61101 Rannu, Tartumaa, Estonia

Piet F. M. VERDONSCHOT, Institute for Forestry and Nature Research, P.O. Box 23, 6700 AA Wageningen, The Netherlands

Hongzhu WANG, Institute of Hydrobiology, Chinese Academy of Sciences, Wuhan Hubei 430072, P. R. China

Wilfred WESTHEIDE, Universität Osnabrück, FB5 Biologie/Chemie, Abt. Spez. Zool., Postfach 4469, D-49069 Osnabrück, Germany

Mark J. WETZEL, Illinois Natural History Survey, Center for Biodiversity, 607 East Peabody Drive, Champaign, Illinois 61820 U.S.A.

Hydrobiologia **406:** xiii–xiv, 1999.
B. M. Healy, T. B. Reynoldson & K. A. Coates (eds), Aquatic Oligochaetes.

In Memoriam Dr Andreina Di Chiara (1937–1997)

Dr Andreina Di Chiara

Dr Andreina Di Chiara was born on December 7th, 1939, in Rome. Later she completed her academic studies in the same city and was awarded the degree of Doctor of Biology after successfully presenting her thesis on the benthic biocoenoses of Lake Monterosi in Italy. In 1967, after moving to Milan with her family, she started work in the Institute of Zoology within the Department of Biology at the University of Milan. Then in 1981 she became a Research Scientist and transferred to the Institute of Ecology, where she taught hydrobiology. From the beginning, she focused her researches on the systematics and ecology of aquatic oligochaetes, particularly the tubificids. She worked on lotic biocoenoses of many Italian rivers, such as the Po, Garigliano and Lambro, and extended her studies to population dynamics, autoecology and reproductive biology of the benthic oligochaete species showing a strong resistance to environmental stresses. These investigations lead to her collaborations with different natural history museums and attending many national and international conferences, most particular was her support of the International Aquatic Oligochaete Symposia.

Andreina Di Chiara was an internationally recognized oligochaetologist and respected by her colleagues. Recently, the *Scientific Committee for the Italian Fauna* invited her to publish a volume on tubificid oligochaetes, but, unfortunately, she did not have enough time to finish this important task. Andreina has always been a friend and a kind colleague, devoid of arrogance, usually smiling and always full of life. Her friends remember her passion for music, literature, theatre and the Arts, and the innumerable intellectual discussions that she loved to start with a provocative question. Her wide knowledge of biology was part of a broader cultural background, rather than the more usual specialized and technical area of ones life. She has gone in silence having fought her disease with dignity and quietness, leaving a space that will forever be Andreina's.

MANUELA PAGANI
Milan, Italy

Papers presented at the International Symposia on Aquatic Oligochaetes by Dr Andreina Di Chiara Paoletti:

1984. & B. Sambugar. Oligochaeta of the middle Po River (Italy): principal component analysis of the benthic data. Hydrobiologia 115: 145-152.

1987. Some notes on the biology of the Asiatic species *Monopylephorus limosus* found in Italy. Hydrobiologia 155:307.

1989. Cohort cultures of *Tubifex tubifex* forms. Hydrobiologia 180: 143-150.

1996. & B. Sambugar. Aquatic Oligochaeta in Italy, with special reference to Naididae. Hydrobiologia 334: 37-49.

Mark J. Wetzel is thanked for providing the photograph.

Hydrobiologia **406:** 1–7, 1999.
B. M. Healy, T. B. Reynoldson & K. A. Coates (eds), Aquatic Oligochaetes.

Past symposium attendees in the group photographs

Group photographs of each of the Second to Sixth International Symposia on Aquatic Oligochaetes with names of the attendees identified using numbers on the respective line drawings.

A group photograph of participants has been taken at every meeting in the International Symposium on Aquatic Oligochaete series. However, the only group photograph published was that taken at the first symposium (Brinkhurst & Cook, 1980).

Since the first symposium in 1979, many people have attended only one or two gatherings and were not known to many of the other attendees. Conversely, some members are well known internationally and have been unable to attend more than one meeting. Current members frequently ask if a particular person attended a past meeting, and what do they look like? Before memories fade too much, it was decided that reproductions of the group photographs from the second to the seventh symposia should be published with the participants names (Figures 1–6). The photographs also provide us with happy reminders of members such as 'Randy' H. R. Baker and Andreina di Chiara Paoletti who have since passed away. Further details of the attendees and events at the respective symposia can be found in the appropriate Proceedings (Brinkhurst & Cook, 1979; Bonomi & Erséus, 1984; Brinkhurst & Diaz, 1987; Kaster, 1989; Reynoldson & Coates, 1994; Coates et al., 1996; current volume). In future it is hoped that the group photograph with identified attendees will be published as part of the Proceedings.

References

Bonomi, G. & C. Erséus, 1984. International Symposium on Aquatic Oligochaete Biology. (2nd held in Pallanza, Italy), Aquatic Oligochaeta, Junk Publ; Dordrecht, The Netherlands: 240 pp.

Brinkhurst, R. O. & D. G. Cook, 1979. Aquatic Oligochaete Biology. Plenum Press, New York: 529 pp.

Brinkhurst, R. O. & R. J. Diaz, 1987. International Symposium on Aquatic Oligochaeta. (3rd held in Hamburg, Germany), Aquatic Oligochaeta, Junk Publ; Dordrecht, The Netherlands: 323 pp.

Coates, K. A., T. B. Reynoldson & T. B. Reynoldson, 1996. Aquatic Oligochaete Biology VI. (6th symposium held in Tjärnö, Sweden, 1994) Kluwer Academic Publ; Dordrecht, The Netherlands: 295 pp.

Kaster, J. L., 1989. Aquatic Oligochaete Biology IV. (4th symposium held in Baton Rouge, La., U.S.A., 1988) Kluwer Academic Publ; The Netherlands: 252 pp.

Reynoldson, T. B. & K. A. Coates, 1994. Aquatic Oligochaete Biology V. (5th symposium held in Tallinn, Estonia, 1991) Kluwer Academic Publ; The Netherlands: 320 pp.

STUART R. GELDER & SUSAN M. PETERSON
Department of Science
University of Maine at Presque Isle
Maine 04769
U.S.A.

Figure 1. Group photograph taken at the Second International Symposium on Aquatic Oligochaetes held in Pallanza, Italy, 1982, with numbers on the line drawing showing the position of the respective named attendees: (1) Olaf Pfannkuche, (2) Simone Sabadini, (3) ?, (4) Barbara Lang, (5) Michel Lafont, (6) Jacques Juget, (7) Olav Giere, (8) ?, (9) Christer Erséus, (10) Marco Ferraguti, (11) Michael Ladle, (12) Ralph Brinkhurst, (13) Tarmo Timm, (14) Benno Wagner, (15) Michael Dzwillo, (16) Pietro Omodeo, (17) Michael Loden, (18) Robert Diaz, (19) Tamara Poddubnaya, (20) Claude Lang, (21) Mark Wetzel, (22) Bent Christensen, (23) Giuliano Bonomi, (24) Barrie Jamieson, (25) Stuart Gelder, (26) Mrs. Jónasson, (27) Petur Jónasson, (28) Deedee Kathman, (29) Klaus Kossmagk-Stephan, (30) Randy Baker, (31) Ludwig Probst, (32) Antanas Grigelis, (33) Krzysztof Kasprzak, (34) Mrs. Zahner, (35) Kathy Coates, (36) Nonna Finogenova, (37) Sandra Casellato, (38) Peter Chapman, (39) Mrs. Wetzel, (40) Rudolf Zahner, (41) Valerie Standen, (42) Göran Milbrink, (43) Brenda Healy.

Figure 2. Group photograph taken at the Third International Symposium on Aquatic Oligochaetes held in Hamburg, Germany, 1985, with numbers on the line drawing showing the position of the respective named attendees: (1) Michael Dzwillo, (2) Yanling Liang, (3) Klaus Kossmagk-Stephan, (4) Peter Chapman, (5) ?, (6) ?, (7) Ralph Brinkhurst, (8) Erich Gnaiger, (9) Sandra Casellato, (10) Doug Lietz, (11) Rosemarie Steinlechner, (12) ?, (13) Olav Giere, (14) Benno Wagner, (15) ?, (16) Göran Milbrink, (17) Marco Ferraguti, (18) Wilfried Westheide, (19) ?, (20) ?, (21) Bob Diaz, (22) Sang Qi, (23) Christer Erséus, (24) ?, (25) Michael Milligan, (26) ?, (27) Jacques Juget, (28) Michael Lafont, (29) ?, (30) Kathy Coates, (31) ?, (32) Jukka Särkkä, (33) Enrique Martinez-Ansemil, (34) Narcisse Giani, (35) Pilar Rodriguez, (36) Piet Verdonschot, (37) Brenda Healy, (38) Stuart Gelder, (39) ?, (40) Andreina Paoletti di Chiara.

Figure 3. Group photograph taken at the Fourth International Symposium on Aquatic Oligochaetes held in Baton Rouge, Louisiana, U.S.A., 1988, with numbers on the line drawing showing the position of the respective named attendees: (1) Narcisse Giani, (2) Robert Diaz, (3) Göran Milbrink, (4) Jacques Juget, (5) Pilar Rodriguez, (6) Brenda Healy, (7) ?, (8) Michael Milligan, (9) Marco Ferraguti, (10) Piet Verdonschot, (11) Sandra Casellato, (12) Roberto Valvasorri, (13) Andreina Paoletti di Chiara, (14) William Fender, (15) Mark Wetzel, (16) Celia Gluzman, (17) Trefor Reynoldson, (18) Jerry Kastner, (19) Olav Giere, (20) Claud Lang, (21) Walter Harman, (22) Kathy Coates, (23) Reinmar Grimm, (24) Ralph Brinkhurst, (25) Christer Erséus, (26) Giulio Lanzavecchia, (27) Erwin Polls, (28) Jukka Särkkä, (29) Michael Loden, (30) Magda de Eguileor, (31) Giuliano Bonomi, (32) Carla Bonacina, (33) Charles Drewes, (34) Stuart Gelder.

Figure 4. Group photograph taken at the Fifth International Symposium on Aquatic Oligochaetes held in Tallinn, Estonia, 1991, with numbers on the line drawing showing the position of the respective named attendees: (1) Alexander Tzetlin, (2) Pietro Omodeo, (3) Brenda Healy, (4) Andreas Anlauf, (5) Emilia Rota, (6) Jukka Särkkä, (7) Ralph Brinkhurst, (8) Mark Wetzel, (9) Guilio Lanzavecchia, (10) Svein-Erik Sloreid, (11) Henn Timm, (12) Rüdiger Schmelz, (13) Olav Giere, (14) Andreina Paoletti di Chiara, (15) Christer Erséus, (16) Michael Milligan, (17) Kathy Coates, (18) Piet Verdonschot, (19) Trefor Reynoldson, (20) Karen Jenderejian, (21) Lena Gustavson, (22) Susanna Hagobian, (23) Giuliano Bonomi, (24) Ljudmila Snimschikova, (25) Patrick Martin, (26) Tamara Dolidze, (27) Claus Lindegaard, (28) Eustolia Lestchova, (29) Elizabeth Dumnicka, (30) Sandra Casellato, (31) Andrea Pasteris, (32) Reinmar Grimm, (33) Carla Bonacina, (34) Doris Rohde, (35) Regine Bönsch, (36) Nonna Finogenova, (37) Tatyana Slepukhina, (38) Tamara Moroz, (39) Henny Graae, (40) Montserrat Real Ortí, (41) Viivi Timm, (42) Tarmo Timm, (43) Marco Ferraguti, (44) Pilar Rodriguez, (45) Goran Milbrink, (46) Robert Diaz, (47) Magda de Eguileor.

Figure 5. Group photograph taken at the Sixth International Symposium on Aquatic Oligochaetes held in Tjärnö, Sweden, 1994, with numbers on the line drawing showing the position of the respective named attendees: (1) Kathy Coates, (2) Mark Siddall, (3) Boris Löhlein, (4) Enrique Martinez-Ansemil, (5) Eugene Burreson, (6) Lena Gustavson, (7) Claud Lang, (8) Mark Wetzel, (9) Piet Verdonschot, (10) Georg Sauter, (11) Michael Milligan, (12) Trond Bremnes, (13) Sergey Afanasyev, (14) Giulio Lanzavecchia, (15) Magda de Eguileor, (16) Emilia Rota, (17) Laslo Molnár, (18) Reinmar Grimm, (19) Andrea Pasteris, (20) Ralph Brinkhurst, (21) Narcisse Giani, (22) Christer Erséus, (23) Marco Ferraguti, (24) Takashi Shimizu, (25) Alex Muir, (26) Ferdinand Sporka, (27) Svein-Erik Sloreid, (28) Donald Stacy, (29) Wilfried Westheide, (30) Robert Diaz, (31) Rüdiger Schmelz, (32) Olav Giere, (33) Agnés Rosso, (34) Svein Manum, (35) Andreina Paoletti di Chiara, (36) Tarmo Timm, (37) Brenda Healy, (38) Carla Bonacina, (39) Nona Finogenova, (40) Elizabeth Dumnicka, (41) Marianne Eckroth, (42) Montserrat Real, (43) Rut Collado, (44) Sandra Casellato, (45) Pilar Rodriguez, (46) Maite Madrid-Martinez, (47) Beatrice Sambugar, (48) Giuliano Bonomi, (49) Tatyana Slepukhina, (50) Patrick Martin, (51) Göran Milbrink, (52) Regine Bönsch, (53) Trefor Reynoldson, (54) Stuart Gelder, (55) Karen Jenderedjian, (56) Viivi Timm, (57) Karin Sindemark, (58) Kerstin Rignéus, (59) Lennart Sandberg, (60) Stephan Lundberg.

Figure 6. Group photograph taken at the Seventh International Symposium on Aquatic Oligochaetes held in Presque Isle, Maine, U.S.A., 1997, by William Duncan, with numbers on the line drawing showing the position of the respective named attendees: (1) Michael Aubele, (2) Ralph Brinkhurst, (3) Marco Ferraguti, (4) Andrea Pasteris, (5) William Moser, (6) Michael Smith, (7) Clive Edwards, (8) Michael Milligan, (9) Elvira Edwards, (10) Charles Drewes, (11) Ivana Teodorovic, (12) Geta Rîsnoveanu, (13) Olav Giere, (14) Rüdiger Schmelz, (15) Patrick Martin, (16) Manuela Pagani, (17) Steve Fend, (18) Kristian Fauchald, (19) Hongzhu Wang, (20) Piet Verdonschot, (21) Stephen Atkinson, (22) Patty Gillis, (23) Jan Seys, (24) Göran Milbrink, (25) Tarmo Timm, (26) Wilfried Westheide, (27) Jan Locke, (28) Günter Purschke, (29) Philipp Egeler, (30) Trefor Reynoldson, (31) Deedee Kathman, (32) Mark Wetzel, (33) Sascha Hallet, (34) Sandra Casellato, (35) Christer Erséus, (36) Akifumi Ohtaka, (37) Leo Shapiro, (38) Giuliano Bonomi, (39) Pilar Rodriguez, (40) Carla Bonacina, (41) Charlotte Rasmussen, (42) Brenda Healey, (43) René Hessling, (44) Alexa Bely, (45) Sue Peterson, (46) Stuart Gelder, (47) Katie Beauchamp, (48) Damhnait McHugh, (49) Billie Kerans, (50) Boris Löhlein.

Hydrobiologia **406:** 9–19, 1999.
B. M. Healy, T. B. Reynoldson & K. A. Coates (eds), Aquatic Oligochaetes.

Retrospect and Prospect: reflections on forty years of study of aquatic oligochaetes

Ralph O. Brinkhurst
205 Cameron Court, Hermitage, TN 37076-1353, U.S.A.

Key words: Annelida, Clitellata, history, taxonomy, ecology

Abstract

Fourteen aquatic oligochaetes were described between 1773 and 1828. Dramatic increases in descriptions occurred in the decades beginning 1900, 1930 and 1960. These can be associated with specific authors, such as W. Michaelsen and S. Hrabě. In the period 1970–1990, some 250 marine species (excluding Enchytraeidae) were described, largely by C. Erséus. Some of the major scholars, such as W. Michaelsen and F. Vejdovský, I. I. Malevič and P. G. Svetlov, founded centers of oligochaete study that survive in Europe today. Others, such as those in the English speaking world, have had to develop in isolation. While numerical and phenetic methods had little influence on microdrile systematics, phylogenetic (parsimony) methods have, since the first publication by B. G. M. Jamieson. Most literature prior to 1970 was taxonomic. Other earlier work documented community structure in relation to pollution. The study of Rybinsk Reservoir (Russia) stimulated a large body of ecological and biological work by T. L. Poddubnaya. The St. Lawrence Great Lakes was also a focus for worm studies until recently. Early physiological work focused on respiration, with studies dating from the 1920s. This has been continued in relation to pollution ecology and energy flow. Physiological work was notably absent from our first proceedings volume and only poorly represented since then. Future prospects are difficult to identify. Developments in ultrastructure (including sperm and muscle cells), molecular biology and parsimony or cladistic phylogenetic methods will move from rather simple initial work that shows the possibilities, to a revisionary phase where the complexities begin to be revealed before it is possible to assess their value. This author has too superficial a level of understanding to predict the outcome, but suspects that agreement among the results of several methods is needed for them to be persuasive. I would hope that multivariate methods of data analysis in pollution studies would prove superior to the development of indices and over-reliance on toxicology. Toxicity studies are of use in comparing risks among products but not in predicting field effects. Benthic biology needs a method for identifying factors that are directly related to worm biology, such as food, rather than physical and chemical factors related to the water column. Surveys, relating communities to chemical and physiographic factors will remain unpredictive. It is easier to forsee some changes in taxonomy that are almost overdue. The haplotaxids should be broken up into several small families, *Haplotaxis* s.s. will be one of them, with a number of highly adapted and very similar species, many currently regarded as synonymous, redefined by careful anatomical study. The lumbriculids and branchiobdellidans may well be combined into a single taxon defined by their unique semi-prosopore male ducts, with lumbriculids defined by their unique muscle structure and the branchiobdellids by many apomorphic characters related to ectocommensal life. The position of Hirudinea remains unresolved, but relationship to this taxon is not excluded. Detailed study of penes in lumbriculids is required. The naidids and tubificids seem to be a monophyly. Existing genera may be hard to analyze phylogenetically because definitions are based on atrial form, which is hard to express as characters. Chaetotaxy in Naididae will prove insufficient and many synonyms will be declared. Capilloventridae and Randiellidae require description of male ducts and gonads. They may provide evidence counter to the octogonadal theory of oligochaete descent, or of polyphyly in the group. Knowledge of phreodrilid diversity will increase. Work will increase in Asia and South America, but sub-Saharan Africa remains under explored. Fish diseases will create more interest in North American worms. New methods, including molecular studies, and renewed interest in ontogeny and detailed anatomical studies, may improve our ability to resolve clusters of taxa with few, if any, apomorphic character states that remain at the base of cladograms. An ability to include sound evidence from vestigial organs and logical arguments on a 'weight of evidence' basis is needed as an adjunct. While molecular studies seem to hold much promise, early studies can prove over simplistic and can provide conflicting hypotheses. We need to complete the review of taxonomy according to parsimony before major terminology changes, such as abandonment of the term Oligochaeta, can gain acceptance.

"On Monday, when the sun is hot,
I wonder to myself a lot:
"Now is it true, or is it not,
That what is which and which is what?"

On Tuesday, when it hails and snows,
The feeling on me grows and grows
That hardly anybody knows
If those are these or these are those."

The Tao of Pooh, B. Hoff, 1982

Introduction

There is a natural progression in the presentations that one gives to scientific meetings. We often begin with a poster, or a small paper tucked into the last afternoon when most people are on their way home. Then we progress to more noticed contributions, until we may be asked to present a summary and conclusions account based on our accumulated wisdom. Eventually, the day comes when one gets asked to present a historical account of one's field and perhaps this serves as a gentle reminder that it is time to stop repeating one's self. It seems that I may now have reached that phase as I try to give you the benefit of my perspective, based on many enjoyable years spent in such good company. This invited contribution, therefore, is more in the nature of an essay than a scientific paper and represents no more than my own opinions, sometimes with little documentation.

I have reached a point where I do not anticipate contributing much more than to complete some taxonomic studies, mostly in association with truly valued younger colleagues. I can see that the lonely position of the few of us in this field forty and more years ago will not be repeated for them. The meeting that I address will surely be one of an ongoing series. My message is basically that a wise scientist develops respect for the past and enthusiasm for the future.

Retrospect: History of biological studies of aquatic oligochaetes

Man's oldest profession is taxonomy, learning to tell food from predator or poison, harmful from beneficial, or simply neutral, fellow organisms. Most of us have lost nearly all trace of the skills that brought survival. The oldest roots of our discipline are also in taxonomy. Several of the well known European species were described before 1828 (Table 1). Descriptions of aquatic species, mainly in the Tubificidae, Naididae, and Lumbriculidae, reached significant peaks in the decades beginning 1900, 1930, and 1960 (Table 2) as determined from the species considered to be valid in 1971, according to Brinkhurst & Jamieson (1971). I rely heavily on this source for my presentation, including some early publication dates that differ slightly from those of other sources. I have omitted Enchytraeidae because of my ongoing inexperience with the family and their superficial treatment in the primary sources I used here. These peaks can be associated with major contributions from specific authors, such as W. Michaelsen in the 1900 decade and S. Hrabě in the 1930 decade (Figure 1). Since then, there has been an explosion of descriptions of marine tubificids in particular. By my count, C. Erséus has described more than 30 species and five genera in the 1970 decade, but more than 200 species in 11 new genera and even a new family in the 1980s, and more later. These numbers exclude enchytraeids in order to make them comparable with the estimates quoted above.

Some of the most influential contributors to our taxonomic literature who began publishing prior to 1970 are recognized in Table 3, with an indication of their publication life spans. I have excluded some authors, such as C. Sperber, who published a limited number of nonetheless influential documents. Some of these, such as F. Vejdovský (Czechoslovakia) and W. Michaelsen (Germany), founded lineages that survive today. The very large school within the former USSR includes over 75 names according to T. Timm. I. I. Malevič in Moscow and P. G. Svetlov in Leningrad were succeeded by N. L. Sokol'skaja and O. V. Čekanovskaja, respectively and these two, with V. V. Izosimov, trained our contemporary colleagues including T. Timm and N. P. Finogenova (Russian translitorations according to T. Timm, personal communication). Many of us, this author included, were self- taught. I owe my start to a small key, published in German by H. Ude in Tierwelt Deutschlands, and the assistance of my mentor, H. B. N. Hynes, who insisted that I learn to read the language.

The Soviet group, if I may call it that, held six All-Union symposia between 1967 and 1987, the proceedings of some of which are available in translation (see Kurashvili, 1983). The very small American group has met several times, most recently attracting a number of European biologists to participate. These informal sessions did not produce printed products. Communication between the founding fathers was ob-

Table 1. A chronological list of some of the oldest aquatic oligochaete species

1767	1776
Stylaria lacustris	*Clitellio arenarius*
1773	1784
Aulophorus furcatus	*Paranais litoralis*
Dero digitata	1821
1774	*Haplotaxis gordioides*
Ophidonais serpentina	1827
Nais barbata	*Chaetogaster limnaei*
Nais elinguis	1828
Tubifex tubifex	*Chaetogaster diastrophus, C. diaphanus*
Lumbriculus variegatus	*Pristina longiseta*

This list includes only Linnean names, not earlier appellations.

viously quite good. F. Smith acknowledges the help of F. Vejdovský in one paper and J. Stephenson dedicated his monograph to W. Michaelsen. Specimens from the work of G. Eisen in California were found in the collections of the Natural History Museum in London, presumably exchanged with F. Beddard. Some joint publications indicate cooperation among authors, but the level of international cooperation seems to have increased in recent years. This can be traced to the effect of these international symposia as well as to easy communication through FAX machines and now E-mail. These methods are especially valuable in exchanges between nations where the postal service seems unreliable at best.

Our taxonomic work was not much influenced by numerical and phenetic taxonomy. There was a considerable effort by R. W. Sims in England to employ these methods in earthworm taxonomy, and B. G. M. Jamieson was involved as early as 1968. Jamieson (1978) introduced phylogenetic or cladistic methods, again mainly with opisthoporous worms and his lead was followed by C. Erséus, K. A. Coates, this author and others from 1987 onwards.

Biology

A great deal of the literature on aquatic oligochaetes published prior to 1970 was taxonomic in nature, or documented distributions of identified taxa, often in relation to pollution assessment. Some exceptions include the use of tubificids in studies of respiration, with the exceptional work of G. Alsterberg in the 1920s. This was followed by authors such as K. Berg (Denmark), H. Munro Fox and R. J. Aston (England) and K. Dausend (Germany). At the first of these symposia on aquatic oligochaetes in Sidney, British Columbia in 1979, many authors were specifically invited to review various subject areas (Brinkhurst & Cook, 1980). I do not know if it was my own bias or a lack of candidates, but physiological work seems to have been omitted. There continues to be a relative shortage of work in this field at our meetings (Table 4). These numbers exclude papers given as abstracts and those that the authors did not publish, so the actual number of presentations at meetings is much higher. For example, there were 56 presentations in 1994, not just the 31 published. When I look through my reprint collection of papers on subjects other than taxonomy (which biases this account, as I have not done a detailed literature search), I find most of them are authored by members of our group and they are referenced in or published in our proceedings (Bonomi & Erséus, 1984; Brinkhurst & Diaz, 1987; Kaster, 1989; Reynoldson & Coates, 1994; Coates et al., 1996). The elegent neurophysical and behavioral work of C. D. Drewes (e.g. Drewes & Fourtner, 1993) appears in only one of our publications. Ultrastructure of oligochaetes was reviewed by Jamieson (1981) and has been the subject of many contributions to these meetings. Many relevant biological papers were written by authors who produce one or two studies on worms and then turn their attention elsewhere. Others have contributed work that combines ecology, physiology and life history studies, such as T. L. Poddubnaya, who published between 1958 and 1980, mostly on Rybinsk Reservoir in Russia. G. Bonomi, C. Bonacina and their colleagues have made extensive studies of cultured populations. Specific sites, such as the St. Lawrence Great Lakes, have created foci for worm work. Au-

Table 2. Numbers of genera and species defined by decade, with cumulative totals, for the oligochaete families Haplotaxidae, Lumbriculidae, Tubificidae, Naididae, Dorydrilidae, Phreodrilidae, and Opistocystidae. Numbers derived from Brinkhurst and Jamieson (1971)

YEAR	<1840	1840	1850	1860	1870	1880	1890
Number							
Genera	6	4	1	3	4	6	3
Species	15	3	6	8	18	13	22
Cumulative							
Genera	6	10	11	14	18	24	27
Species	15	18	24	32	50	63	85
	1900	1910	1920	1930	1940	1950	1960
Number							
Genera	10	2	1	10	4	1	8
Species	82	17	28	72	36	26	74
Cumulative							
Genera	37	39	40	50	54	55	63
Species	167	184	212	284	320	346	420

thors include J. K. Hiltunen, D. R. Spencer (Spencer, 1980), P. L. McCall, T. B. Reynoldson, D. S. White, D. R. Barton, their associates and many others. Marine biological work has been ably reviewed by Giere & Pfannkuche (1982).

Prospects

It is, of course, much more difficult to talk about future possibilities than to review the past. When the manuscript for my book on lake benthos was reviewed, one scientist said "Why does the author not tell us all what we should be doing?" and the other said "Why does the author keep telling us what to do?". The editors and I left the work unchanged! While it may be difficult for my colleagues to believe, I actually feel quite humble about my ability to exhort anyone to do anything other than to persue their own interests and opportunities with vigor and an open mind. As my own circumstances caused my work on aquatic oligochaetes to revert to taxonomic and anatomical research after leaving the facilities and colleagues available in Toronto, I am not well equiped to advise anyone about future prospects for ultrastructure, sperm morphology, molecular biology, or developments in phylogenetic analysis. I have lately been exposed to an indication that some of the early publications on molecular systematics in our field may be too superficial. We should expect that to some degree. Surely in every discipline, new methods are often applied before we know some of the limitations involved. One trap we could avoid is ignoring 'within taxon' variation when trying to evaluate 'between taxon' relationships by using several representatives of each taxon involved in whatever study we make. Using standard and often unrepresentative single taxa to represent major groups such as families, or even higher taxa, is an obvious pitfall. I believe it is worth publishing some work 'prematurely' simply to draw attention to new possibilities. My own brief venture into the taxonomy of marine tubificids is an example – I even confused some fragmentary material of two species as belonging to a single taxon in a first paper. However, that paper did help alert at least American marine biologists to the tip of what proved to be a major iceberg! Subsequent more careful work with good material by others has greatly expanded that field to the point where this author bowed out of it, but the 'premature' recognition of the Phallodrilinae (Brinkhurst & Jamieson, 1971) has withstood the test of time.

Figure 1. Some notable 'ancestors'. Top row, from left: F. Vejdovský, L. Cernosvitov, W. Michaelsen. Bottom row: J. Stephenson, S. Hrabě. (Thanks to B. G. M. Jamieson for some of the above photographs.)

Table 3. Some major contributors to aquatic oligochaete taxonomy (of families identified in Table 2) to 1970, with dates of active publication

Beginning before 1900:				
Grübe 1844–80	Leidy 1850–80	Eisen 1878–10	Beddard 1888–20	Benham 1890–50?
Smith 1895–19	Bretscher 1896–05	Michaelsen 1899–35		
Begin 1900–1949:				
Mrazek 1900–27	Stephenson 1907–32	Lastockin 1918–55	Malevic 1925–58	Izosimov 1937–72
Cernosvitov 1926–45	Hrabě 1927–84	Yamaguchi 1936–53	Marcus E. and E du B. 1942–65	
Begin 1950–1970:				
Čekanovskaya 1952–75	Sokol'skaya 1951–83	Juget 1957–94	Jamieson 1957–	Liang 1958–
Omedeo 1958–	Timm 1958–	Brinkhurst 1960–	Harman 1961–84	Finogenova 1962–

Many other contributors could be added, such as F. Vejdovský and C. Sperber, but these authors produced important but a small number of publications and their inclusion would create the need for a great many other additions.

Table 4. Numbers of papers published in the proceedings of the Aquatic Oligochaete Symposia by topic

SYMPOSIUM	I	II	III	IV	V	VI
Year	1979	1982	1985	1988	1991	1994
Topic						
Taxonomy and Evolution	5	8	12	6	3	4
Genetics	0	1	0	0	1	1
Physiology/behavior	0	2	3	2	0	2
Life cycle/reproduction	1	4	2	6	1	2
Production/population dynamics	3	4	3	0	8	3
Pollution/communities	7	9	7	6	12	7
Faunistics	10	7	7	0	8	9
Ultrastructure	0	1	4	5	5	4
Other	0	0	4	1	0	0
Number of papers published	26	36	42	26	38	31
Nationalities represented	13*	17	12	10*	20	21

* = held in North America

Predictions concerning taxonomy

Some taxonomic problems do seem to be predictable. A few are considered here.

Haplotaxidae

The haplotaxids are clearly not a monophyletic group and I am sure that someone will propose making a distinct family for the predatory *Haplotaxis sensu stricto*. More detailed work recording the nature and distribution of the dorsal chaetae, where present, may well provide characters that separate the *H. gordioides* complex into separate taxa. When we examine specimens of *H. ichthyophagus*, its great size alone suggests its separate identity. The form of the pharynx may vary if superficial examination of the exterior of the prostomium and peristomium can be used as an indication (Brinkhurst & Marchese, 1987; Brinkhurst & McKey-Fender, 1991). Absence of mature specimens must be overcome by more attention to other anatomical features and new types of evidence.

Lumbriculidae

The lumbriculids also provide a fertile field for new work. The structure of the penes is odd and poorly understood. In many, it seems the penis is an extension of atrial lining cells and so may be anatomically distinct from pendent penes found elsewhere.

A phylogenetic analysis of the lumbriculids based on generalized generic characters (Brinkhurst, 1989) proved very revealing, as it set up the hypothesis that the semiprosopore condition (also the commonest condition) was plesiomorphic to all other male duct arrangements in the family. This in turn provides a synapomorphy for Lumbriculidae plus Branchiobdellida. However, while the latter is clearly monophyletic, the lumbriculids within a resultant merged taxon are not, unless their muscular ultrastructure remains unique after study of other clitellate taxa (Eguileor et al., 1987). Attempts to follow that generic-level phylogenetic analysis by (unpublished) studies of all species included in sectors of the resultant cladogram have so far failed to support monophyly for several existing genera (*Trichodrilus, Stylodrilus, Lumbriculus* and its several supposed synonyms and *Rhynchelmis*). Cladograms that differ from the expected or traditional pattern, like the generic-level analysis, are valuable. The confused results of the species-group analysis, caused by supposed homoplasious characters, send us back to the organisms for more detailed work, or to alternative methods, even though they are unhelpful as they stand.

The lumbriculids have always been recognized as having a very inconsistent male reproductive system, although in terms of frequency of various combinations this seems a little exaggerated (Brinkhurst, 1991). Recent work on lumbriculids from the western states of North America suggest the possibility that there is a rich post glacial fauna of closely related taxa in isolated cool groundwater and montane habitats.

The origin of leeches

The Euhirudinea may be traceable to a separate root within this prosopore assemblage, as it can be argued that their male reproductive system can be traced to a prosoporous ancestor. Ontogenetic study of the development of the male system in leeches might substantiate the assumption that we all make, that the enclosed testes in leeches, situated in segments behind the male pore, originated in the segment bearing the male ducts and pore. However, Acanthobdellida appears to possess a male pore that is anything but prosopore. In its present configuration, the distinct oligochaete-like male duct appears potentially opithoporous (ontogenetic studies might determine the relative positions of testes, male funels and atrium-like structures). This could be derived from a semi-prosoporous state by reduction of the posterior testes and ducts, just as the Euhirudinea can be derived from the semiprosoporous state by the loss of the anterior pair set of those structures, but these are assumptions at this stage. Acanthobdellida exhibit a series of synapomorphies with Euhirudinea, but have a few segments bearing chaetae and have lost the prostomium, peristomium and maybe more anterior segments. The number of ganglia in the nervous system does not correspond with the number of annuli (Purschke et al., 1993). All these facts may help to resolve an apparent difference in the fixed number of segments between this taxon and leeches. The branchiobdellidans seem to be more certainly related to the lumbriculids than to the leeches because of the shared semiprosoporous condition of the stem forms of both taxa. This subject will be reviewed in detail elswhere in these proceedings, with a report on discussions at the last two conferences. That paper will substantiate differences between this account and that presented by Purschke et al. (1993).

Tubificidae and Naididae

The general recognition that the tubificids and naidids may be a single monophyly also needs to be resolved. A difficulty here is that the tubificid genera may also prove intractable to phylogenetic analysis until we can express the characteristic shapes of the male ducts in appropriate terms. Again, analyses at the generic level seem to provide useful answers within various subfamilies, but these are based on assumptions about generic characters, discussed below. Perhaps analyses of type species alone should be used to isolate small sectors of a cladogram and then all species within the subgroup of genera so identified could be analyzed to establish generic limits. There is a growing suspicion and certainly my opinion, that chaetotaxy is insufficient to differentiate species of Naididae and that many synonyms remain to be identified. There is conflicting evidence regarding polymorphic forms of species such as *T. tubifex*. Experiments prove that hair and pectinate chaetae can be rapidly replaced by simple bifid chaetae, but some field populations of the variant named (as a species, subspecies or variant according to source) *blanchardi* can be stable and genetically controlled even where conductivity of the water is not elevated. Various populations of *T. tubifex* have been tested in relation to whirling disease in North American trout, and there seems to be a great deal of variation in regard to this as well as to size and other morphological characteristics. We should expect to see more intense work on this species in regard to this economically significant question (Brinkhurst, 1996).

Capilloventridae, Randiellidae and others

A number of very small clitellate and non-clitellate worms have been described or reinvestigated recently. They form the basis for the definition of very small families. Two European terrestrial forms are enigmatic and one, *Hrabeiella periglandulata* (Pizl & Chalupský, 1984) may be related to oligochaetes (Rota, 1998).

Resolving the reproductive anatomy of Narapidae, Capilloventridae and Randiellidae and newly discovered forms yet to be described may provide evidence that challenges the concept of an ancestral form of aquatic oligochaete with eight pairs of gonads and numerous gonopores, including several pairs of spermathecae. Some authors see some of these very small, interstitial, taxa as descendents of very early oligochaete forms, in contrast to the evolutionary pattern I have described in a variety of publications. My views are heavily influenced by the concept that a small earthworm-like form evolved from an unsegmented worm as an adaptation to constant burrowing. The reproductive system of such an animal had four pairs of gonads and four pairs of very similar gonoducts. Advances in the efficiancy of gamete storage and copulation then led to a reduction in the number of reproductive segments until we reach the tubificid condition. Multiple and elaborate chaetae are then an adaptation to a secondary aquatic existence. Another alternative is that the whole microdrile assemblage is polyphyletic and so both hypotheses can be valid. However, a logical scenario to justify a shift from a streamlined reproductive system with a minimum number of gonads and gonopores associated with adequate storage for gametogenesis to the more complex octogonadal state with multiple pores, rather than the reverse, has yet to be presented by oponents of the octogonadal theory. Most proponents of phylogenetic (or cladistic, or parsimony) methods refuse to indulge in such speculation in the name of objectivity, but there is a 'weight of evidence' approach that provides, in my humble opinion, equally valuable insights given that the logic is demonstrated.

Phreodrilidae

The diversity among Phreodrilidae is now being explored in Australia and new interest in South America may contribute more. The interesting old records from Cape Province, South Africa and Sri Lanka that I had hoped to be able to pursue will need investigation. So will the single record from Bolivia.

Zoogeography

In terms of other zoogeographical coverage, work in Asia may be stimulated by the careful work of A. Ohtaka and in Australia by A. Pinder for as long as they can be supported, a problem for taxonomists everywhere. Sub-Saharan Africa remains largely *terra incognita*. In America, a number of new colleagues are being stimulated to take up oligochaete research by the outbreak of whirling disease in trout, caused by a myxozoan that has *Tubifex* as an alternate host (Brinkhurst, 1996). Fish diseases caused by various myxozoans (which may be Ctenophora!) are carried by oligochaetes from several families, again to be discussed elsewhere in these proceedings.

Phylogeny

Phylogenetic studies continue to intrigue some of us, even if our jigsaw puzzle has lost so many species that reconstructing the past is almost impossible. The new methods available to us have stimulated a lot of work in this area. One of the great difficulties of working with oligochaetes is that homoplasy is definitely a very large problem. Some of these obvious cases of homoplasy are listed in Table 5. These lead to such inconsistency in data matrices that there are many plausible alternative models. Many of us now use phylogenetic methods that rely on the parsimony principle. This is probably the largest change in our science since these meetings began. These methods have many strengths, but we should not be blind to the problems that still exist when using them, unless we wish to stifle further development. Willmer (1995) indicated some of the limitations of both cladistic and molecular methods in a paper that is part of a very valuable volume. Like cluster analysis in ecological data, variations in the precise method used can be valuable in exploring the data set for robust information.

Evolution still proceeds by means of natural selection, so far as we are aware. While present phylogenetic methods cannot incorporate this fact, perhaps future modifications will. This might permit logical selection of alternative scenarios by informed weighting. The following is an example of a situation in which I think logic can be used to determine character polarity, or 'which came first?'

The anterior set of gonads, male funnels and vasa deferentia (the GI complex) in lumbriculids is often reduced to non-functional elements in genera such as *Rhynchelmis*. The situation varies, with anterior testes small or absent, lack of anterior male funnels or very small ones and/or non-patent vasa deferentia. There are several lumbriculids in which the parts of the male duct are present without the associated gonad. It appears that there are species in which a progressive reduction of the anterior set of structures is evident. This situation fits the concept of vestigial organs, here being represented by functionless traces and I deny anyone to suggest that the reverse scenario (partly formed gonads and gonoducts leading to the development of a second full set) is possible. While the identification of selective versus non-selective characters, which dates from Darwin's time, is clearly unacceptable (who are we to say what confers a selective advantage in subtle characters) it remains true that extreme selection towards very specialized habits does produce convergence. This may be obvious (flight in birds, bats and insects) or not. Such arguments should at least cause careful examination of what are apparently obvious homologies.

In analyses of taxa above the species level, the plesiomorphic character state of a stem form may be used to permit comparison with other taxa, but derived forms of that taxon may exhibit a convergent apomorphy. For example, in analyzing clitellate taxa, it is reasonable to assume that Enchytraeidae are derived from a bisetate ancestor (Coates, 1989), which separ-

Table 5. Examples of some homoplasious characters for clitellate taxa as they are currently defined. The first five are clearly not homologous characters. Other characters require detailed study

Character	Comments
gills	*Branchiura* (Tubificidae), *Dero, Branchiodrilus* (Naididae).
modified pharynx	*Agriodrilus, Phagodrilus* (Lumbriculidae), *Chaetogaster* (Naididae), *Haplotaxis* (Haplotaxidae).
stalked prostates	Found in Tubificinae, Telmatodrilinae, and Phallodrilinae, differing in number.
spermatozeugmata	Specific spermatozeugmata in Tubificinae, other forms in Telmatodrilinae, and marine tubificids.
anterior attachment organ	Branchiobdellida, Hirudinea, structure and position and mode of action differ
pendent penes	Found in a variety of Tubificidae, Lumbriculidae, Phreodrilidae, Branchiobdellida, Hirudinea
eversible pseudopenes	Found in a variety of Phreodrilidae and Tubificidae
large grooved genital chetae (spermathecal and/or penial)	Found in some species of *Phreodrilus* (Phreodrilidae), some genera in all subfamilies of Tubificidae – different subfamilies, not a monophyly, *Rhynchelmis* (Baikal lumbriculid species). Similar chetae in one or two Naididae. *Kathrynella* (Alluroididae), *Dendrobaena* (Lumbricidae).
hair-like penial chetae	*Haplotaxis* (Haplotaxidae), *Lumbriculus* (Lumbriculidae), *Rhyacodrilus, Nootkadrilus, Inanidrilus* (Tubificidae), *Randiella* (Randiellidae), *Capilloventer* (Capilloventridae).
hair chetae	Form with support chaetae in most Phreodrilidae may differ from those found in most naidid genera and commonly, but variably within genera, of Tubificidae. Capilloventridae have ventral hairs.
proboscis on prostomium	Several Phreodrilidae, Lumbriculidae, Naididae, Propappidae and Opistocystidae.
'eyes'	Several Naididae, varies within a genus, not consistent within species.
pectinate/palmate chaetae	Widespread in Tubificinae and Rhyacodrilinae (Tubificidae), even as an intraspecific variant, also in Naididae (*Dero, Nais*).
spermatheca	Gut connection in some Enchytraeidae, *Rhyacodrilus* (Tubificidae), *Eclipidrilus* (Lumbriculidae).
median male pores	Found in *Eclipidrilus, Tatriella* (Lumbriculidae), *Monopylephorus, Bothrioneurum, Aulodrilus, Limnodriloides, Smithsonidrilus* (Tubificidae), *Slavina, Pristina* (Naididae), Branchiobdellida, Hirudinea.

ated the taxon from the multisetate tubificid - naidid assemblage according to Brinkhurst (1994). However, derived enchytraeids can have multiple chaetae. Conversely, marine tubificids may have few or even paired chaetae, although Brinkhurst coded the entire tubificid-naidid-opistocystid complex as multichaetate assuming that the marine forms were derived. The analysis actually suggested the reverse, as Heterodrilinae and Phallodrilinae were basal within that section of the cladogram, an unlooked-for result!

The difficulty here is, perhaps, the attempt to make an exact science out of an art form. After all, parsimony is simply used as a convention for deciding between alternate hypotheses. It is not a universal law. The major problem I find with phylogenetic methods is that there is always a small 'bush' of taxa with largely plesiomorphic character states at the root of the cladogram. If there are ancestors, there must be logical roots to monophylies.

While we wrestle with the problems created by taxa that are apparently paraphyletic, we have to be able to refer to study groups by some acceptable terminology. Thus, I try to use 'oligochaetes' in place of Oligochaeta until someone is smart enough to divide up that taxon into true monophylies. The term 'Non-branchiobdellidan, non-hirudinean Clitellata' is too cumbersome.

Phylogenies based on specific character sets (ultrastructure, molecular data, sperm, anatomy) may not be comparable. Should these character sets from each discipline be combined in a single analysis? Such combined analyses will require cooperation among various specialists. The taxonomic literature is replete with examples of conflicting phylogenies of one taxon in which different data sets were used, or weighted very differently by their respective authors. The non-specialist is then left with two conflicting opinions which are impossible to resolve.

One of the great advantages of phylogenetic methods is that the process is more open than trying to follow the arguments of at least some authors. Character state matrices, when presented, allow readers to assess the allocation of character states. This should allow duplication of earlier work, but even that is not always possible. For example, Siddall & Burreson (1996) were unable to duplicate a published cladogram of mine. The problem may be that I did not specify all of the many possible alternative analytical method options used. In my result (Brinkhurst, 1994) there were only 12 alternative trees and most of the variation involved small rearrangements among four microdrile taxa, the megadriles and the hypothetical ancestor. The re-analysis, with characters ordered, provided five alternative trees and the new consensus is actually quite similar to the original scheme. Using unordered character states produced 310 alternatives, resulting in a meaningless consensus tree, one which would have sent me back to the drawing board! The lengths and consistency indices of the new trees are not presented for comparison. The difference may lie in the use of weighting of multistate characters to bring them to equal weight to binary characters, as suggested in the original PAUP manuals, a technique I use consistently, but which may be applied quite differently in more recent editions of the method.

What I am attempting to show here is that there is no single 'correct' solution to a phylogenetic study, simply one dependent upon many method options. The result is even more significantly affected by decisions regarding homology, as demonstrated elsewhere in these proceedings (Brinkhurst, 1999).

Molecular taxonomy

Molecular methods may provide the ultimate panacea. Or not! Even in a very limited situation, the relationship between the two species of Panda and the Asian bears, Schaller (1993) tells us that two analyses produced opposite conclusions about the status of pandas. I do not have sufficient understanding of such methods to evaluate how fair such a comparison of results might be. How much more difficult will be our work with invertebrates. Will enough attention be paid to variation within a taxon before a few 'type' species are seized and used to represent large complex groups? What would have happened if Kim et al. (1996) and Moon et al. (1996) had used a lumbriculid instead of a lumbricid in their molecular study? We await the outcome of parallel contemporary studies with interest and, one hopes, with an open mind. Willmer (1995) suggests that developmental markers and genome architecture show promise where cladistics and molecular work is limited.

Biological studies

The future of biological studies on aquatic oligochaetes will still be determined by the interests of a small cadre of biologists and information derived in the context of broader studies such as pollution surveys. Whirling disease research proposals indicate a widespread adherence to the concept of limiting variables, such as temperature, substratum type, flow and other physical attributes. Surely it is the interaction among species, both other tubificids and taxa from other major groups, that determines the distribution of worms, together with the available food supply. Pollution biology in North America, at least, will continue to try and establish tolerance limits and so-called metrics. In other regions, the already large literature of surveys with comparisons between chemical and physiographic variables and worms communities will continue despite the low significance values or percent variance explained by multivariate analyses. The use of meaningful variables, such as the abundance of specific bacteria used as food, is too complex, expensive and difficult in the absence of aquatic microbiologists. Benthic ecology will make a quantum leap once we can typify soft sediments in a way that is significant to the organisms inhabiting them.

Finale

I, for one, am excited at the prospect of the steadily improving quality of our work, the promise of new technologies and the quantum improvement in means of communication, that I am convinced that I leave my almost life-time study in very good hands. Above all, I hope for all our sakes that the spirit of fun and cooperation that has characterized members of this group continues. We do not need any Confucianist, Dedicated scholars....

"What's that?" Pooh interrupted. "What's what?" I asked. "What you just said - the Confusionist, Desiccated scholar" "Well, let's see. The Confusionist, Desiccated scholar is one who studies Knowledge for the sake of Knowledge and keeps

what he learns to himself or to his own small group, writing pompous and pretentious papers that no one else can understand, rather than working for the enlightenment of others. How's that?" "Much better" said Pooh.

The Tao of Pooh, B. Hoff, 1982.

Acknowledgements

I wish to thank B. G. M. Jamieson and T. Timm for their personal contributions to this manuscript. I also wish to thank S. R. Gelder for his invitation to attempt this task and for his continued support and friendship, as well as his help with the final drafts. Various reviewers have also contributed much to the manuscript. The list of friends and colleagues who have freely contributed to the many projects I have been involved with for about 40 years is too long to list individually, but I would like to take this opportunity to express my gratitude to them all, especially to Professor H. B. N. Hynes, without whose stimulus I would not be in this field.

References

Bonomi, G. & C. Erséus, 1984. Aquatic Oligochaeta. Proceedings of the second International symposium on aquatic oligochaete biology, held in Pallanza, Italy, September 21–24, 1982. Developments in Hydrobiology 24, Dr W. Junk, Dordrecht: 240 pp.

Brinkhurst, R. O. ,1989. A phylogenetic analysis of the Lumbriculidae (Annelida, Oligochaeta). Can. J. Zool. 67: 2731–2739.

Brinkhurst, R. O., 1991. Ancestors. Mitt. hamb. zool. Mus. Inst. 88: 97–110.

Brinkhurst, R. O., 1994. Evolutionary relationships within the Clitellata: an update. Megadrilogica 5: 109–112.

Brinkhurst, R. O., 1996. On the role of tubificid oligochaetes in relation to fish disease with special reference to the Myxozoa. Ann. Rev. Fish. Disease 6: 29–40.

Brinkhurst, R. O., 1999. Lumbriculids, branchiobdellidans and leeches, a review of progress. Hydrobiologia 406: 281–290.

Brinkhurst, R. O. & D. G. Cook, 1980. Aquatic Oligochaete Biology. Plenum Press, New York: 529 pp.

Brinkhurst, R. O. & R. J. Diaz, 1987. Aquatic Oligochaeta. Proceedings of the third International symposium on aquatic Oligochaeta held in Hamburg, Germany September 29 - October 4, 1985. Developments in Hydrobiology 40, Dr W. Junk, Dordrecht: 323 pp.

Brinkhurst, R. O. & B. G. M. Jamieson, 1971. Aquatic Oligochaeta of the World. Univ. Toronto Press, Toronto, Canada: xi + 860 pp.

Brinkhurst, R. O. & M. Marchese, 1987. A contribution to the taxonomy of the aquatic Oligochaeta (Haplotaxidae, Phreodrilidae, Tubificidae) of South America. Can. J. Zool. 65: 3154–3165.

Brinkhurst, R. O. & D. McKey-Fender, 1991. The anatomy of the pharynx of two predatory aquatic oligochaetes. Can. J. Zool. 69: 669–675.

Coates, K. A., 1989. Phylogeny and origins of Enchytraeidae. Hydrobiologia 180: 17–33.

Coates, K. A., T. B. Reynoldson & T. B. Reynoldson, 1996. Aquatic Oligochaete Biology VI. Proceedings of the VI International symposium on aquatic oligochaetes held in Strömstad, Sweden, September 5–10, 1994. Developments in Hydrobiology 115. Kluwer Academic Publishers, Dordrecht: 295 pp.

de Eguileor, M., G. Lanzavecchia, R. Valvassori & P. Lanzavecchia, 1987. Unusual mode of lumbriculids' helical muscles: comparison with body-wall muscles in other microdriles. Hydrobiologia 155: 135–144.

Drewes, C. D. & C. R. Fourtner, 1993. Helical swimming in a freshwater oligochaete. Biol. Bull. 185: 1–9.

Giere, O. & O. Pfannkuche, 1982. Biology and ecology of marine Oligochaeta. Oceanogr. mar. Biol. Ann. Rev. 20: 173–308.

Jamieson, B. G. M., 1978. Phylogenetic and phenetic systematics of the opisthoporous Oligochaeta (Annelida: Clitellata). Evol. Theory 3: 195–233.

Jamieson, B. G. M., 1981. The ultrastructure of the Oligochaeta. Academic Press, London: 462 pp.

Kaster, J. L. , 1989. Aquatic Oligochaete Biology. Proceedings of the 4th International symposium on aquatic oligochaete biology, held in Baton Rouge, Louisiana, 1988. Developments in Hydrobiology 51. Kluwer Academic Publishers, Dordrecht: 252 pp.

Kim, C. B.,S. Y. Moon, S. R. Gelder & W. Kim, 1996. Phylogenetic relationships of annelids, molluscs and arthropods evidenced from molecules and morphology. J. Mol. Evol. 43: 207–215.

Kurashvili, B. E., 1983. Papers presented at the fourth all-union symposium, Tbilisi, 5–7 October. Metsniereba Publishing House, Tbilisi, USSR: 136 p. (Canadian Translation of Fisheries and Aquatic Sciences 5091, 1984, National Research Council, Ottawa, Ontario, Canada KiA 0S2).

Moon, S. Y., C. B. Kim, S. R. Gelder & W. Kim, 1996. Phylogenetic positions of the aberrant branchiobdellidans and aphanoneurans within the Annelida as derived from 18S ribosomal RNA gene sequences. Hydrobiologia 324: 229–236.

Purschke, G., W. Westheide, D. Rhode & R. O. Brinkhurst, 1993. Morphological reinvestigation and phylogenetic relationship of *Acanthobdella peledina* (Annelida, Clitellata). Zoomorph. 113: 91–101

Reynoldson, T. B. R. & K. A. Coates, 1994. Aquatic oligochaete biology V. Proceedings of the 5th oligochaete symposium held in Tallin, Estonia, 1991. Developments in Hydrobiology 95. Kluwer Academic Press, Dordrecht: 320 pp.

Rota, E., 1998. Morphology and adaptations of *Parergodrilus* Reisinger and *Hrabeiella* Pizl and Chalupsky, two enigmatic soil-dwelling annelids. Ital. J. Zool. 65: 75–84.

Schaller, G. B. , 1993. The last Panda. University of Chicago Press, Chicago.

Spencer, D. R. , 1980. The aquatic Oligochaeta of St. Lawrence Great Lakes Region. In R. O. Brinkhurst & D. G. Cook (eds), Aquatic Oligochaete Biology, Plenum Press, New York: 115–164.

Siddall, M. E. & E. M. Burreson, 1996. Leeches (Oligochaeta?:Euhirudinea), their phylogeny and the evolution of life-history strategies. Hydrobiologia 334: 277–285.

Willmer, P. G. , 1995. Modern approaches to the phylogeny of body cavities. In G. Lanzavecchia, R. Valvassori & D. C. Carnevali, body cavities: function and phylogeny. Selected symposia and monographs 8. Mucchi Editore, Modena: 280 pp.

Hydrobiologia **406:** 21–31, 1999.
B. M. Healy, T. B. Reynoldson & K. A. Coates (eds), Aquatic Oligochaetes.

Zoogeography of branchiobdellidans (Annelida) and temnocephalidans (Platyhelminthes) ectosymbiotic on freshwater crustaceans, and their reactions to one another *in vitro*

Stuart R. Gelder
Department of Science, University of Maine at Presque Isle, Maine 04769, U.S.A.

Key words: Annelida, branchiobdellidans, Platyhelminthes, temnocephalidans, biogeography, interactions

Abstract

The zoogeographical distribution of branchiobdellidans (Annelida) and temnocephalidans (Platyhelminthes), ectosymbiotic on freshwater crustaceans, is reviewed and revised. In the areas where the Holarctic branchiobdellidans and Gondwanian temnocephalidans overlap, distributions of species and their respective hosts are reviewed in detail. Specimens of the branchiobdellid, *Branchiobdella astaci*, and the scutariellid temnocephalidan, *Bubalocerus pretneri*, were allowed to meet *in vitro* on the dorsal carapace of a crayfish, *Austropotamobius pallipes*. This usually resulted in the scutariellid being ingested by a branchiobdellid. Specimens of *B. astaci* were placed in contact with the host of *B. pretneri*, *Troglocaris* sp., a cave-dwelling shrimp and the branchiobdellid immediately moved away from the shrimp.

Introduction

Members of the Branchiobdellida, or crayfish worms (Gelder, 1996), are clitellate annelids which have an ectosymbiotic association primarily with freshwater crayfish (Holt, 1986; Gelder & Brinkhurst, 1990). The Temnocephalida are ectosymbiotic turbellarian Platyhelminthes also found on freshwater crayfish, other decapod crustaceans, and occasionally vertebrates (Jennings, 1971; Williams, 1981; Cannon, 1986). The first specimens of temnocephalidans were collected from crayfish near Santiago in Chile by Claudius Gay about 1846. This information was sent in a letter to De Blainville and subsequently quoted by Moquin-Tandon (1846) who assumed the worms to be 'branchiobdellid leeches', and accordingly named them *Branchiobdella chilensis*. However, branchiobdellidans and temnocephalidans have evolved independently to fill the same ectosymbiotic niche on their freshwater crayfish hosts, but in disjunct Holarctic and largely Gondwanian regions, respectively.

The taxonomic positions of the two types of ectosymbionts are still being debated but each taxon forms a monophyletic clade of approximately ordinal rank within its respective phylum. For the purpose of this study, the taxonomy of the annelid Branchiobdellida follows that proposed by Holt (1986), and that of the turbellarian Temnocephalida summarized by Cannon (1986). The terms 'branchiobdellidan', coined by Holt (1986), and the comparable 'temnocephalidan', are used in the common sense for the higher ranks in the respective taxa, thus reserving 'branchiobdellid' and 'temnocephalid' specifically for common usage associated with members of the respective families. Although temnocephalidan is used in this paper for the reasons given, the preferred form by workers on these turbellarians is 'temnocephalans' (Cannon & Sewell, 1994).

The distributions of the two taxa are given using the freshwater faunal realms of Sclater and Wallace (Humphries & Parenti, 1990: 13), including more precise geographical descriptors where necessary. The treatise by Banarescu (1990) on the zoogeography of freshwater animals of the world is, and will remain, the standard reference text on this subject for many years. Unfortunately, the information on the distributions of the two ectosymbiotic taxa (Banarescu, 1990: 383, Figures 5/6 and 387, Figures 5/7) contains a number of inaccuracies and some currently known locations are omitted. The distributions of the ectosymbionts are no longer solely endemic, but include increasingly larger numbers of sightings in new countries due to the commercial introductions of their hosts (Franzén, 1962; Avenant-Oldewage, 1993; Oki et al., 1995). The names of crayfish and other hosts have been cited as given in the source publication where possible, with any modern nomenclature following that of Hobbs (1989).

Given the known distributions of branchiobdellidans and temnocephalidans, it is difficult to accept that individuals from the two groups have not confronted each other while sharing a common host species or in the process of one crustacean host feeding on another. This raises an interesting question as to which one of the two opportunistic carnivores (Jennings, 1968a; Jennings & Gelder, 1979) would be successful in a challenge for the freshwater crustacean host.

There were two aims of this study. The first, was to collate and up-date the distributional information on the two taxa in a brief, general review that would include new records and give details regarding the two taxa in the areas of geographical overlap. The second aim was to report observations made when a representative from each of the two ectosymbiotic taxa came into physical contact with one another and their respective hosts *in vitro*.

Zoogeography of branchiobdellidans

Branchiobdellidans have been reported in the Nearctic, and in two disjunct areas in the Palaearctic realms (Figure 1). Many reports of branchiobdellidans in the Nearctic show them in habitats from the southern Canadian border region (Gelder & Hall, 1990; Gelder, 1991) into Mexico (Holt, 1973), however, the most southerly sighting has occurred in Costa Rica (Holt, 1964), in the northern Neotropical realm. For references to the branchiobdellidans in the Nearctic realm see Holt (1986) and Gelder (1996).

The Palaearctic realm contains two disconnected areas of branchiobdellidan distribution, one in Europe and the other in the Orient (Figure 1). The European area extends east from western France almost to the Ural Mountains in Russia, and from a southern limit bounded by southern France-northern Italy-the Balkans-Ukraine to a northern boundary in southern Sweden-southern Finland-adjacent Russia. For details of references from this region see Karaman (1970), Kosarov et al. (1972), Boshko (1983) and Nesemann (1994). The British Isles were believed to be free of branchiobdellidans until Leeke & Price (1965) found *Branchiobdella astaci* Odier, 1823 on *Austropotamobius pallipes* (Lereboullet, 1858) near Reading in England. They suggest that the crayfish may have been imported from mainland Europe, where the branchiobdellidan-crayfish association is common, and subsequently released into the local English waters. However, no branchiobdellidans were found on any of the over 200 *A. pallipes* collected from the Aire and Ouse river basins and northern Humberside in northern England from 1966 to 1976 by the author. *Austropotamobius pallipes* is also found in Spain (Lowery & Holdich, 1988: 286), but branchiobdellidans have never been reported from the Iberian peninsula. Attempts in June 1996 by the author failed to find specimens of *A. pallipes* in streams east of Gernika, and in the stream tributaries of the Altube Erreka in the Orozko region, Basque Province, Spain. However, while searching in the main course of the Altube Erreka, specimens of the introduced North American Signal crayfish *Pacifastacus leniusculus* (Dana, 1852) were observed and three were collected for examination. Each crayfish carried a healthy and large population, over 40 specimens, of *Xironogiton victoriensis* Gelder & Hall, 1990. Similarly, specimens of *A. pallipes* and *Astacus astacus* (Linnaeus, 1758) were sought in Galicia Province, Spain (Lowery & Holdich, 1988: 286) in June 1996 by the author. Only 32 specimens of *A. pallipes* were collected from one location after searching many sites in the Miño River drainage north of Lugo, and none of these supported branchiobdellidans. *Procambarus (Scapulicambarus) clarkii* (Girard, 1852) was imported from North America into central and southern Spain and now specimens are spreading northwards (Lowery & Holdich, 1988: 287). However, no reports have been published as to whether these crayfish support imported branchiobdellidans.

There are no reports of branchiobdellidans in central and northern Asia, specifically Siberia. The oriental distribution of branchiobdellidans consists of areas along the Amur River in southeastern Russia (Pierantoni, 1912; Timm, 1991), Hokkaido and northern Honshu Islands, Japan (Yamaguchi, 1934; Gelder, 1987), Korea (Subchev et al., 1991) and Manchuria, China (Liu & Zhang, 1983). Further isolated collections in Guangdong (Liu, 1984), Yunnan and Sichwan (Liang, 1963) Provinces of China have extended the southern limits of branchiobdellidans in the Oriental realm. All of these records of the ectosymbiont are from endemic species of crayfish or other crustaceans. An exception is the report of *Cambarincola okadai* Yamaguchi, 1933, on "crayfish formerly transferred

Figure 1. The world-wide distributions of branchiobdellidan annelids and temnocephalidan platyhelminthes is shown using the respective shading for multiple reports and symbols for isolated reports.

from America into Lake Chûzenji, Nikko" (Yamaguchi, 1934: 191).

Zoogeography of temnocephalidans

Temnocephalidans have been reported from the Neotropical, Ethiopian, Australian, and Oriental realms, plus the aberrant Balkan and Caucasus areas in the Palaearctic (Figure 1). For details on the distribution of temnocephalidans, most of the source references are cited in Jennings (1971), Schaefer (1971) and Matjasic (1990). The primary interest of this review is the distribution of temnocephalidans in the areas of overlap with branchiobdellidans, and since a number of early distributional records have been omitted from the recent literature reviews, these are included here.

Only species of *Temnocephala* have been reported in the Neotropical realms, and they are found in three areas. The largest area includes the central temperate region centered on the Plate River drainage and extends westwards into Chile (Jennings, 1971). More recently, Ponce de León (1979), described a new species from Uruguay making it the second from that country. The remaining two areas are in Venezuela and Central America (Jennings, 1971).

The most taxonomically diverse region for temnocephalidans consists of the Australian, Oriental and Ethiopian realms, but only Madagascar in the latter (Matjasic, 1990). The location of the monotypic Actinodactyllidae is restricted to Tasmania and Victoria, Australia. Except for the genera *Temnocephala* and *Dactylocephala*, members of the Temnocephalidae are restricted to the Australian realm. Recent taxonomic studies on temnocephalidans in this realm have resulted in a number of new genera and species being described (Cannon, 1993; Cannon & Sewell, 1995) with others to follow. The monotypic *Dactylocephala* appears to be restricted to Madagascar, while the genus *Temnocephala* which is found in the Australian and Neotropical realms (Banarescu, 1990: 383), and in addition extends across the Oriental realm from southeastern China into central India (Figure 1). It appears that *Temnocephala semperi* Weber, 1890 was reported on the islands of Aru, Celebes, Java and Sumatra, and accounts for virtually all of the records in this northern region. Previously, Semper (1872) had found specimens on the islands of Luzon and Mindanao in the Philippines. On the mainland of southeast Asia, Rohde (1966) obtained specimens from Malaysia and Gravely (1913) collected *T. semperi* from Tavoy and

Dawna Hills of Tennasserin, Mayamar on the border with Thailand, and in Yunnan Province, China. The most northern record so far of this species was found by Dr C. J. Shen at Foochow, Fukien Province (=Fuzhou, Fujian Province), in China (Lee, 1936). To the west, *T. semperi* were collected by Major Godwin-Austen in the Dafla Hills (Wood Mason, 1875), Abor and Manipur Hills (Gravely, 1913), with additional material "collected by Dr. H. S. Roa . . . from Jaintia Hills" (Chauhan & Ramakrishna, 1954), all in Assam, India. Chauhan and Ramakrishna (1954) reviewed the known distribution, and specimens of *T. semperi* in the collection of the Zoological Survey of India, Calcutta, India, and included the first record of the species from the central plateau region of India by Shir D. D. Mukherjee with material from the River Narmada, "probably [at] Hoshangabad," Madhya Pradesh. The most northerly report of a temnocephalidan in the Oriental realm is, "of *T. minor* . . . first observed in the beginning of April, 1985" following, "about 20 specimens of marron . . . introduced into [the] Ibusuki Branch of the KPFES [Kagoshima Prefectural Fisheries Experimental Station] from Western Australia, in June, 1984" (Oki et al., 1995: 73). This facility is located on the southern tip of Kyúshu Island, Japan. The collector of *Temnocephala minor* Haswell, 1887 at Ibusuki, Mr M. Ninagawa, reported that the ectosymbionts, "may reattach to the body of the Japanese freshwater crab, *Geothelphusa dehaani*" (Oki et al., 1995).

Members of the temnocephalidan Scutariellidae in the Oriental realm extend from Sri Lanka to Japan. *Monodiscus macbridei* Fernando, 1952, *Monodiscus parvus* Plate, 1914 and *Caridinicola platei* Fernando, 1952, are extensively cited from Sri Lanka (Matjasic, 1990), however, references to the Indian temnocephalidan records tend to be superficial. Annandale (1912: 247) reported *Caridinicola indica* Annandale, 1912, from the River Madanaddi and adjacent canals at Cattack, and Sambalpar, Orissa, together with a collection by B. L. Chudhuri from the River Ganges near Rajmehal, Bengal, India. About the same time, Gravely (1913: 231), reported collections of *C. indica* from "Dibrugarth by Mr. Kemp" and "the Assam-Bhutan frontier by Dr. Annandale", both in Assam, India. One often cited and imprecise record actually appears as a footnote and was never elaborated, "Since this was written Mr. Gravely has obtained specimens of *Caridinicola* on *Caridina sumatrensis* in the Western Ghats [India]" (Annandale, 1912: 245). In a review of the taxonomy of Japanese scutariellids (Kawakatsu et al., 1989: 45, 46), *C. indica, Caridinicola ceylonica* Monticelli, 1914 and *C. platei* in Japan, India and Sri Lanka were reduced to junior synonyms of *Scutariella indica* (Annandale, 1912), and other records of *C. indica* and *C. platei* in the Indo portion of the region became junior synonyms of *Paracaridinicola indicus* (Plate, 1914). *Caridinicola indica* was first reported in Lake Biwa, by Kemp (1917) followed by other reports in Japan (Kawakatsu et al., 1989). Recently this species in Japan was recognized as distinct (*Caridinicola japonica* Matjasic, 1990) along with *Caridinicola sinica* Matjasic, 1990 in Guanxi Zhuang Province, China (Matjasic, 1990: 11). In 1998, Kawakatsu (pers. comm.) compared and evaluated the type descriptions of *C. japonica* and *C. sinica* with those of specimens from the Indo-oriental region (Kawakatsu et al., 1989). He concluded that all specimens reported from Japan were *Scutariella japonica* (Matjasic, 1990), and that *C. sinica* should be considered a junior synonym of *S. japonica* for the time being.

Temnocephalidans in the Balkans are all members of the Scutariellidae and were found on shrimps mainly from troglodytic habitats (Matjasic, 1990). In the Caucasus region, two scutariellids, *Troglocaridincola mrazeki* Joffe & Djanashvili, 1981 and *Scutariella georgica* Joffe & Djanashvili, 1981, were reported on the blind cave prawn, *Troglocaris anophthalmus* Kollar, 1887, in the west of the Republic of Georgia. Subsequent collections of *T. mrazeki* were made from the lake in the Motion cave near Coatis, Georgian Republic (Joffe et al., 1997). These Georgian records are consistent with a scutariellid distribution along the southern boundary of the Palaearctic realm.

The Ethiopian realm remains a distributional anomaly, as neither branchiobdellidans nor temnocephalidans have been found in Africa, but the monotypic temnocephalid, *Dactylocephala madagascarensis* Vayssière, 1892, does occur in Madagascar. Matjasic (1990: 45), "failed to find [temnocephalidans] on the prawns in western Africa." Similarly, no members of either ectosymbiotic taxa were found on freshwater crabs collected from streams south of Kolwasi, Shaba Province, or from outside Kindu, Maniema Province, Zaire, in 1974 by the author. While these observations are consistent with past reports where the ectosymbionts are absent, modern commercial ventures have introduced large quantities of a variety of crustaceans into the continent. The importation of the Louisiana red swamp crayfish into many African countries (Huner & Barr, 1991) may have already established colonies of branchiobdellidans on the continent's endemic crustaceans. Specimens of *Temnocephala chaerapis* Hett, 1925 (corrected identification is *T. minor*, L. R. G. Cannon, pers. comm.) living on the 'marron', *Cherax tenuimanus* (Smith, 1912), have been reported in South Africa, following the crustacean's importation from Australia for aquaculture (Avenant-Oldewage, 1993). The ectosymbiont was also observed to colonize specimens of the indigenous South African crab, *Potamonantes warreni* Calman, 1918, which proved to be an acceptable, alternate host.

Symbionts' areas of distributional overlap

The three areas in which the taxa overlap are northern Neotropical (Central America), Oriental/Palaearctic (eastern Asia) and the western Palaearctic relicts (Balkans and Caucasus) (Figure 1). In these areas, the symbionts are given with their respective hosts for a better assessment of possible competition.

There are four reports of temnocephalidans in the northern Neotropical (Central America) realm and these are taken in geographical order proceeding north. Specimens of *Temnocephala brenesi* Jennings, 1968 were removed from the branchial chambers of the giant shrimp, *Macrobrachium americanum* Bate, 1868, collected from the Rio Virile and its tributaries the Quebrada El Cacao, near Las Ventanas and Nuestro Amo in Alajuela Province, and the Rio Barranca east of Barranca in Puntarenas Province, Costa Rica (Jennings, 1968b). Kraus (1954: 19) studied *Temnocephala brevicornis* Monticelli, 1899, from the freshwater crab, *Pseudothelphusa* sp., collected at the hacienda San José, northeast of Metapan, in Santa Ana Province, El Salvador. The first report of temnocephalidans in Central America came from material supplied by Prof. L. Bouvier to Vayssière (1898). Specimens of *Temnocephala mexicana* Vayssière, 1898 were removed from the crayfish, *Cambarus digueti* Bouvier, 1897 [= *Procambarus (Procambarus) digueti*] that were collected from 'Rivers in Mexico.' The importance of this study has been overlooked in subsequent reviews as Vayssière (1898: 227) also reported that significant numbers of unidentified branchiobdellidans were found along with the temnocephalids on the host. A similar observation was made by Hobbs and Villalobos (1958) when they found branchiobdellidans and temnocephalids on the freshwater crab, *Pseudothelphusa lamellifrons* Rathbun, 1898 from Arroyo Tapalapan, Santiago Tuxtla, Veracruz, Mexico. The branchiobdellidan material was subsequently examined and a new species, *Cambarincola carcinophilus* Holt, 1973, was described. Under the section 'Materials examined,' Holt (1973: 14) stated that "Temnocephalids were associated with branchiobdellids at . . . Arroyo de Zapoapan de Cabana, Veracruz," Mexico. It is assumed that the two named locations refer to the same site or area. In this same area, the branchiobdellidan had two hosts, *Pseudothelphusa veracruzana* Rodriguez & Smalley, 1970, and *Procambarus ruthveni zapoapensis* Villalobos, 1954, but it is not clear if both carried temnocephalids.

Sixteen species of branchiobdellidans have been reported from 'Mesoamerica' by Holt (1973). However, *Pterodrilus mexicanus* Ellis, 1919, was not found in his survey and was cited from the type description only for completeness. The hosts of the branchiobdellidans were surprisingly diverse and consisted of 15 species of astacid crayfish, three species of freshwater crab and two species of isopod. Specimens of *Bdellodrilus illuminatus* (Moore, 1894), a widely distributed species across the Nearctic, were removed from *Procambarus digueti* by Dieter Enkerlin at Rancho Nueva España, near Jacona, Michoacan Province, Mexico (Holt, 1973: 9). This is the only species and site recorded for a branchiobdellidan in western Mexico. The majority of the remaining branchiobdellidans were collected from the region with easterly-flowing drainages that enter the Gulf of Mexico. *Cambarincola ellisi* Holt, 1973, *Cambarincola hoffmani* Holt, 1973, *Cambarincola jamapaensis* Holt, 1973, *Cambarincola miradenus* Holt, 1973, *Cambarincola olmecus* Holt, 1973, *Cambarincola susanae* Holt, 1973, *Sathodrilus veracruzicus* Holt, 1973, and *Sathodrilus villalobosi* Holt, 1973, were each found on from one to four species of the crayfish, *Procambarus.* In addition, *Cambarincola nanognathus* Holt, 1973, *Cambarincola toltecus* Holt, 1973, and *Sathodrilus prostates* Holt, 1973, were also found on crayfish as well as the crabs *P. veracruzana* and *Potamocarcinus nicaraguensis* Rathbun, 1893. However, *C. carcinophilus* and *Cambarincola smalleyi* Holt, 1964 were restricted to the crabs *P. veracruzana* and *Pseudothelphusa tumimanus* Rathbun, 1898, respectively, and *Cambarincola acudentatus* Holt, 1973 to two isopods, *Speocirolana bolivari* (Rioja, 1953) and *Speocirolana pelaezi* (Bolivar & Pieltain, 1950). Of these, two species of branchiobdellidans are of particular importance due to their extended southern distribution. *Cambarincola nanognathus* was found on the crab, *P. veracruzana*, from the Rio Tapalapa, Santiago Tuxtla, and the crayfish, *Procambarus vazquezae* Villalobos, 1954, from Laguna de Catemarco, Playa Azul, both in Veracruz

Province, southeastern Mexico, and from Isleta de Granada, Lago de Nicaragua, Nicaragua, on the freshwater crab, *P. nicaraguensis*, collected by G. Alvilez (Holt 1973: 23). The single record of *C. smalleyi* on the freshwater crab, *P. tumimanus*, in Costa Rica (Holt, 1964) remains the most southerly collection of a branchiobdellidan in the Central America.

So far there have been no records of temnocephalidans and branchiobdellidans together on the same host within the Oriental realm. However, the two types of ectosymbionts have both been reported in China and Japan. The Chinese temnocephalid in the potential overlap region is *T. semperi* on the freshwater crab, *Potamon* sp. at Fuzhou, Fujian Province (Lee, 1936), and the scutariellid, *C. sinica*, on the prawn *Caridina* sp. from the Lijiang River, Guanxi Zhuang Province, China (Matjasic, 1990: 11); both locations are in the southeast of China. Gravely (1913: 229) wrote, "Specimens ... of *Potamon andersonianum* from Yunnan ... bear eggs closely resembling those evidently belonging to this species [*T. semperi*] on crabs from the Abor country and Dawna Hills." This remark has subsequently been repeated and interpreted to mean that *T. semperi* occurs in Yunnan Province, however, it remains to be substantiated.

The only branchiobdellidans reported in southern China are the monotypic *Caridinophila unidens* Liang, 1963, on *Caridina yunnanesis* Yu, Shou-chie, 1938, from Lake Er-Hai [=Erh hai], Yunnaun Province, and *Holtodrilus truncatus* (Liang, 1963) collected from *Neocaridina denticulata sinensis* (Kemp, 1918) at Sichwan, Honan [=Hunan] Province by Liang (1963) and in the Changjiang River near Shaoguan, Guangdong Province by Liu (1984). These records provide the most southerly limits of branchiobdellidan distribution in the Oriental realm. Based on the present information, it is questionable whether both branchiobdellidans and temnocephalidans are present in Yunnan Province. However, the potential certainly exists for both symbiont taxa to meet in the southeastern areas of China.

Temnocephalidans in Japan consist of the imported *T. minor* on the Australian marron located at the southern tip of Kyúshu, (Oki et al., 1995) and the endemic scutariellid, *S. japonica* (=*C. indica*) on Honshû, Japan (Kawakatsu et al., 1989). Kemp (1917) reported *C. indica* on the small shrimp, *Paratya compressa compressa* de Haan, 1849, from Omatus (Omatus pond adjacent to Lake Biwa), and corrected the host's genus name from *Xiphocaridina*. Subsequent records of this symbiosis in Lake Biwa by Kobayashi (1935) and Honjô (1937) continued to use the junior synonym *Xiphocaridina*. In addition, Honjô (1937) described *X. compressa* as living most of the year in deep water and returning to the littoral zone from the end of March to the beginning of July for its eggs to hatch and the juveniles to mature. Such a migration is in contradiction to 20 years of observations of *P. compressa compressa* remaining in the littoral zone throughout the year (Dr Machiko Nishino, Lake Biwa Research Institution, Japan, pers. comm.). The same association of *C. indica* on *P. compressa* was observed by Yanada (1955) near Okazaki City, Aichi Prefecture, about 100 km east of Lake Biwa. The occurrence of *C. indica* on *P. compressa improvisa* (see Nishino, 1981: 202) has been reported in four Prefectures on northern Honshû: Fukushema and Miyagi Fukushema (Kamita, 1970), Yamagata (Umemoto, 1936) and Aomori (Negishi, 1935). Branchiobdellidan species on the endemic crayfish, *Cambaroides japonicus* (de Haan, 1849), in Japan are restricted to the islands of Hokkaido and the northern area of Honshû (Yamaguchi, 1934, Ohtaka, pers. comm.). The Honshû record consists of *Cirrodrilus aomorensis* (Yamaguchi, 1934: 196) from the northernmost Prefecture of Aomori on Honshû. Therefore a definite area of geographical proximity of the endemic branchiobdellidans and scutariellids exists. In addition, the introduction of the American *Cambarincola okadai* into Lake Chûzenji, Tochigi Prefecture (Yamaguchi, 1934: 191) artificially placed a branchiobdellidan close to the scutariellid area in central Honshû.

The European area of overlap is determined by the restricted geographical distribution of the scutariellid temnocephalidans which extend from Slovenia and the adjacent area in Italy (Gasparo et al., 1984), south through Croatia, Bosnia and Herzegovina, and into Montenegro and Macedonia (Matjasic, 1990: 47). The scutariellids in this area are predominantly troglodytes, or cave-dwellers, on atyaeid shrimps, although gammarid amphipods have proven to be acceptable, alternative hosts (Matjasic, 1990: 41,43,47). However, *Scutariella didactyla* Mrazek, 1907 is the exception being found on the shrimp, *Atyaëphyra desmaresti* (Millet, 1834), from shallow-water vegetation near the mouth of the river Moraca prior to entering Lake Skadar in Montenegro (Mrazek, 1907). In the 1950s, Matjasic (1990: 48) visited the northern shores of Lake Skadar and its tributaries in the vicinity of Viripaza, and collected *A. desmaresti* carrying *S. didactyla*. Additional specimens of *S. didactyla* were sent to Matjasic (1990: 48) after collection from the same type of habitat and host-species in Lake Dojran, Macedonia.

Georgévitch (1955) described a large number of branchiobdellidans found on crayfish from Lake Dojran; for the modern nomenclature of these species see Gelder (1996). Previously, Moszyński (1938) had reported branchiobdellidans on *Potamobius astacus* [= *Astacus astacus*] in the River Ibar near Kosovska Mitrovica, Kosovo, Yugoslavia. The author visited the northern shore of Lake Skadar in Montenegro in August, 1966, but did not obtain any shrimps. However, approximately 50 crayfish, probably *Austropotamobius torrentium* according to Horton H. Hobbs Jr., were collected from the river in the village of Rijeka Crnojevica located west of Lake Skadar. These crayfish did not have any scutariellids, but did have specimens of *Branchiobdella hexodonta* Gruber, 1883 in the gill chambers and *B. pentodonta* on the ventral thorax and abdomen. As part of his review of branchiobdellidans in Europe, Karaman (1970), reported branchiobdellidans from all parts of Yugoslavia as the country was then defined. Finally, Gelder et al. (1994) reported *B. astaci*, *Branchiobdella parasita* Henle, 1835, *Branchiobdella balcanica* Moszyński, 1938 and *B. pentodonta* on crayfish from drainage ditches, streams and lakes in the Ljubljana and Bled areas in Slovenia. No underground crayfish or shrimps were collected for examination. The subterranean distribution of scutariellids in Slovenia (Matjasic, 1990: 50, 51) occurs in the southern half of the country. It is not known if a search for scutariellids has been conducted in the northern region.

Observations on in vitro exposures of a representative species of a branchiobdellidan to a temnocephalidan and their respective hosts

Troglocaris sp., a cave-dwelling shrimp, were collected in the Grotta di Comarie, VG 4221, 15 m down, near the village Doberdo del Lago, Province of Gorizia, Italy, by Dr F. Gasparo and Prof. A. Minelli on 17th June 1989. Six specimens, all about 25 mm long, supported a total of ten mature specimens of the scutariellid temnocephalidan, *Bubalocerus pretneri* Matjasic, 1958, on their carapaces and anterior legs. Three specimens of *Austropotamobius pallipes* were collected from a tributary of the Rosandra River near S. Dorligo della Valle, Trieste Province, Italy, by Dr F. Gasparo on 20th June 1989 (Gelder et al., 1994). The crayfishes carried six specimens of *Branchiobdella astaci* on their carapaces.

Three specimens of *B. astaci*, ranging from 5.0 to 6.0 mm long, on the dorsal carapace of a 35 mm long specimen of *A. pallipes*, swept their heads around in typical food-searching motions. A 2.0 mm long specimen of the scutariellid *B. pretneri* was placed about 2.0 mm beyond the reach of one of the branchiobdellids. The scutariellid transferred itself from the mounted needle point onto the dorsal carapace without any hesitation and then moved around its immediate area with characteristic looping motions. When the scutariellid contacted the branchiobdellid, both paused, and then the branchiobdellid sucked the smaller worm into its pharynx. A moment later, the posterior adhesive disc of the scutariellid detached from the carapace surface and ingestion was completed. A second scutariellid followed the same fate. A third specimen was partially ingested but did not release its posterior adhesive hold on the carapace and succeeded in pulling itself out of the branchiobdellid's pharynx. Immediately the scutariellid moved rapidly under the protection of the rostrum without having encountered a branchiobdellid. The fourth scutariellid moved rapidly from its point of contact with the carapace to the rostrum. The branchiobdellids were removed from the crayfish and a thorough search of the crayfish's surface, including the branchial chambers, and dish was made but neither of the two escaped scutariellids were found. However during the dismemberment of the crayfish, five mature specimens of *Branchiobdella italica* Canegallo, 1928, each about 2.0 mm long, were discovered in the protective folds of the ventral cephalothorax and rostrum regions. These specimens probably explain why the two scutariellids that escaped were not recovered.

In another set of observations, three *B. astaci* was placed separately in contact with the dorsal surface of a *Troglocaris* sp.. On each occasion, when the shrimp's cuticle touched the branchiobdellid's peristomium, the latter made violent twisting body movements and moved away from the shrimp. After the *in vitro* observations finished, the shrimps were dissected and 12 specimens of a second scutariellid, *Troglocaridicola (Jeanbaeria) vilkae* Matjasic, 1990, were found deep in the hosts' branchial chambers.

From these limited observations three conclusions can be made. The scutariellids found the crayfish an acceptable host, but the branchiobdellids were violently opposed to touching the cuticle of the shrimp, host of the scutariellid. The scutariellids were kept from colonizing the crayfish probably due to the predatory behavior of the branchiobdellids. The escape of one partially ingested scutariellid indicates that these ectosymbionts may have a defense mechanism that under certain circumstances can deter a branchiobdellid from completing ingestion.

Discussion

The Scutariellidae and Temnocephalidae in the Temnocephalida are clearly separate families (Cannon, 1986). The scutariellids on atyaeid shrimps are distributed primarily along the southern margin of the Palaearctic with the Oriental and Ethiopian realms, although those in Sri Lanka do pose a slight anomaly, while the temnocephalids have a Gondwanian distribution. The order Branchiobdellida does not show a similar distributional regionalisms of the five families (Holt, 1986). However, the genera of branchiobdellidans do show a clear endemic separation into those reported in the Palaearctic and Nearctic realms respectively. Unfortunately there is no single morphological character to validate such a supra-generic arrangement. This could be seen as an additional reason for arguing that branchiobdellidans only warrant the rank of family as accepted by Stephenson (1930) and others.

The common host taxa of the branchiobdellidans and temnocephalidans are the freshwater crayfish in the Infraorder Astacidea, which consists of three families, Astacidae, Cambaridae and Parastacidae (Hobbs, 1988: 55). The parastacids have a Gondwanian distribution and this matches the distribution of the temnocephalids. The first two families are Holarctic and generally reflect the distribution of the branchiobdellidans. However, members of the Astacidae are restricted to the western regions of both the Nearctic and Palaearctic realms with the Cambaridae being similarly restricted to both of the eastern regions, including Mexico. A satisfactory explanation for this distribution remains elusive. Although some branchiobdellidan genera show an east or west polarization on the respective crayfish in each realm, others do not.

Regions of branchiobdellidan and temnocephalidan overlap

Branchiobdellidans have been reported from the Republic of Ukraine (Boshko, 1983), but not from the adjacent Republic of Georgia, even though crayfish inhabit both countries (Hobbs, 1988). Joffe & Djanashvili (1981) reported two species of the scutariellid *Troglocaridincola* in Georgia on cave shrimps, and so it is possible that another area of overlap similar to that in the Balkans may exist. In spite of the geographical overlap of the branchiobdellids and scutariellids in the Balkans they have never been reported on the same host or host species. Separate host species and their different respective habitats offers an obvious explanation; branchiobdellidans live on surficial crayfish (Georgévitch, 1957) and scutariellids on trogloditic shrimps (Matjasic, 1990). However, the habitat factor appears incidental when scutariellid-carrying shrimps and branchiobdellidan-carrying crayfish were both found in the shallow waters of Lake Dorjan, Macedonia (Georgévitch, 1955; Matjasic, 1990). It is difficult to accept that the respective ectosymbiont-bearing shrimps and crayfishes in Lake Dorjan have not come into physical contact with each other on repeated occasions. Such contact would be more than adequate for the ectosymbionts to change hosts. The fact that this phenomenon has not been observed is consistent with the experimental observations reported in the current study.

Branchiobdellidans in southern China are beyond the southern range of crayfish and have adopted the shrimp, *Neocaridina* sp. as their new host. The northward extension of the range of *T. semperi* has brought the ectosymbiont into southern China where it is found on the freshwater crab, *Potamon* sp. Although it is possible that the two ectosymbionts could meet, there is no reason to believe that they have. The situation in Aomori Prefecture, northern Honshû, Japan, is different to that found in the Balkans where the ranges of both endemic scutariellids on shrimps and branchiobdellidans on crayfish are separated by a significant distance.

The second area of overlap is in the Neotropical realm's northern region. From the data already cited, there appears to be ample opportunity for the two taxa to come into physical contact on a variety of host species. Contrary to the predicted results based on the confrontational branchiobdellid-scutariellid interactions, a sympatric association between the branchiobdellidans and temnocephalids is expected based on previous observations (Vayssière, 1898). It is, therefore, surprising that such a sharing of a common host is limited to one actual observation, and a second, possible association (Hobbs & Villalobos, 1958).

Host tolerance of ectosymbionts

The observations described on the reaction of the ectosymbionts to the two host species indicated that preference in the associations was determined only by the ectosymbionts. It is generally assumed that the host is unable to affect or prevent the ectosymbiont's colonization. However, specimens of *P. (S.) clarkii*, *Cambarus bartonii* (Fabricius, 1798), *Orconectes virilis* Hagen, 1870, and *Macrobrachium* sp. have been observed scraping their exoskeleton and 'combing' their gills, with appropriate appendages (Gelder, unpub. obs.). From these observations, very few places on the body were effectively out-of-reach of the cleaning behaviour. The hosts have sensory setae on the body surface, and so the presence and movement of the ectosymbionts will be monitored by the host. Although most crayfish appear to ignore their ectosymbionts, Gelder & Smith (1986) observed a crayfish, *O. virilis*, to scrape off successfully branchiobdellidans on its carapace. Jones & Lester (1993) observed that the temnocephalid, *Diceratocephala boschmai* Baer, 1953, lived and deposited cocoons on five species of *Cherax*, but did not survive on *Macrobrachium* spp. because of the host's grooming behaviour. In a subsequent study in the laboratory, Jones & Lester (1996) showed that grooming behaviour was an important factor in reducing the temnocephalid populations on *Cherax* spp. However, the absence of temnocephalidans from *Macrobrachium* sp. was due to grooming, 60%, but the remaining 40% left the host for unknown reasons (Jones & Lester, 1996). A different interaction was reported by Jennings (1968a) who observed established populations of *T. brenesi* and their cocoons along with accumulated detritus on the gills of *M. americanum* in Costa Rica. It would help to know what factors cause a host to be tolerant of the ectosymbionts. A study of these factors could help to predict whether or not an exotic ectosymbiont is likely to find an acceptable endemic host. Specimens of *T. minor* have been introduced into South Africa and Japan (Avenant-Oldewage, 1993; Oki et al., 1995) and in both cases the temnocephalids have found acceptable endemic hosts. The future impact of these introduced specimens on other endemic organisms has not been considered.

Multiple ectosymbiont species interactions on a single host

It is usual for more than one species of branchiobdellidan to be found on the same host. Hobbs et al. (1967: 8) reported six species on a single crayfish in Virginia, and the author has removed seven species representing five genera from a crayfish collected in Tennessee. Both of these observations noted a microhabitat preference for some of the species, i.e. *Bdellodrilus illuminatus* in the gill chambers, *Xironogiton instabilis* (Moore, 1894) on the chelae, and *Cambarincola holti* Hoffman, 1963 on the ventral thoracic surface of the host. Conversely, other species i.e. *Cambarincola fallax* Hoffman, 1963, and *Pterodrilus annulatus* Gelder, 1996, could be found anywhere on the exoskeleton, excluding the gills, as long as the area was not occupied by another species. Comparable multi-species populations of scutariellids with different microhabitats on their shrimp hosts have been reported in the Balkan region (Matjasic, 1990). Similarly temnocephalidans do approach the diversity of the North American branchiobdellidans on a single host with six species, each representing a different genus, on *Cherax quadricarinatus* (von Martens, 1868) in Australia (Cannon & Sewell, 1994). The report of an unidentified branchiobdellidan and *T. mexicana* living together in the gill chambers of a crayfish in Mexico (Vayssière, 1898: 234) is at first surprising, especially given that both taxa are opportunistic carnivores, and following the *in vitro* observation made in the present study. Vayssière also reported that both ectosymbionts feed on the epizoonts and organic particles carried into the gill chambers by the water currents. This apparent sharing of the food resource is comparable to the sympatry documented on multi-species populations of both ectosymbiotic taxa on their respective hosts. Under these conditions, maybe the Mexican observations are not too surprising.

Resource partitioning based on subtle differences in the food and feeding strategies of the respective ectosymbiont species is probably the most important criterion in these harmonious associations. Coupled with food preferences is a defense or deterrence mechanism to prevent or reduce predation. The best hope of finding the basis for a successful sympatric association is to rediscover and study the branchiobdellidan-temnocephalid association found in Mexico by Vayssière (1898).

Acknowledgements

Parts of this work have been supported by the 1995 LIBRA Professorship, Faculty Development grants and Mini-Grant funds from the University of Maine at Presque Isle, U.S.A. to S.R.G. from 1985 to the present, and Consiglio Nazional della Ricerche (Rome) Italy as a Bilateral Project to Dr Marco Ferraguti, Dipartimento di Biologia, Universita degli Studi di Milano, Italy (1989–1996). Grateful thanks are expressed to Dr Marco Ferraguti, Dr Falvio Gasparo and Prof. A. Minelli in Italy, Prof. Dr Boris Sket, University of Ljubljana, and Ms Natalija Budihna and colleagues, Fisheries Research Intitute, Ljubljana, Slovenia, Drs Pilar Rodriguez and Maite Madrid, Universidade del Pais Vasco, Bilbao, Spain, Dr Enrique Martinez-Ansemil and colleagues, Universidade de Coruña, A Coruña, Spain, Dr Machiko Nishino, Lake Biwa Research Institute, Shiga, Japan, and Prof. Masaharu Kawakatsu, Fuji Women's University, Sapporo, Japan, for assistance in collecting specimens, hospitality or providing information. Constructive criticism of the manuscript was provided by Drs Andrea M. Gorman, University of Maine at Presque Isle, Seth Tyler, University of Maine, U.S.A., and Lester R. G. Cannon, Queensland Museum, South Brisbane, Australia. This paper is dedicated to Dr Joseph B. Jennings, University of Leeds, U.K., for his patient support over the years and who, inadvertently, introduced me to the branchiobdellidans in 1966.

References

Annandale, H., 1912. *Caridinicola*, a new type of Temnocephaloidea. Rec. Ind. Mus. 7: 243–253.

Avenant-Oldewage, A., 1993. Occurrence of *Temnocephala chaeropsis* on *Cherax tenuimanus* imported into South Africa, and notes on its infestation of an indigenous crab. S. Afr. J. Sci. 89: 427–428.

Banarescu, P., 1990. Zoogeography of Fresh Waters. AULA-Verlag, Wiesbaden, Germany, 1617 pp.

Boshko, E. G., 1983. Small worms dwelling on river crayfish of some water bodies in the Ukraine. Vestnik zool. 7: 13–20.

Cannon, L. R. G., 1986. Turbellaria of the World. A Guide to Families & Genera. Queensland Museum, South Brisbane, Australia, 136 pp.

Cannon, L. R. G., 1993. New temnocephalans (Platyhelminthes): ectosymbionts of freshwater crabs and shrimps. Mem. Queensl. Mus. 33: 17–40.

Cannon, L. R. G. & K. B. Sewell, 1994. Symbionts and biodiversity. Mem. Queensl. Mus. 36: 33–40.

Cannon, L. R. G. & K. B. Sewell, 1995. Craspedellinae Baer, 1931 (Platyhelminthes: Temnocephalida) ectosymbionts from the branchial chamber of Australian crayfish (Crustacea: Parastacidae). Mem. Queensl. Mus. 38: 397–418.

Chauhan, B. S. & G. Ramakrishna, 1954. *Temnocephala semperi* Weber, 1890 from the Narmada River, with a note on other Temnocephalid material in the Zoological Survey of India, Calcutta. Rec. Indian Mus. 51: 421–425.

Franzén, Å., 1962. Notes on the morphology and histology of *Xironogiton instabila* (Moore, 1893) (Fam. Branchiobdellidae) with special reference to the muscle cells. Zool. Bidr. Upps. 35: 369–383.

Gasparo, F., A. Minelli & P. Brandayr, 1984. First report of temnocephalids in Italy. Boll. zool. ital. 51(suppl.): 51.

Gelder, S. R., 1987. Observations on three species of branchiobdellid (Annelida: Clitellata) worms from eastern Asia. Hydrobiologia 155: 15–25.

Gelder, S. R., 1991. New location reports of Nearctic branchiobdellidans (Annelida: Clitellata). Can. Field-nat. 105: 390–391.

Gelder, S. R., 1996. A review of the taxonomic nomenclature and a checklist of the species of the Branchiobdellae (Annelida: Clitellata). Proc. biol. Soc. Wash. 109: 653–663.

Gelder, S. R. & R. O. Brinkhurst, 1990. An assessment of the phylogeny of the Branchiobdellida (Annelida: Clitellata) using PAUP. Can. J. Zool. 68: 1318–1326.

Gelder, S. R., G. B. Delmastro & M. Ferraguti, 1994. A report on branchiobdellidans (Annelida: Clitellata) and a taxonomic key to the species in northern Italy, including the first record of *Cambarincola mesochoreus* on the introduced American red swamp crayfish. Boll. Zool. 61: 179–183.

Gelder, S. R. & L. A. Hall, 1990. Description of *Xironogiton victoriensis* n. sp. from British Columbia, Canada, with remarks on other species and a Wagner analysis of *Xironogiton* (Clitellata: Branchiobdellida). Can. J. Zool. 68: 2352–2359.

Gelder, S. R. & R. C. Smith, 1986. Distribution of branchiobdellids (Annelida: Clitellata) in northern Maine, U.S.A. Trans. am. microsc. Soc. 106: 85–88.

Gravely, F. H., 1913. Temnocephalidae. Rec. indian Mus. 8: 229–232.

Georgévitch, J., 1955. Sur les Branchiobdellides des écrevisses du Lac Dojran. Acta Mus. Macedonici Sci. Nat., Skopje 2: 199–221.

Georgévitch, J., 1957. Les Branchiobdellides de Yougoslavie. Bull. Acad. Serbe Sci. Cl. Sci. Math. Nat. 5: 5–25.

Hobbs, H. H. Jr., 1988. Crayfish distribution, adaptive radiation and evolution. In D. M. Holdich & R. S. Lowery (eds), Freshwater Crayfish: Biology, Management and Exploitation. Timber Press, Portland (Oregon), U.S.A.: 52–82.

Hobbs, H. H. Jr., 1989. An illustrated checklist of the American crayfishes (Decapoda: Astacidae, Cambaridae, and Parastacidae). Smithson. Contr. Zool. 480: 1–236.

Hobbs, H. H. Jr., & A. Villalobos. 1958. The exoskeleton of a freshwater crab as a microhabitat for several invertebrates. Virginia J. Sci. 9: 395–396.

Hobbs, H. H. Jr., P. C. Holt & M. Walton, 1967. The crayfishes and their epizootic ostracod and branchiobdellid associates of the Mountain Lake, Virginia, region. Proc. U.S. nat. Mus. 123: 1–84.

Holt, P. C., 1964. A new branchiobdellid (Annelida) from Costa Rica. Tulane Stud. Zool. 12: 1–4.

Holt, P. C., 1973. A summary of the branchiobdellid (Annelida: Clitellata) fauna of Mesoamerica. Smithson. Contr. Zool. 142: 1–40.

Holt, P. C., 1986. Newly established families of the order Branchiobdellida (Annelida: Clitellata) with a synopsis of the genera. Proc. biol. Soc. Wash. 99: 676–702.

Honjô, I., 1937. Physiological study on the neuromuscular system of lower worms. 1. *Caridinicola*. Mem. Coll. Sci. Kyoto Univ. ser. B, 12: 187–210.

Humphries, C. J. & L. R. Parenti, 1990. Cladistic biogeography. Clarendon Press, Oxford, U. K. 98 pp.

Huner, J. V. & J. E. Barr, 1991. Red swamp crayfish: biology and exploitation. Louisiana Sea Grant College Program, 128 pp.

Jennings, J. B., 1968a. Feeding digestion and food storage in two species of Temnocephalid flatworms (Turbellaria: Rhabdocoela). J. Zool. Lond. 156: 1–8.

Jennings, J. B., 1968b. A new Temnocephalid flatworm from Costa Rica. J. nat. Hist., Lond. 156: 117–120.

Jennings, J. B., 1971. Parasitism and commensalism in the Turbellaria. Adv. Parasit. 9: 1–32.

Jennings, J. B. & S. R. Gelder, 1979. Gut structure, feeding and digestion in the branchiobdellid oligochaete *Cambarincola macrodonta* Ellis 1912, an ectosymbiote of the freshwater crayfish *Procambarus clarkii*. Biol. Bull. 156: 300–314.

Joffe, B. I. & R. A. Djanashvili, 1981. First record of temnocephalids in the territory of the U.S.S.R. Parasitol. (Leningrad) 24: 473–481.

Joffe, B. I., I. V. Solovei, N. A. Watson & L. R. G. Cannon, 1997. Structure and evolution of the pharynx in the Temnocephalida (Platyhelminthes). Can. J. Zool. 75: 205–226.

Jones, T. C. & R. G. J. Lester, 1993. Aspects of the biology and pathogenicity of *Diceratocephala boschmai* (Platyhelminthes: Temnocephalida), an ectosymbiont on the Redclawed crayfish, *Cherax quadricarinatus*. Aust. J. mar. freshwat. Res. 44: 927–933.

Jones, T. C. & R. G. J. Lester, 1996. Factors influencing populations of the ectosymbiont *Diceratocephala boschmai* (Platyhelminthes: Temnocephalida), on the redclaw crayfish *Cherax quadricarinatus* maintained under laboratory conditions. Aquaculture 143: 233–243.

Kamita, H., 1970. Studies on the freshwater shrimps, prawns and crawfishes of Japan. Sonoyama Shoten, Matsue, Japan, 213 pp. (in Japanese).

Karaman, S. M., 1970. Beitrag zur Kenntnis der europaischen Branchiobdelliden (Clitellata, Branchiobdellidea). Int. Rev. Hydrobiol. 55: 325–33.

Kawakatsu, M., N. Nunomura & H. Suzuki, 1989. Corrections and emendations of specific names of several Japanese species of turbellarians (Acoela, Tricladida-Maricola, Polycladida-Cotylea, and Temnocephalida-Scutariellidata) and Nemertinea (Enopla-Hoplonemertea-Monostylifera). Bull. Fuji Women's Coll. No. 27, ser. II: 41–51.

Kemp, S., 1917. Notes on Crustacea Decapoda in the Indian museum. XI Atyidae of the genus *Paratya* (=*Xiphocaridina*). Rec. Ind. Mus. 13: 293–306.

Kobayashi, H., 1935. Observation on *Caridinicola indica*, a Temnocephalen parasite of *Xiphocaridina compressa* in Lake Biwa. Bot. et Zool. 3: 2124–2128.

Kosarov, G., P. Michailova & M. A. Subchev. 1972. Studies on the Branchiobdellidae (Oligochaeta, Annelida) of Bulgaria. Ann. Univ. Sofia Kliment Ochridski- Fac. Biol. 64: 77–89.

Kraus, O., 1954. *Temnocephala brevicornis* en El Salvador. Comun. Inst. Trop. Invest. Cient. 3: 19–20.

Lee, L. Y., 1936. *Temnocephala semperi* Weber on crab *Potamon* sp. South China. Contr. Inst. zool. Acad. Peiping 3: 127–129.

Leeke, C. J. & A. Price, 1965. *Branchiobdella astaci* Odier (Oligochaeta, Annelida). Reading Nat. 17: 18–19.

Liang, Y.-L., 1963. Studies on the aquatic Oligochaeta of China. I. Descriptions of new naids and branchiobdellids. Acta Zool. sinica 15: 560–570.

Lui, S.-C., 1984. Descriptions of two new species of the genus *Stephanodrilus* from Northeast China and notes on *St. truncatus* Liang from Guangdong Province (Oligochaeta: Branchiobdellidae). Acta Zootaxon. sinica 9: 351–355.

Liu, S.-C. & D.-C. Zhang, 1983. Three new species of the genus *Branchiobdellida* (Oligochaeta: Branchiobdellidae) from China. Acta Zootaxon. sinica 8: 246–251.

Lowery, R. S. & D. M. Holdich, 1988. *Pacifastacus leniusculus* in North America and Europe, with details of the distribution of introduced and native crayfish species in Europe. In D. M. Holdich & R. S. Lowery (eds), Freshwater Crayfish: Biology, Management and Exploitation. Timber Press, Portland (Oregon), U.S.A.: 283–308.

Matjasic, J., 1990. Monograph of the family Scutariellidae (Turbellaria, Temnocephalidea). Slovenska Akademija Znanosti in Umetnosti, Ljubljana, 166 pp.

Moquin-Tandon, A., 1846. Monographie de la famille des Hirudines. J.-B. Bailliere, Paris, 448 pp.

Moszyński, A., 1938. Oligochtes parasites de l'crevisse (*Potamobius astacus* L.) de la Yougoslavie. Bull. Soc. Sci. Skopje 18: 69–75.

Mrazek, A., 1907. Ein europischer Vertreter der Gruppe Temnocephaloidea. Sitz.. K. Bhm. Ges. wiss. Prag 11: 1–7.

Nesemann, H., 1994. The epizoic leeches found on crayfishes of the basin of the Upper Danube (Austria, Germany) with an identification key to the European species (Clitellata, Branchiobdellida). Lauterbornia 19: 79–93.

Negishi, H., 1935. Notes on *Caridina indica* Annandale, found in the vicinity of Hirosaki in Aomori Prefecture. Botany zool., Tokyo 4: 949–952 (in Japanese).

Nishino, M., 1981. Brood habits of two subspecies of a freshwater shrimp, *Paratya compressa* (Decapoda, Atyidae) and their geographical variations. Jap. J. limnol. 42: 201–219 (in Japanese).

Oki, I., S. Tamura, M. Takai & M. Kawakatsu, 1995. Chromosomes of *Temnocephala minor*, an ectosymbiotic turbellarian on Australian crayfish found in Kagoshima Prefecture, with karyological notes on exotic turbellarians found in Japan. Hydrobiologia 305: 71–77.

Pierantoni, U., 1912. Monografia dei Discodrilidae. Ann. Mus. Zool. R. Univ. Napoli 3: 1–27.

Ponce de León, R., 1979. Especies americanas de Temnocephalidea Benham (Platyhelmintha), 1. Descripcion de *Temnocephala rochensis* n. sp. de la camara paleal de *Pomacea canaliculata* (Lamarck). Rev. Biol. Uruguay 7: 39–43.

Rohde, K., 1966. A Malayan record of *Temnocephala sempari*, ectocommensal on the freshwater crab *Potamon rafflesi* Roux. Med. J. Malaya 20: 356.

Schaefer, C. W., 1971. Observations on temnocephalid hosts and distributions. Sonder. Z. zool. Syst. Evolut. 9: 139–143.

Semper, C., 1872. Zoologische Ajhoismen II. über die Gattung *Temnocephala*. Z. wiss. Zool. 22: 307–310.

Stephenson, J., 1930. The Oligochaeta. Clarendon Press, Oxford, pp. 978.

Subchev, M. A., L. S. Stanimirova & T. Tomek, 1991. Distribution of branchiobdellidans (Annelida: Clitellata) on the Korean Peninsula. Acta Zool. bulg. 41: 12–17.

Timm, T., 1991. Branchiobdellida (Oligochaeta) from the farthest South-East of the U.S.S.R. Zool. Scr. 20: 321–331.

Umemoto, H., 1936. Notes on *Caridina indica* found in Tsuruoka District of Yamagata Prefecture. Botany zool., Tokyo 3: 1375–1376 (in Japanese).

Vayssière, A., 1898. Description du *Temnocephala mexicana*, nov. sp. Ann. Fac. Sci. Marseille 8: 227–235.

Williams, J. B., 1981. Classification of the Temnocephala Platyhelminthes. J. nat. Hist., Lond. 15: 277–300.

Wood-Mason J., 1875. Note on the Geographical Distribution of the 'Temnocephala chilensis' of Blanchard. Ann. Mag. nat. Hist. 15: 336–337.

Yamaguchi, H., 1934. Studies on Japanese Branchiobdellidae with some revisions on the classification. J. Fac. Sci. Hokkaido Univ. series VI, Zool. 3: 177–219.

Yanada, T., 1955. Annual changes on the parasitic rate of *Caridinicola* on a freshwater shrimp, *Paratya compressa improvisa*. Biol. Bull., Okazaki 5: 105–109 (in Japanese).

Hydrobiologia **406:** 33–47, 1999.
B. M. Healy, T. B. Reynoldson & K. A. Coates (eds), Aquatic Oligochaetes.

Studies on the aquatic oligochaete fauna in Lake Biwa, central Japan. II. Records and taxonomic remarks of nine species

Akifumi Ohtaka[1] & Machiko Nishino[2]
[1]*Department of Natural Science, Faculty of Education, Hirosaki University, Hirosaki 036-8560, Japan*
[2]*Lake Biwa Research Institute, Otsu 520-0806, Japan*
E-mail: ohtaka@cc.hirosaki-u.ac.jp; Tel/Fax: [+81] 172 39 3369

Key words: aquatic Oligochaeta, Lake Biwa, Japan, taxonomy, zoogeography

Abstract

At least 41 taxa of aquatic oligochaetes including five new records have been recorded from the ancient Lake Biwa in Japan. The lake is characterized by eurybathyal occurrence of *Branchiura sowerbyi.* The posterior gill filaments of *B. sowerbyi* develop normally in lagoons adjacent to the lake, while in the lake they weakly develop or often disappear at littoral and sublittoral depths shallower than 30 m, and they almost completely disappear in the profundal zone. The profundal population of *B. sowerbyi* also differs from other populations in eastern Asia in having fewer and shorter hairs, and simple-pointed crotchets, or bifid crotchets with parallel teeth, in dorsal bundles of chaetae. This suggests that an intralacustrine speciation is progressing in the profundal population. An aberrant modification is found in the pectination of dorsal crotchets in *Tubifex tubifex* occurring at deep profundal depths of the lake. The ventral chaetae in segment II of *Teneridrilus mastix* are rarely modified in Lake Biwa. This requires cancellation of the modification of chaetae in II as a diagnostic characteristic of the genus *Teneridrilus. Limnodrilus amblysetus,* previously known only from China, is recorded from Lake Biwa. Descriptions and taxonomic remarks are provided for *Haemonais waldvogeli* (new to Japan), *Pristina longiseta, Stephensoniana trivandrana* and *Ripistes parasita* (new to Lake Biwa), and *Ophidonais serpentina,* based on Lake Biwa material.

Introduction

Lake Biwa, located in central Japan, is a tectonic lake having a long and complicated history which can be traced back to ca. 4 million years ago (Research Group of the Natural History of Lake Biwa, 1994). Based on its antiquity, Biwa is regarded as an ancient lake, harboring 57 endemic species, for example highly diverse pleurocerid gastropods (Nishino & Watanabe, in press).

The oligochaete fauna of Lake Biwa was first reported by Stephenson (1917). However, the faunal composition of this group of animals was left poorly known until the 1980s because of a long period without any further taxonomic study. Recently, Ohtaka & Nishino (1995) gave a checklist for 35 oligochaetes from the lake. Although their list covered all the species previously recorded from the lake, the taxonomic status of several taxa were still left unclear. Furthermore, several additional species have been collected from the lake in the most recent surveys. As a second report of faunal studies on Lake Biwa oligochaetes, we herein provide records of nine species with some taxonomic and faunal discussions.

Material and methods

Lake Biwa is composed of a small, shallow and eutrophic southern basin (4 m mean depth; hereafter called the southern lake) and a large, deep and mesotrophic northern basin (103.6 m maximum depth; hereafter called the northern lake). The bottom of the northern lake is bathymetrically divided into littoral (<10 m), sublittoral ($10 \ll 30$ m) and profundal (> 30 m) zones. Oligochaete specimens used in the present study were collected from various sites in the lakes and the adjacent lagoons by the authors, and by the colleagues whose names are given in the text. They were observed and illustrated under a light microscope after they were fixed in 10% formalin solution, and mounted whole in Canada Balsam or polyvinyl lactophenol. Anatomical observation was made on dissected specimens and specimens cut serially (sections 8–10 μm thick) and stained with haematoxylin and eosin. A detailed observation of chaetal morphology was made by scanning electron microscopy. In addition, type material of *Teneridrilus mastix*, deposited in the United States National Museum of Natural History (USNM), and specimens from other localities were examined for comparison with Lake Biwa forms.

Systematic account

Family Naididae

Subfamily Naidinae

Ophidonais serpentina (Müller, 1773)

Ophidonais serpentina (Müller). Sperber, 1948, p. 100, Figures 10, 26, pl. VI, Figure 1; Brinkhurst, 1971, p. 327, Figure 7.4F–H; Ohtaka & Iwakuma, 1993, p. 252, Figures 2–3.

Material examined: Ten immature individuals, a ditch at Otsu, leading to the southern lake, mud with dense submerged vegetation, 21st October, 1992. Sixty-two immature individuals, off Shimosakamoto (the southern lake), 3 m depth, mud with submerged vegetation, 2nd August, 1995.

Remarks: Living worm crawls and never swims. Among 15 individuals from Shimosakamoto (the southern lake) examined, three lacked dorsal needles throughout the segments, and the remainder also had lost them in a variable number of segments. The absence of dorsal chaetae in some segments of this species were also reported from another Japanese lake (Ohtaka & Iwakuma, 1993). This species was once recorded from Lake Biwa (Ohtaka & Nishino, 1995).

Haemonais waldvogeli Bretscher, 1900 (Figure 1A–D)

Haemonais waldvogeli Bretscher. Sperber, 1948, p. 154, Figures 18c, 27b; Brinkhurst, 1964, p. 212, Figure 4a; 1971, p. 356, Figure 7.11M–P; 1986, p. 72; Naidu, 1962, p. 522, Figure 10a–c.

Haemonais laurentii Stephenson. Stephenson, 1923, p. 79, Figure 27; Chen, 1940, p. 41, Figure 9.

Material examined: One immature individual, off Hayasaki (the northern lake), 10 m depth, sand, 26th February, 1993.

Description: In fixed state, body 5.2 mm long, maximally 0.3 mm wide, segments 32. No eyes. Dorsal chaetal bundles beginning in XIV, each consisting of one short (110–142 μm long), smooth hair and one curved, bifurcate crotchet. Dorsal crotchets (Figure 1A) 92–104 μm long, with nodulus distally and with long teeth; upper tooth much longer and as thick as or a little thinner than lower. Ventral chaetae all bifurcate; those in anterior segments (Figure 1B) 2–3 per bundle, 90–98 μm long, with nodulus medially to proximally, upper tooth longer and thinner than lower; those in middle and posterior segments (Figure 1C, D) 2–3 per bundle, 76–84 μm long, becoming a little thicker than in anterior segments, with nodulus distally and with upper tooth shorter and much thinner than lower.

Remarks: All chaetal characteristics found in the present specimen coincide well with those in the previous descriptions. This is the first recorded occurrence of this species in Japan.

Pristina longiseta Ehrenberg, 1828 (Figure 2)

Pristina longiseta Ehrenberg. Kondo, 1936, p. 388, pl. 24, Figure 18; Chen, 1940, p. 46, Figure 12; Yamaguchi, 1953, p. 285, Figure 5, pl. VII, Figure 1; 1965, p. 538, Figure 167.1–6; Rodriguez, 1987, p. 39, Figures 2–3.

Pristina longiseta longiseta Ehrenberg. Sperber, 1948, p. 236, pl. XXI, Figures 2, 6; Naidu, 1963, Figure 34 A–K; Brinkhurst, 1971, p. 402, Figures 7.21J, 7.25E–I; Hiltunen & Klemm, 1980, p. 13.

Pristina longiseta sinensis Sperber, 1948, p. 237.

Pristina longiseta leidyi Smith. Sperber, 1948, p. 237; Brinkhurst, 1971, p. 403; Hiltunen & Klemm, 1980, p. 13.

Pristina longiseta bidentata Cernosvitov. Sperber, 1948, p. 238; Brinkhurst, 1971, p. 404, Figure 7.25 J–M.

Pristina leidyi Smith. Harman & McMahan, 1975, p. 171. Figure 1–4; Harman, 1982, p. 295; Brinkhurst, 1986, p. 67.

Material examined: One immature individual, Ohura bay (the northern lake), 0.5 m depth, 6th November, 1994, collected by Y. Masuda. Five immature specimens, a shallow pond with dense vegetation, Wakasakkanai, Soya, Hokkaido, 29th October, 1984. Seven immature individuals, a water tank, Odate, Akita Pref. 13th June, 1997.

Remarks: This species is common in Japan, but this is the first record from Lake Biwa. All the Japanese material examined had serrated hairs in II as well as

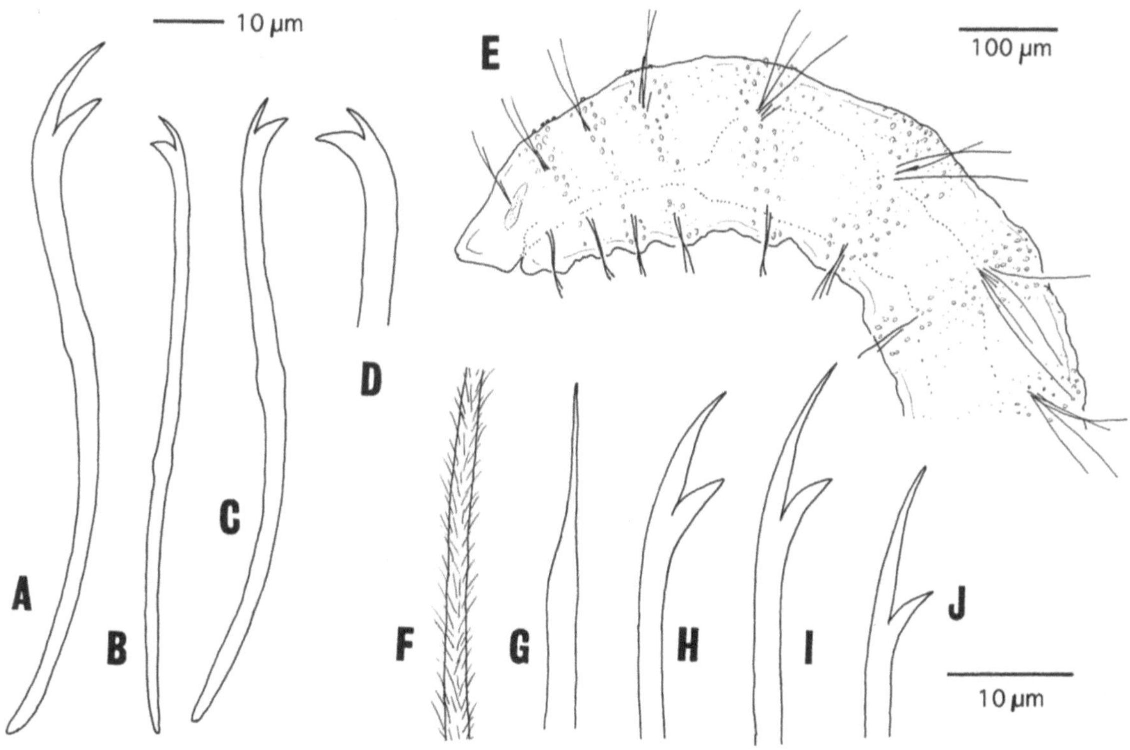

Figure 1. Haemonais waldvogeli and *Stephensoniana trivandrana* from Lake Biwa. (A–D) *Haemonais waldvogeli*: (A) dorsal chaeta in XV; (B) ventral chaeta in III; (C) the same in XVIII; (D) the same in a posterior segment. (E–J) *Stephensoniana trivandrana*: (E) anterior part of body; (F) a part of hair chaeta; (G) distal end of needle chaeta in an anterior segment; (H) distal end of ventral chaeta in III; (I) the same in XV; J, the same in a posterior segment.

in IV and onwards and bifid needles with short and diverged teeth, as Yamaguchi (1953, 1965) once noted and illustrated. In the Japanese material, the ventral chaetae in II are 4–6 per bundle, 56–64 μm long, being more numerous and a little longer than those in III (2–4 per bundle, 56–60 μm long) and in the following segments (3–5 per bundle, 49–54 μm long); the ventral chaetae in III are often thickened. These characteristics of the Japanese material cover both features of *P. longiseta* and *P. leidyi*, which supports the opinion that both should be merged into the single species *P. longiseta* (Rodriguez, 1987; Paoletti & Sambugar, 1996).

The density of serration in dorsal hairs has been used as an infraspecific criterion (Sperber, 1948). However, it is variable even within a single individual and not significant for taxonomic discrimination. In the Japanese material, dorsal hairs are devoid of serration proximally, while the teeth of serration are more than 5 μm apart medially and become closer toward the distal end, where they are less than 1 μm apart (Figure 2).

Stephensoniana trivandrana (Aiyer, 1926) (Figure 1E-J)

Stephensonia trivandrana (Aiyer). Aiyer, 1929, p. 27, Figure 7, pl. I, Figures 2–4.

Stephensoniana trivandrana (Aiyer). Cernosvitov, 1938, p. 539, Figures 1–4; Sperber, 1948, p. 208, Figure 28c; Naidu, 1963, p. 201, Figure 28a–d; Brinkhurst, 1971, p. 389, Figure 7.21B–F; 1986, p. 99; Brinkhurst et al., 1990, p. 908, Figure 1I.

Material examined: Four immature individuals, off Ayame-hama (the northern lake), 10 m depth, mud, 30th June, 1990, 31st July, 1991. One immature individual, center of the southern lake, 4.3 m depth, mud, 24th February, 1995. One immature individual, off Shimosakamoto (the southern lake), 3 m depth, muddy sand with submerged vegetation, 2nd August, 1995. Two immature individuals, center of Lake Kitaura, 6 m depth, mud, 3rd August, 1983, 20th January, 1984.

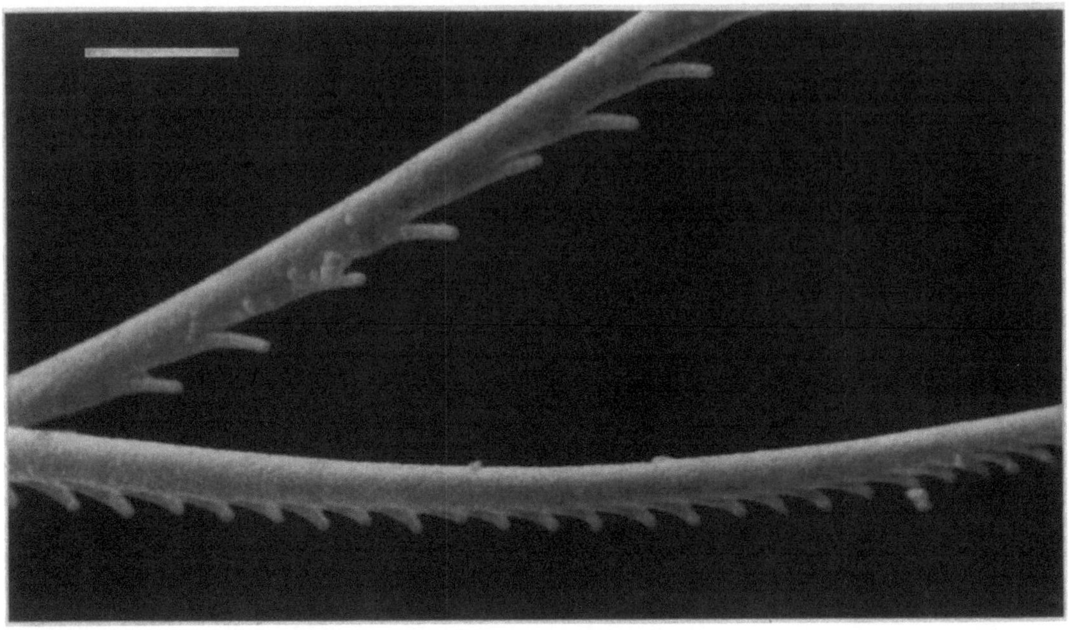

Figure 2. SEM photomicrograph of distal parts of hair chaetae in a specimen of *Pristina longiseta* from Wakasakkanai, Hokkaido, Japan. The bottom hair shows a more distal part than the top one. Scale, 5 μm.

Description of Lake Biwa material: Body 2.0–2.5 mm long, 0.18 mm wide in fixed state. Segments of an intact and immature specimen 32. No eyes. Foreign matter adhering to body wall, concentrated in middle part of each segment (Figure 1E). Prostomium conical. Dorsal chaetal bundles beginning in II, consisting of finely hispid hairs (Figure 1F), and needles without nodulus and tapering distally with simple pointed end (Figure 1G). Dorsal hairs 2–4 per bundle anteriorly, 1–2 per bundle posteriorly, 120–160 μm long. Dorsal needles 3–4 per bundle anteriorly, 2–3 per bundle posteriorly, 40–60 μm long. Ventral chaetae bifurcate with nodulus proximally; those in anterior segments 1–3 per bundle, upper tooth about twice longer and as thick as lower (Figure 1H, I), those in posterior segments 1–2 per bundle, the length hardly changing but upper tooth becoming much (more than three times) longer than lower (Figure 1J) . Stomach suddenly widening in VI.

Remarks: Newly recorded from Lake Biwa. The dorsal hairs of Lake Biwa specimens are invariably hispid. Such hispid hairs were also confirmed in other Japanese specimens from Lake Kitaura, although they have not been described before. Other characteristics examined in Japanese material agree well with those in the previous descriptions.

Subfamily Stylarinae

Ripistes parasita (Schmidt, 1847)

Ripistes parasita (Schmidt). Sperber, 1948, p. 143, Figures 15H–K, 16, 26A, pl. 9, Figures 4–7; Brinkhurst, 1971, p. 351, Figure 7.10L–N; Hiltunen & Klemm, 1980, p. 19; Ohtaka, 1985, p. 119, Figure 2H–L.

Ripistes rubra Lastockin. Yamaguchi, 1940, p. 384, Figures 3–4.

Material examined: One immature individual, off Shimosakamoto (the southern lake), 3m depth, 2nd August, 1995.

Remarks: This is the first record of this species from Lake Biwa.

Family Tubificidae

Subfamily Tubificinae

Tubifex tubifex (Müller, 1774) (Figures 3–4)

Tubifex tubifex (Müller). Brinkhurst, 1971, p. 444, Figures 8.1A–D, 8.3G–J; Holmquist, 1983, p. 189, Figures 1–4.

Tubifex hattai Nomura, 1926, p. 193, Figures 1–18; Yamaguchi, 1965, p. 541, Figure 177. 1–4.

Tubifex sp. Ohtaka, 1993, p. 28; Ohtaka & Nishino, 1995, p. 173, Figure 3.

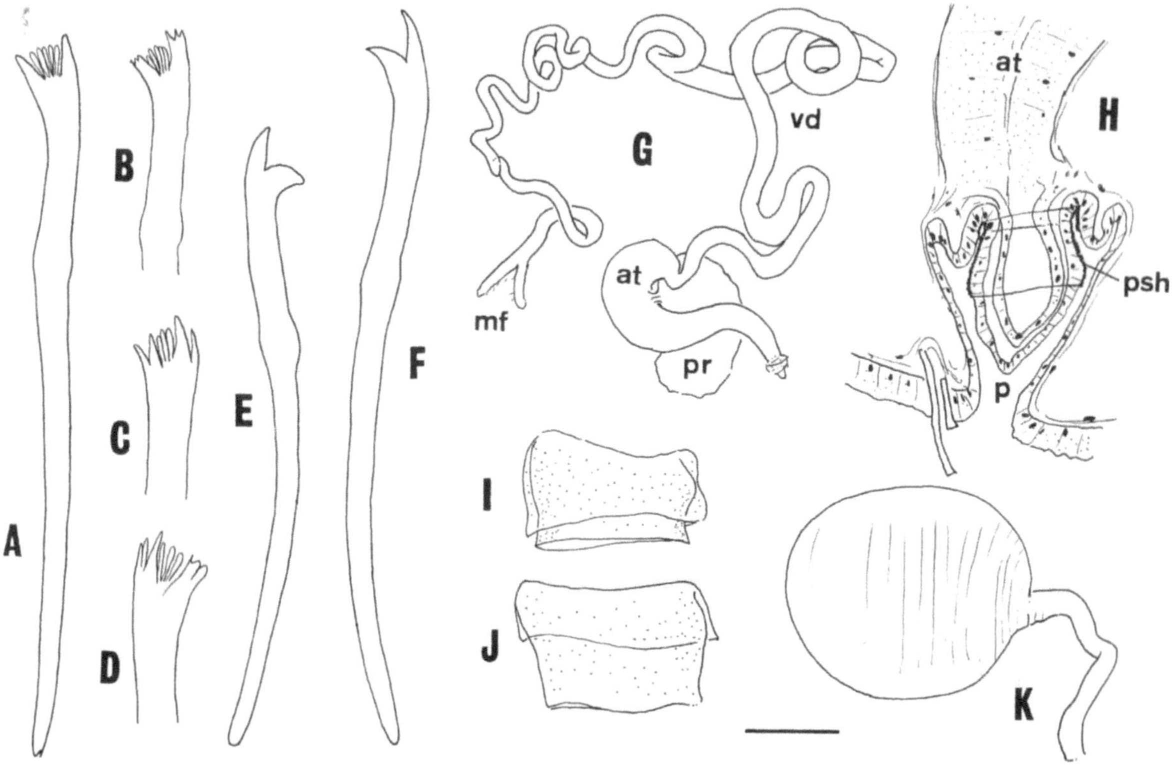

Figure 3. Tubifex tubifex from Lake Biwa. (A) dorsal pectinate chaeta in VII; (B–D) variations of distal pectination in anterior dorsal chaetae; (E) dorsal chaeta in XXX; (F) ventral chaeta in VI; (G) male duct; (H) ectal part of atrium with penis, sagittal section; (I–J) penis sheaths; (K) spermatheca. Abbreviations: at – atrium; mf – male funnel; p – penis; pr – prostate gland; psh – penis sheath; vd – vas deferens. Scale: (A–F) 20 μm; (G) and (K) 200 μm; (H) 65 μm; (I) and (J) 50 μm.

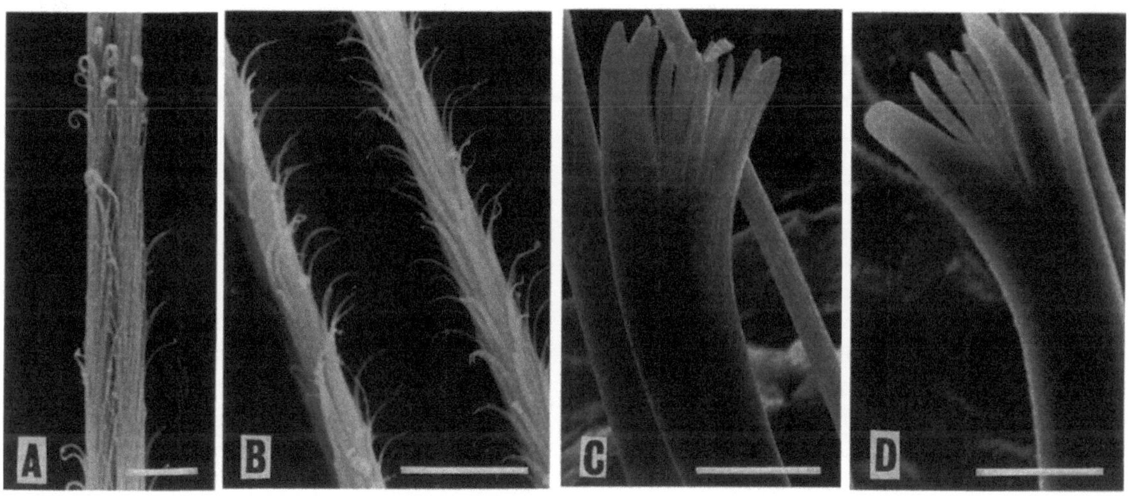

Figure 4. SEM photomicrographs of chaetae in *Tubifex tubifex* from profundal of Lake Biwa. (A) and (B) parts of hair chaetae in anterior segments; (C) and (D) modifications of dorsal pectinate chaetae in anterior segments. Scales, 5 μm.

Material examined: Thirty-five mature individuals, off Wani (the northern lake), 70–74 m depth, 14th October, 1982, 25th October, 1993, 13th April, 1993; Seven mature individuals, off Hayasaki (the northern lake), 90 m depth, 14th February, 1995.

Description of Lake Biwa material: In mature and fixed state, body 10–23 mm long, 0.4–0.6 mm wide in anterior segments except clitellum where it measures up to 0.8 mm wide. Segments 65–110. Dorsal chaetal bundles composed of hairs and pectinate chaetae. Dorsal hairs slightly to heavily hispid distally (Figure 4A, B), 1–3 per bundle, 250–560 μm long preclitellarly, while 0–1 per bundle, 180–200 μm long postclitellarly. Dorsal pectinate chaetae in preclitellar segments (Figure 3A) 2–5 per bundle, 90–144 μm long, with upper tooth as long as or a little longer and as thick as lower, and with several distinct club-shaped intermediate teeth; those in postclitellar segments (Figure 3E) 1–3 per bundle, 110–130 μm, upper tooth thinner and intermediate teeth becoming absent. Aberrant modification of distal teeth often found in anterior dorsal pectinates (Figures 3B–D, 4C, D). Ventral chaetae in preclitellar segments (Figure 3F) 2–4 per bundle, 95–142 μm long, upper tooth a little longer and thinner than lower, sometimes with short intermediate spine; those in postclitellar segments 2–3 per bundle, 110–124 μm long, upper tooth nearly as long as and thinner than the lower.

Clitellum whitish, occupying from 1/2 X to end of XII. Vasa deferentia very long (up to 3900 μm) and coiled in XI–XIII (Figure 3G), composed of narrow and ciliated proximal part (22–26 μm wide) and wide and non-ciliated distal part (40–45 μm wide), entering atria subapically. Atria tubular and bent proximally and distally, 450–600 μm long,100–120 μm wide in maximum; inner epithelium thick and glandular. Prostate glands nearly as large as atria, connected with atria at the concave side of proximal portion of atria through a short stalk. Penes in penial poach about 100 μm long, protrusible, opening in XI ventrally (Figure 3H). Penis sheath very thin, granular and plastic, usually tub-shaped (Figure 3I, J). Spermathecae in X (Figure 3K); ampullae ovoid in shape, 500 μm long by 380 μm and ducts thin and about 450 μm long, well marked off from ampullae.

Remarks: This is the most dominant taxon in the macrozoobenthos of the profundal region of the northern lake, whereas so far the species has never been recorded from the littoral zone of the northern lake or from the shallow southern lake (Nishino et al., unpubl.). This species was formerly recorded under the name *Tubifex* sp. (Ohtaka, 1993; Ohtaka & Nishino, 1995); the present examination reveals that it is ascribable as *Tubifex tubifex* on the basis of chaetal and genital morphology. However, as already noted by Ohtaka & Nishino (1995), the Lake Biwa form is noticeable in having an aberrant modification of distal teeth in anterior dorsal pectinate chaetae. The modification involves a split or multiplication of lateral teeth (Figures 3B–D, 4C,D), and the frequency of occurrence in the preclitellar bundles was found to be 50% in mature worms, whereas the modification was rare in young ones. Similar chaetal abnormalities in *Tubifex tubifex* were demonstrated to be induced by mercury and salinity stress by Chapman & Brinkhurst (1987, Figure 2h, i), and has also been reported for another tubificid, *Potamothrix hammoniensis*, from a mercury-polluted Swedish lake (Milbrink, 1983). In the profundal of Lake Biwa, however, the concentration of mercury in the sediment is lower than 0.3 mg kg^{-1} dry mud (Shiga Prefectural Institute of Public Health & Environmental Science, 1990) and the electric conductivity in the bottom water is lower than 125 μS cm^{-1} at 25 °C (Kumagai et al., 1995), which is much lower than the level for induction of chaetal abnormarities recorded by the above authors. Thus the modification of dorsal chaetae in the Lake Biwa form is caused by a more sensitive nature against these environmental factors, or by some other, unknown mechanism.

Teneridrilus mastix (Brinkhurst, 1978) (Figures 5–6)

Ilyodrilus mastix Brinkhurst, 1978, p. 2171, Figure 3; Erséus & Qi, 1985, p. 193, Figure 1.

Teneridrilus mastix (Brinkhurst). Holmquist, 1985, p. 357, Figures 18, 21C, 22D, 23J, 24O; Brinkhurst, 1986, p. 175; Erséus et al., 1990, p. 840.

Teneridrilus sp. Ohtaka & Usman, 1997, p. 152, Figure 5.

Material examined: One mature and 25 immature individuals, off Hayasaki (the northern lake), 10–90 m depth, 26th January, 29th September, 1992, 26th February, 1993. Ten mature and 10 immature individuals, off Ayamehama (the northern lake), 10 m depth, 3rd April, 1990. Twenty-five immature individuals, off Wani (the northern lake), 3–76 m depth, 27th January, 25th October, 1992. Nineteen immature individuals, Lake Suwa, Nagano prefecture (exact locality unknown), 8th December, 1983, 11th November, 1984, 9th, 19th, 23rd November, 1985, collected by K. Takada. USNM 55375 (holotype) and USNM 55375

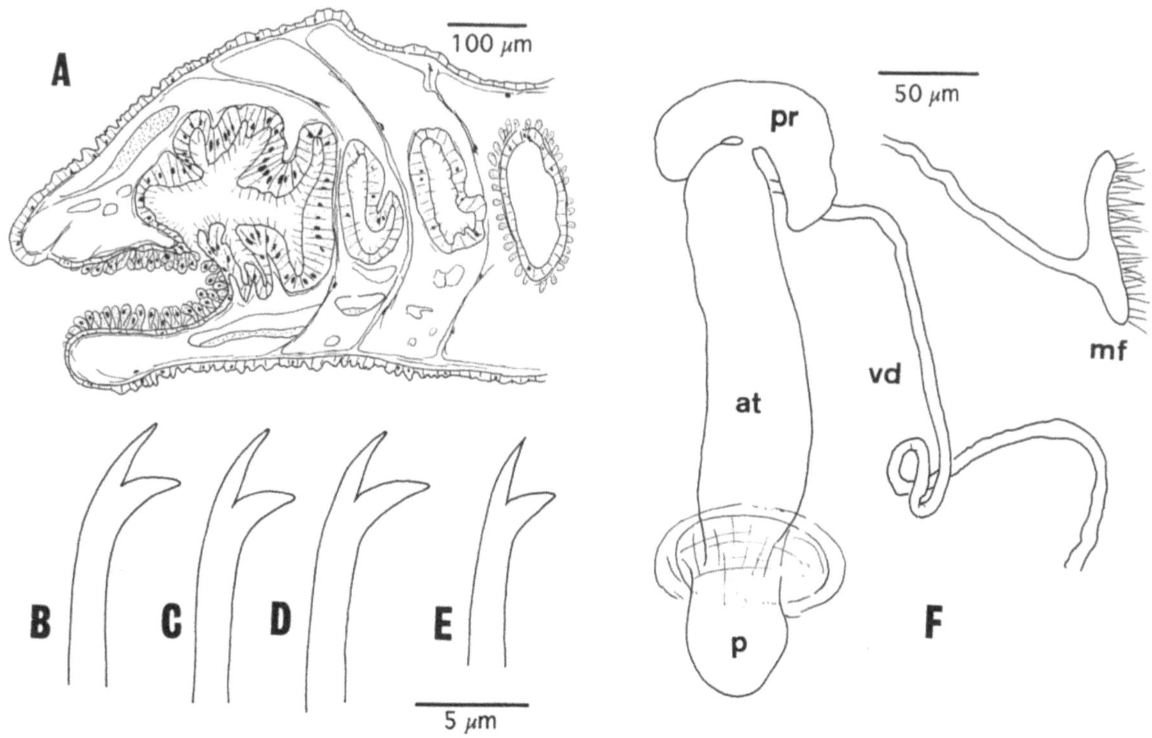

Figure 5. Teneridrilus mastix from Lake Biwa. (A) sagittal section of anterior part of body; (B–D) distal end of ventral chaetae in II; (E) the same in a posterior segment; (F) male duct. Abbreviations as in Figure 3.

(paratype), Fraser River, B. C., Canada, 1972, collected by R. O. Brinkhurst. Sixteen immature and one mature individuals, the type locality (R. O. Brinkhurst collection). One mature and 12 immature individuals, Pearl River, People's Republic of China, July, 1988, January, 1989, collected by Qi Sang (R. O. Brinkhurst collection).

Description of Lake Biwa material: In mature and fixed state, body 10–25 mm long, becoming slender posteriorly. Segments 56–80. No coelomocytes. Prostomium small. Mouth enlarged, becoming narrower when fixed. Buccal cavity large and densely covered with pear-shaped gland cells (Figure 5A). Pharynx in II and III, large, folded and eversible, devoid of pharyngeal glands. Oesophagus in IV and V. Intestine beginning in VI, covered with chloragogen cells, becoming wider in VIII–XI. Dorsal chaetal bundles consisting of hairs and bifurcate crotchets except in II and III (occasionally only in II) where hair chaetae absent. Hair chaetae short and bayonet-shaped, almost smooth or finely hispid distally, 2–5 per bundle, 84–124 μm long anteriorly, 2–4 per bundle, 64–100 μm long posteriorly. Dorsal and ventral crotchets with nodulus at 1/4 from distal end; upper tooth almost as long as but thinner than lower in II and III (Figure 5B–D), while upper tooth a little longer and thinner than lower in the following segments (Figure 5E). Dorsal crotchets 2–6 per bundle, 56–80 μm long anteriorly, 2–3 per bundle, 60–78 μm posteriorly. Ventral chaetae 2–7 per bundle, 56–74 μm long anteriorly and 2–4 per bundle, 50–66 μm long posteriorly. No modified genital chaetae.

Clitellum from X to XII, thin. Testes in X, very small. Ovaries in XI, large. Sperm sac and spermathecae absent. Ovisac large in XI–XIII. Male ducts (Figure 5F) in XI, but often missing. Male funnels on 10/11, 80 μm in diameter. Vasa deferentia thin (8–10 μm in diameter) and hardly winding, entering atria almost apically. Atria elongate tubular with thick wall, 200 μm long and 50 μm wide medially. Prostate glands small (90 μm in maximum width), connected with atria near apical end. Penes protrusible without penis sheaths, 70 μm long, opening ventrally.

Remarks: *Teneridrilus mastix* has been recorded from Pacific North America (Brinkhurst, 1978, 1986), China (Erséus & Qi, 1985), and two central Japan-

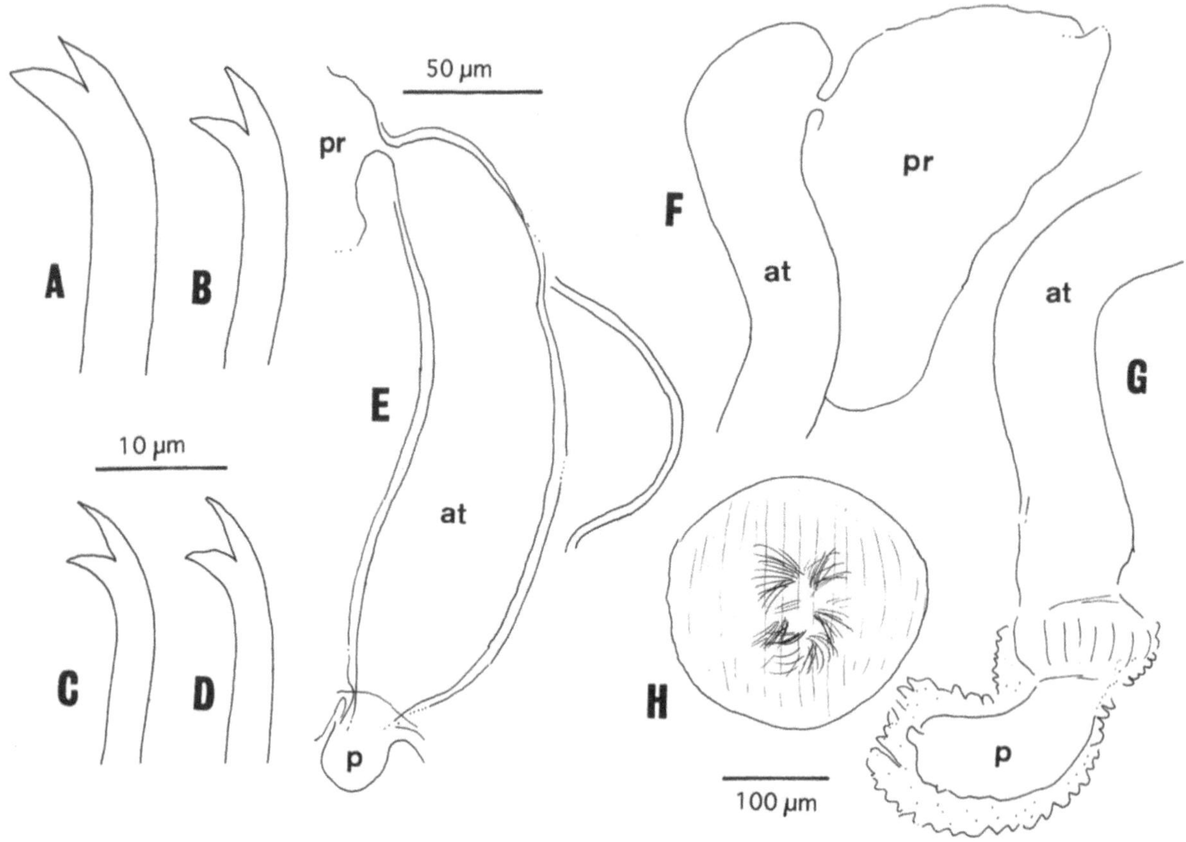

Figure 6. Teneridrilus mastix from Fraser River, B.C., Canada (A–E) and from Pearl River, China (F–H). (A) distal end of ventral chaeta in II; (B) the same in III; (C) the same in IV; (D) the same in V; (E–G) ectal parts of male ducts; (H) spermatheca. Abbreviations as in Figure 3.

ese lakes, L. Biwa and L. Suwa (Ohtaka & Nishino, 1995; present study). The present examination of the type material and topotypes (from Fraser River, B. C., Canada) confirmed that their chaetae are stouter in II than in the following segments, and that, in II also, the upper teeth are shorter than the lower (Figure 6A–D). The chaeta of II in the topotypes are usually single as in the original description (Brinkhurst, 1978), however, 2 chaetae per bundle were often found (5 out of 15 specimens examined). Chinese specimens from Pearl River coincide with the topotypes in the number and distal form of chaetae of II, on the other hand, they have strongly hispid hairs which has not been reported in the topotypes. One Chinese specimen was mature; its spermathecal ampullae are spherical with sperm as loose bundles while the ducts were not detected (Figure 6H); atria are cylindrical with large prostates and large protruded penes (Figure 6F, G). Consequently, the Chinese form resembles the Canadian one in chaetal characteristics, as already noted by Erséus & Qi (1985), while it differs from the latter in having hispid hairs, larger penes and spermathecae. The present Japanese form is, on the other hand, different from the Canadian and Chinese specimens in that the chaetae of II are not modified, while it is intermediate between the latter two forms with regard to the degree of hispid condition of hairs and the size of the penes (Figures 5F, 6E, G). *Teneridrilus* sp. recorded from Sumatra (Ohtaka & Usman, 1997) closely resembles the Japanese form in the number and distal form of chaetae in II, but its hair chaetae are not hispid. We regard all the above forms compared as a single taxon *T. mastix* because the diagnostic characteristics overlap with each other. Erséus et al. (1990) proposed the modification of chaetae in II and the eversible pharynx as synapomorphies for the genus, but the former character should be cancelled because Japanese and Sumatran materials never show such a modification. This may also apply to the taxonomy of *Varichaetadrilus* species without hair chaetae; they also have a modification of chaetae in anterior segments (Brinkhurst, pers. comm.). The modifications in anterior ventral chaetae may be caused by chemical conditions of water, as in the Spanish form of another tubificid, *Tubifex tubifex* (*Tubifex* sp. in Rodriguez & Armas, 1983), whose anterior ventral chaetae were enlarged related to very high mineralized waters (Rodriguez, 1986).

All Japanese material examined have no spermatheca. In addition, the male ducts are also often missing, although the clitellum and ova are fully developed. Among 50 mature specimens from Lake Biwa examined, only six had male ducts. This species may be able to reproduce facultatively by parthenogenesis as suggested by Erséus & Qi (1985).

Limnodrilus amblysetus Brinkhurst et al., 1990 (Figure 7)

Limnodrilus amblysetus Brinkhurst et al., 1990, p. 910, Figure 3a.

Material examined: One immature individual, off Wani (the northern lake), 5m depth, mud, 25th October. 1992. One immature individual, off Ayame-hama (the northern lake), 10 m depth, mud, 30th January, 1992. Three immature individuals, off Karasaki (the southern lake), 1.5 m depth, 14th February, 1994. One immature individual, off Shina (the southern lake), 2.5 m depth, 14th February, 1994. Five immature individuals, off Shimosakamoto (the southern lake), 2.0–2.5 m depth, 23rd October, 1992, 14th February, 1994. Three mature individuals, Lake Chenhu, Hanyang, Hubei Province, People's Republic of China, 8th June, 1983 (paratypes) (R.O. Brinkhurst collection).

Description of Lake Biwa material: In immature and fixed state body up to 50 mm long, segments up to 325. Posterior part of body slender, thread-like. Cuticular covering and body wall thick. Prostomium bluntly conical. Segments II–V biannulate, each anterior annulus shorter than posterior one. Both ventral and dorsal chaetal bundles consisting of simple-pointed or bifurcate crotchets only. Chaetae in anterior segments 3–7 per bundle, 110–150 μm long with maximum size in V–VII, distal end simply pointed in II and III (Figure 7A), bearing minute upper tooth from IV to VII on (Figure 7B, C). Chaetae in middle segments 2–4 per bundle, 90–110 μm long, with upper tooth shorter and thinner than lower (Figure 7D); in posterior segments 1–2 per bundle, 44–70 μm long, upper tooth becoming almost as long as but still thinner than lower (Figure 7E). Nodulus of all chaetae situated about 1/3 from distal end. Chloragogen cells from VI on, thickly covering gut. No coelomocytes.

Remarks: The present species was previously known only from China. According to the original account of *L. amblysetus* (Brinkhurst et al., 1990), all chaetae are bifid or bluntly simple-pointed, and the nodulus of anterior chaeta is situated in a slightly proximal position. However, re-examination of the paratypes (from Hubei Province) reveals that chaetae in some most anterior segments are sharply simple-pointed and all chaetae have a distal nodulus (Figure 7F, G). In the paratypes, the chaetae attain their maximum size in V. Thus, the chaetal conditions of the paratypes coincides well with the present Lake Biwa material (Figure 7). This species resembles the congeners, *L. grandisetosus* Nomura and *L. udekemianus* Claparède, in having slender posterior body, but their chaetal forms are different from each other and diagnostic.

Subfamily Rhyacodrilinae

Branchiura sowerbyi Beddard, 1892 (Figures 8–10)

Branchiura sowerbyi Beddard, 1892, p. 325, pl. XIX, Figures 1–15; Chen, 1940, p. 90, Figure 27; Yamaguchi, 1953, p. 296; 1965, p. 541, Figure 180.1–3; Ohtaka, 1993, p. 35; Ohtaka & Nishino, 1995, p. 176.

Kawamuria japonica Stephenson, 1917, p. 89, Figure 105.

Material examined: Over 1000 immature and mature individuals, many sites covering various areas in Lake Biwa, August 1988–January 1996. Sixteen mature and 29 mature individuals, two lagoons adjacent to Lake Biwa, 0.3–0.5 m depth, 9th March, 1994, 11th May, 1996. Two mature and 4 immature individuals, a pond in Sapporo, 23rd May, 1983. Three mature individuals, Niitsu, Niigata Pref., 19th May, 1996, N. Hayakawa collected. Two mature and 6 immature individuals, center of Lake Suwa, 6 m depth, 11th June, 1980. Twenty immature individuals, center of Lake Kitaura, 6.5 m depth, 5th January, 1985. Three immature individuals, a pond in Akashi, Hyogo Pref., 8th October, 1982. Four immature individuals, a pond in Takamatsu, Kagawa Pref., 11th October, 1982. Two immature individuals, Kanna River, Okinawa, 8th August, 1990, M. Tsuchiya collected. Four immature individuals, Bao-an Lake, Hubei Prov., People's Republic of China, 8th April, 1987 (R. O. Brinkhurst collection). Twelve immature individuals, paddy fields at Padang, Sumatra, Indonesia, 4th August, 1988.

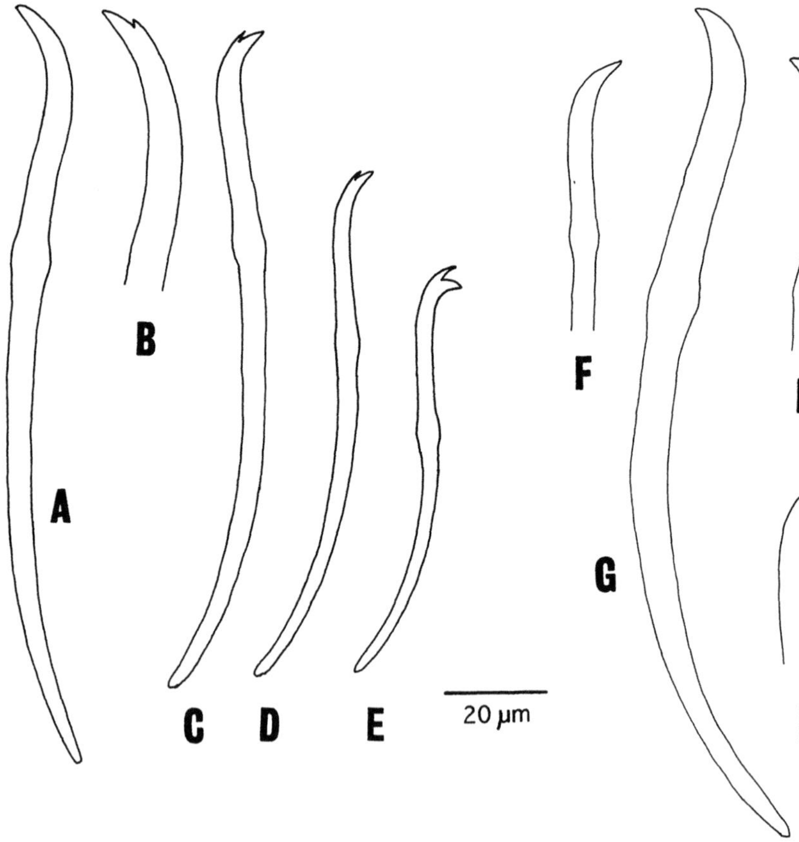

Figure 7. Limnodrilus amblysetus from Lake Biwa (A–E) and Lake Chenhu, Hanyang, Hubei Province, P.R. China (paratype) (F–J). (A) dorsal chaeta in III; (B) ventral chaeta in IV; (C) dorsal chaeta in VIII; (D) the same in XVI; (E) ventral chaeta in LXXXVI; (F) ventral chaeta in II; (G) the same in III; (H) the same in VI; (I) the same in a posterior segment. (B) (F) (H) and (I) showing only distal ends.

Remarks: In Lake Biwa, the profundal population of *Branchiura sowerbyi* lacks posterior gills (Ohtaka & Nishino, 1995). The present examination reveals that development of gill filaments is corelated with water depth in the lake (Table 1; Figure 8). In the profundal region of the northern lake with depths deeper than 30 m, almost all individuals were devoid of gill filaments (Figure 8A) and only a few individuals had vestigial gills (Figure 8B). In the sublittoral and littoral regions of the northern lake, worms with little developed gills occurred, but more than half of the individuals examined were still devoid of gills (Figure 8C, D). In the shallow southern lake, worms with gills surpassed those without gills, however, the number and length of gill filaments were few and short (Figure 8E). On the other hand, all worms in the adjacent lagoons had developed gills (Figure 8F,G). These observations support the opinion presented by the previous authors (Chen, 1940; Yamaguchi, 1953; Ohtaka & Nishino, 1995), that *Kawamuria japonica* Stephenson, which is devoid of posterior gill filaments and was originally described from profundal of Lake Biwa (Stephenson, 1917), is an infraspecific variant of *Branchiura sowerbyi* presented. The form lacking gills has never been recorded from sites other than Lake Biwa.

Chaetal morphology was examined for specimens from different localities within Lake Biwa and elsewhere. A hispid condition on the dorsal hairs was confirmed in all the material, and the distal form of the ventral chaetae did not vary much among populations, in which the upper teeth were almost absent or much shorter than lower teeth (Figure 10A). On the other hand, considerable variation was detected between the populations in the number and length of dorsal hairs, and in the distal form of dorsal crotchets

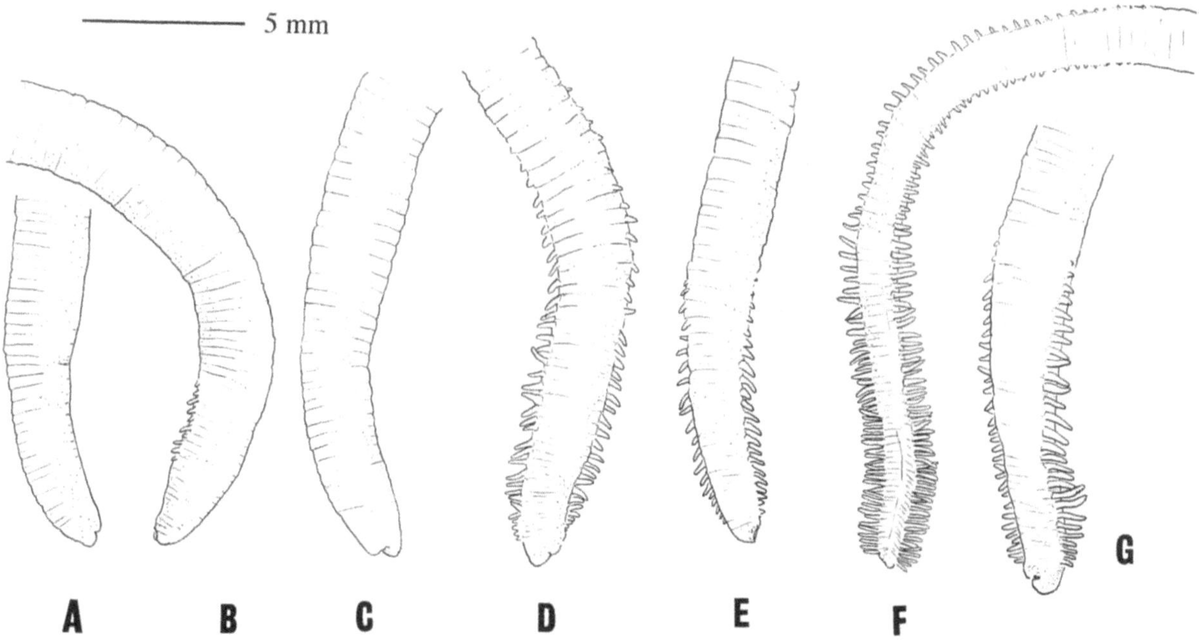

Figure 8. Posterior ends of *Branchiura sowerbyi* from Lake Biwa and the adjacent lagoon. (A) the northern lake, 90 m depth; (B) the same, 70 m depth; (C) the same, 30 m depth; (D) the same, 5 m depth; (E) the southern lake, 4 m depth; (F) and (G) an adjacent lagoon.

(Figures 9, 10B–D). The dorsal hairs in preclitellar segments were 1–5 per bundle (usually 2–3), and the longest hair was 2–3 times longer than the crotchets in the same bundle, in all specimens examined except those from Niigata and profundal Lake Biwa, which had fewer (0–2 per bundle) and distinctly shorter hairs (not exceeding twice the length of the crotchets). As for the distal ends of the dorsal crotchets, a bifurcate tip with a much shorter upper tooth was the most common condition in the Japanese populations, but pectinations or split lower teeth were often found in specimens from some Japanese localities including the southern lake of Lake Biwa (Figures 9B–K, 10C). Pectinations were the most prominent condition in Chinese and Sumatran material (Figures 9L, M, 10D). On the other hand, the distal ends of the dorsal chaetae in the profundal Lake Biwa form were simple-pointed or cleft with parallel teeth (Figures 9A, 10B). Such short hairs and simple-pointed dorsal chaetae as in the present material were previously described and figured in the original description of *Kawamuria japonica* from profundal Lake Biwa (Stephenson, 1917). Consequently the profundal population of *B. sowerbyi* in Lake Biwa is different from those of other localities, not only in terms of absence of gills, but also regarding chaetal characteristics. This suggests that speciation is progressing in the profundal population of Lake Biwa (also see 'Discussion'). However, it is also probable that the differences in chaetal forms between local populations depend on chemical condition of waters, as previously discussed for *Tubifex tubifex* and *Teneridrilus mastix*. To estimate the degree of differentiation in the profundal Lake Biwa population, critical comparison is needed for genital organs that could effectively isolate populations.

Discussion

At least 41 taxa of aquatic oligochaetes have so far been recorded from Lake Biwa (Ohtaka & Nishino, 1995; present study). They include at least 23 naidids (*Chaetogaster diastrophus, C. diaphanus, C. limnaei, Amphichaeta* sp., *Paranais* sp., *Specaria josinae, Uncinais uncinata, Ophidonais serpentina, Nais communis, N. bretscheri, N. pardalis, N. barbata, Slavina appendiculata, Vejdovskyella simplex, Arcteonais lomondi, Ripistes parasita, Stylaria lacustris, Haemonais waldvogeli, Branchiodrilus hortensis, Dero* spp., *Stephensoniana trivandrana, Pristina aequiseta, P. longiseta*), 15 tubificids (*Tubifex tubifex, Teneridrilus mastix, Limnodrilus hoffmeisteri, L. claparedianus, L. profundicola, L. grandisetosus,*

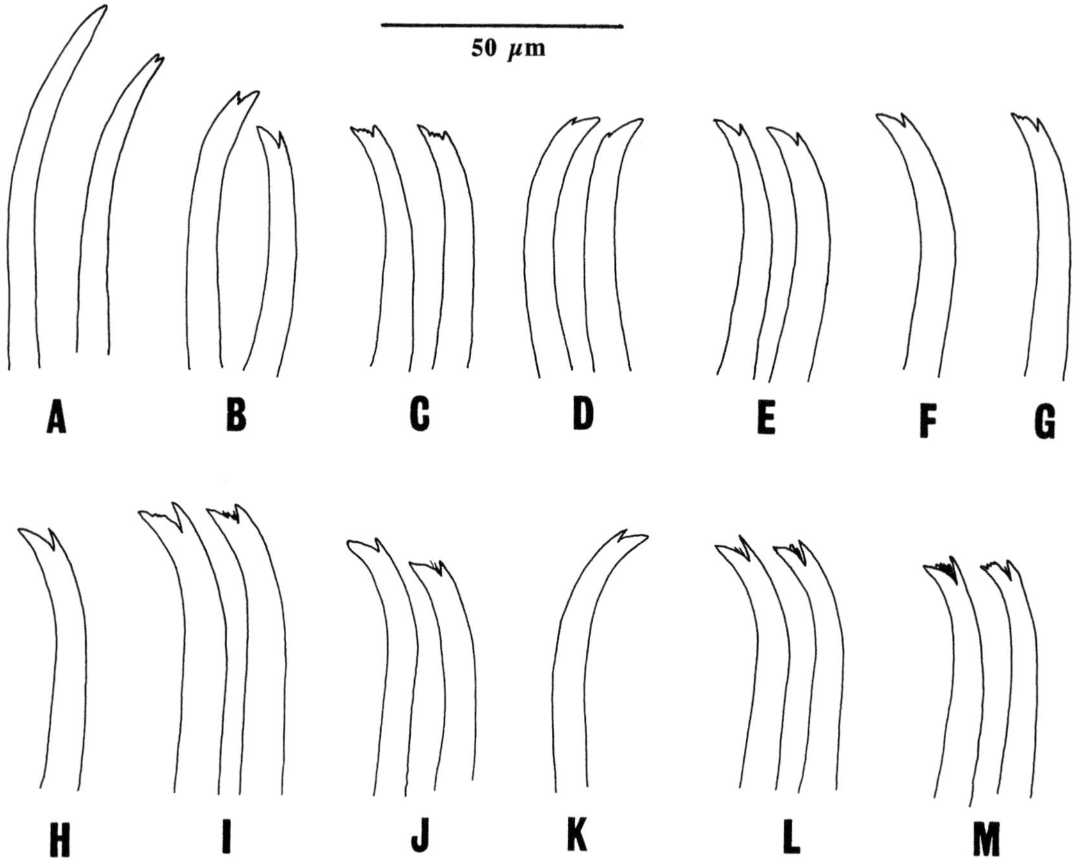

Figure 9. Distal ends of dorsal crotchets in *Branchiura sowerbyi* at various localities. (A) the northern lake of Lake Biwa, 70 m depth; (B) the same, 10 m depth; (C) the southern lake of Lake Biwa, 4 m depth; (D) a lagoon adjacent to Lake Biwa; (E) Sapporo, Japan; (F) Niigata, Japan; (G) Lake Suwa, Japan; (H) Lake Kitaura, Japan; (I) Akashi, Hyogo Pref., Japan; (J) Takamatsu, Kagawa Pref., Japan; (K) Kanna R., Okinawa, Japan; (L) Bao-an Lake, Hubei Province, China; (M) Padang, Sumatra Is., Indonesia.

L. amblysetus, Embolocephalus yamaguchii, Aulodrilus limnobius, A. japonicus, A. pigueti, Rhyacodrilus hiemalis, R. sp., *Bothrioneurum vejdovskyanum, Branchiura sowerbyi*), one biwadrilid (*Biwadrilus bathybates*), and poorly studied lumbriculids and enchytraeids. The fauna mainly consists of widely distributed species as well as some potentially Japanese (*R. hiemalis, B. batybates*), circum-Pacific (*T. mastix*) and East Asian species (*L. grandisetosus, L. amblysetus, V. simplex*).

Only one endemic oligochaete species (*E. yamaguchii*) has been recorded from Lake Biwa, and the endemicity is low if compared to that of other ancient lakes (Martin, 1996). According to Martin, this could be due mainly to the relatively short and complicated history of the lake, in which a sedimentary basin can be traced back to 4 Ma, but which was filled up with fluvial sediments twice during its history. However, this is partly wrong because of a misunderstanding with regard to the paleo-history of Lake Biwa. There were at least five old lakes called the Paleo-Lake Biwa Group, all located apart from the present Lake Biwa (Research Group for Natural History of Lake Biwa, 1994). The deep basin of Lake Biwa was produced about 0.4 Ma, and it has never been dried up since then (Horie, 1987; Meyers et al., 1993). There have not yet been any reports of any oligochaete fossils, although a lot of fossils of desmids, diatoms, protozoans, sponges, platyhelminthes, molluscs, crustaceans, chironomids and fishes have been found from the Paleo-Lake Biwa (Research Group for Natural History of Lake Biwa, 1994). Martin (1996) suggested that the disappearance of the precursors of Lake Biwa is responsible for the low extant diversity, but no evidence has been available for oligochaetes. A

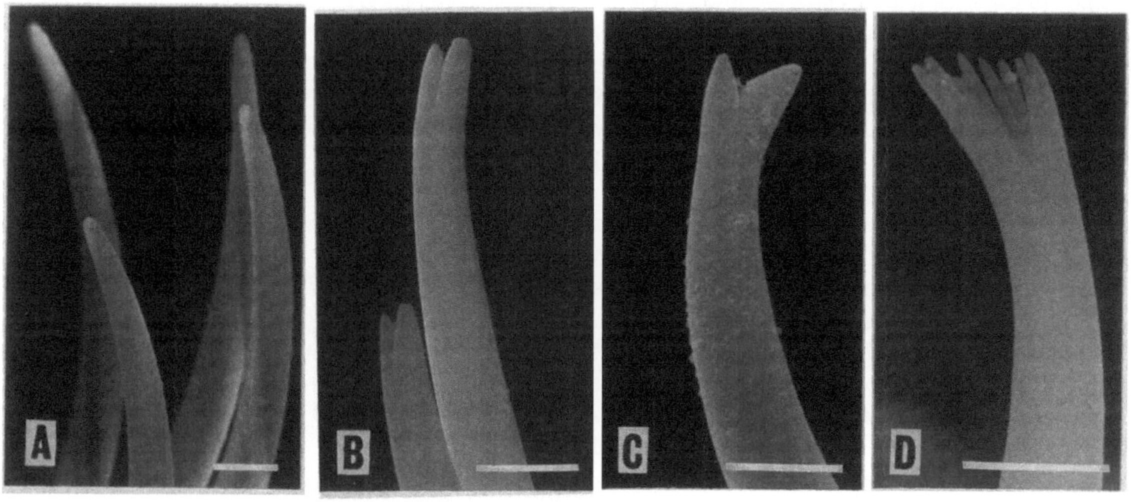

Figure 10. SEM photomicrographs of distal ends of chaetae in anterior segments of *Branchiura sowerbyi*. (A) ventral chaetae (off Wani, 70 m depth, Lake Biwa); (B) dorsal chaetae (off Wani, 70 m depth, Lake Biwa); (C) dorsal chaeta (Niitsu, Niigata); (D) dorsal chaeta (Padang, Sumatra). Scales, 5 μm.

Table 1. Development of posterior gill filaments in *B. sowerbyi* in Lake Biwa and the adjacent lagoons. Specimens collected from 1990 to 1996 were combined

Locality	Depth (m)	No. specimens	Without gills %	With gills %	No. gills paired Mean	No. gills paired Range
The northern lake	$\geq$ 90	152	100	0	-	-
	90 > $\geq$ 70	440	99	1	7.3	6–10
	70 > $\geq$ 50	147	99	1	8.0	8
	50 > $\geq$ 30	52	100	0	-	-
	30 > $\geq$ 10	51	53	47	14.0	7–26
	< 10	179	69	31	24.1	6–60
The southern lake	$\leq$ 5.5	52	15	85	17.1	5–56
Adjacent lagoons	< 1	24	0	100	68.0	38–105

careful seach for oligochaete fossils in the Paleo-Biwa Group is needed.

It is noticeable that Lake Biwa is the only ancient lake located in the zoogeographical Sino-Indian region (Timm, 1980). The presence of *L. amblysetus* in Lake Biwa suggests that there is a close faunal relationship between Lake Biwa and the continental China, and common occurrence of *Branchiura sowerbyi* in the lake could also be due to the geographical location of the lake, as the Sino-Indian region is the original home of this peregrine species (Timm, 1980). Although infraspecific variability of *B. sowerbyi* has not so far been subject to much attention, because of the species' other distinct features. The present study has revealed that the chaetae vary much between populations. This implies that local speciation is progressing, and the profundal population in Lake Biwa might be the most prominent instance. The less productive and monomictic condition of the northern lake maintains the high oxygen content in the bottom waters throughout the year. The gill-lacking form of *B. sowerbyi* found there may have originated intralacustrily to adapt for such an oxygen rich environment. *Branchiura sowerbyi* is a thermophilous tubificid, and the reproduction usually occurs in spring in the temperate region (Ohtaka & Kikuchi, 1997). On the other hand, continuous reproduction was found in the profundal of Lake Biwa (unpubl.), the bottom water of which is almost constantly within 6–8 °C throughout the year. As pointed out before (Ohtaka &

Nishino, 1995), such temperatures are lower than the threshold value for embryonic development (10 °C) estimated by Bonacina et al. (1994). Consequently, it is highly probable that the profundal population of *B. sowerbyi* has specialized not only morphologically but also physiologically to be able to live in the cool environment.

Lake Biwa used to be oligotrophic, but it has suffered from eutrophication in the last 30 years (LBRI & ILEC, 1988), and faunal and floral changes have been reported (Negoro, 1981; Hamabata & Kira, 1987; Maehata, 1987; Matsuda & Uenishi, 1992; Nishino et al., 1998). However, our knowledge of the oligochaete fauna in Lake Biwa is mostly restricted to knowledge developed during recent years, and faunal changes have not been detected. Nevertheless, some distributional changes can be noted for some species. That is, Stephenson (1917) recorded unidentified lumbriculids and a biwadrilid (*Biwadrilus batybates*) from the profundal of the lake. However, our repeated surveys since 1986 have not been able to confirm their presence in such deep habitats (Nishino et al., unpubl.). A similar situation was encountered for an endemic glossiphoniid leech, *Ancyrobdella biwae* Oka, which has never been recorded since the original record from the profundal of Lake Biwa in 1915 (Nishino, 1993). The disappearance of such profundal fauna might be caused by some recent environmental changes in the lake.

Acknowledgements

We are grateful to Dr R. O. Brinkhurst (Aquatic Resources Center, TN, U.S.A.) for the loan of oligochaete specimens and for his invaluable taxonomic advice, to Dr K. Fauchald and Ms L. Ward (Smithsonian Institution, Washington, DC, U.S.A.) for their support in studying USNM type specimens. We are also indebted to Dr M. Tsuchiya (Ryukyu University, Okinawa), Mr. Y. Masuda (Kawasaki Medical School, Kurashiki) and Mses K. Takada (Tokyo) and N. Hayakawa (Niigata) who provided us oligochaete specimens used in the present study.

References

Aiyer, K. S. P., 1929. An account of the Oligochaeta of Travancore. Rec. Ind. Mus. 31: 13–76.

Beddard, F. E., 1892. A new branchiate Oligochaeta (*Branchiura sowerbyi*). Quart. J. Microsc. Sci. 33: 325–341.

Bonacina, C., A. Pasteris, G. Bonomi & D. Marzuoli, 1994. Quantitative observations on the population ecology of *Branchiura sowerbyi* (Oligochaeta, Tubificidae) Hydrobiologia 278: 267–274.

Brinkhurst, R. O., 1964. Studies on the North American aquatic Oligochaeta. I. Naididae and Opistocystidae. Proc. Acad. Nat. Sci. Philad. 116: 195–230.

Brinkhurst, R. O., 1971. Part 2. Systematics, 7. Family Naididae, 8. Family Tubificidae. In R. O. Brinkhurst & B. G. M. Jamieson (eds), Aquatic Oligochaeta of the World. Oliver and Boyd. Edinburgh: 304–625.

Brinkhurst, R. O., 1978. Freshwater Oligochaeta in Canada. Can. J. Zool. 56: 2166–2175.

Brinkhurst, R. O., 1986. Guide for the freshwater aquatic microdrile oligochaetes of North America. Canad. Spec. Publ. Fish. Aquat. Sci. 84, Ottawa: 259 pp.

Brinkhurst, R. O., S. Qi & Y. Liang, 1990. The aquatic Oligochaeta from the People's Republic of China. Can. J. Zool. 68: 901–916.

Chapman, P. M. & R. O. Brinkhurst, 1987. Hair today, gone tomorrow: induced chaetal changes in tubificid oligochaetes. Hydrobiologia 155: 45–55.

Chen, C., 1940. Taxonomy and faunal relations of the limnic Oligochaeta of China. Contr. Biol. Lab. Sci. Soc. China, Zool. 14: 1–131

Cernosvitov, L., 1938. Oligochaeta. In R. Washbourn & R. F. Jones (eds), Report of the Percy Sladen Expedition to Lake Hulech. Ann. Mag. nat. Hist. ser. 11: 2: 533–550.

Erséus, C. & S. Qi, 1985. Two aberrant Tubificidae (Oligochaeta) from Pearl River in the People's Republic of China. Hydrobiologia 127: 193–196.

Erséus, C., J. K. Hiltunen, R. O. Brinkhurst & D. W. Schloesser, 1990. Redefinition of *Teneridrilus* Holmquist (Oligochaeta: Tubificidae), with description of two new species from North America. Proc. biol. Soc. Wash. 103: 839–846.

Hamabata, E. & T. Kira, 1987. Submerged plants in Lake Biwa. The Natural History of Japan. 1(7): 24–29 (in Japanese).

Harman, W. J., 1982. The aquatic Oligochaeta (Aeolosomatidae, Opistocystidae, Naididae) of Central America. Southwest. Nat. 27: 287–298.

Harman, W. J. & M. L. McMahan, 1975. A reevaluation of *Pristina longiseta* (Oligochaeta: Naididae) in North America. Proc. biol. Soc. Wash. 88: 167–178.

Hiltunen, J. K. & D. J. Klemm, 1980. A guide to the Naididae (Annelida: Clitellata: Oligochaeta) of North America. United States Environmental Protection Agency Research Report, Environmental Monitoring Series 600/4-80-031: 48 pp.

Holmquist, C., 1983. What is *Tubifex tubifex* (O. F. Müller) (Oligochaeta, Tubificidae)? Zool. Scr. 12: 187–201.

Holmquist, C., 1985. A revision of the genera *Tubifex* Lamarck, *Ilyodrilus* Eisen, and *Potamothrix* Vejdovský & Mrázek (Oligochaeta, Tubificidae) with extensions to some connected genera. Zool. Jb. Syst. 112: 311–366.

Horie, S. (eds), 1987. History of Lake Biwa. Its transition as discovered in an extremely long core of 1,400 m obtained in Lake Biwa. Institute of Paleolimnology and Paleoenvironment on Lake Biwa, Kyoto University: 242 pp.

LBRI (Lake Biwa Research Institute) & ILEC (International Lake Environment Commitee)(eds), 1988. Biwako (Lake Biwa). Data Book of World Lake Environments – A Survey of the State of World Lakes-. ILEC, Otsu: 1–12.

Kondo, M., 1936. A list of naidiform Oligochaeta from the waterworks plant of the city of Osaka. Annot. Zool. Japon. 15: 382–393.

Kumagai, M., S. Nakano & F. Maruo (eds), 1995. Data Report (II). Lake Biwa Institute, Otsu: 241 pp.

Maehata, S., 1987. On the large mouth bass in Lake Biwa – from the viewpoint of its food habit. Freshwater Fish, 13: 44–49 (in Japanese).

Martin, P., 1996. Oligochaeta and Aphanoneura in ancient lakes: a review. Hydrobiologia 334: 63–72.

Matsuda, M. & M. Uenishi, 1992. A newly introduced mussel, *Limnoperna lacustris* (Martens) in Lake Biwa, Shiga Prefecture, Japan (Mollusca; Mytilidae). Annual Rep. Biwako-Bunkakan. 10: 45. (in Japanese).

Meyers, P. A., K. Takemura & S. Horie, 1993. Reinterpretation of late Quaternary sediment chronology of Lake Biwa, Japan, from correlation with marine glacial-interglacial cycles. Quat. Res. 39: 154–162.

Milbrink, G., 1983. Characteristic deformities in tubificid oligochaetes inhabiting polluted bays on Lake Vänern, Southern Sweden. Hydrobiologia 106: 169–184.

Naidu, K.V., 1962. Studies on the freshwater Oligochaeta of South India. I. Aeolosomatidae and Naididae, Part 3. J. Bombay Nat. Hist. Soc. 59: 520–546.

Naidu, K.V., 1963. Studies on the freshwater Oligochaeta of South India. I. Aeolosomatidae and Naididae, Part 5. J. Bombay Nat. Hist. Soc. 60: 201–227.

Negoro, K., 1981. On the disorder of plankton calendar (plankton periodicity) in Lake Biwa, followed by the increased eutrophication. Water Temp. Res. 25: 15–19, (in Japanese).

Nishino, M., 1993. Hirudinea. In M. Nishino (ed.), Handbooks of Zoobenthos in Lake Biwa III. Porifera, Platyhelminthes, Annelida, Tentaculata and Crustacea. Lake Biwa Research Instutute, Otsu, Japan: 41–45, (in Japanese).

Nishino, M., M. Fukui & T. Nakajima, 1998. Dense Mats of *Thioploca*, gliding filamentous sulfur-oxidizing bacteria in Lake Biwa, central Japan. Wat. Res. 32: 953–957.

Nishino, M. & N. Watanabe, in press. An overview of endemic species of Lake Biwa. Proceedings of the International Conference on Ancient Lakes, 1997.

Nomura, E., 1926. On the aquatic oligochaete *Tubifex hattai*, n. sp. Sci. Rep. Tokoku Imp. Univ. 1: 193–228.

Ohtaka, A., 1985. Taxonomical revision of three Japanese *Limnodrilus* species (Oligochaeta, Tubificidae). Proc. Jap. Soc. syst. Zool. 30: 18–35.

Ohtaka, A., 1993. Oligochaeta. In M. Nishino (ed.), Handbooks of zoobenthos in Lake Biwa III. Porifera, Platyhelminthes, Annelida, Tentaculata and Crustacea. Lake Biwa Research Instutute, Otsu, Japan: 18–41 (in Japanese).

Ohtaka, A. & T. Iwakuma, 1993. Redescription of *Ophidonais serpentina* (Müller, 1773) (Naididae, Oligochaeta) from Lake Yunoko, central Japan, with record of the oligochaete composition in the lake. Jap. J. Limnol. 54: 251–259.

Ohtaka, A. & M. Nishino, 1995. Studies on the aquatic oligochaete fauna in Lake Biwa, central Japan. I. Checklist with taxonomical remarks. Jap. J. Limnol. 56: 167–182.

Ohtaka, A. & H. Kikuchi, 1997. Composition and abundance of zoobenthos in the profundal region of Lake Kitaura, central Japan, during 1980–1985, with special reference to oligochaetes. Publ. Itako Hydrobiol. Stn. 9: 1–14.

Ohtaka, A. & R. Usman, 1997. Records of tubificid oligochaetes from Padang, west Sumatra, Indonesia, with description of a new species of *Aulodrilus* Bretscher. Species Diversity 2: 145–154.

Paoletti, A. & B. Sambugar, 1996. Aquatic Oligochaeta in Italy, with special reference to Naididae. Hydrobiologia 334: 37–49.

Research Group for Natural History of Lake Biwa, 1994. The Natural History of Lake Biwa. Yasaka Shobo Inc. Tokyo, 340 pp. (in Japanese).

Rodriguez, P., 1986. Nuevos resultados acerca de la fauna de oligoquetos acuáticos del País Vasco y cuenca alta del Ebro. 1. Haplotaxidae, Naididae y Tubificidae. Munibe 38: 75–80.

Rodriguez, P., 1987. The variability of setae of *Pristina longiseta* Ehrenberg (Oligochaeta, Naididae). Hydrobiologia 155: 39–44.

Rodriguez, P. & J. C. Armas, 1983. Contribution à la connaissance de la faune d'Oligochètes aquatiques du pays basque et zones limitrophes. Annls Limnol. 19: 93–100.

Shiga Prefectural Institute of Public Health & Environmental Science, 1990. Report of survey on the bottom sediments condition of Lake Biwa for 1986–1988. Shiga Prefectural Institute of Public Health & Environmental Science, 203 pp. (in Japanese).

Sperber, C., 1948. The taxonomical study of the Naididae. Zool. Bidr. Upps. 28: 1–296.

Stephenson, J., 1917. Zoological results of a tour in the Far East. Aquatic Oligochaeta from Japan and China. Mem. Asiat. Soc. Bengal 6: 85–99.

Stephenson, J., 1923. Oligochaeta. The Fauna of British India, including Ceylon and Burma. London 518 pp.

Timm, T., 1980. Distribution of aquatic oligochaetes. In R. O. Brinkhurst & D. G. Cook (eds), Aquatic Oligochaete Biology, Plenum: 55–77.

Yamaguchi, H., 1940. Oligochaeta of Manchoukuo. In T. Kawamura (ed.), Report of Limnobiological Survey of Kwantung and Manchoukuo: 382–394 (in Japanese).

Yamaguchi, H., 1953. Studies on the aquatic Oligochaeta of Japan VI. A systematic report, with some remarks on the classification and phylogeny of the Oligochaeta. J. Fac. Sci. Hokkaido Univ. Ser. VI, Zool. 11: 277–342.

Yamaguchi, H., 1965. Oligochaeta. In New Illustrated Encyclopedia in the Fauna of Japan, 1. Hokuryukan Inc., Tokyo: 537–546 (in Japanese).

Hydrobiologia **406:** 49–55, 1999.
B. M. Healy, T. B. Reynoldson & K. A. Coates (eds), Aquatic Oligochaetes.

Monopylephorus camachoi nov. sp., a new rhyacodriline worm (Tubificidae: Clitellata) from the Coiba Island, on the east Pacific Coast of Panama

Pilar Rodriguez
Dpto. Biología Animal y Genética. Facultad de Ciencias. Universidad del País Vasco. Apdo. 644. Bilbao, 48080. Spain
Tel: 34-94-6012712 (ext: 2899). Fax: 34-94-4648500; E-mail: ggprorop@lg.ehu.es

Key words: Monophylephorus, new species, Rhyacodrilinae, Central America

Abstract

A new species of freshwater tubificid worm is described from the East Pacific Coast of Central America. *Monopylephorus camachoi* nov. sp. is mainly characterised by the presence of a single spermatheca with a short bulbous muscular duct and a pair of relatively elaborated male ducts, each consisting of a short vas deferens, a prostate-covered tubular atrium that opens apical-laterally to a muscular ovoid cavity interpreted as a pseudopenis, separated by a constriction from an elongated penial bursa. Pseudopenes and penial bursae have thick muscular walls arranged in two perpendicular layers. Both male ducts open into a mid-ventral copulatory bursa in segment XI. Setae in ventral bundles of both spermathecal (segment X) and postatrial (segment XII) segments are enlarged.

Introduction

Following the revision of Baker & Brinkhurst (1981) and the description of *M. moleti* Brinkhurst & Marchese (1987), the genus *Monopylephorus* Levinsen, 1884 comprises a total of nine species. In this paper, the author follows the synonymies proposed by Baker & Brinkhurst (1981) for *M. rubroniveus* Levinsen, 1884. *M. rubroniveus* and *M. parvus* Ditlevsen, 1904 are cosmopolitan species, being found on both Atlantic and Pacific coasts of North America, the Atlantic coast of South America, in Europe, Asia and South Africa (only *M. parvus*). On the Pacific Northeastern American coast, both species have been reported from British Columbia in West Canada, *M. parvus* at Saturna Island and *M. rubroniveus* at Prince Rupert, Powell River, Port Alice, Patricia Bay and Nuchatlitz Inlet (data from Baker & Brinkhurst, 1981). *M. rubroniveus* has also been reported in central California (Fields 1985), Massachusetts (Moore 1905, as *M. glaber*), West Texas, Louisiana and Florida (Loden 1980, as *M. helobius*), and the bay of Montevideo (Marcus 1952, as *M. corderoi*). Likewise, *M. parvus* has also been reported in Delaware (U.S.A.) (Baker & Brinkhurst 1971), the Gulf of Mexico (Milligan, 1996) and Brazil (São Paulo and Cananeia, Marcus 1965). The remaining species of *Monopylephorus* reported in America appear to have a more restricted geographical distribution. Thus, *M. cuticulatus* Baker & Brinkhurst (1981) is known in British Columbia and Alaska; *M. evertus* Baker & Brinkhurst 1981 has been found in the Atlantic U.S.A. at Delaware, Massachusetts and South Carolina; *M. irroratus* (Verrill, 1873) was confirmed in Massachusetts by Baker & Brinkhurst (1981) although previously known in France and the Baltic Sea (Europe); finally, *M. moleti* is known from inland waters in Santa Fe, Argentina (Brinkhrust & Marchese, 1987). Figure 1 shows the distribution of *Monopylephorus* species on the American continents. The new species *M. camachoi* on the Pacific Coast of Panama raises to seven the number of *Monopylephorus* species reported in America, so far.

Description of site

Santa Cruz River is situated on North Coiba Island, 500 m upstream from the river mouth, coordinates 17NMU158414, 100 m above sea-level. Flowing

Figure 1. Distribution of *Monopylephorus* species in the American continent. The American species are indicated on their distribution area with their specific names, while the records of the cosmopolitan species *M. rubroniveus* (*) and *M. parvus* (▲) are indicated by symbols with reference source into brackets. (See text for explanation of the synonyms and authors that reported the species).

water with low mineral content (pH: 7.5–7.8; conductivity: 126–239 μS/cm, alkalinity: 56–99 mg/l). Chlorides were moderately high (Cl^-: 5.0–8.9 mg/l). Data has been taken from García-Valdecasas et al. (1997).

Material and methods

Samples were taken by the kick method, 21 August, 1994, by Dr García-Valdecasas and Dr Camacho. Worms were fixed in 4% formaline and preserved in 70% ethyl alcohol. Specimens have been deposited at the National Museum of Natural Sciences (MNCN) in Madrid (Spain) and the University of the Basque Country (UPV/EHU) in Bilbao.

Taxonomy

Monopylephorus camachoi nov. sp. (Figures 2 & 3)

Holotype: MNCN 16.03/3024, one dissected specimen stained in Ehrlich's haematoxylin, mounted in Canada balsam.

Paratypes: MNCN 16.03/3024, one whole-mounted specimen stained in Ehrlich's haematoxylin and mounted in Canada balsam. Two serially sectioned specimens stained in haematoxylin -eosin. All from type locality.

Other material: Two immature and two mature individuals preserved in ethyl alcohol 70% are deposited in the MNCN. Two mature individuals (one dissected, one preserved in ethyl alcohol 70%) are in author's collection. All from type locality.

Type locality: River Santa Cruz. Coiba Island, Panama, in the Pacific Ocean. 21 August, 1994.

Diagnosis: 17 to 51 segments. Diameter of the body at segment VIII, 150–225 μm. Prostomium rounded (67–95 μm long). Nucleate granular coelomocytes ovoid, 7–10 μm longer axis, abundant in some individuals, but scarce in others. Epidermal layer of the body wall up to 4.8 μm thick, on the dorsal side of segment VIII, and up to 6.4 μm on the ventral side. Longitudinal muscular layer up to 4.8 μm on dorsal side, and up to 19 μm on ventral side at segment VIII. Chloragogen cells cover the gut from segment VI, backwards.

Only bifid setae, 3–5 setae per bundle in anterior segments and down to 2–3 setae per bundle in the posterior part of the body. Anterior setae are 40 μm long in II; 50–59 μm long in the following anterior segments, 58–64 μm long in the median region of the body, and 48–54 μm backwards. Width of somatic setae ca. 2 μm. Teeth in the anterior setae are about the same length in dorsal bundles, the upper tooth being slightly shorter in segment II and longer than the lower in the following ventral bundles. Setal teeth increase in length gradually and get thinner from II to IX (Figure 2 A–C and 3 C). Posterior setae with upper tooth about equal or shorter than lower. In segment X, two enlarged setae in ventral bundles 60 μm long, 2.7 μm thick, with upper tooth about 1.5 times longer and about the same thickness as the lower (Figure 3-D). Ventral setae absent in segment XI. In three of the examined mature individuals, ventral setae of segment XII are present in a number of one or two per bundle and are like those of segment X. Enlarged setae in segments X and XII are associated with setal glands

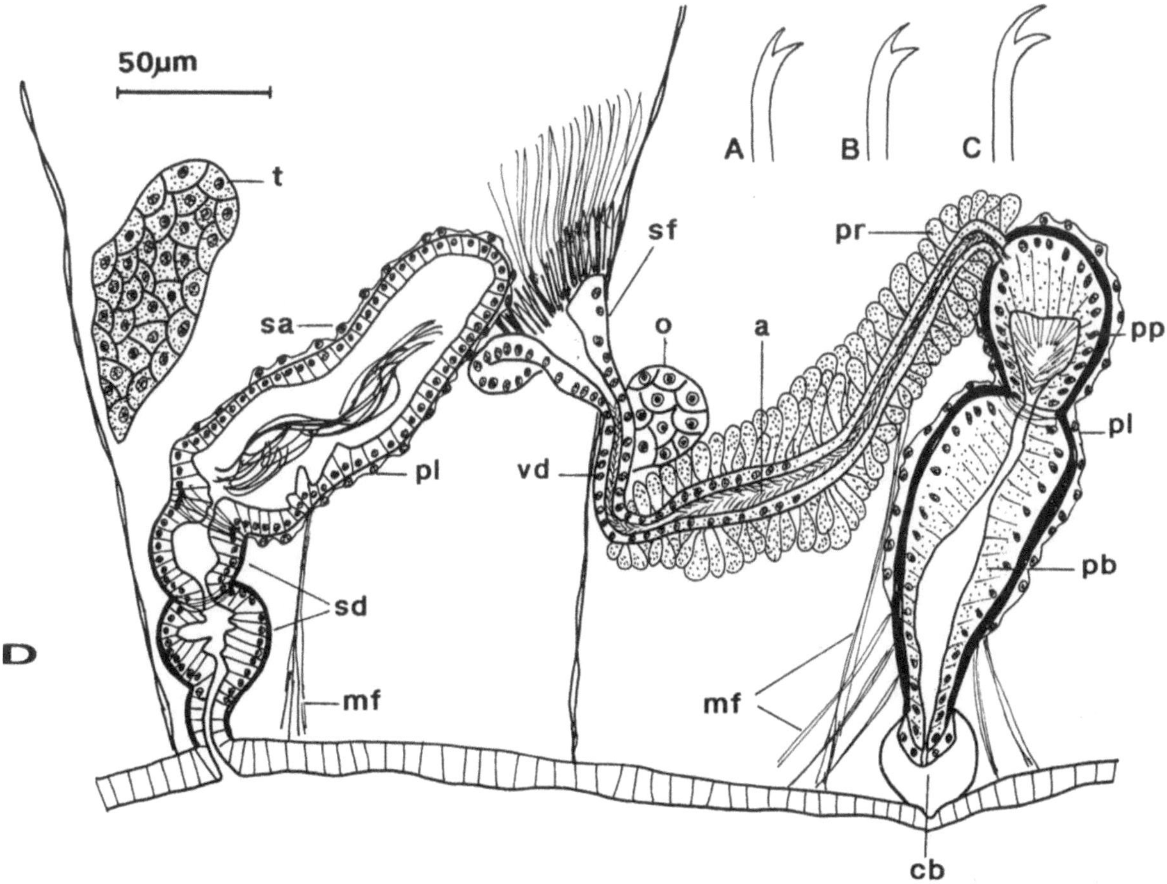

Figure 2. Setal types and anatomy of the male genitalia in *Monopylephorus camachoi* nov. sp. (A) Seta of segment II. (B) Most common type of somatic seta. (C) Ventral seta of segment VIII. (D) Lateral view of the male duct, gonads and spermatheca. *Abbreviations*: a: atrium, cb: copulatory median bursa, mf: muscular fibers, o: ovary, pb: penial bursa, pl: peritoneal layer, pp: pseudopenis, pr: prostatic cells, sa: spermathecal ampulla, sd: spermathecal duct, sf: sperm funnel, t: testis, vd: vas deferens.

which are more developed than those of the remaining setal bundles.

Clitellum non-prominent (up to 6.4 μm high), from behind setae of segment X to the end of segment XII. Sperm sacs extending forward to IX and backwards to XII. Egg sac up to XIV. One single spermathecal ventral pore that opens on the mid-line of the body, in the anterior part of segment X. One pair of male pores discharge close together into a single median copulatory bursa, that opens on the mid-ventral line of segment XI. One pair of testes in X and one pair of ovaries in XI.

One pair male ducts, each consisting of a sperm funnel opening to segment X, situated ventrally on septum X/XI. The male funnel leads to a short vas deferens (about 30 μm long, 11–12 μm diameter) that enters apically the ciliated tubular atrium (167–175 μm long), wider in its ental part (15–17 μm diameter) than in its ectal part (10–13 μm diameter), covered by a diffuse layer of prostatic cells (up to 31 μm high) and lined by a thin (less than 1 μm) muscular layer. The atrial cells show nuclei in the proximal part of the channel, not clearly seen in the rest of the duct. Ejaculatory duct absent. Atrium discharges laterally into the apical part of a muscular ovoid sac containing a protrusible pseudopenis (in the sense of Baker & Brinkhurst, 1981), 50–60 μm long and 37–43 μm maximum diameter, formed by ciliated cells lined by circular and longitudinal muscular fibres, surrounded by a peritoneal layer. Finally, the pseudopenial sac separates by a constriction from an elongated penial bursa (125–131 μm long and 52–64 μm maximum diameter), formed by a glandular epithelium lined with fine circular, followed by thick longitudinal, muscular fibres (up to ca. 4 μm thick) and a peritoneal layer. Both penial bursae open into a

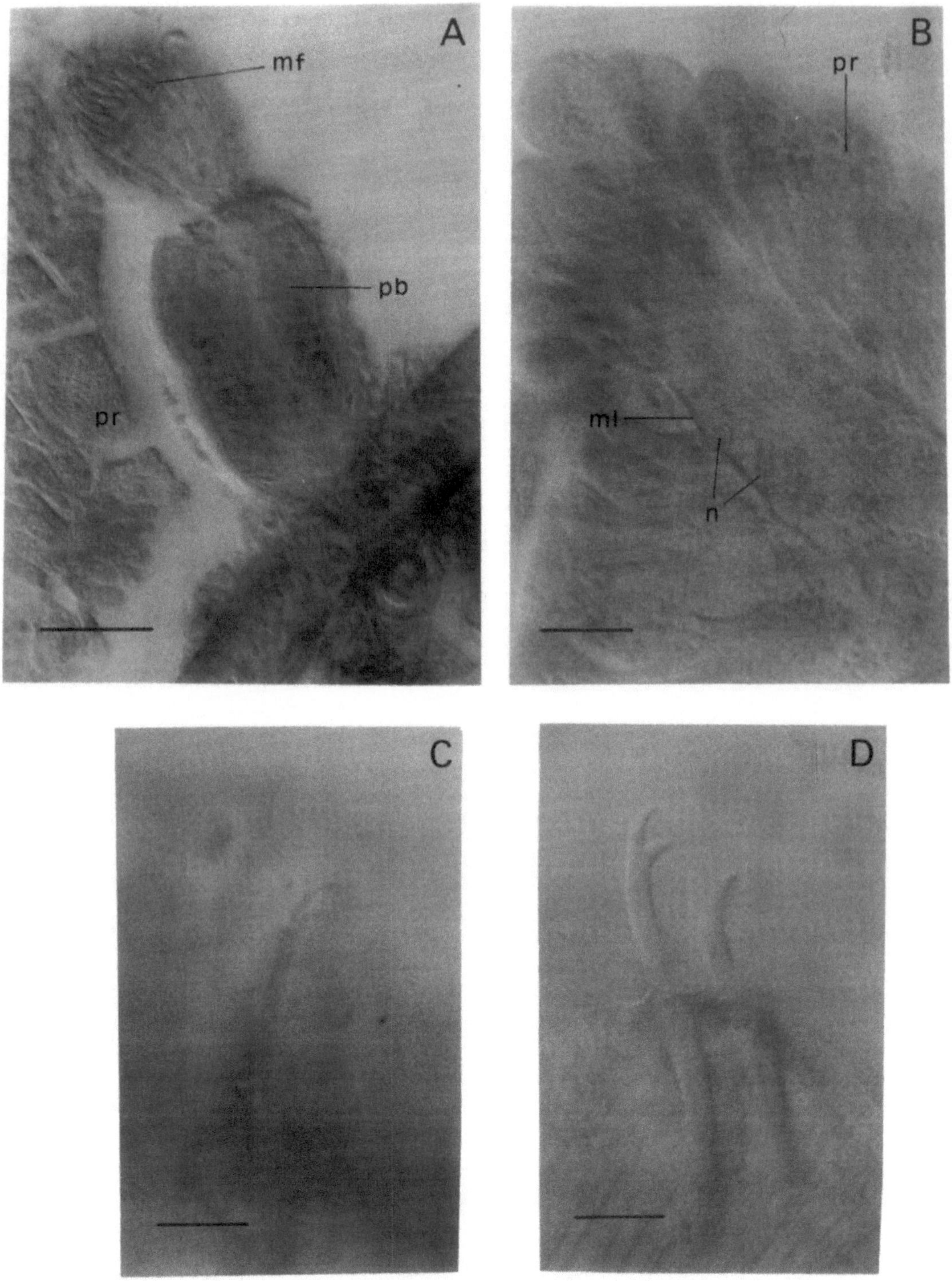

Figure 3. Details of the anatomy of *Monopylephorus camachoi* nov. sp. (A) Pseudopenis and penial bursa of the male duct. mf: circular and longitudinal muscular fibres, over the pseudopenis, pb: penial bursa, pr: prostate. Scale bar equal to 30 μm. (B) Proximal section of the atrium, showing nuclei (n) in the epithelial cells, and muscular lining (ml), covered by prostatic cells (pr). Scale bar equal to 10 μm. (C) and (D) ventral seta of segment IX and X, respectively. Scale bar equal to 10 μm.

single median copulatory bursa in the ventral side of segment XI.

One unpaired spermatheca is present, usually situated on the left side of segment X, formed by an elongated ampulla (120–160 μm long, and 32–65 μm maximum diameter) and a tripartite spermathecal duct, about 90 μm long, separated from the ampulla by a constriction of circular muscular fibres. Spermathecal duct comprises one ental section (22–32 μm long, 30–40 μm wide) separated by a constriction of a bulbous region (30–37 μm long, 42–50 μm wide), followed by a short narrow ectal duct (20–30 μm long, 12–15 μm wide) that opens to the spermathecal pore. Wall of spermathecal ampulla consisting of an inner epidermal lining (7–11 μm high) and a fine outer muscular layer (ca. 1 μm), covered with peritoneal cells. Lumen of ampulla filled by sperm in a loop-shaped mass. Wall of the spermathecal duct formed by an inner epithelial lining covered by a muscular layer, 1–2 μm thick, thicker in the bulbous part (ca. 2.5 μm). The external cuticle of the tegument enters internally by the spermathecal duct as far as the bulbous region.

Etymology: The species is affectionately dedicated to Dr Ana Camacho, who has always encouraged me in the study of subterranean aquatic worms.

Habitat characteristics: *M. camachoi* is found in clean freshwater, close to the estuary, as a component of a community formed by Odonata, Trichoptera, Ephemeroptera, Plecoptera, and others (García-Valdecasas et al., 1997). *Monopylephorus* is typically regarded as a genus which inhabits brackish waters; however, it is a recognisable fact that some species are euryhalines and therefore can be also found in freshwater (Marcus, 1952), commonly associated with high conductivity due to pollution or salt beds in the watershed (Brinkhurst & Marchese, 1987). Thus, *M. rubroniveus*, *M. limosus* and *M. moleti* have been reported in freshwater habitats (e.g., Erséus & Paoletti 1986; Brinkhurst & Marchese, 1987; Timm, 1997; respectively). The habitat characteristics are unclear for *M. kermadecensis* (Baker and Brinkhurst, 1981).

Distribution: Coiba Island, Panama, Pacific coast of Central America.

Discussion

Regarding the presence of conspicuous, usually abundant coelomocytes, a diffuse prostate covering the atrium and the general structure and position of the reproductive apparatus, the new species is a rhyacodriline tubificid. It is not simply an interpretation of the different parts of the male duct of the new taxon and there are few works that give light on this question (Hrabě 1939; Baker & Brinkhurst, 1981). The definitions by Baker & Brinkhurst (1981) of vas deferens (naked epithelial tube, without muscular lining and devoid of prostatic cells), atrium (prostate-covered tube, with or without a muscular layer), ejaculatory duct (the part of the male duct that follows the atrium, without muscular lining, that can appear occasionally distended by the accumulation of sperm) and pseudopenis, have been followed in the present work. However, a study of the development of the male duct and of compared anatomy of different clitellate families is insufficient, to adequately define homologous parts of the reproductive male apparatus. The ontogenesis of the male duct in *Monopylephorus* is now being studied (C. Erséus, personal communication) and the different male duct sections could be redefined for the genus in the near future.

The genus has last been revised by Baker & Brinkhurst (1981) and Finogenova (1982). *Monopylephorus* Levinsen, 1884 was established for tubificids with rudimentary or without vas deferens, entering apically to a narrow-tubular prostate-covered atrium, that leads to an ejaculatory duct which enters by a protrusible pseudopenis into a muscular penial sac (Baker & Brinkhurst, 1981). *M. camachoi* n. sp. fits in general terms the definition of the genus, however, the section after the prostate-covered atrium cannot be interpreted as an ejaculatory duct due to the existence of thick muscular layers. Therefore, this region has been interpreted as a pseudopenis because of its thick muscular wall and the presence of long epithelial cells in the ental part that could have a protrusible nature. The pseudopenis is separated by a constriction from an elongated muscular penial bursa and the functioning of these two distal parts of the male duct is not easy to ascertain. Therefore, the absence of an ejaculatory duct, the junction of the atrium lateral to the pseudopenis, and a thick muscular wall organised in both longitudinal and circular layers at the pseudopenis and penial bursa are new for the genus *Monopylephorus*. Only in *M. moleti*, there has not been described any section of the male duct as an ejaculatory duct (the taxonomic position of this species is discussed below). Among the *Monopylephorus* species, *M. limosus* (Hatai, 1878) also shows two separate muscular sections after the ejaculatory duct (the so-called pseudopenis and male bursa, in Erséus & Paoletti, 1986).

The presence of only one spermatheca is already known in *Monopylephorus parvus* Ditlevsen, 1904 and *M. moleti* Brinkhurst & Marchese, 1987. The new taxon is also close to these species by the presence of bifid setae and a median copulatory male bursa. However, *M. parvus* also has single-pointed setae in posterior segments, as has been commonly reported (Ditlevsen, 1904; Marcus, 1965; Brinkhurst & Baker, 1981; Erséus, 1984). The spermathecal ampulla of *M. parvus* is subject to geographical variation, and Erséus (1984) suggested the existence of two probable subspecies depending on the size and form of spermathecae: the north Atlantic populations with shorter untwisted ampulla, and the tropical and subtropical forms with long twisted ampulla. *M. camachoi* n.sp. shows no single-pointed setae nor twisted spermathecal ampulla, despite its tropical distribution. A bulbous spermathecal duct reinforced by musculature is known in other species of the genus *M. aucklandicus* (Benham, 1909), but it is not present in *M. parvus*. The structure of the spermatheca in *M. moleti* is not described but authors emphasized its complex histology (Brinkhurst & Marchese, 1987). *M. limosus* also shows muscular spermathecal duct divided into sections separated by constrictions, that resembles the sections in *M. camachoi*.

The presence of modified genital setae is unknown in the genus *Monophylephorus*, except for *M. moleti*. The taxonomic position of this species is uncertain, as it presents modified penial setae, similar to those of the genus *Rhizodrilus*, with their tips protruding into the so-called copulatory bursa, where a short atrium (not as tubular as in other *Monopylephorus* species), with prostatic cells covering only one side, enters laterally in a way that is also reminiscent of some *Rhizodrilus* species.

The new species shows enlarged ventral setae in pre- and post-atrial segments, of the same shape as the somatic ones, probably representing the most primitive way of differentiation of the genital setae. Erséus (1984) and Milligan (1996) have described specimens of *M. parvus* from southern China and the Gulf of Mexico, respectively, that also show enlarged ventral setae in a few segments close to the clitellar region. In the Rhyacodrilinae, the presence of modified setae in the postatrial segment is not common, being known in *Protuberodrilus tourenqui* Giani & Martinez-Ansemil, 1979. Contrarily, the presence of modified spermathecal setae is also known in several rhyacodriline taxa, such as some species of the genus *Rhyacodrilus* (see Giani & Rodriguez, 1988), *Rhizodrilus* (see *R. pacificus* Finogenova, 1982), *Ainudrilus* (see Erséus, 1990) and the recently described genus *Stochidrilus* Martinez-Ansemil et al., 1997.

Acknowledgements

I am endebted to Dr Ana Camacho and Dr García-Valdecasas from the National Museum of Natural Sciences of Madrid for placing the collection at my disposal; to Mrs. Bidaurrazaga for technical assistance in histological sections; to the laboratory of Pathological Anatomy of the S.I.M.A. where the histological sections have been stained. I thank Dr Brinkhurst and Dr Milligan for critically reviewing the manuscript.

References

Baker, H. R. & R. O. Brinkhurst, 1981. A revision of the genus *Monopylephorus* and redefinition of the subfamilies Rhyacodrilinae and Branchiurinae (Tubificidae: Oligochaeta). Can. J. Zool. 59: 939–965.

Benham, W. B., 1909. Report on Oligochaeta of the subantarctic islands of New Zealand. In C. Chilton & J. Mackay (eds), The Subantarctic Islands of New Zealand, 1: 251–294.

Brinkhurst, R. O. & M. Marchese, 1987. A contribution to the taxonomy of the aquatic Oligochaeta (Haplotaxidae, Phreodrilidae, Tubificidae) of South America. Can. J. Zool. 65: 3154–3165.

Ditlevsen, A., 1904. Studien an Oligochaeten. Z. wiss. Zool.77: 398–480.

Erséus, C., 1984. The marine Tubificidae (Oligochaeta) of Hong Kong and Southern China. Asian Mar. Biol. 1: 135–175.

Erséus, C., 1990. Marine Oligochaeta of Hong Kong. Proc. Second Int. Mar. Biol. Workshop: (ed.). In B. Morton, The Marine Flora and Fauna of Hong Kong and Southern China. Hong Kong Univ. Press. 1986: 259–335.

Erséus, C. & A. Paoletti, 1986. An Italian record of the aquatic oligochaete *Monopylephorus-limosus* (Tubificidae), previously known only from Japan and China. Boll. zool. 53: 115–118.

Fields, W. C. Jr., 1985. New distribution records for *Monopylephorus rubroniveus* (Oligochaeta: Tubificidae). Trans. am. microsc. Soc.104: 285–286.

Finogenova, N. P.,1982. Revision of the genus *Monopylephorus* (Oligochaeta, Tubificidae). Zool. Zh. 61: 5–17.

García-Valdecasas, A., A. I. Camacho, P. Aranzadi & E. Bello, 1997. Las Aguas Dulces de la Isla de Coiba (Panama). In S. Castroviejo, (eds), Flora y Fauna del Parque Nacional de Coiba (Panama). Inventario Preliminar. AECI: 107–125.

Giani, N. & E. Martinez-Ansemil, 1979. Description d'une nouvelle espèce de Tubificidae du Nord-Ouest de l'Espagne, *Protuberodrilus tourenqui* n.g., n.sp. Ann. Limnol. 15: 291–297.

Giani, N. & P. Rodriguez, 1988. Description de quelques espèces nouvelles de Tubificidae (Oligochaeta) des grottes et sources karstiques de la Péninsule Ibérique. Stygologia 4: 122–137.

Hatai, S., 1898. On *Vermiculus limosus*, a new species of Aquatic Oligochaeta. Annot. Zool. Japon. 2: 103–111.

Hrabě, S., 1939. Über die ontogenetische Entwicklung des männlichen Ausführungsganges bei einigen Oligochaeten. Sb. Kl. Prir. Brno 3: 56–65.

Levinsen, G. M. R., 1884. Systematisk-geografisk Oversigt over de nordiske Annulata. Vidensk. Medd. Dan. Natur-hist. Foren Khobenhaven 43(1883): 92–350.

Loden, M., 1980. A new euryhaline species of *Monopylephorus* (Oligochaeta: Tubificidae) from the southern United States. Bull. mar. Sci. 30: 600–603.

Marcus, E. du B.-R., 1952. A new tubificid from the Bay of Montevideo. Commun. zool. Mus. Hist. nat. Montevideo 56 (1949): 1–6.

Marcus, E. du B.-R., 1965. Naidomorpha aus brasilianischem Brackwasser. In E. Titschack (ed.), Beitrag neotrop. Fauna. 4. 61–83.

Martinez-Ansemil E., Sambugar B. & N. Giani, 1997. Ground oligochaetes from Southern -Europe. I. A new genus and new species of Rhyacodrilinae (Tubificidae) with a redescription of *Tubifex pescei* (Dumnicka) comb. n. Ann. Limnol. 33: 33–44.

Milligan, M. R. 1996. Identification Manual for the Aquatic Oligochaeta of Florida: Volume II. Florida Department of Environmental Protection: 239 pp.

Moore, J. P., 1905. Some marine Oligochaeta of New England. Proc. Acad. nat. Sci. Philad. 57: 373–399.

Timm, T., 1997. Freshwater Oligochaeta of some urban watercourses in the Russian Far East. Int. Rev. ges. Hydrobiol. 82: 437–467.

Verrill, A. E., 1873. Report on the inverterbate animals of Vineyard Sound. Rep. U.S. Comm. Fish. 324: 622–624.

Hydrobiologia **406:** 57–66, 1999.
B. M. Healy, T. B. Reynoldson & K. A. Coates (eds), Aquatic Oligochaetes.

Records of Enchytraeidae (Clitellata) from the People's Republic of China

Hongzhu Wang, Zhicai Xie & Yanling Liang
Institute of Hydrobiology, Chinese Academy of Sciences, Wuhan Hubei 430072, P. R. China

Key words: Enchytraeidae, Oligochaeta, terrestrial, aquatic, new species, new records, China

Abstract

Eleven species of terrestrial and aquatic Enchytraeidae are reported from southeastern China. *Fridericia multisegmentata* and *Enchytraeus athecatus* are new to science, while most of the others are recorded from the country for the first time.

Introduction

Owing to taxonomical difficulties and the influence of the opinion that enchytraeids are rare in the Oriental realm (Stephenson, 1923), taxonomists of microdrile oligochaetes in China paid more attention to Naididae and Tubificidae than to Enchytraeidae before the 1980s. The first recorded species, *Henlea ventriculosa* (d'Udekem), was reported from Tibet by Stephenson (cited in Cernosvitov, 1941). Until 1990, only three further species, *Fridericia bulbosa* (Rosa) (Chen, 1959), *Sinenchytraeus glacialis* Liang & Hsü (Liang et al., 1979), and *Isosetosa minxianensis* Xu, Zhang & Jiang, 1989 were recorded from the country. In 1992, Liang & Xie reported eight species belonging to four genera from Mt. Tianmushan in Zhejiang Province, *viz. Achaeta parva* Nielsen & Christensen, *A. indica* Prabhoo, *Hemienchytraeus bifurcatus* Nielsen & Christensen, *H. theae* Prabhoo, *Enchytraeus* sp., *Fridericia carmichaeli* Stephenson, *F. bulboides* Nielsen & Christensen and *F. callosa* (Eisen). Regarding marine enchytraeids, 11 species of five genera were recorded (Erséus, 1990, 1992a,b; Erséus et al., 1990).

In recent years, a preliminary investigation of Enchytraeidae in China was carried out. More than 20 species belonging to eight genera were found. Among them, four species of *Hemienchytraeus, viz. H. planisetosus* Xie, Wang & Liang sp.nov., *H. brachythecus* Xie, Wang & Liang sp.nov., *H. loksai* Dózsa-Farkas and *H. stephensoni* (Cognetti), will be published in another paper (Xie et al., 1999). This paper reports 11 species including *Fridericia multisegmentata* sp. nov. and *Enchytraeus athecatus* sp. nov.

Methods

Worms were extracted from samples by the wet-funnel method (O'Connor, 1962). Observations were based on living material and whole-mounted specimens, preserved in 10% formalin, stained by alum cochineal or borax carmine and mounted in Canada balsam. Measurements of internal organs refer to whole-mounted specimens and drawings were made from the mounts using a camera lucida, except Figure 2F and Figure 4B. The types and other specimens are deposited in the Specimen Room of Invertebrates, Institute of Hydrobiology, Chinese Academy of Sciences (CAS).

Sampling sites

Sampling sites are located in the following regions (Figure 1): Fengqiu (35.1° N, 114.4° E), Henan Province; Xi'an (34.3° N, 108.9° E), Shaanxi Province; Wuhan (30.5° N, 114.3° E), Hubei Province; Changsha (24.2° N, 116.1° E), Mt. Hengshan (27° 10′–20′ N, 112° 34′–114° 44′ E) and Hengyang (26.9° N, 112.6° E), Hunan Province; Lingchuan (25.4° N, 110.3° E), Yangshuo (24.8° N, 110.5° E), Zhongshan (24.5° N, 111.3° E) and Tengxian (23.4°

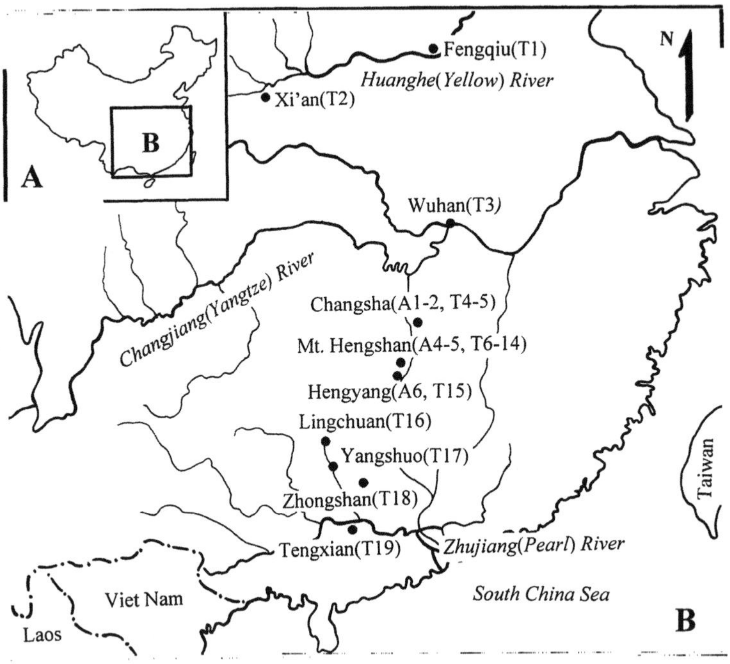

Figure 1. Distribution of sampling sites. A. China, B. sampling area.

N, 110.9° E), Guangxi Autonomous Region. These sites are as follows:

Aquatic sites
A1: Ditches around Taozihu pond, Changsha, April 12, 1991.
A2: Xiangjiang River, Changsha, April 12, 1991.
A3: Brooks on Mt. Hengshan, May 9, 1991.
A4: Rice fields on Mt. Hengshan, May 9, 1991.
A5: Nanyue Reservoir at the foot of Mt. Hengshan, May 9, 1991.
A6: Leishui River, Hengyang, May 21, 1991.

Terrestrial sites
T1: Fengqiu, villages and their vicinities, September 13–23, 1991.
T2: Xi'an, parks, September 5, 1991.
T3: Luojiashan, Wuhan, forest soil, July, 1994.
T4: Juzizhoutou in the middle of Xiangjiang River in Changsha, mossy soil, April 13,1991.
T5 : Yuelushan, Changsha, with the highest peak 296 m above sea level, forest soil, April 17, 1991 and November 13, 1994.
T6: Zhurongfeng, Mt. Hengshan, 1290 m above sea level, forest soil, April 24, 1991.
T7: Tianzhufeng, Mt. Hengshan, 1061 m, grass soil, April 25, 1991.
T8: Cangjingdian, Mt. Hengshan, 1050 m, forest soil, April 25, 1991.
T9: Huangdiyan, Mt. Hengshan, ca. 950 m, mossy soil, April 24, 1991.
T10: Tiefusi, Mt. Hengshan, 840 m, forest soil, April 29, 1991.
T11: Yehoushuyuan, Mt. Hengshan, 750 m, forest soil, April 28, 1991.
T12: Banshanting, Mt. Hengshan, 620 m, forest soil, April 26, 1991.
T13: Yubanqiao, Mt. Hengshan, 290 m, forest soil, April 27, 1991.
T14: Forest soil near Nanyue Reservoir at the foot of Mt. Hengshan, April 9, 1991.
T15: Xiangshancun, Hengyang, forest soil, May 21, 1991.
T16: Lingchuan, vegetable patch, September 7, 1992.
T17: Yangshuo, on the bank of Lijiang River, September 10, 1992.
T18: Zhongshan, various habitats, July 21–25, 1993.
T19: Tengxian, various habitats, July 10–13, 1993.

List of species

(#: new genus record for China, *: new species record for China.)

Mesenchytraeus sp. #
Achaeta brevivasa Graefe, 1980*
Achaeta cf. *indica* Prabhoo, 1960
Hemifridericia parva Nielsen & Christensen, 1959 #*
Henlea perpusilla Friend, 1911*
Fridericia cf. *alba* Moore, 1894*
Fridericia bulboides Nielsen & Christensen, 1959
Fridericia multisegmentata sp. nov.
Enchytraeus athecatus sp. nov.
Enchytraeus cf. *christenseni* Dózsa-Farkas, 1992*
Marionina riparia Bretscher, 1899*

Systematic descriptions

***Mesenchytraeus* sp.** (immature)

Localities: A3, T5 (13.XI.94), T6, T7, T8, T9, T10, T11, T12.

***Achaeta brevivasa* Graefe, 1980 (Figure 2 C, D, E, F)**

Length 1.4–2 mm (fixed), 2 mm (living); segments 17–25. Head pore small. No chaetal follicles. Clitellum over XII–1/2XIII, with gland cells in transverse rows. Brain about twice as long as width, anterior concave, posterior rounded. Septal glands all united dorsally; one pair of secondary glands behind septum 5/6. Gut dilatation gradual in VII. Oesophageal appendages one pair, confined to V and without internal canal (Figure 2C). Dorsal vessel originating in VII. Nephridia from 6/7 onwards with anteseptal parts ca. 1/2 as long as postseptal; efferent ducts terminal (Figure 2E). Coelomocytes abundant and round, with attached filaments (Figure 2F); cytoplasm transparent, nuclei 1/4–1/2 of the cell, round or irregular, sometimes absent. Sperm funnels about twice as long as width, with collar slightly narrower than funnel (Figure 2D). Vasa deferentia short, confined to XII. Penial bulbs crescent-shaped. No seminal vesicles. Spermathecae dumb-bell-shaped, opening laterally and confined to V, length shorter than body width. Ampullae irregularly oval, with globules on surface and small lumen.

Material examined: ten whole mounts and three live specimens from T3, T5 (17.IV.91), T8.

Remarks: The morphological characters of our specimens conform quite well to the original description. *A. camerani* is similar to *A. brevivasa* but is larger (length=6–10mm, segments=32–35), possesses two pairs of secondary septal glands in V–VI, a dorsal vessel was originating in VIII, and coelomocytes had larger nuclei (Graefe, 1980).

Distribution: Europe; Hubei Province and Hunan Province, China

***Achaeta* cf. *indica* Prabhoo, 1961 (Figure 2A, B)**

Achaeta indica Prabhoo. Liang & Xie, 1992.

Fixed length 2–3 mm; segments 20–30. No chaetal follicles. Clitellum over XII–XIII; gland cells scarce and small, arranged in transverse rows. Brain 1.5 times as long as wide, anterior end concave, posterior rounded. Septal glands all united dorsally. Oesophageal appendages one pair, confined to V. Gut dilatation sudden in VII (Figure 2A). Dorsal vessel originating in VII. Coelomocytes round with numerous granules. Efferent ducts of all nephridia terminal. Sperm funnels 2–4 times as long as wide, with collar slightly narrower than funnel (Figure 2B). Seminal vesicle present. Spermathecae opening ventro-laterally and extending to VII (Figure 2A); ampullae oval, ca. 60 μm in length, 26 μm in maximum width; ectal ducts ca. 160 μm long and 8 μm wide, dilated near ectal pore and forming a small spindle-shape lumen.

Material examined: four whole mounts and two live specimens from T14.

Remarks: Although individuals from the type position (India) were longer (length 7 mm, segments 31–36), and possessed oesophageal appendages extending into VI, the present material agrees with the original description (Prabhoo, 1961) in many characters, especially spermathecal morphology. The authors are of the opinion that it should be identified as *A. indica* for the time being.

Distribution: India; Zhejiang Province and Hunan Province, China.

***Hemifridericia parva* Nielsen & Christensen, 1959 (Figure 3 A)**

Length=1.6 mm (fixed), 2 mm (living); segments=21–22. Dorsal chaetae 2 per bundle; ventral chaetae 3 per bundle (2 in II) in the anteclitellar, 2 per bundle postclitellar. Clitellum over XII–1/2XIII; gland cells oval and transversely arranged; chaetae present in XII. Posterior of brain deeply incised. No septal glands unite dorsally; secondary glands one pair in IV (Figure 3A). Dorsal vessel originating in XII–XIII. Nephridia from 6/7 onwards with efferent duct terminal. Coelomocytes of two types. Sperm funnels pear-shaped, 1.5 times as long as width, with inconspicuous collar. No seminal vesicles. Spermathecae small, connecting jointly with oesophagus (Figure 3A).

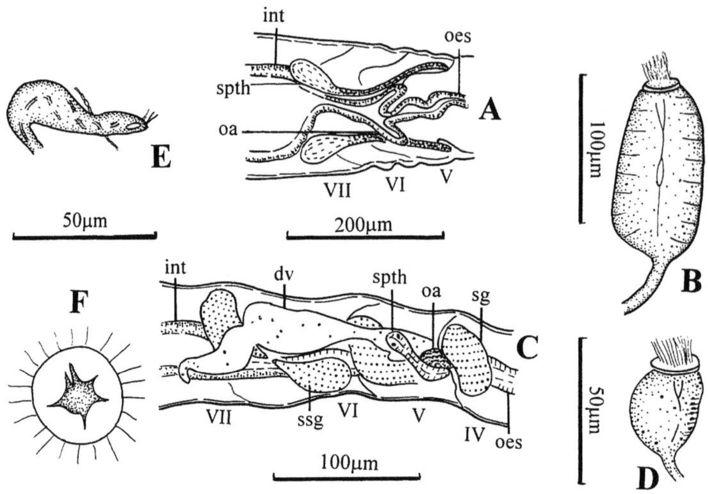

Figure 2. (A–B) *Achaeta* cf.*indica*, (C–F) *Achaeta brevivasa*; (A) dorsal view of IV–VII, (B, D) sperm funnel, (E) nephridium, (F) coelomocyte (freehand drawing). Abbreviations: dv, dorsal vessel; int, intestine; oa, oesophageal appendage; oes, oesophagus; sg, septal gland; ssg, secondary septal gland; spth, spermatheca.

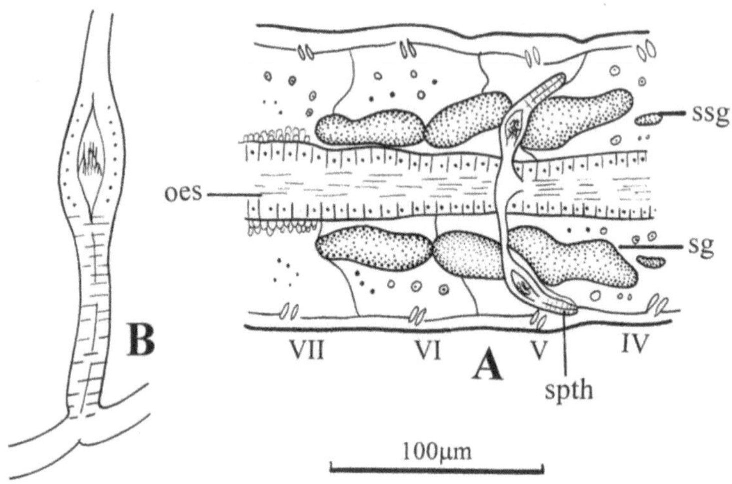

Figure 3. (A) *Hemifridericia parva*, (B) *Henlea perpusilla*, (A) dorsal view of IV–VII, (B) spermatheca, Abbreviations: oes, oesophagus; sg, septal gland; spth, spermatheca; ssg, secondary septal gland.

Material examined: three specimens from T6.

Remarks: The specimens from the type locality (Denmark) were described as possessing 2 chaetae per bundle ventrally in XI (Nielsen & Christensen, 1959), whereas our specimens have 3 in the same position. The presence of secondary septal glands in IV was not mentioned in the original description.

Distribution: Europe; Hunan Province, China.

***Henlea perpusilla* Friend, 1911 (Figure 3 B)**

Henlea perpusilla Friend. Nielsen & Christensen, 1959.

Length 7.4 mm(fixed), 8.5 mm (living); segments 36. Chaetae 2–7 per bundle. Clitellum over XII–1/2XIII, with gland cells irregularly arranged. Posterior of brain slightly incised. Three oesophageal appendages, dorsal pair short and unbranched, ventral piece branching into two long twigs. No intestinal diverticula. Dorsal vessel originating in posterior of VIII. Sperm funnels cylindrical, 2–3 times as long as width. Spermathecal ampullae spindle-shaped, without glands at ectal pore (Figure 3B).

Material examined: one whole mount and two live specimens from T7.

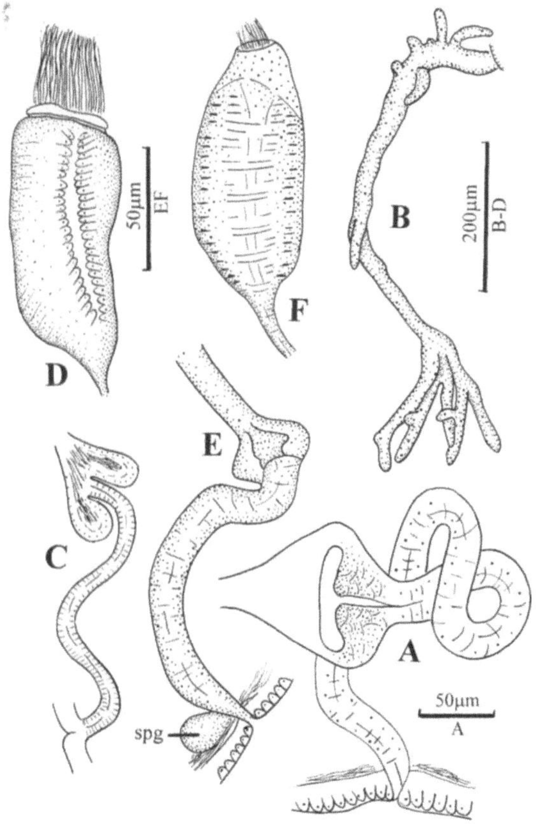

Figure 4. (A) *Fridericia* cf. *alba*, (B–D) *Fridericia multisegmentata* sp. nov., (E, F) *Fridericia bulboides*; (A, C, E) spermatheca, (B) peptonephridium on one side, (D, F) sperm funnel. Abbreviation: spg, gland at spermathecal pore.

Distribution: Europe, Asia (Russia; Hunan Province, China), N. America, Antarctica (South Georgia Island).

***Fridericia* cf. *alba* Moore, 1895 (Figure 4 A)**
Length 11–12 mm (fixed), 16 mm (living); width 0.3–0.5 mm; segments 52–59. Head pore at 0/I, dorsal pores from VII onwards. Chaetae from II onwards, 2–4 in dorsal and ventral bundle, mostly 4 anteriorly (from II to ca. XXI) and 2 posteriorly; chaetae absent in XII; outer chaetae ca. 62 μm in length, 6μm in maximum width. Epidermal gland cells transversely arranged, ca. 14 rows per segment (when whole-mounted), more conspicuous in the anteclitellar region. Clitellum over XII–2/3XIII, with gland cells irregularly arranged; medioventral gland cells absent or scant anteriorly and in a small posterior patch.

Brain 140 μm long, 50–60 μm wide anteriorly and 90 μm wide posteriorly, anterior end triangularly convex, and posterior rounded or truncated. Peptonephridia one pair, straight and short, extending to IV or anterior of V, unbranched or with 1–2 twigs. Septal glands with dorsal connection in first and second pairs. Gut dilatation gradual. Dorsal vessel originating in XIX. Chylus cells present in 3 postclitellar segments, e.g. from XIV to XVI, or from XVII to XIX. Chloragogen cells from ca. VII onwards. Nephridia from 6/7 onwards with 4 pairs in front of clitellum, anteseptal parts ca. 1/2 as long as postseptal; efferent ducts originating from anterior of postseptal parts. No copulatory glands enveloping ventral nerve cord. Nucleate coelomocytes numerous, oval to round, with regular outline, with granules evenly distributed and without refractile globules, ca. 20 μm long and 10–20 μm wide; small anucleate corpuscles scarce, 5–7 μm long and ca. 4 μm wide (when whole-mounted). Detached chaetae present in coelome.

Sperm funnels cylindrical, ca. 220 μm long and ca. 90 μm wide, with distinct collar as wide as funnel, mostly distorted when preserved. Vasa deferentia irregularly coiled. Penial bulbs one pair, hemispherical, 60 μm long, 36 μm wide, 30 μm in height, opening ventrolaterally at mid XII. One large seminal vesicle in X–XI. No egg sacs, 1–2 eggs present.

Spermathecae in V, communicating with dorsolateral region of oesophagus separately. Ampullae onion-shaped, with transversely scattered spermatozoa, 120 μm long, 96–100 μm in maximum width. Ectal ducts ca. 360 μm long and 26 μm wide, without glands at ectal pore. Ental ducts ca. 100 μm long and 40 μm wide (Figure 4A).

Material examined: five whole mounts and two live specimens from T7, T12, T16.

Remarks: The distinctive characters of *F. alba* are large size, short peptonephridia and simple spermathecae. In these respects, our specimens are similar to those from the type locality (Philadelphia, USA). However, the original material had chaetae were 2–7 per bundle, mostly 4, the brain slightly emarginate posteriorly, and the dorsal vessel originating in XXII. Male genital organs and nephridia were not described by Moore (1895).

Distribution: N. America; Hunan Province and Guangxi Autonomous Region, China.

***Fridericia bulboides* Nielsen & Christensen, 1959 (Figure 4 E, F)**

Fridericia bulboides Nielsen & Christensen. Liang & Xie, 1992.

Length 6 mm (fixed), 9 mm (living); segments 28–

42. Chaetae 2–4 in dorsal and ventral bundle, mostly 4 in the anteclitellar and 2 in postclitellar regions. Epidermal gland cells 5–6 rows per segment. Clitellum over XII–2/3XIII, with gland cells irregularly arranged and well developed everywhere. Peptonephridia unbranched and very long, much coiled in IV–V, extending to VI–VII. Dorsal vessel originating in XVII. Chylus cells present in region XIII–XIV. Sperm funnels cylindrical, with tall collar (Figure 4F). No seminal vesicles. Spermathecal ampullae onion-shaped, with a large gland at ectal pore (Figure 4E).

Material examined: seven whole mounts and two live specimens from T6, T7, T11, T14.

Remarks: The characteristics of our specimens coincide with the original description, except that our specimens have larger glands at spermathecal pore and a more posterior origin of dorsal vessel (XIV–XVI).

Distribution: Europe, Turkey, N. America, Greenland, Canada, Australia, Siberia; Zhejiang Province and Hunan Province, China.

***Fridericia multisegmentata* sp. nov. Wang, Xie & Liang (Figure 4 BCD)**

Holotype: Mature, whole mount, collected from Cangjingdian, Mt. Hengshan, 1050 m, broad-leaf forest dominated by beech (Fagaceae), yellow-brown soil, April 25, 1991 (T8). T8.

Paratypes: eight whole mounts (2 dissected), eight in 10% formalin all from T8.

Etymology: Named *multisegmentata* for the large number of segments.

Description: Fixed length 20–28 mm (holotype: 27 mm); width 0.7–0.8 mm; segments 73–80 (holotype: 80). Body whitish. Head pore at 0/I, dorsal pores from VII. Chaetae 2–4 in dorsal and ventral bundle, mostly 4 anteriorly (from II to ca. XX) and 2 posteriorly, disposed in pairs; straight, simple-pointed, with ental hook; outer chaetae ca. 101 μm in length, 10 μm in maximum width; chaetae absent in XII. Epidermal gland cells numerous, irregularly arranged. Clitellum over XII–XIII, gland cells irregularly arranged; mediventral gland cells absent in the anterior region (in front of the posterior border of male pores) between male pores.

Brain 126 μm long, 100 μm wide anteriorly and 132 μm wide posteriorly, anterior slightly concave and posterior truncated. Peptonephridia arising ventrolaterally, branching into 6 and 7 twigs respectively in the anterior and posterior parts, one twig present in the middle part (Figure 4B). Septal glands with dorsal connection in first and second pairs; no secondary glands. Gut dilatation gradual. Dorsal vessel originating in XXV–XXVI (holotype: XXVI). Chylus cells from XI to ca. XVII. Chloragogen cells from VII. Nephridia from 6/7 with 5 pairs in front of clitellum, anteseptal parts ca. 1/3 as long as postseptal; efferent ducts originating from terminal regions of postseptal parts. No copulatory glands enveloping ventral nerve cord. Coelomocytes of two types, both oval to round, with regular outline. Nucleate coelomocytes numerous, with granules evenly distributed and with several refractile nucleoli, 30–40 μm long and 22–26 μm wide; small anucleate corpuscles scarce, 7–8 μm long and 6–7 μm wide (when whole-mounted).

Sperm funnels cylindrical, 338 μm long and 152 μm wide, with raised collar narrower than funnel (Figure 4D). Vasa deferentia confined to XII, irregularly coiled. Penial bulbs one pair, hemispherical, 116–132 μm long, 97–101 μm wide, 77–97 μm in height, opening ventrolaterally at mid XII. One large seminal vesicle in the anterior of XI. No egg sacs. 1–2 eggs present.

Spermathecae in V, having conical ampulla and 2 diverticula with almost cylindrical outline and lumen (Figure 4C). Spermatozoa scattered throughout the common lumen of ampulla and diverticula. Ampullae ca. 70 μm in length (from connection with ectal duct to beginning of ental duct); paired diverticula 39–43 μm in length (from distal end to connection with ectal duct) and ca. 40 μm in width. Each ampulla communicating with dorsolateral region of oesophagus separately by a short ental duct. Ectal ducts 480–500 μm long and 27 μm wide, without glands at ectal pore.

Remarks: The new species differs from all known *Friderica* species by the following combination of characters:

1. the large number of segments,
2. a maximum of 4 chaetae in a bundle,
3. spermathecae with 2 diverticula and without glands at ectal pore (Table 1).

Nevertheless, there is some resemblance to *Fridericia perrieri* (Vejdovský), *Fridericia agricola* Moore and *Fridericia jordanensis* Bell concerning spermathecae and peptonephridia. Besides the different chaetae, segments and sperm funnel, the present species differs from the first two by the morphology of spermathecal lumina and nephridia, and differs from the last by the more posterior origin of the dorsal vessel.

Distribution: Hunan Province, China.

***Enchytraeus athecatus* sp. nov. Wang, Xie & Liang (Figure 5)**

Table 1. Comparison of *Fridericia multisegmentata* sp. nov. with allied species

	F. multisegmentata sp. nov.	*F. perrieri* (Vejdovsky, 1878)[1)]	*F. agricola* Moore, 1895[2)]	*F. jordanensis* Bell, 1962
Length (mm)	20–28	10–25	20–25	10
Segments	73–80	33–64	65–72	48–58
Chaetae/bundle	2–4	4–8	2–4 (5)	4–6
Clitellum	XII–XIII	XII–1/2XIII	XII–1/2XIII	absent
Posterior of brain	truncated	round	round	round
Peptonephridia	much branched	branched	much branched	with several terminal twigs
Origin of dorsal vessel	XXV–XXVI	XVI–XXI	?	XV–XVIII
Efferent duct of nephridia	terminal	ventral	ventral	terminal or ventral
Sperm funnel (length:width)	2.2:1	2.5–3:1	4.5:1	1:1
Seminal vesicle	well developed	poorly developed or absent	well developed	well developed
Spermathecae	lumen of ampulla and 2 diverticula inseparable; no ectal glands	lumen of ampulla and 2 diverticula separable; canal of ectal duct coiled before reaching ampulla; no ectal glands	lumen of ampulla and 2 diverticula separable; no ectal glands	lumen of ampulla and 2 diverticula inseparable; no ectal glands
Distribution	China	Europe, E. Africa	USA, Brazil,	Jordan

[1)] data from Nielsen & Christensen, 1959; Möller, 1971. [2)] from Moore, 1895; Bittencourt, 1974.

Holotype: Whole mount, collected from Juzizhoutou in the middle of Xiangjiang River in Changsha, red, mossy soil attached to the bank, April 13, 1991 (T4).

Paratype: one whole mount, also from T4.

Other localities: A4, T1, T2, T15, T16, T17, T18, T19.

Etymology: Named *athecatus* for the absence of spermathecae.

Description: Fixed length 4.7–4.8 mm (holotype: 4.7 mm); width 0.2–0.3 mm; segments 30. Head pore present in dorsal center of prostomium. Chaetae from II, mostly 2 per bundle; 1 additional chaeta (smaller or not) occasionally present ventrally from II to XIII (present in both holotype and paratype); straight, simple-pointed, without nodulus and with ental hook; 48–52 μm in length, 2–4 μm in width (Figure 5A); chaetae absent in XII. Clitellum over XII–1/2XIII, with gland cells irregularly arranged; medioventral gland cells absent between male pores.

Brain trapezoidal, posterior round, 96 μm in length, 36–48 μm in width (Figure 5B). Pharyngeal plate in II–III well developed. Peptonephridia unbranched, extending to IV (Figure 5C). A pair of small postpharyngeal bulbs attached to pharynx and situated dorsal to peptonephridia, with numerous nuclei. Septal glands without dorsal connection, or with thin connection in first pair (holotype); no secondary glands. Gut dilatation gradual. No oesophageal or intestinal appendages. Dorsal vessel originating in clitellar region. Chloragogen cells from VI. Nephridia from 6/7, with 5 pairs in front of clitellum; anteseptal parts small and postseptal elongated; efferent ducts originating from terminal of postseptal parts (Figure 5D). Coelomocytes oval and nucleate, 12–20 μm long and 9–16 μm wide (when whole-mounted). Detached chaetae few in coelom.

Sperm funnels sub-spherical, length 72 μm, maximum width 62 μm; collars narrower than funnels, 10 μm in length, 34 μm in width (Figure 5E). Vasa defer-

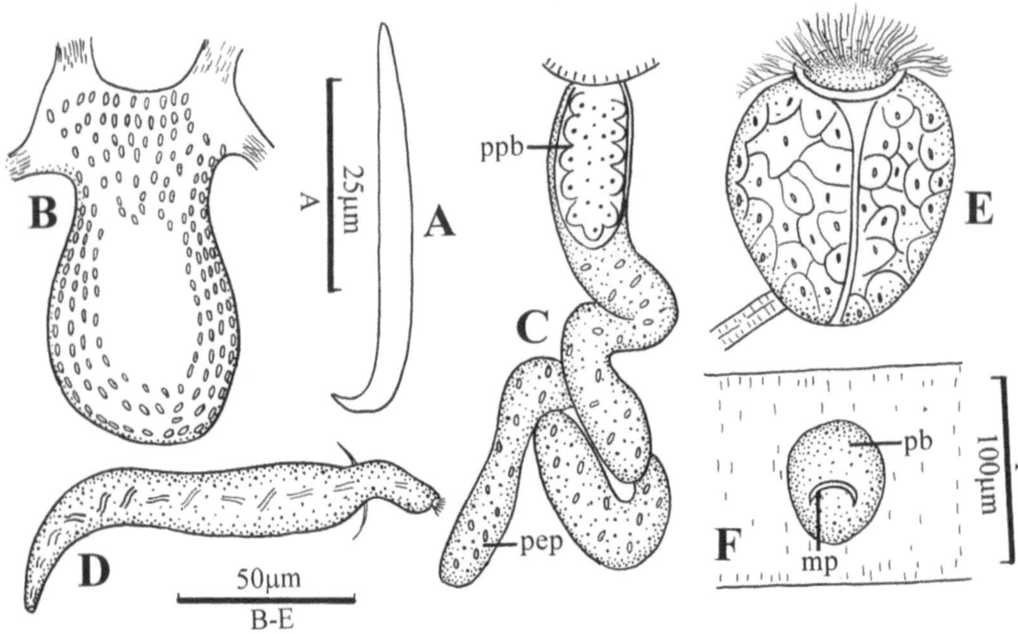

Figure 5. Enchytraeus athecatus sp. nov. (A) chaeta (VIII), (B) brain, (C) peptonephridium and post-pharyngeal bulb on one side, (D) nephridium(21/22), (E) sperm funnel, (F) penial bulb. Abbreviations: mp, male pore; pep, peptonephridium; pb, penial bulb; ppb, post-pharyngeal bulb.

entia confined to XII, irregularly coiled. Penial bulbs, large and spherical, 56–64 μm in maximum length, 64–80 μm in width, opening ventrolaterally at mid XII (Figure 5F). Seminal vesicles oval to round, one pair from posterior of X to anterior of XI (holotype) or in anterior of XI (paratype), 80–140 μm in length and 44–80 μm in maximum width. No egg sacs; 1–2 eggs present. Spermathecae absent.

Remarks: The new species resembles *Enchytraeus florentinus* Bell and *Enchytraeus varithecatus* Bouguenec & Giani in the absence of spermathecae. However, the chaetae of *E. florentinus* are curved, the degree of curvature varying considerably in the same worm and its penial bulb is reduced to a small cluster of gland cells. The present species also differs from *E. varithecatus* by having one more pair of anteclitellar nephridia, regularly oval coelomocytes, larger penial bulb and a small number of mature eggs at a time (Table 2).

Distribution: Henan Province, Shaanxi Province, Hunan Province and Guangxi Autonomous Region, China.

Enchytraeus cf. _christenseni_ Dózsa-Farkas, 1992

Enchytraeus minutus (nec Tauber) Nielsen & Christensen, 1961

Enchytraeus christenseni Dózsa-Farkas, 1992

Length 2–3 mm (fixed), 5 mm (living); segments 27–29. Chaetae 2–4 per bundle. Epidermal gland cells scarce. Clitellum over XII–1/2XIII, with gland cells in transverse rows; medioventral gland cells absent in area between male pores. Brain slightly incised posteriorly. Peptonephridia unbranched, with irregular outline. No septal glands united dorsally. Dorsal vessel originating in XIII. Efferent ducts of nephridia originating from posterior of postseptal part. Coelomocytes abundant and oval. Sperm funnels cylindrical or pear-shaped, 2–3 times as long as wide, with conspicuous collar as wide as funnel. Penial bulbs large, 48–56 μm in length, 40–44 μm in width, ca. 20 μm in height. Seminal vesicles small, one pair in XI. Spermathecal ampullae spherical, ca. 22 μm in diameter; ectal ducts covered with glands, 1.5–2 times as long as ampullae, without glands at ectal pore.

Material examined: nine whole mounts and three live specimens from A2, A3, T13.

Distribution: Europe; Hunan Province, China.

Marionina riparia Bretscher, 1899 (Figure 6)

Marionina riparia Bretscher. Nielsen & Christensen, 1959.

Table 2. Comparison of *Enchytraeus athecatus* sp. nov. with allied species

	E. athecatus sp. nov.	*E. florentinus* Bell, 1947	*E. varithecatus* Bouguenec & Giani, 1987
Length (mm)	4.7–4.8	2–3	6–7.5
Segments	30	27–30	25–30
Chaetal shape	straight	curved	straight
Dorsal chaetae/bundle	2	2	2
Ventral chaetae/bundle	2 (3)	3	2
Coelomocytes	oval	oval	irregularly oval to spindle-shaped
Posterior of brain	round	slightly incised	truncated or slightly incised
Anteclitellar nephridia	5 pairs	?	4 pairs
Sperm funnels (length:width)	1.2:1	2:1	1.3:1
Penial bulbs	large	degenerate	small and compact
Eggs	1–2	?	1–6
Spermathecae	absent	absent	sometimes absent
Distribution	China	Italy	France

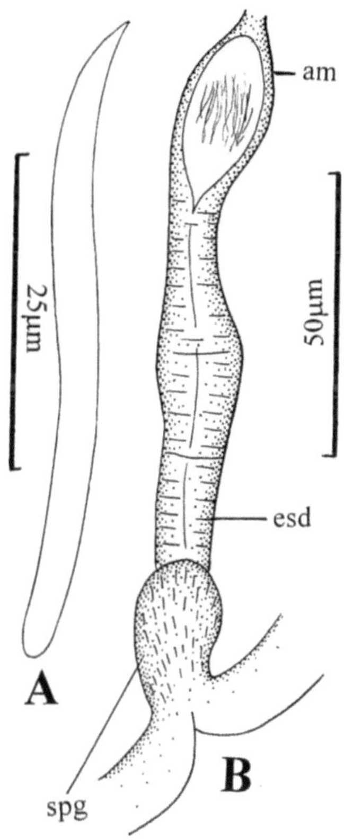

Figure 6. Marionina riparia, (A) chaeta (IV), (B) spermatheca. Abbreviations: am, ampulla; esd, ectal duct of spermatheca; spg, gland at spermathecal pore.

Fixed length 3 mm; segments 26. Body reddish when living. Chaetae sigmoid (Figure 6A), 2–4 per bundle, mostly 3. Epidermal gland cells 2–4 rows per segment, reddish when living. Clitellum over 1/2XI–1/2XIII, with gland cells irregularly arranged. Brain deeply incised posteriorly. Only first pair of septal glands united dorsally. Dorsal vessel originating in XIII, with reddish blood when living. Efferent ducts of nephridia originating from posterior of postseptal parts. Coelomocytes spindle-shaped and abundant. Sperm funnels cylindrical, twice as long as width, with collar narrower than funnel. Vasa deferentia confined to XII. Penial bulbs large. No seminal vesicles. The connection between spermatheca and oesophagus rather thin. Spermathecal ampullae oval, ca. 26 μm in length, ca. 18 μm in maximum width. Ectal ducts ca. 90 μm long and 10 μm wide, with a gland at ectal pore(Figure 6B).

Material examined: five whole mounts and two live specimens from A1, A3, A4, A5, A6.

Remarks: In specimens described by Nielsen & Christensen (1959), the spermathecal ampullae are larger than those of our specimens and the clitellar gland cells are arranged in transverse rows.

Distribution: Europe, N. Africa, Turkey; Hunan Province, China.

Acknowledgement

The project was supported by a grant for systematic and evolutionary biology, CAS and by National Natural Science Foundation of China. The latter also financed the first author to participate in the Aquatic Oligochaete Symposium. We are deeply indebted to Drs Ruediger M. Schmelz, Emilia Rota and Brenda Healy for their great help and valuable comments on the manuscript, and to Ms Yin Zheng for drawing our figures in ink.

References

Cernosvitov, L., 1941. Oligochaeta from Tibet. Proc. zool. Soc., Lond. 111: 281–287.

Chen, Y., 1959. Icones faunarum sinicarum, pars. Annelida, app. Myriapoda. Science Press, Beijing: 78pp. (in Chinese).

Dózsa-Farkas, K., 1992. List of enchytraeid synonyma. Newsletter on Enchytraeidae 3: 16–35.

Erséus, C., 1990. Marine Oligochaeta of Hong Kong. In B. Morton (ed.), The Marine Flora and Fauna of Hong Kong and Southern China II. Hong Kong University Press, Hong Kong: 259–335.

Erséus, C., 1992a. Hong Kong's marine Oligochaeta: A supplement. In B. Morton (ed.), The Marine Flora and Fauna of Hong Kong and Southern China III. Hong Kong University Press, Hong Kong: 157–180.

Erséus, C., 1992b. Oligochaeta from Hoi Ha Wan. In B. Morton (ed.), The Marine Flora and Fauna of Hong Kong and Southern China III. Hong Kong University Press, Hong Kong: 909–917.

Erséus, C., D. Y. Sun, Y. L. Liang & B. Sun, 1990. Marine Oligochaeta of Jiaozhou Bay, Yellow Sea coast of China. Hydrobiologia 202: 107–124.

Graefe, U., 1980. Systematische Untersuchungen an der Gattung *Achaeta* (Enchytraeidae, Oligochaeta). 1. Beschreibung von *Achaeta brevivasa* sp. n. und *Achaeta camerani* (Cognetti). Mitt. hamb. zool. Mus. Inst. 77: 35–39.

Liang, Y. L., C. F. Hsü & T. N. Chang, 1979. A new genus and species of Enchytraeidae from Tibet. Acta Zootax. Sinica 4: 312–315. (in Chinese with English abstract)

Liang, Y. L. & Z. C. Xie, 1992. Chapter 7. Annelida, Mollusca and Tardigrada, 1.Oligochaeta, i. Oligochaeta Plesiopora. In W. Y. Yin et al. (eds), Subtropical Soil Animals of China. Science Press, Beijing: 194–201 (in Chinese).

Moore, J. P., 1895. Notes on American Enchytraiedae. I. New species of *Fridericia* from the vicinity of Philadelphia. Proc. Acad. nat. Sci. Phila. 1895: 341–345.

Nielsen, C. O. & B. Christensen, 1959. The Enchytraeidae, critical revision and taxonomy of European species. Nat. Jutl. 8–9: 1–160

Nielsen, C. O. & B. Christensen, 1961. The Enchytraeidae, critical revision and taxonomy of European species. Supplement 1. Nat. Jutl. 10: 1–23.

O'Connor, F. B., 1962. The extraction of Enchytraeidae from soil. In P. W. Murphy (ed.), Progress in soil zoology. Butterworths, Lond: 279–285.

Prabhoo, N. R., 1961. Studies on Indian Enchytraeidae (Oligochaeta: Annelida). I. Description of three new species. J. Zool. Soc. India 12: 125–132.

Stephenson, J., 1923. Oligochaeta. The fauna of British India, including Ceylon and Burma. Taylor and Francis, London: 518 pp.

Xie, Z. C., H. Z. Wang & Y. L. Liang, 1999. Studies on the Enchytraeidae (Oligochaeta, Annelida) of China. I. On new species and new records of the genus *Hemienchytraeus*. Acta Hydrobiol. Sin. 22(4).

Xu, Z. F., D. N. Zhang & J. M. Jiang, 1989. A new genus and species of Enchytraeidae from Gansu Province (Oligochaeta: Plesiopora). Acta Zootax. Sinica 14: 153–156 (in Chinese with English abstract).

Hydrobiologia **406:** 67–81, 1999.
B. M. Healy, T. B. Reynoldson & K. A. Coates (eds), Aquatic Oligochaetes.

Distribution of freshwater oligochaetes in the west and east coastal regions of the North Pacific Ocean

Tarmo Timm
Võrtsjärv Limnological Station of the Institute of Zoology and Botany, EE-61101 Rannu, Tartumaa, Estonia

Key words: Oligochaeta, freshwater, zoogeography, Far East, North America

Abstract

Faunistic data available from 10 different regions were reviewed: Middle and South China (55 species of freshwater oligochaetes recorded); Amur Basin, Manchuria and Primor'e (86); Korean Peninsula (20); Japan, Sakhalin and Kuril Islands (77); Kamčatka Peninsula (42); Čukči Peninsula including the Anadyr Basin and Korjak Mountains (41); Alaska Peninsula (31); Southern Alaska and British Columbia (62); Washington, Oregon and California (119); Mexico (25). A total of 306 valid species of freshwater oligochaetes includes the branchiobdellidans, freshwater and amphibiotic enchytraeids and lumbricids, as well as some brackish-water representatives of freshwater genera. Among them, 61 species of Naididae (56 on the Asian, 35 on the American side; 30 species common for both coasts); 68 of Tubificidae (40, 45; 17); 2 of Propappidae (2, 0; 0); 43 of Enchytraeidae (16, 27; 0); 43 of Lumbriculidae (29, 17; 3); 4 of Haplotaxidae (3, 2; 1); 1 of Biwadrilidae (1, 0; 0); 5 of Lumbricidae (4, 1; 0); 79 of the five branchiobdellidan families (36, 43; 0). Endemicity at the species or even generic level on either side of the Pacific appears to occur in over 2/3 of the total number of species, being prevalent in Lumbriculidae, Enchytraeidae, Haplotaxidae, Biwadrilidae and Branchiobdellida, less expressed in Tubificidae and rare in Naididae. Most of the species common to both coasts of the North Pacific are widely distributed in the Holarctic, or appear cosmopolitan. A weak faunal exchange appears to have happened in the Bering Strait region (*Alexandrovia ringulata, Tasserkidrilus americanus, Styloscolex opisthothecus, Lumbriculus olgae*).

Introduction

Oligochaete research in most countries of the North Pacific region started comparatively recently. In the 19th century, a vast majority of the faunistic studies on aquatic oligochaetes covered only a few European countries, and only single collections of expedition material from overseas were treated. Eisen's papers (1879, 1881, 1885, 1910) serve as a rare example of research actively carried out in a Pacific coast region. In the first half of the 20th century, the first monographic studies appeared in China (Chen, 1940) and Japan (Yamaguchi, 1934, 1953, etc.). The importance of the contributions of Sokol'skaja (e.g. 1958, 1961, 1972, 1983a,b) to oligochaete research in all regions of the Russian Far East in the second half of the century should be stressed. Liang (e.g. 1958, 1962, 1963, 1987) and Brinkhurst et al. (1990) considerably improved the knowledge of freshwater oligochaetes in China, Ohtaka (e.g. 1985) and Ohtaka & Nishino (1995) in Japan, Morev (e.g. 1982, 1986) in the northeastern and Timm (e.g. 1995, 1997) in the southeastern corner of Russia. Altman (1936), Brinkhurst (e.g. 1964, 1965, 1978, 1986) and Brinkhurst & Cook (1966) presented a good overview of the U.S. and Canadian fauna, which was completed by Holmquist (e.g. 1974a,b, 1976, 1979) with her account of the Alaska Peninsula. Goodnight (e.g. 1940), Holt (e.g. 1977) and Holt & Opell (1993) greatly contributed to the knowledge of branchiobdellidans in the U.S.A., as did Liu (e.g. 1964) in China. The first communication on the free-living freshwater oligochaetes of Korea was recently presented by Brinkhurst et al. (1994), while the commensal branchiobdellidans of this region were studied already by Yamaguchi (1934) and Subchev (1986). In Mexico, all freshwater oligochaete groups except for the branchiobdellidans (surveyed by Holt, 1973) are so far poorly studied (Brinkhurst & Marchese, 1989). A number of minor communications cannot be surveyed within the limits of this paper.

Thus, a broad picture of the freshwater oligochaete fauna occurring along the coasts of the North Pacific is now available. The west and east coasts are separated from each other by a vast space of ocean, except in the narrow Bering Strait (which dried up repeatedly in the geological past; see Morev, 1986) in the Arctic. It is now possible to compare the composition and probable exchange of the fauna of both sides of the North Pacific Ocean, the more so because heavy transoceanic shipping began later here, and remained less intense, than over the North Atlantic, where it is often hard to establish the origin of freshwater worms at present distributed both in North America and Europe.

Material and methods

Using all available literature sources, I have drawn up maps of distribution for all species of aquatic oligochaetes and terrestrial microdriles. It is impossible to list the vast number of publications here; the most important of them are mentioned in the Introduction. Only freshwater and amphibious taxa present in the countries (states, provinces) bordering on the North Pacific are included, together with some brackish-water representatives of freshwater genera (e.g. *Nais borutzkii*), amphibious genera of the generally terrestrial family of Enchytraeidae (*Mesenchytraeus*, *Cernosvitoviella* and *Cognettia*) and some earthworms (Lumbricidae) noted in various lists of hydrofauna. I have preferred valid names, avoiding synonyms as well as apparently artificial lumpings. In many cases when the status or limits of a taxon seemed doubtful (e.g. in case of *Rhyacodrilus sodalis* or *Embolocephalus nikolskyi*), my choice was inevitably subjective.

The study area comprises ten regions, six of them on the Asian, and four on the American side (Figure 1).

1. Middle and South China, including the lowland portions of the Huanghe, Yangtze and Pearl River basins with their environs.
2. The Amur basin together with the rest of Manchuria and Primor'e.
3. Korean Peninsula.
4. Japan Islands together with Sahalin and the Kuriles, belonging to Russia.
5. Kamčatka Peninsula.
6. Čukči Peninsula together with the Korjak Highlands, the Anadyr and Upper Kolyma basins.
7. Alaska Peninsula including the northwestern corner of Canada up to the Mackenzie delta.
8. Southern Alaska and British Columbia.
9. States of Washington, Oregon and California in the U.S.A.
10. Mexico.

The species list with distribution data is given in Table 1. The total number of species, and their number in separate groups, were counted for all regions under study. The number of species common to both west and east coasts of the North Pacific, and the number of species recorded on one side only, were established, including the taxa endemic in these regions as well as the more widely distributed ones (Table 2).

Results

Oligochaete fauna in general

The number of species known from the Asian side of the study area is 187, and from the American side 171. The slight preponderance for the western coast may be accounted for by its more complicated topography. Zones with moderate climate contain the most diverse oligochaete fauna on both sides: the Amur basin s.l. with 86 species and large islands (79 species) on the Asian side, and the U.S. coastal states with 119 species on the American side. The fauna of Korea (with only 20 recorded species), located at the same latitudes, is still poorly studied. The number of species decreases both northwards (to 31 on the Alaskan Peninsula), where the Naididae become scanty and Branchiobdellida are entirely lacking, and southwards (55 in Middle and South China), where the climate is too warm for most Lumbriculidae, Tubificidae and amphibious Enchytraeidae. The low number of species (25) in Mexico can certainly be attributed to the scarcity of studies on all aquatic oligochaete groups except for Branchiobdellida in this country.

A total of 306 freshwater oligochaete species have been recorded in the study area so far, only 52 of them (about 17%) being common to both sides of the Pacific, whereas 209 species (about 69%) are endemic either for one or several neighbouring regions within the area.

Naididae

The Asian side is considerably better represented by naidid species than the American coast: 56 and 35 species of the total 61, respectively. Twenty-six naidid species are known from the Asian side only, including

Table 1. Distribution of freshwater oligochaetes on the western and eastern coasts of the North Pacific M. & S. China – Middle & South China; Amur, Prim. – Amur basin, Primor'e & NE China; Jap., S., Kuriles – Japan, Sakhalin & Kuril Islands; Kam. Pen. – Kamčatka Peninsula; Čukči Pen. – Čukči Peninsula, incl. the Anadyr & Kolyma basins; Alas. Pen. – Alaska Peninsula; S. Al., B.C. – Southern Alaska & British Columbia; Was., O., Ca. – Washington, Oregon & California

Species	M. & S. China	Amur, Prim.	Korea	Jap., S., Kuriles	Kam. Pen.	ČukČi Pen.	Alas. Pen.	S. Al., B.C.	Was., O., Ca.	Mexico
Naididae:										
Stylaria lacustris (Linnaeus, 1767)		x		x	x	x	x	x	x	
Stylaria fossularis Leidy, 1852	x	x		x		x			x	
Ripistes parasita (Schmidt, 1847)		x		x		x				
Arcteonais lomondi (Martin, 1907)		x		x			x	x		
Vejdovskyella comata (Vejd., 1883)		x		x	x			x	x	
Vejdovskyella intermedia (Br., 1896)								x		
Vejdovskyella simplex Liang, 1958	x	x		x						
Slavina appendiculata (Udek., 1855)	x	x		x	x			x	x	
Dero digitata (Müller, 1773)	x	x	x	x				x	x	
Dero dorsalis Ferronière, 1899	x									
Dero indica Naidu, 1962				x						
Dero nivea Aiyer, 1929	x							x		
Dero obtusa Udekem, 1855		x							x	
Aulophorus furcatus (Müller, 1773)	x	x							x	
Aul. heptabranchionus Liang, 1958	x									
Aulophorus moghei N. et Sr., 1980									x	
Aulophorus pectinatus Steph., 1931	x									
Aulophorus tonkinensis (Vejd., 1894)	x									
Aulophorus varians Liang, 1958	x									
Branchiodrilus cleistochaeta D., 1957			x							
Branchiodrilus hortensis (St., 1910)	x	x		x						
Allonais gwaliorensis (Steph., 1920)	x									
Allonais paraguayensis (Mich., 1905)	x									
Allonais pectinata (Steph., 1910)	x									
Nais barbata Müller, 1773		x		x	x	x		x		
Nais behningi Michaelsen, 1923		x						x	x	
Nais borutzkii Sokol'skaja, 1964		x		x	x					
Nais bretscheri Michaelsen, 1899		x		x						
Nais communis Piguet, 1906	x	x		x	x	x		x	x	
Nais elinguis Müller, 1773		x		x	x	x		x		
Nais inflata Liang, 1963	x									
Nais pardalis Piguet, 1906	x	x		x	x			x		
Nais pseudobtusa Piguet, 1906								x	x	
Nais simplex Piguet, 1906	x				x			x	x	
Nais variabillis Piguet, 1906	x	x		x	x			x	x	
Specaria fraseri Brinkhurst, 1978								x		
Specaria josinae (Vejdovský, 1883)		x		x	x	x		x	x	
Piguetiella amurensis Sok, 1958		x								
Piguetiella michiganensis Hilt., 1967		x								
Haemonais waldvogeli Bret., 1900	x	x	x							
Ophidonais serpentina (Müller, 1773)				x				x	x	
Uncinais uncinata (Oersted, 1842)		x		x	x	x		x	x	
Homochaeta naidina Bretscher, 1896				x	x					
Stephensoniana trivandrana (A., 1926)	x			x						x
Paranais birsteini Sokol'skaja, 1971					x			x		

Table 1. Continued

Species	M. & S. China	Amur, Prim.	Korea	Jap., S., Kuriles	Kam. Pen.	Čukči Pen.	Alas. Pen.	S. Al., B.C.	Was., O., Ca.	Mexico
Paranais frici Hrabě, 1941	x	x						x	x	
Paranais litoralis (Müller, 1784)	x			x		x		x	x	
Amphichaeta asiatica Liang, 1958	x									
Amphichaeta leydigi Tauber, 1879		x							x	
Amphichaeta raptisae (Chapm., 1981)								x		
Chaetogaster diaphanus (Gr., 1828)	x	x		x	x	x			x	
Chaetogaster diastrophus (Gr., 1828)	x	x		x					x	
Chaetogaster langi Bretscher, 1896	x			x						
Chaetogaster limnaei Baer, 1827	x	x		x					x	
Pristina aequiseta Bourne, 1891	x	x		x		x		x	x	
Pristina biserrata Chen, 1940	x	x								
Pristina longiseta Ehrenberg, 1828	x	x		x					x	
Printina proboscidea Beddard, 1896	x									
Pristinella acuminata (Liang, 1958)	x									
Pristinella jenkinae (Steph., 1932)		x		x				x	x	
Pristinella sima (Marcus, 1944)	x									
Tubificidae:										
Rhyacodrilus brevidentatus Br., 1965									x	
Rhyacodrilus coccineus (Vejd., 1875)	x	x		x	x	x	x	x	x	
Rhyacodrilus falciformis Bret., 1901								x		
Rhyacodrilus hiemalis Ohtaka, 1995					x					
Rhyacodrilus komarovi Timm, 1990		x								
Rhyacodrilus levanidovae Sok., 1973					x					
Rhyacodrilus montanus (Brink., 1965)								x	x	
Rhyacodrilus sinicus (Chen, 1940)	x									
Rhyacodrilus sodalis (Eisen, 1897)								x	x	
Rhyacodrilus suputensis Timm, 1990		x								
Rhyacodrilus svetlovi Sok., 1976						x				
Bothrioneur. vejdovskyanum Št., 1888	x	x	x	x	x			x	x	
Branchiura sowerbyi Beddard, 1892	x	x	x	x				x	x	x
Monopylephorus limosus (Hatai, 1898)	x			x						
Monopyleph. rubroniveus Lev., 1884	x	x	x	x				x		
'Ilyodrilus' asiaticus Chen, 1940	x									
Alexandrovia ringulata (Sok., 1961)					x	x	x			
Telmatodrilus mcgregori Eisen, 1900									x	
Telmatodrilus vejdovskyi Eisen, 1879									x	
Tubifex minor Sokol'skaja, 1961		x								
Tubifex tubifex (Müller, 1774)	x	x	x	x				x	x	
Lophochaeta paucipilifer Holm., 1985							x			
Tasserkidrilus americanus (Br., 1966)					x	x	x		x	
Tasserkidrilus harmani (Loden, 1979)								x	x	
Tasserkidrilus hrabei (Sok., 1973)					x					
Spirosperma acapillatus (L. & S., 1953)		x		x	x					
Embolocephalus alaskensis H., 1979							x			
Embolocephalus beetoni (Br., 1965)									x	
Emboloc. chukotensis (Morev, 1975)						x				

Table 1. Continued

Species	M. & S. China	Amur, Prim.	Korea	Jap., S., Kuriles	Kam. Pen.	Čukči Pen.	Alas. Pen.	S. Al., B.C.	Was., O., Ca.	Mexico
Emboloceph. kurenkovi (Sok., 1961)					x					
Emboloc. nikolskyi (L. & Sok., 1953)	x	x		x	x					
Emboloc. nomurai (Yam. & Ok., 1940)				x						
Emboloceph. oregonensis (Br., 1965)									x	
Emboloceph. yamaguchii (Br., 1971)				x						
Quistadrilus multisetosus (Sm., 1900)									x	x
Haber amurensis (Sok. & Hr., 1969)		x								
Limnodrilus amblysetus (Brinkh. 1990)	x									
Limnodrilus claparedeanus R., 1868	x	x	x	x				x	x	
Limnodrilus grandisetosus N., 1932	x	x		x						
Limnodrilus hoffmeisteri Clap., 1862	x	x	x	x				x	x	x
Limnodrilus profundicola (Verr., 1871)		x		x	x	x		x	x	x
Limnodrilus silvani Eisen, 1879									x	
Limnodrilus udekemianus Clap., 1862	x	x	x	x	x			x	x	x
Isochaetides freyi Brinkhurst, 1965								x		
Isochaetides suspectus (Sok., 1964)				x						
Varichaetadr. angustipenis (Br., 1966)									x	
Varichaetadrilus minutus (Br., 1965)									x	
Varichaetadrilus pacificus Br., 1981							x	x	x	
Varichaetadr. potatiensis (Sok., 1983)						x				
Psammoryctides californianus B., 1965									x	
Psammoryctid. convolutus Lod., 1978									x	x
Arctodrilus wulikensis Br. & K., 1983							x			
Potamothrix bavaricus Oschm., 1913									x	
Ilyodrilus fragilis Eisen, 1879									x	
Ilyodrilus frantzi Brinkhurst, 1965								x	x	
Ilyodrilus perrierii Eisen, 1879									x	
Ilyodrilus sokolskayae Brinkh., 1970				x						
Ilyodrilus templetoni (Southern, 1909)				x	x			x	x	
Kopilrok flexipenis Holmquist, 1985							x			
Kopilrok sagavanirktoki Holmq., 1985							x			
Aulodrilus americanus Br. & C., 1966				x				x	x	
Aulodrilus japonicus Yamag., 1953				x						
Aulodrilus limnobius Bretscher, 1899		x		x				x	x	
Aulodrilus pigueti Kowalewski, 1914	x	x		x				x	x	
Aulodrilus pluriseta (Piguet, 1906)	x	x						x	x	x
Teneridrilus calvus Ers. & Br., 1990									x	
Teneridrilus columbiensis (Br., 1985)									x	
Teneridrilus mastix (Brinkhurst, 1978)	x			x				x	x	
Propappidae:										
Propappus arhynchotus Sok., 1972		x			x					
Propappus volki Michaelsen, 1916		x								
Enchytraeidae:										
Mesenchytraeus armatus (Lev., 1884)					x					
Mesenchytraeus arcticus Bell, 1962							x			
Mesenchytraeus asiaticus Eis., 1904						x				

Table 1. Continued

Species	M. & S. China	Amur, Prim.	Korea	Jap., S., Kuriles	Kam. Pen.	Čukči Pen.	Alas. Pen.	S. Al., B.C.	Was., O., Ca.	Mexico
Mesenchytraeus atriaphorus Al., 1936									x	
Mesenchytraeus beringensis E., 1904					x					
Mesenchytr. chromophorus Al., 1936									x	
Mesenchytraeus chaunus Piper, 1982						x				
Mesenchytr. crenobius Timm, 1994		x								
Mesenchytr. diverticulatus Piper, 1982						x				
Mesenchytr. diplobulbosus Bell, 1949									x	
Mesenchytr. eastwoodi Eisen, 1904									x	
Mesenchytr. fontinalis Eisen, 1904									x	
Mesenchytr. franciscanus Eis., 1904									x	
Mesenchytraeus fuscus Eisen, 1904									x	
Mesenchytraeus gelidus Welch, 1916									x	
Mesenchytraeus grandis Eisen, 1904								x		
Mesenchytr. grebnitzkyi Mich., 1901					x					
Mesenchytr. harrimani Eisen, 1904							x	x	x	
Mesenchytraeus hydrius Welch, 1919									x	
Mesench. kontrimavichusi Pip., 1982						x				
Mesenchytraeus kincaidi Eisen, 1904							x			
Mesench. konyamensis Mich., 1916						x				
Mesenchytraeus macnabi Bell, 1942									x	
Mesenchytraeus maculatus E., 1904							x		x	
Mesenchytraeus magnus Altm., 1936									x	
Mesenchytraeus minimus Altm., 1936									x	
Mesenchytr. monothecatus Bell, 1945									x	
Mesenchytraeus nanus Eisen, 1904							x			
Mesenchytraeus obscurus Eis., 1904							x			
Mesenchytraeus orcae Eisen, 1904							x	x		
Mesenchytraeus pedatus Eisen, 1904									x	
Mesenchytraeus penicillus Eis., 1904							x			
Mesenchytraeus setchelli Eisen, 1904							x			
Mesenchytr. solifugus (Emery, 1898)							x	x	x	
Mesenchytraeus svetae Piper, 1982						x				
Mesenchytraeus tundrus Piper, 1982						x				
Mesenchytr. unalaskae Eisen, 1904							x			
Mesenchytraeus vegae Eisen, 1904								x		
Mesenchytr. vshivkovae Timm, 1994		x								
Cernosvitoviella pensau Timm, 1994		x								
Cognettia glandulosa (Mich., 1888)						x				
Cognettia lapponica Nurminen, 1965						x				
Lumbricill. kamtschatkanus (M., 1929)					x					
Lumbriculidae:										
Lumbriculus ambiguus Holmq., 1976							x			
Lumbriculus genitosetosus H., 1976							x			
Lumbriculus illex Timm & Rodr., 1994		x								
Lumbriculus inconstans (Sm., 1905)							x	x	x	
Lumbriculus japonicus Yamag., 1936				x						
Lumbriculus kolymensis Morev., 1982						x				

Table 1. Continued

Species	M. & S. China	Amur, Prim.	Korea	Jap., S., Kuriles	Kam. Pen.	Čukči Pen.	Alas. Pen.	S. Al., B.C.	Was., O., Ca.	Mexico
Lumbriculus mukoensis Yam., 1953				x						
Lumbriculus multiatriatus Yam., 1937				x						
Lumbriculus olgae Sokol'skaja, 1976						x	x			
Lumbriculus sachalinicus S., 1967				x						
Lumbriculus variegatus (Müller, 1774)	x	x		x	x	x	x	x	x	x
Eumuliercula casta (Timm, 1995)		x								
Eumuliercula emendata (Timm, 1995)		x								
Lamprodrilus isossimovi Morev, 1982						x				
Lamprortus orientalis Rodriguez, 1994			x							
Trichodrilus itchaensis Sokol., 1973					x	x				
Kincaidiana freidris Cook, 1966									x	
Kincaidiana hexathexa Altman, 1936									x	
Stylodrilus californianus Rodr., 1996									x	
Stylodrilus chukotensis (Sok., 1975)					x	x				
Stylodrilus heringianus Clap., 1862						x		x		
Stylodrilus sowaliki (Holmquist, 1976)							x			
Stylodrilus suputensis Timm, 1995		x								
Stylodrilus tschaunensis Morev, 1982						x				
Stylodr. wahkeenensis R. & C., 1996									x	
Hrabea ogumai Yamaguchi, 1936				x						
Rhynchelmoides alaskana (H., 1976)							x			
Rhynchelmoides brooksi (H., 1976)							x			
Rhynchelmoides elrodi S. & D., 1918							x	x	x	
Rhynchelmoides malevici Sok., 1983					x					
Rhynchelmoides orientalis (Y., 1936)				x						
Sutroa rostrata Eisen, 1888									x	
Eclipidrilus frigidus Eisen, 1881									x	
Styloscolex distinctus Morev, 1984						x				
Styloscolex japonicus Yamag., 1937				x						
Styloscolex levanidovi (Sokol., 1977)						x				
Styloscolex macer Sokol'skaja, 1976						x				
Styloscolex opisthothecus Sok., 1969					x		x			
Styloscolex sokolskajae Morev, 1978						x				
Styloscolex tetrathecus Burow, 1931		x				x				
Styloscolex tubulatus Timm, 1994						x				
Phagodrilus macnabi McKey-F., 1988									x	
Kurenkovia magna Sokol'skaja, 1969					x					
Haplotaxidae:										
Haplotaxis gastrochaetus Yam., 1953				x						
Haplotaxis gordioides (Hartm., 1821)		x		x	x				x	
Haplotaxis ichthyophagus Gat., 1971									x	
Heterochaetella glandularis Y., 1953				x						
Biwadrilidae										
Biwadrilus bathybates (Steph., 1917)				x						

Table 1. Continued

Species	M. & S. China	Amur, Prim.	Korea	Jap., S., Kuriles	Kam. Pen.	Čukči Pen.	Alas. Pen.	S. Al., B.C.	Was., O., Ca.	Mexico
Lumbricidae:										
Dendrodrilus rubidus (Savigny, 1826)					x					
Eiseniella nordenskjoldi (Eisen, 1879)					x					
Eiseniella koreana Zicsi, 1972			x							
Eiseniella tetraedra (Savigny, 1826)								x	x	
Allolobophora parva Eisen, 1874		x		x	x					
Branchiobdellidae:										
Branchiobdella cheni Liu, 1964		x								
Branchiobdella domina Timm, 1991		x								
Branchiobdella kobayashii Y., 1934		x	x							
Br. macroperistomium Liu & Z., 1983		x								
Branchiobdella minuta Pier., 1912		x								
Branch. monodonta Liu & Z., 1983		x								
Branchiobdella orientalis Yam., 1934		x	x							
Branchiobdella teresae Subch., 1986			x							
Sinodrilus heterorchis (Liu & Z., 1983)		x	x							
Cirrodrilus aequiannulus (Liu, 1984)		x								
Cirrodrilus anodontus (Liu, 1964)		x								
Cirrodrilus aomorensis (Yam., 1934)				x						
Cirr. breviformis (Liu & Zhang, 1964)		x								
Cirrodrilus chosen (Yamaguchi, 1934)		x	x							
Cirrodrilus cirratus Pierantoni, 1905				x						
Cirrodrilus ezoensis (Yamag., 1934)				x						
Cirrodrilus fimbriatus Timm, 1991		x								
Cirrodr. heteroglandularis (Liu, 1964)		x								
Cirrodrilus homodontus (Yam., 1932)				x						
Cirrodrilus inukaii (Yamaguchi, 1934)				x						
Cirrodrilus japonicus (Pierant., 1912)				x						
Cirrodrilus kawamurai (Yam., 1934)		x	x							
Cirr. liaoningensis (Liu & Ch., 1964)		x								
Cirrodrilus makinoi (Yamag., 1934)				x						
Cirrodr. megalodentatus (Yam., 1934)				x						
Cirrodrilus minimus (Liu, 1964)		x								
Cirrodrilus nipponicus (Yamag., 1932)				x						
Cirrodrilus peristomialis (Liu, 1964)		x								
Cirrodrilus pugnax Timm, 1991		x								
Cirrodr. quadritentacularis (Liu, 1984)		x								
Cirrodrilus sapporensis (Pier., 1906)				x						
Cirrodrilus suzukii (Yamaguchi, 1934)		x	x							
Cirrodrilus uchidai (Yamaguchi, 1932)				x						
Uglukodrilus hemophagus (Holt, 1977)									x	
Xironogiton cassiensis Holt, 1974									x	
Xironogiton fordi Holt, 1974									x	
Xironogiton instabilis (Moore, 1893)									x	
Xironogiton kittitasi Holt, 1974								x	x	
Xironogiton occidentalis Ellis, 1920									x	
Xironog. victoriensis Geld. & H., 1990								x	x	

Table 1. Continued

Species	M. & S. China	Amur, Prim.	Korea	Jap., S., Kuriles	Kam. Pen.	Čukči Pen.	Alas. Pen.	S. Al., B.C.	Was., O., Ca.	Mexico
Caridinophilidae:										
Caridinophila unidens Liang, 1963	x									
Bdellodrilidae:										
Bdellodrilus illuminatus (Moore, 1893)										x
Hidejiodrilus koreanus (Pier., 1912)			x							
Cambarincolidae:										
Cambarincola barbarae Holt, 1981									x	
Cambarinc. carcinophilus Holt, 1973										x
Cambarincola ellisi Holt, 1973										x
Cambarincola fallax Hoffman, 1963								x	x	
Cambarincola gracilis Robinson, 1954								x	x	
Cambarincola hoffmani Holt, 1973										x
Cambarincola jamapaensis Holt, 1973										x
Cambarincola macrocephala G., 1943									x	
Cambarin. mesochoreus Hoffm., 1963									x	
Cambarincola micradenus Holt, 1973										x
Cambarincola montanus (G., 1940)								x	x	
Cambarinc. nanognathus Holt, 1973										x
Cambarincola olmecus Holt, 1973										x
Cambarincola pamelae Holt, 1984									x	
Cambarincola serratus Holt, 1981									x	
Cambarinc. speocirolanae Holt, 1984										x
Cambarinc. shoshone Hoffman, 1963									x	
Cambarincola susanae Holt, 1973										x
Cambarincola toltecus Holt, 1973										x
Ceratodrilus ophiorhysis Holt, 1960									x	
Magmatodrilus obscurus (G., 1940)									x	
Oedipodrilus cuetzalanae Holt, 1984										x
Pterodrilus mexicanus Ellis, 1920										x
Sathodrilus attenuatus Holt, 1981									x	
Sathodrilus chehalisae Holt, 1981									x	
Sathodrilus dorfus Holt, 1977									x	
Sathodrilus inversus (Ellis, 1920)									x	
Sathodrilus lobatus Holt, 1977									x	
Sathodrilus norbyi Holt, 1977									x	
Sathodrilus prostates Holt, 1973										x
Sathodrilus shastae Holt, 1981									x	
Sathodrilus veracruzicus Holt, 1968										x
Sathodrilus villalobosi Holt, 1968										x
Sathodrilus wardinus Holt, 1981									x	
Triannulata magna Goodnight, 1940									x	
Xironodrilidae:										
Holtodrilus truncatus (Liang, 1963)	x									
Total number of the valid species In the whole North Pacific area 306	55	86	20	78	42	41	31	62	119	25

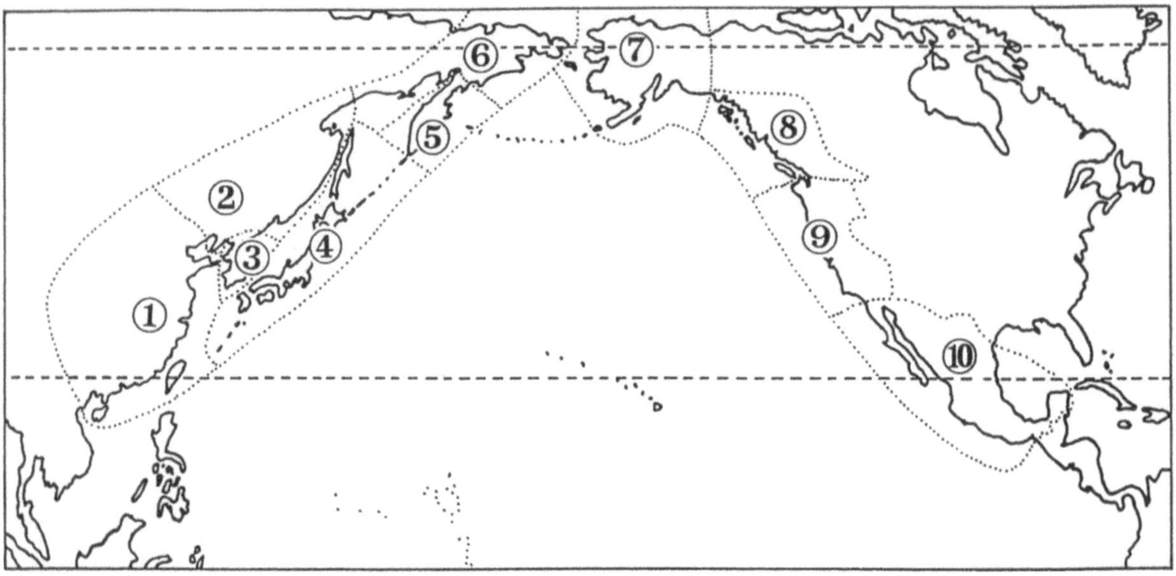

Figure 1. Study area and its division into regions. (1) Middle and South China, (2) the Amur basin with adjacent territories, (3) Korea, (4) Japan with other adjacent islands, (5) Kamčatka, (6) Čukči Peninsula with adjacent territories, (7) Alaska Peninsula, (8) Southern Alaska and British Columbia, (9) western coastal states of the USA, (10) Mexico.

Table 2. Number of species in the different families on both coasts of the North Pacific

Family	Whole area	Asian side				American side				Common for both sides
		All species	On this side only	Endemic	Widely distribut.	All species	On this side only	Endemic	Widely distribut.	
Naididae	**61**	**56**	26	8	18	**35**	5	2	3	**30**
Tubificidae	**68**	**40**	23	20	3	**45**	28	18	10	**17**
Propappidae	**2**	**2**	2	1	1	–	–	–	–	–
Enchytraeidae	**43**	**16**	16	13	3	**27**	27	27	–	–
Lumbriculidae	**43**	**29**	25	24	1	**18**	14	13	1	**4**
Haplotaxidae	**4**	**3**	2	2	1	**2**	1	1	–	**1**
Biwadrilidae	**1**	**1**	1	1	–	–	–	–	–	–
Lumbricidae	**5**	**4**	4	1	3	**1**	1	–	1	–
Branchiobdellida (5 families)	**79**	**36**	36	36	–	**43**	43	43	–	–
Total	**306**	**187**	135	106	30	**171**	119	104	15	**52**

eight possible endemics (*Vejdovskyella simplex, Aulophorus heptabrachionus, A. varians, Nais borutzkii, N. inflata, Piguetiella amurensis, Amphichaeta asiatica* and *Pristinella acuminata*). The rest are more widely distributed forms but include, e.g. the originally African *Branchiodrilus cleistochaeta* found now in Korea by Brinkhurst et al. (1994), as well as *Piguetiella michiganensis*, common in the eastern part of North America but not recorded from its western coast (Figure 2). Only five species were confined to the American side, including two endemics, *Specaria fraseri* and *Amphichaeta raptisae* (but also *Nais pseudobtusa* and *Vejdovskyella intermedia*, widely distributed in the western Palearctic). The difference between the Asian and the American side with respect to naidids seems to reflect an insufficient knowledge of these worms, especially in Mexico where the occurrence of many widely distributed species of *Aulophorus* and *Allonais*, characteristic of the tropical and subtropical zones, can be expected. No

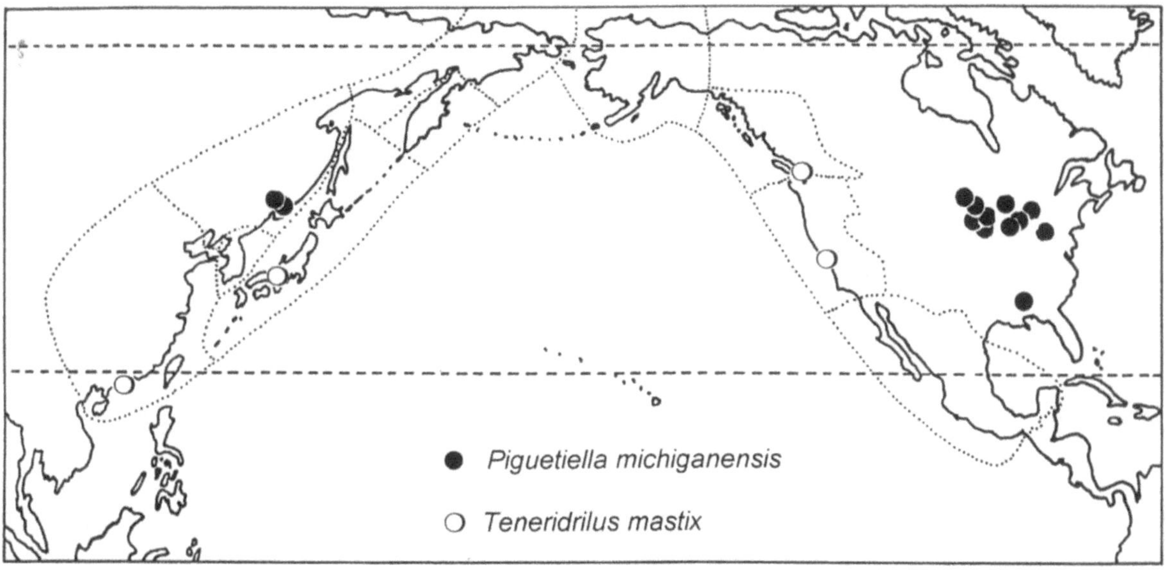

Figure 2. Distribution ranges of *Piguetiella michiganensis* and *Teneridrilus mastix.*

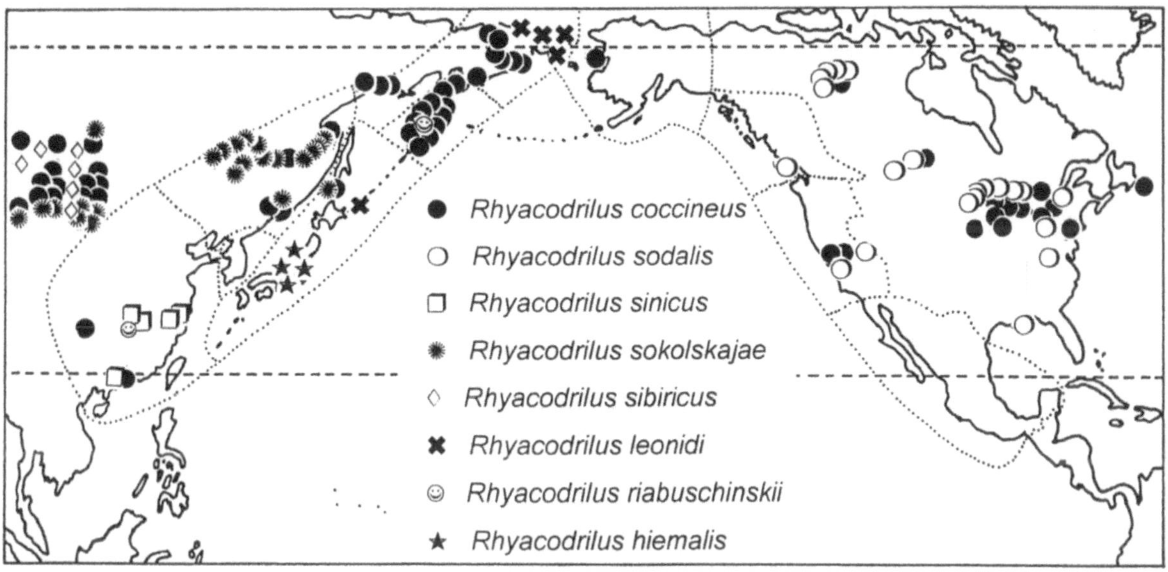

Figure 3. Distribution of the *Rhyacodrilus coccineus* group in the North Pacific area.

Transberingian contacts are observed in the case of the Naididae.

Tubificidae

These worms are almost evenly distributed on both sides of the North Pacific with slight odds in favour of America (40 and 45 species, respectively, of the total of 68). Only 17 species are common to both sides. Out of 23 species specific for the Asian side, 20 can be considered endemic (but including also *Monopylephorus limosus*, apparently introduced to Italy, see Erséus & Paoletti (1986); and *Embolocephalus kurenkovi* with an enigmatic identification from Israel by Pascar-Gluzman & Dimentman (1984)), while on the American side the corresponding numbers are 28 and 18, respectively. On the American side, some widely distributed tubificids seem to have been recently introduced, e.g. the originally European *Rhyacodrilus falciformis* and *Potamothrix bavaricus*. Two endemic

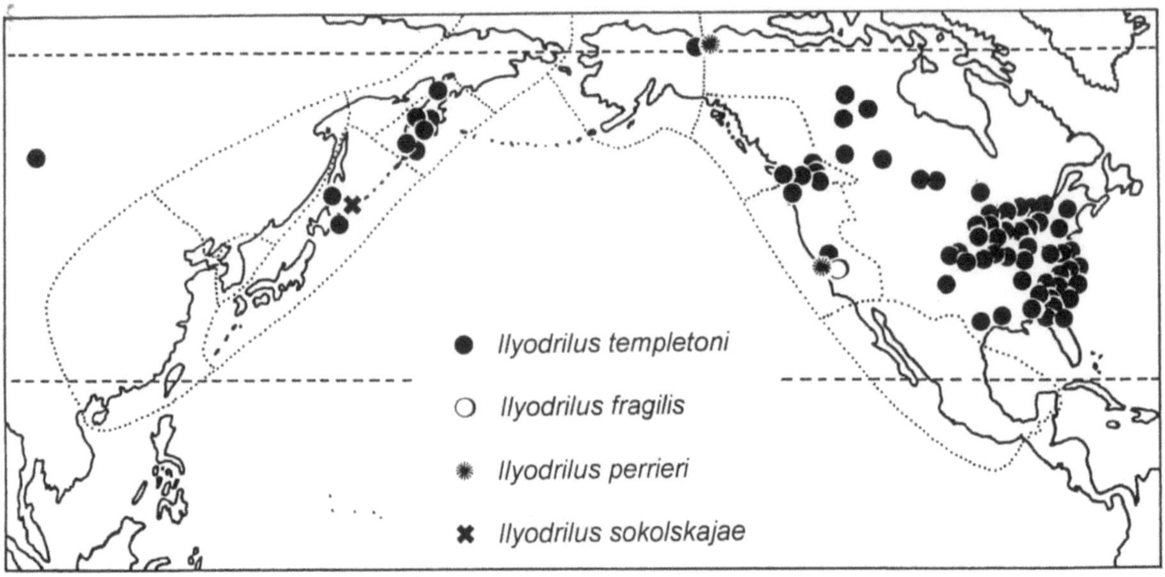

Figure 4. Distribution of the *Ilyodrilus templetoni* group in the North Pacific area.

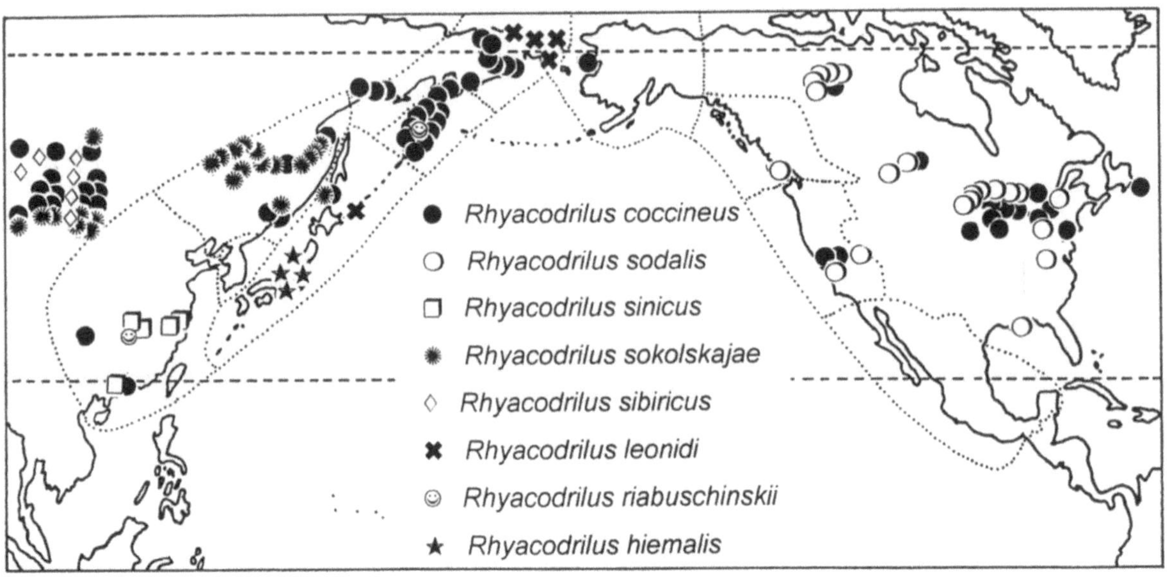

Figure 5. Distribution of *Alexandrovia ringulata* and *Tasserkidrilus americanus* in the North Pacific area.

genera, *Arctodrilus* Brinkhurst & Kathman, 1983 and *Kopilrok* Holmquist, 1985, are described from the American side. Some probably autochthonous clusters of related species or subspecific units are known on both sides, such as the *Rhyacodrilus coccineus-riabuschinskii-sinicus-hiemalis* group (including the disputable *R. leonidi* Sokol'skaja, *R. sokolskajae* Semernoj, and *R. sibiricus* Semernoj but also the American *R. sodalis*) in East Asia (Figure 3) and the *Ilyodrilus templetoni-perrieri-fragilis* group in North America (*I. templetoni* occurring also in East Asia, along with the endemic *I. sokolskayae*) (Figure 4). The presence of *Alexandrovia ringulata* in western Alaska, as well as *Tasserkidrilus americanus* in Kamčatka (and maybe also *Aulodrilus americanus* on Hokkaido Island, according to a personal communication by A. Ohtaka) are signs of a direct species exchange over the Beringian region (Figure 5). *Teneridrilus mastix* is endemic for both coasts of the Pacific but occurs in more southern regions (Qi & Erséus, 1985; Ohtaka

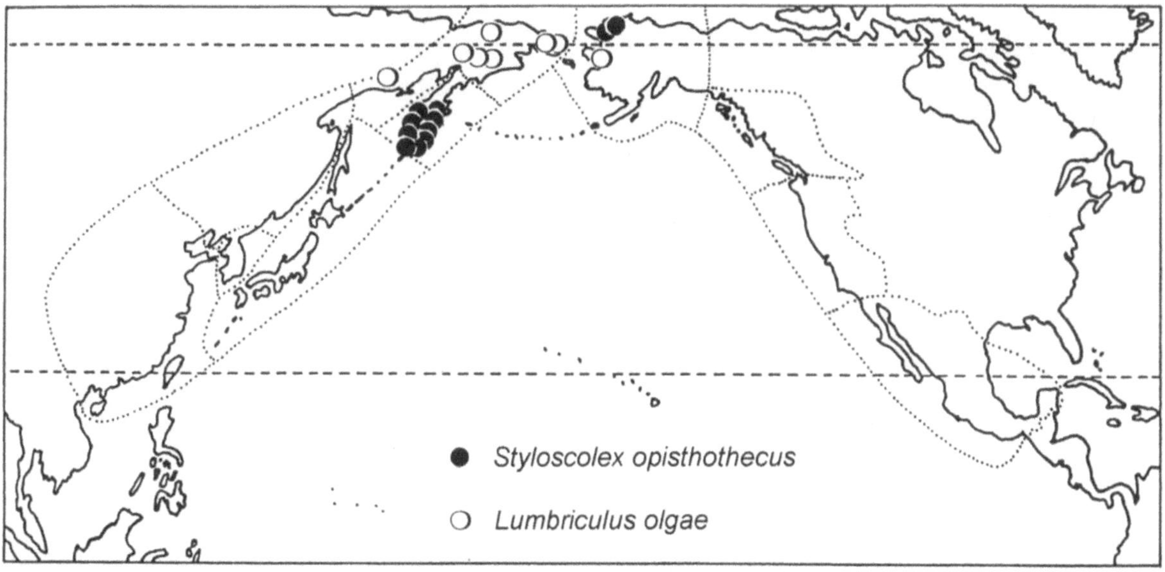

Figure 6. Distribution ranges of *Lumbriculus olgae* and *Styloscolex opisthothecus*.

& Nishino, 1995) (Figure 2). In general, tubificids are considerably less peregrine than naidids, and this is reflected in their higher degree of endemicity. The slightly larger number of tubificid species found on the American side may ensue from the more intensive water quality studies carried out in the U.S.A. and Canada during recent decades.

Lumbriculidae

Among 43 lumbriculid species, 29 were found on the Asian and 18 on the American side, with only 4 species in common (*Lumbriculus variegatus, L. inconstans, L. olgae* and *Styloscolex opistothecus*). The vast majority of lumbriculids are endemic (24 for the Asian and 13 for the American regions under study). In many cases, endemicity reaches the generic level, as for *Eumuliercula* Timm, 1997; *Hrabea* Yamaguchi, 1936; *Kurenkovia* Sokol'skaja, 1969; and *Lamprortus* Rodriguez in Brinkhurst et al., 1994 on the Asian side, as well as for *Kincaidina* Altman, 1936; *Phagodrilus* McKey-Fender, 1988; and *Sutroa* Eisen, 1888 on the American side.

Lumbriculids are cold-preferring animals distributed almostly exclusively in moderate and cold climate. However, their high endemicity does not favour any wide circumpolar distribution. The invasion of Alaska by *Styloscolex opisthothecus* and *Lumbriculus olgae*, evidently originating from Asia, seems to be the only Transberingian contact (Figure 6). The considerably larger number of species on the Asian side can be related to its geographical diversity (presence of many large islands and peninsulae). However, many faunistic treasures are probably still to be discovered in the mountain regions of North America.

Enchytraeidae

These mostly amphibious species from the genera studied proved strongly endemic: 13 species out of 16 encountered on the Asian side and all 27 American ones. The larger number of species in America can be particularly attributed to intensive research on the genus *Mesenchytraeus* (Eisen 1910, Welch 1919, and others).

Smaller families

The Propappidae (with two species) and Biwadrilidae (with one species) are known from the Asian side only, as are three of the four species of Haplotaxidae. High endemicity was observed in all these groups. The selection of generally terrestrial lumbricids is somewhat accidental here, and so is their larger number of species found on the Asian side. However, an endemic amphibious species, *Eiseniella koreana*, was described from East Asia, while its peregrine congener *E. tetraedra* has been recorded from the waterbodies of the American side only.

Branchiobdellida

All five families can be treated together here. The study comprises 36 Asian and 43 American species without any common species or even genus except for the dubious *Cambarincola okadai*, apparently introduced to Japan from North America (Yamaguchi, 1933) . All species are endemic as are most of the genera and even some families (Caridinophilidae in Asia, and Cambarincolidae in North America). The probable cause of this is the symbiotic way of life of the worms on crayfish whose distribution is confined to fresh waters in moderate climates.

Conclusions

Both the Asian and American coastal regions of the North Pacific show the same pattern of species diversity of freshwater oligochaetes, with the maximum number of species at middle latitudes: in the Amur basin s.l., in Japan, with the neighbouring islands, and in the western coastal states of the U.S.A. These regions offer the most favourable conditions for all the families considered and are also the most thoroughly studied. The species numbers decrease both northwards and southwards from this zone; in Mexico the phenomen is related to the poor knowledge of many groups. The number of species common to both North Pacific coasts proved very low, only about 17%, while more than two thirds of the fauna consist of endemic taxa. The level of endemicity is different in separate groups: it is low among the Naididae, higher among the Tubificidae, even higher among the Lumbriculidae, Enchytraeidae and some smaller families, and reaches a maximum among the families of Branchiobdellida. Few cases of recent intercontinental fauna exchange over the Bering Strait can be traced, and there is even less evidence of other immediate intercontinental invasions over the Pacific. The Holarctic or cosmopolitan species now common to both sides of the Pacific may have reached this area from the western Palaearctic or from the eastern part of North America, too.

Acknowledgements

I am greatly indebted to all my colleagues for the reprints of their papers, sent to me during four decades, thus enabling the compilation of distribution maps of the Oligochaeta. I am grateful to Mrs Ester Jaigma who kindly checked my English text, and to Dr C. Erséus for his careful review of the submitted manuscript.

References

Altman, L. C., 1936. Oligochaeta of Washington. Univ. Washington Publ. in Biology 4: 1–37.

Brinkhurst, R. O., 1964. Studies on the North American aquatic Oligochaeta I: Naididae and Opistocystidae. Proc. Acad. nat. Sci. Philad. 116: 195–230.

Brinkhurst, R. O., 1964. Studies on the North American aquatic Oligochaeta II: Tubificidae. Proc. Acad. nat. Sci. Philad. 117: 117–172.

Brinkhurst, R. O., 1978. Freshwater Oligochaeta in Canada. Can. J. Zool. 56: 2166–2175.

Brinkhurst, R. O., 1986. Guide to the freshwater aquatic microdrile oligochaetes of North America. Can. spec. Publ. Fish. aquat. Sci. 84: 1–259.

Brinkhurst, R. O. & D. G. Cook, 1966. Studies on the North American aquatic Oligochaeta III: Lumbriculidae and additional notes and records of other families. Proc. Acad. nat. Sci. Philad. 118: 1–33.

Brinkhurst, R. O. & M. R. Marchese, 1989. Guide to the freshwater aquatic Oligochaeta of South and Central America. Colección CLIMAX 6, Asociación Ciencias Naturales del Litoral, Santo Tomé: 180 pp.

Brinkhurst, R. O., Qi Sang & Liang Yanling, 1990. The aquatic Oligochaeta from the People's Republic of China. Can. J. Zool. 68: 901–916.

Brinkhurst, R. O., P. Rodriguez, T.-S- Chon & T.-S. Kwon, 1994. A new genus of Lumbriculidae (Oligochaeta) from Korea. Can. J. Zool. 72: 1960–1966.

Chen, Y., 1940. Taxonomy and faunal relations of the limnitic Oligochaeta of China. Contributions from the biological Laboratory of the Science Society of China, zool. ser. 14, Shanghai: 131 pp.

Eisen, G., 1879. Preliminary report on genera and species of Tubificidae. Bihang k. svenska Vet.-Akad. Handl. 5(16): 1–25.

Eisen, G., 1881. Eclipidrilidae and their anatomy. Nova Acta Regiae Soc. Sci. upsal. 11(4): 1–10.

Eisen, G., 1885. Oligochaetological researches. Annual Report of the Commissioner of Fish and Fisheries for 1883, Washington 11: 879–964.

Eisen, G., (1904) 1910. Enchytraeidae of the west coast of North America. Harriman Alaska Series, Smithsonian Institution, Washington 12: 1–126.

Erséus, C. & A. Paoletti, 1986. An Italian record of the aquatic oligochaete *Monopylephorus limosus* (Tubificidae), previously known only from Japan and China. Boll. Zool. 53: 115–118.

Goodnight, C. J., 1940. The Branchiobdellidae (Oligochaeta) of North American crayfishes. Illinois biol. Monogr. 177(3): 1–75.

Holmquist, C., 1974a. A species of *Styloscolex* Michaelsen (Oligochaeta, Lumbriculidae) found in two lakes in northern Alaska. Int. Rev. ges. Hydrobiol. 59: 49–55.

Holmquist, C., 1974b. On *Alexandrovia onegensis* Hrabe from Alaska, with a revision of the Telmatodrilinae (Oligochaeta, Tubificidae). Zool. Jb. Syst. 101: 249–268.

Holmquist, C., 1976. Lumbriculids (Oligochaeta) of northern Alaska an northwestern Canada. Zool. Jb. Syst. 103: 377–431.

Holmquist, C., 1979. Revision of the genus *Peloscolex* (Oligochaeta, Tubificidae). 2. Scrutiny of the species. Zool. Scr. 8: 37–60.

Holt, P. C., 1973. A summary of the branchiobdellid (Annelida: Clitellata) fauna of Mesoamerica. Smiths. Contr. Zool. 142: 1–40.

Holt, P. C., 1977. An emendation of the genus *Sathodrilus* Holt 1968 (Annelida: Branchiobdellida), with the description of four new species from the Pacific drainage of North America. Proc. biol. Soc. Wash. 90: 116–131.

Holt, P. C. & B. D. Opell, 1993. A checklist of and illustrated key to the genera and species of the Central and North American Cambarincolidae (Clitellata: Branchiobdellida). Proc. biol. Soc. Wash. 106: 251–295.

Liang, Y., 1958. On some new species of Naididae from Nanking including remarks of certain known species. Acta hydrobiol. sin. 7: 41–58.

Liang, Y., 1962. Records of naidids and tubificids in the North China. Acta hydrobiol. sin. 12: 14–26.

Liang, Y., 1963. Studies on the aquatic Oligochaeta of China. 1. Description of new naids and branchiobdellids. Acta zool. sin. 15: 560–570.

Liang, Y., 1987. Preliminary study of the aquatic Oligochaeta of the Changjiang (Yangtze). Hydrobiologia 155: 195–198.

Liu, S., 1964. A second report of Branchiobdellidae in Liaoning Province with descriptions of three new species. Acta zool. sin. 16: 602–610.

Morev, A. P., 1982. Novye vidy semejstva Lumbriculidae (Oligochaeta) iz vodoemov Severo-Vostoka SSSR. Zool. Žurnal 61: 663-670 [new species of lumbriculids from the waterbodies of northwestern SSSR. In Russian].

Morev, A. P., 1986. Stanovlenie presnovodnyh faunističeskih kompleksov oligohet Beringii. In: Biogeografija Beringijskogo sektora Subarktiki, Vladivostok: 52-57 [origin of the freshwater faunistic complexes of Oligochaeta in the Beringia. In Russian].

Ohtaka, A., 1985. Taxonomical studies of the Japanese Naididae (Annelida, Oligochaeta) 1. Four unrecorded species in small genera. J. Fac. Sci. Hokkaido Univ. Zool. 24: 113–121.

Ohtaka, A. & M. Nishino, 1995. Studies on the aquatic oligochaete fauna in Lake Biwa, Central Japan. I. Checklist with taxonomic remarks. Jap. J. Limnol. 56: 167–182.

Pascar-Gluzman, C. & C. Dimentman, 1984. Distribution and habitat characteristics of Naididae and Tubificidae in the inland waters of Israel and the Sinai Peninsula. Hydrobiologia 115: 197–205.

Qi, S. & C. Erséus, 1985. Ecological survey of the aquatic oligochaetes in the Lower Pearl River (People's Republic of China). Hydrobiologia 128: 39–44.

Sokol'skaja, N. L., (1957) 1958. Presnovodnye maloščetinkovye červi bassejna Amura. Trudy amurskoj ihtiologičeskoj Ékspedicii 1945–1949 gg. 4, Moskva: 287-358 [freshwater Oligochaeta of the Amur basin. In Russian].

Sokol'skaja, N. L., 1961. Materialy po faune presnovodnyh maloščetinkovyh červej bassejna Amura (po sboram kompleksnoj Sovetsko-Kitajskoj Amurskoj ékspedicii 1957 i 1958 gg.). Sbornik Trudov zoologičeskogo Muzeja moskovskogo gosudarstvennogo Universiteta 8: 79–101 [materials on the oligochaete fauna of the Amur basin, on the collections of a Chinese-Soviet expedition in 1957 and 1958. In Russian].

Sokol'skaja, N. L., 1972. K faune vodnyh Oligochaeta Dal'nego Vostoka SSSR. Trudy vsesojuznogo gidrobiologičeskogo Obščestva 17: 50–65 [on the freshwater oligochaete fauna of the Soviet Far East. In Russian].

Sokol'skaja, N. L., 1983a. Presnovodnye maloščetinkovye červi (Oligochaeta) Kamčatki i Korjakskogo nagor'ja. Sbornik Trudov zoologičeskogo Muzeja moskovskogo gosudarstvennogo Universiteta 20: 22–119 [freshwater Oligochaeta of Kamčatka and the Korjak Highland. In Russian].

Sokol'skaja, N. L., 1983b. K faune tubificid i ljumbrikulid (Oligochaeta: Tubificidae, Lumbriculidae) Čukotskogo poluostrova. Sbornik Trudov zoologičeskogo Muzeja moskovskogo gosudarstvennogo Universiteta 20: 120–143 [on the fauna of Tubificidae and Lumbriculidae of the Čukči Peninsula. In Russian].

Subchev, M. A., 1986. On the Korean branchiobdellids (Annelida, Clitellata) with a description of a new species – *Branchiobdella teresae* sp. n. Acta zool. bulg. 31: 60–66.

Timm, T., 1995. Aquatic Oligochaeta from the farthest Southeast of Russia. IV. Lumbriculidae. Proc. estonian Acad. Sci. Biol. 44: 78–91.

Timm, T., 1997. Freshwater Oligochaeta of some urban watercourses in the Russian Far East. Int. Rev. ges. Hydrobiol. 82: 437–467.

Welch, P. S., 1919. Further studies on North American Mesenchytraeids (Oligochaeta). Trans. am. microsc. Soc. 38: 175–188.

Yamaguchi, H., 1933. Description of a new branchiobdellid, *Cambarincola okadai* n. sp., parasitic on American crayfish transferred into a Japanese lake. Proc. imp. Acad. [of Japan] 9: 191–193.

Yamaguchi, H., 1934. Studies on Japanese Branchiobdellidae, with some remarks on the classification. J. Fac. Sci. Hokkaido Univ. Zool. 3: 177–219.

Yamaguchi, H., 1953. Studies on the aquatic Oligochaeta of Japan. VI. A systematic report, with some remarks on the classification and phylogeny of the Oligochaeta. J. Fac. Sci. Hokkaido Univ. Zool. 11: 277–342.

Hydrobiologia **406:** 83–86, 1999.
B. M. Healy, T. B. Reynoldson & K. A. Coates (eds), Aquatic Oligochaetes.

Enzymatic variability in sympatric species of *Tubifex* (Clitellata: Tubificidae)

Andreina Paoletti Di Chiara, Manuela Pagani* & Barbara Leoni
Department of Biology, via Celoria 26, I-20133 Milan, Italy
*(*author for correspondence)*

Key words: *Tubifex* spp., enzymatic variability, genotypic differences, polyploidy

Abstract

The globally distributed species *Tubifex tubifex* Müller, 1774 (Clitellata; Tubificidae) is sometimes considered to include several different morphological forms, each recognizable by their chaetal structure. The 'blanchardi' form is a distinct species: sympatric populations of *Tubifex tubifex* and *Tubifex blanchardi* Vejdovsky, 1871 show complete reproductive isolation. In an attempt to assess the genotypic differences between these closely related species, a comparative allozymic survey was carried out. More than 400 individuals of *T. tubifex* (235 specimens) and *T. blanchardi* (203 specimens), from a mixed community of tubificid species in a highly polluted river were examined, using as markers two enzymatic loci: phosphoglucose isomerase (PGI) and phosphoglucose mutase (PGM). Both species are polymorphic at the two loci, with at least one unshared allele for each locus, and show different electromorph patterns. Moreover, the occurrence of multiple banded patterns suggests a polyploid condition in both *T. tubifex* and *T. blanchardi.*

Introduction

The cosmopolitan genus of oligochaete, *Tubifex*, includes many species that are widely distributed, both in clean and polluted waters. In particular, the species *Tubifex tubifex* Muller, 1774 is very common in Italy, usually found associated with species such as *Limnodrilus udekemianus, Limnodrilus hoffmeisteri, Limnodrilus profundicola* and *Monopylephorus limosus*, forming complex tubificid-dominated communities (Paoletti & Rusconi, 1985). The definition of *T. tubifex* has encompassed diverse morphological forms, in particular showing variation in chaetal structure. *Tubifex blanchardi* (Vejdovsky, 1891), which has only forked chaetae in dorsal bundles, has been subsumed by this broad definition of *T. tubifex* and Chapman & Brinkhurst (1987) demonstrated that environmental stress often induces a reduction of the hair and pectinate chaetae in *T. tubifex*, producing 'blanchardi-like' individuals.

However, other data support the recognition of *T. blanchardi* as a distinct species (Vejdovsky, 1891). In the Lambro River, Italy, a very polluted tributary of the Po River, *T. tubifex* and *T. blanchardi* are sympatric, with *T. blanchardi* the more numerous. Studies comparing these sympatric populations of *T. tubifex* and *T. blanchardi* demonstrated that even though they have similar life-cycles, *T. blanchardi* is less resistant to environmental variability of some abiotic factors, such as temperature (Paoletti, 1989). Moreover, while *T. tubifex* can reproduce parthenogenetically, *T. blanchardi* seems unable to reproduce in the absence of a sexual partner. In addition, crossbreeding experiments showed complete reproductive isolation between the two species (Paoletti, 1989). As a consequence, we conclude that *T. tubifex* and *T. blanchardi* are distinct species.

In an attempt to assess the degree of genotypic difference between these closely related species, a comparative allozymic survey was started on the Italian worm populations.

Materials and methods

More than 400 mature individuals, 235 of *T. tubifex* and 203 of *T. blanchardi* from the mixed Lambro community were examined on a light microscope

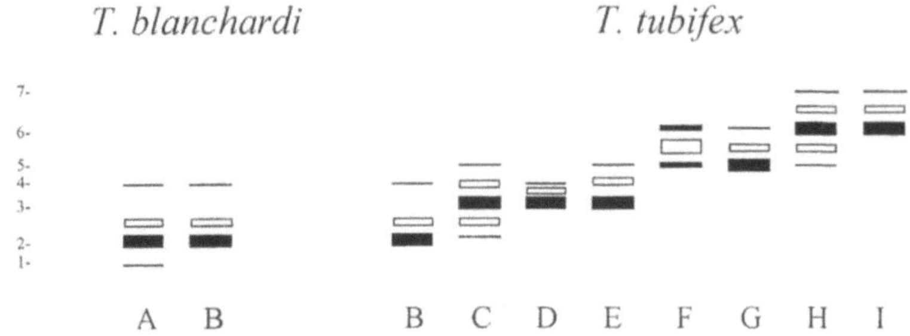

Figure 1. Electrophoretic patterns for the dimeric enzyme phosphoglucose isomerase (PGI) of *T. tubifex* and *T. blanchardi*. Alleles are numbered (1–7) according to speed of migration through the gel.

to separate the species. Each worm was processed for electrophoresis: the postclitellar portion was amputated, frozen in a sterile tube in liquid nitrogen (−196 °C), and later homogenized in sample buffer (0.05 M Tris HCl pH 8.+ 0.02% bromophenol).

A few enzyme systems were screened (APH, ME, PGI and PGM), but only two loci appeared highly polymorphic and presented good staining resolution: phosphoglucose isomerase (PGI, EC 5.3.1.9) and phosphoglucose mutase (PGM, EC 2.7.5.1). Moreover, preliminary tests showed that PGM analysis required a five fold concentration of the tissue extract with respect to PGI (3 μl sample buffer *vs.* 15 μl).

Electrophoresis was performed using a cellulose acetate system (Labometrics, Milan, Italy), following the method of Richardson et al. (1986). Cellulose gels were pre-soaked 15 min in running buffer (0.02 M Tris glycine pH 8.5), then samples were loaded and electrophoresis was carried out at 220 V for 1h in a refrigerator (4 °C). About 1 μl of the extract was loaded in each sample slot.

Staining techniques for the two enzymes were: PGI - 3 ml 0.1M Tris HCl pH 8, 20 mg F-6-P, 7.8 mg NADP, 6 mg MTT, 1.2 mg PMS, 15 μl G-6-PDH; PGM - 10 ml 0.05 M Tris HCl pH 7.4, 100 mg G-1-P, 20 mg NADP, 20 mg MTT, 6 mg PMS, 40 mg $MgCl_2$, 4 μl G-6-PDH. Both enzymes were stained in agar at 38 °C.

The resulting electromorphs were listed according to speed of migration through the gel and different alleles were identified.

To evaluate the degree of genotypic difference between *T. tubifex* and *T. blanchardi* a binary matrix was built using presence/absence data of different alleles, for both enzymes we tested and we calculated the Lance and Williams dissimilarity coefficient (SPSS, 1989), that ranges from 0 to 1.

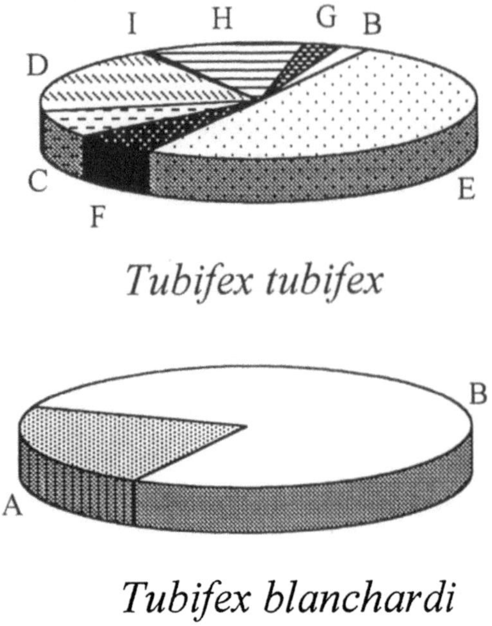

Figure 2. Frequencies of PGI electromorphs. *T. blanchardi*: A=22%; B=78%. *T. tubifex*:B=2%; C=8%; D=18%; E=50%; F=6%; G=3%; H=12%; I=1%.

Results and discussion

We detected nine different electromorphs of PGI, two in *T. blanchardi*, and eight in *T. tubifex*, with seven different alleles. Only one electromorph, B, was shared by the two species (Figure 1) and this was the dominant electromorph in *T. blanchardi* (78%), but was infrequent in *T. tubifex* (2%) (Figure 2). Electromorphs E, D and H occurred in 50%, 18% and 12%, respectively, of *T. tubifex*, whereas, the other five electromorphs were less frequent (C= 8%; F= 6%; G= 3%; I= 1%). Only 28% of the alleles were shared by *T. tubifex* and *T. blanchardi*. Electromorph A, detected in 22% of the individuals, was exclusive to *T. blanchardi* and showed the unique allele 1. The unique alleles

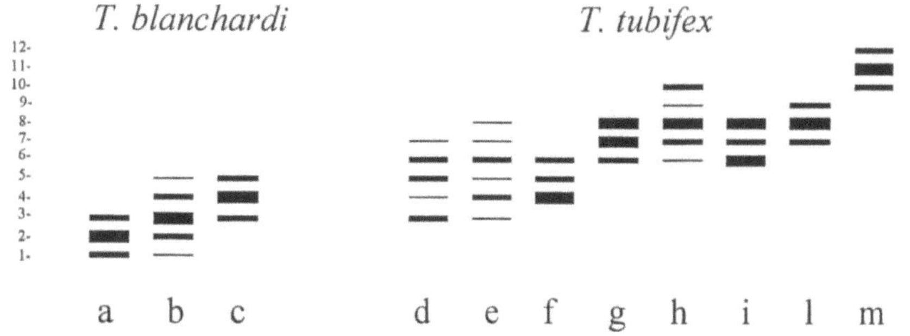

Figure 3. Electrophoretic patterns for the monomeric enzyme phosphoglucose mutase (PGM) of *T. tubifex* and *T. blanchardi*. Alleles are numbered (1–12) according to speed of migration through the gel.

of *T. tubifex* were: allele 3 (shared by electromorphs C, D and E), allele 5 (present in electromorphs C, E, F, G and H), allele 6 (electromorphs F, G, H and I) and finally allele 7 (shared by electromorphs H and I).

For PGM, we detected 11 different electromorphs, eight in *T. tubifex* and three in *T. blanchardi*, with 12 different alleles, and no shared patterns (Figure 3). Electromorph d occurred in 34% of *T. tubifex*, electromorphs f and e occurred in 18% and 14%, respectively, other electromorphs, g–m, had frequencies of 6–9%. In *T. blanchardi* electromorph b was found in 50% of the population and electromorphs a and c were each found in 24% of the population (Figure 4). Alleles 1 and 2 were exclusive to *T. blanchardi* (electromorphs a and b), whereas 25% of alleles (3, 4 and 5) were shared by the two species. Any other alleles (6–12) were unique to *T. tubifex*. The calculation of the Lance and Williams coefficient (based on presence/absence of different alleles) confirmed a quite high degree of dissimilarity between *T. tubifex* and *T. blanchardi* either for PGI (0.57) or PGM (0.60) enzymes.

The electrophoretic data for PGI and PGM loci identified 32 different genotypes in *T. tubifex* and 6 genotypes in *T. blanchardi*. These results confirm the high level of enzymatic variability already recorded for tubificids. Weider (1992) found 13 alleles at the PGM locus and 10 alleles at the PGI locus in five species of *Limnodrilus*, while Anlauf (1997) and Anlauf & Neumann (1997) detected eight alleles for PGI and six alleles for PGM in *T. tubifex* populations from 20 different limnetic and lotic sites.

Since PGI is usually a dimeric enzyme (consisting of two polypeptide chains), it shows a single banded pattern in homozygotes and a triple banded pattern in heterozygotes. In heterozygotes, the two external (slowest and fastest) bands are different alleles and the symmetrical central band would represent the het-

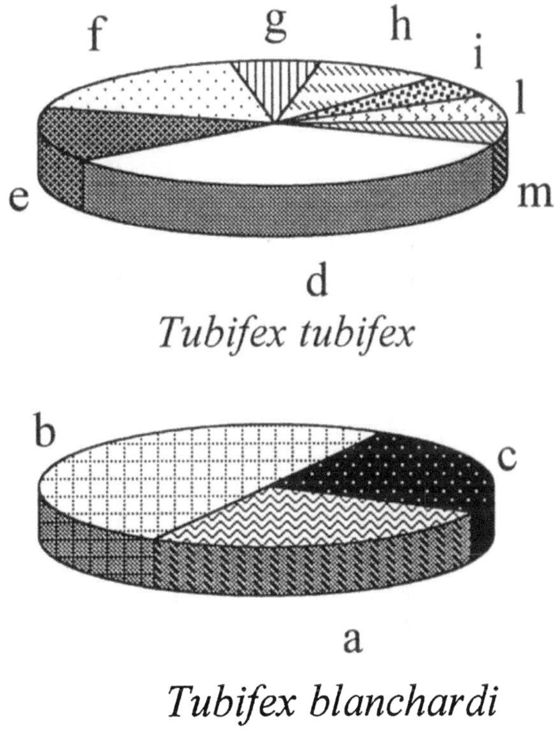

Figure 4. Frequencies of PGM electromorphs. *T. blanchardi*: a=26%; b=50%;. c=24%. *T. tubifex*:d=34%; e=14%; f=18%; g=6%; h=9%; i=6%; l=7%; m=6%.

eropolymer. PGM, is usually a monomeric enzyme (consisting of one polypeptide chain) and shows only one band in homozygotes and two bands in heterozygotes. In our PGI patterns a maximum of five bands occurred in a single electromorph (Figure 1). A larger number of bands in a single electromorph, in the light of highly conservative enzyme structure, suggests the presence of more alleles coding for the enzyme. More alleles might be due to a multiple loci system coding for different proteins with the same enzyme activity. In this case, the resulting banding pattern would appear characterized by separate zones of different relative mobilities (Avise & Kitto, 1973; Taggart et al., 1981) and our PGI pattern did not agree with this interpretation (Figure 1). The recorded symmetry of our multiple-band pattern indicates polyploidy, with gene duplication followed by subsequent divergence.

Polyploids have been reported for many tubificids. Christensen (1980) recorded different degrees of ploidy (3x, 4x, 5x and 6x) in *T. tubifex* and Weider (1992) found tri- and tetra-allelic electromorph banding patterns in *Limnodrilus udekemianus*. Anlauf (1994, 1997) reported a diploid condition in the *T. tubifex* populations he studied but could not exclude occasional occurrence of polyploidy, as some individuals showed tetrameric banding patterns.

On the basis of simple calculations (Allendorf & Utter, 1976; Taggart et al., 1981) using the PGI data, it is possible to hypothesize a tetraploid condition in both populations of *Tubifex* we studied, with a minimum of two and a maximum of three alleles for each electromorph.

In the PGM pattern, however, some electromorphs showed five or six bands, for both species (Figure 3). Since in monomeric enzymes, as PGM, all bands are representative of direct allelic expression, the tetraploid hypothesis has to be rejected. A minimum ploidy of six is necessary to obtain such a complex pattern, if we exclude a multiple loci interpretation. Different levels of ploidy have been observed in single populations of oligochaetes (Christensen et al., 1976; Christensen, 1980; Coates, 1995). Fifty-seven percent of *T. tubifex* and 45% of *T. blanchardi* showed electrophoretic genotypes for PGM in accord with a ploidy level certainly higher than 4x, but chromosome counts, or some other analyses of cellular DNA quantities, would be required to determine whether different ploidy levels are present in our populations.

In conclusion, our *T. tubifex* population shows a very high level of enzymatic polymorphism at both the investigated loci. *T. blanchardi* has less enzymatic polymorphism. Only one PGI genotype and 26% of the total alleles for both enzymes we found are shared between *T. tubifex* and *T. blanchardi*, suggesting they are closely related but separate species. Finally, our enzymatic data suggest a polyploid condition in both species. Further studies of enzymes and karyological surveys are needed to determine the effective genetic distance between *T. tubifex* and *T. blanchardi* in the sympatric Italian population and to confirm the ploidy levels.

References

Allendorf, F. W. & F. M. Utter, 1976. Gene duplication in the family *Salmonidae* III. Linkage between two duplicated loci coding for aspartate aminotransferase in the cutthroat trout (*Salmo clarki*). Hereditas 82: 19–24.

Anlauf, A., 1994. Some characteristics of genetic variants of *Tubifex tubifex* (Müller, 1774) (Oligochaeta: Tubificidae) in laboratory cultures. In T. B. Reynoldson & K. A. Coates (eds), Aquatic Oligochaete Biology V Kluwer Academic Publishers, Belgium. Hydrobiologia 278: 1–6.

Anlauf, A., 1997. Enzyme variability of *Tubifex tubifex* (Müller) (Oligochaeta, Tubificidae) and seven other tubificid species. Arch. Hydrobiol. 139: 83–100.

Anlauf, A. & D. Neumann, 1997. The genetic variability of *Tubifex tubifex* (Müller) in 20 populations and its relation to habitat type. Arch. Hydrobiol. 139: 145–162.

Avise, J. C. & G. B. Kitto, 1973. Phosphoglucose isomerase gene duplication in the bony fishes: and evolutionary story. Biochem. genetics 8: 113–132.

Chapman, P. M. & R. O. Brinkhurst, 1987. Hair today, gone tomorrow: induced chaetal changes in tubificid oligochaetes. Hydrobiologia 155: 45–55.

Christensen, B., 1980. Animal Cytogenetics. 2. Annelida. Gebrüder Borntraeger, Berlin, 81 pp.

Christensen, B., U. Berg & J. Jelnes, 1976. A comparative study on enzyme polymorphism in sympatric diploid and triploid forms of *Lumbricillus lineatus* (Enchytraeidae, Oligochaeta). Hereditas 84: 41–48.

Coates, K. A., 1995. Widespread polyploid forms of *Lumbricillus lineatus* (Müller) (Enchytraeidae: Oligochaeta): comments on polyploidism in the enchytraeids. Can. J. Zool. 73: 1727–1734.

Paoletti, A., 1989. Cohort cultures of *Tubifex tubifex* forms. In J. L. Kaster (ed.), Aquatic Oligochaete Biology Kluwer Academic Publisher, Belgium. Hydrobiologia 180: 143–150.

Paoletti, A. & C. Rusconi, 1985. Preliminary data on the biology of a tubificid community in a small stream. S. It. E. Atti 5: 277–280 (in Italian).

Richardson, B. J., P. R. Baverstock & M. Adams, 1986. Allozyme electrophoresis. A handbook for animal systematics and population studies. Academic Press, Sydney, 410 pp.

SPSS, 1989. Statistical Package for Social Sciences. Release 8.0.0., 1997. SPSS Inc., Illinois, U.S.A.

Taggart, J., A. Fergusson & F. M. Mason, 1981. Genetic variation in Irish populations of brown trout (*Salmo trutta* L.): electrophoretic analysis of allozymes. Comp. Biochem. Physiol. 69B: 393–412.

Vejdovský, F., 1891. Note sur un *Tubifex* d'Algérie. Mem. Soc. Zool. de France 4: 596–603.

Weider, L. J., 1992. Allozymic variation in tubificid oligochaetes from the Laurentian Great Lakes. Hydrobiologia 234: 79–85.

Hydrobiologia **406:** 87–99, 1999.
B. M. Healy, T. B. Reynoldson & K. A. Coates (eds), *Aquatic Oligochaetes.*

Terrestrial polychaetes – models for the evolution of the Clitellata (Annelida)?

Günter Purschke
Spezielle Zoologie, Fachbereich Biologie/Chemie, Universität Osnabrück, D-49069 Osnabrück, Germany. E-mail: purschke@biologie.uni-osnabrueck.de; Phone: [+49] 541 2857; Fax: [+49] 541 2587

Key words: phylogeny, Polychaeta, Clitellata, Oligochaeta, Parergodrilidae, *Hrabeiella*

Abstract

The origin and the position of the Clitellata in the phylogenetic system are the object of vigorous scientific debate. Since most extant Clitellata are either terrestrial or limnetic, the question arises which characters are specific adaptations to these environments and which are plesiomorphies taken from the annelid stem species. A promising approach is to study whether clitellate characters are present in terrestrial polychaetes as well. In fact, the terrestrial polychaetes *Hrabeiella periglandulata* Pizl & Chalupský, 1984 and *Parergodrilus heideri* Reisinger, 1925 have many features in common with the Clitellata, as does *Stygocapitella subterranea* Knöllner, 1934, which inhabits sandy beaches at the transition between the marine and terrestrial realms. These features include a prostomium lacking appendages, chaetae that are simple and short, an absence of parapodia, an epidermis without kinocilia, sensory cells with cilia that project only slightly beyond the cuticle or not at all, direct development, direct sperm transfer and eggs laid in cocoons. The nuchal organs so characteristic of polychaetes are either displaced into the body and reduced (*S. subterranea*, *H. periglandulata*) or absent (*P. heideri*). In *H. periglandulata*, the nervous system and the foregut also closely resemble those of the oligochaetes. These polychaetes are certainly not closely related and clearly do not belong to the Clitellata, as shown by their lack of a clitellum, the different structure and position of the reproductive organs and ultrastructural differences in the sperm. Hence their similarities to clitellates have most probably arisen by convergence. A hypothesis is presented explaining the evolution of the terrestrial polychaetes as a model for that of the Clitellata, documenting a primary terrestrial origin of the latter.

Introduction

The phylogenetic system of the higher taxa of the Annelida is one of the most unsatifactorily resolved problems in invertebrate systematics. This problem appears of topical interest and is the object of vigorous scientific debate (Rouse & Fauchald, 1995, 1997; Eibye-Jacobsen & Nielsen, 1996; McHugh, 1997; Westheide, 1997; Purschke, 1997; Kojima, 1998). One question in these discussions is whether there exists a monophyletic taxon Polychaeta, the occurrence of nuchal organs being their sole autapomorphy (Rouse & Fauchald, 1995, 1997) or whether the polychaetes form a paraphyletic assemblage of primarily marine annelids (McHugh, 1997; Westheide, 1997). This question leads to two hypotheses. In the first hypothesis, the Clitellata are the sister group of the monophyletic Polychaeta or form at least a separate clade (Rouse & Fauchald, 1995, 1997). Consequently, many clitellate characters, such as a small prostomium lacking appendages, numerous homonomous segments without parapodia and only a few simple chaetae as well as the absence of nuchal organs, would have to be regarded as plesiomorphies taken from the stem species of both clitellates and polychaetes. From the second hypothesis (McHugh, 1997; Westheide, 1997; Kojima, 1998) follows an opposite interpretation: the Clitellata are highly derived annelids forming a subordinate taxon within the paraphyletic polychaetes. Thus, most characters typical of the Clitellata are interpreted as derived character states – including the absence of nuchal organs, which is regarded as a loss (Purschke, 1997; Westheide, 1997).

Since most extant Clitellata are either terrestrial or limnetic, the question arises which characters are specific adaptations to these environments rather than plesiomorphies taken from their marine annelid stem species without any changes. A promising approach is to study whether such clitellate-like characters are present in true terrestrial polychaetes as well. Apart from a few air-breathing Nereidae inhabiting the upper shore of mangrove swamps (Storch & Welsch, 1972), *Hrabeiella periglandulata* Pizl & Chalupský, 1984 and the parergodrilid *Parergodrilus heideri* Reisinger, 1925 are the only known non-clitellate annelids which have successfully colonized the terrestrial environment (Rota & Lupetti, 1996; Rota, 1997, 1998). The importance of the other parergodrilid, *Stygocapitella subterranea* Knöllner, 1934, in comparative analyses has so far not completely been recognized. Therefore, this species, known as a marine interstitial polychaete (Schmidt, 1970), is included in this study. It inhabits the upper eulittoral of sandy beaches, the transition region between the marine and terrestrial realms. The Parergodrilidae and *Hrabeiella* have many features in common with the Clitellata, although they are related neither to each other nor to the Clitellata (see Rota, 1998), and, thus, similarities among these taxa are convergent. It is discussed whether the convergently evolved characters may give us a better understanding of the evolution of the annelids in general and whether they can be taken as a model for the evolution of the Clitellata.

Materials and methods

Specimens of *Hrabeiella periglandulata* Pizl & Chalupský, 1984 (Figure 1E) were collected from well-drained, fresh garden soil in Osnabrück (Germany), those of *Parergodrilus heideri* Reisinger, 1925 (Figure 1C) from thick layers of leaf litter in a beech forest near Göttingen (Germany), and those of *Stygocapitella subterranea* Knöllner, 1934 (Figure 1A) in the upper intertidal area of sandy beaches on the North Sea island of Sylt (East Coast, List, Germany). Other localities from which *S. subterranea* was obtained are: Kristineberg (Sweden), Helgoland (Düne, Germany), Mellum (Germany), Weser-Estuary (Dedesdorf, Germany), Seignosse le Penon (Biarritz, France), Nahant (Canoe Beach, Massachusetts, U.S.A.), Quadra Island (British Columbia, Canada), Sidney (British Columbia, Canada), Coos Bay (Horsefall Beach, Oregon, U.S.A.). The animals were extracted in the laboratory either by a modified version of O'Connor's wet-funnel method (Jans & Römbke, 1989) or by the $MgCl_2$-method (Westheide & Purschke, 1988). Samples were inspected and sorted with a dissecting microscope. Immediately after sorting, the live material was examined by light microscopy.

For electron microscopy, specimens were fixed either in 2.5% glutaraldehyde in 0.1 M phosphate buffer at pH=7.3 (*P. heideri*, *H. periglandulata*) or a phosphate buffered (pH=7.2, 0.05 M) mixture of sucrose, picric acid, formaldehyde and glutaraldehyde (see Westheide & Purschke, 1988) (*S. subterranea*). After rinsing in the appropriate buffer specimens were postfixed in 1% OsO_4 and dehydrated in a graded ethanol series. For scanning electron microscopy they were then critical point dried, mounted on aluminium stubs and sputter-coated with gold. The specimens were examined in Cambridge Stereoscan 250S and Zeiss DSM 962 microscopes. For transmission electron microscopy, specimens were embedded in an Epon-Araldite mixture. Ultrathin sections were cut with a diamond knife (Reichert Ultracut E microtome), stained with aqueous solutions of uranyl acetate and lead citrate (Leica Ultrostrain I and II) in a Leica Ultrostainer and examined in Zeiss EM 109 or EM 902A microscopes.

The zonation pattern of *S. subterranea* was analyzed for a beach at the east coast of Sylt (List, North Sea). Samples (100 cm^3 volume) were taken on a transect through the beach perpendicular to the water line from the beginning of the beach slope far beyond high tide level (up to 22 m) at intervals of 2 m. Samples were taken monthly between 1 h before and 1 h after low tide. Sediment cores were taken from the top to 30 cm below ground-water level, cut into pieces of 10 cm length and 100 cm^3 volume. Duplicate sets of samples were analyzed; depending on the actual ground water and high tide level each set consisted of up to 127 samples. For quantitative extraction of meiofauna the $MgCl_2$-technique was used followed by three washes with sea water (100 cm^3 sample, 80 μm sieve). Cocoons were collected from these samples and reared in the laboratory in Petri dishes with sea water, moist sand or 1.5% agar at the temperature of the habitat until hatching.

Results

Distribution and habitat

In the beach chosen for quantitative sampling, *Stygocapitella subterranea* occupied a wide range of the beach slope but the zonation pattern was clearly correlated with the high tide level throughout the year

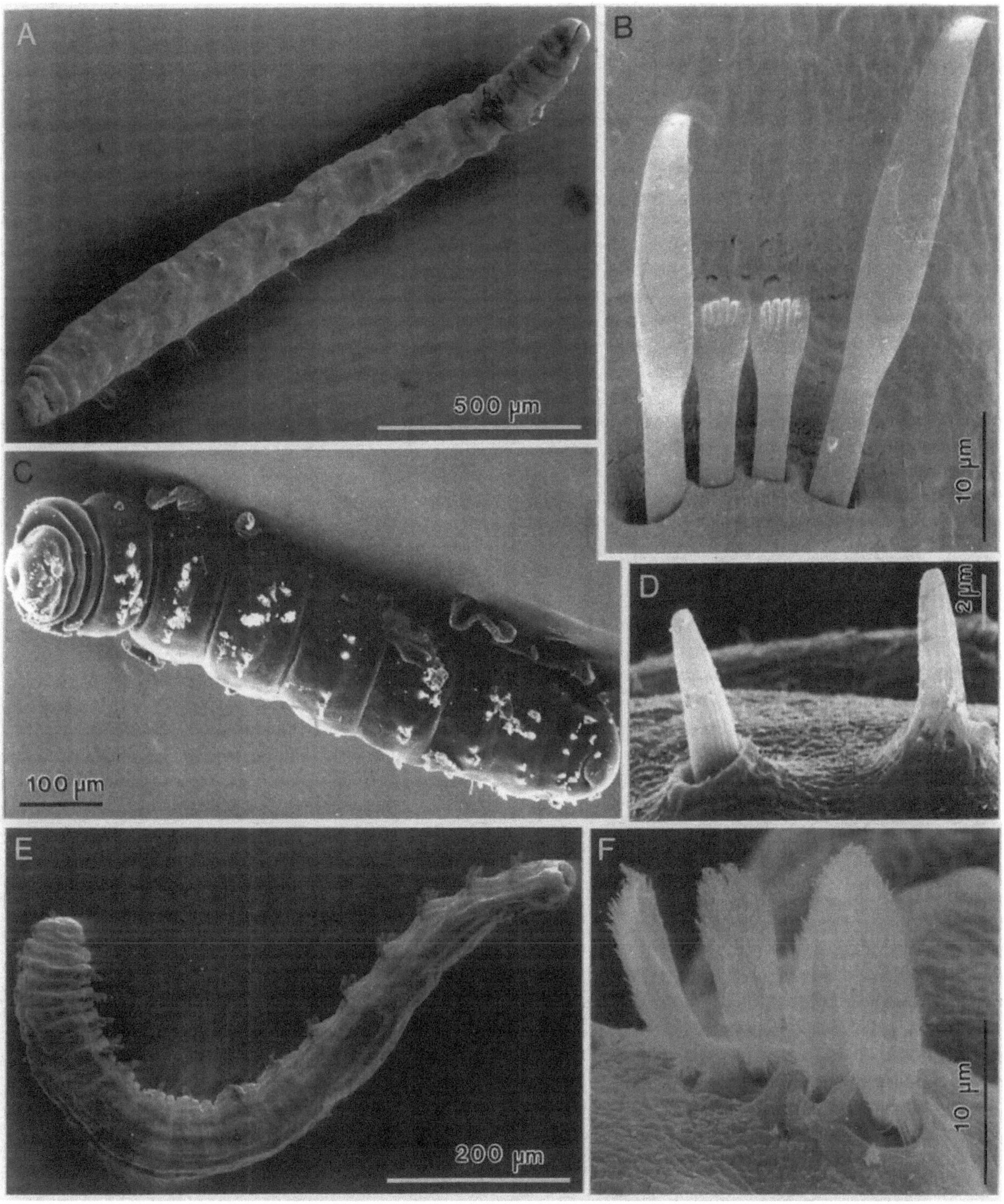

Figure 1. Habitus (anterior to left) and chaetae of the species investigated. SEM micrographs (A–B). *Stygocapitella subterranea.* (A) Entire male specimen, ventral view. (B) Bundle of chaetae. (C–D) *Parergodrilus heideri.* (C) Female, ventral view. (D) Group of chaetae. (E–F) *Hrabeiella periglandulata.* (E) Entire specimen, ventro-lateral view. (F) Two bundles of chaetae, each chaeta with a shovel-like tip bearing numerous fibres on one side.

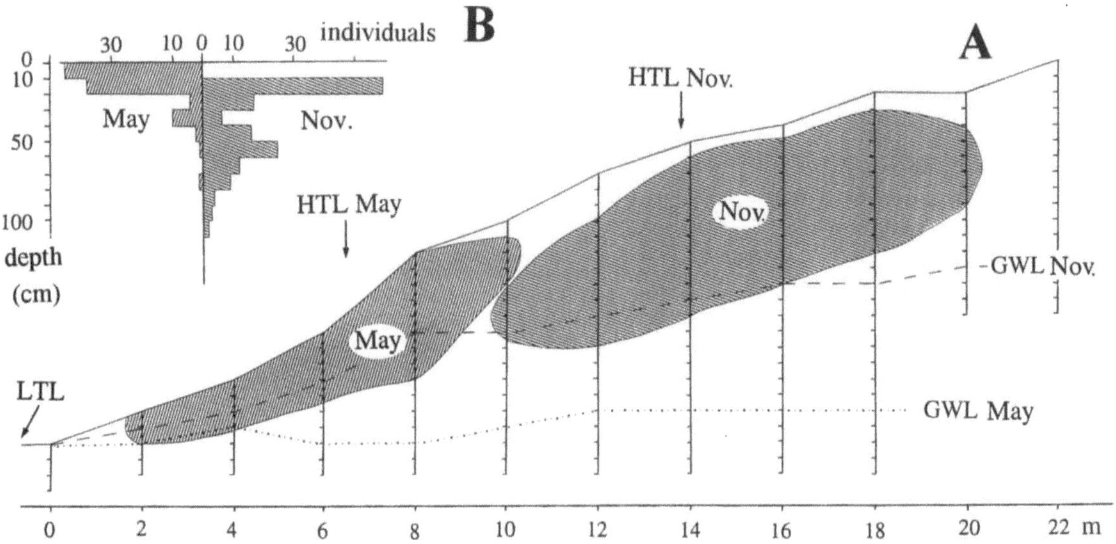

Figure 2. Stygocapitella subterranea. (A) Zonation pattern at low tide in a sandy beach on the island of Sylt for two representative months in spring and winter. Hatched areas represent inhabited regions. Each vertical line represents a core of sediment divided into pieces 10 cm long with volume 100 cm^3. LTL – low tide level, HTL – high tide level; GWL – ground water level. (B) Vertical distribution and abundance in spring and winter.

(Figure 2A). About 50% of the individuals were found close to the high tide level. High numbers of individuals were recorded at the sites just below and above high tide level. A maximum of 69 individuals were counted in one sample of 100 cm^3. In November, the area of colonization was extended and more individuals were found above high tide level. In winter months (November through February), not more than 10 individuals were found in one sample. Vertically, the uppermost layers were preferred (Figure 2B). For instance, in May about 90% of the population occupied the upper 20 cm. In winter (November), the 0–10 cm layer was avoided but again a maximum close to the surface was observed. This upper zone was extended to a depth of 20 cm in January. At all times, only occasional individuals (1–2 individuals per 100 cm^3) could be found just above and below the ground water level. Total abundance was between 225 (July) and 93 (February).

Reproductive biology and genital organs

The Parergodrilidae are gonochoristic and reproduction occurs throughout the year. In *Stygocapitella*, the population consisted of 70–90% mature females and males in a ratio of about 1:1 throughout the time of observation and 10–30% juveniles. The smallest juveniles had 4 chaetigers and were 0.65–0.78 mm long. New chaetigers appear successively until the final number of 10 chaetigers has been developed. Usually the anlagen of the genital organs can be recognzied under the light microscope in late stages with 9 chaetigers. In *Parergodrilus* a total of 340 individuals were observed, only two of which were males. About 10% of the individuals were juveniles with 4–7 chaetigers (adult females have 8 chaetigers).

The genital organs of the Parergodrilidae have been described in detail on the light microscopic level (Reisinger, 1925, 1960; Karling, 1958). Usually there is only one mature oocyte of comparatively large size at a time (300–400×150 μm in *S. subterranea* and 120–150 μm diameter in *P. heideri*), although vitellogenic oocytes may be present in each ovary (Figure 3A, B). In the females of both species, there is a complex of gland cells behind the ovaries (Figure 3A, B). In *S. subterranea* the glands are situated in chaetiger 10 and the following achaetigerous segment (Figure 3A) and open behind the genital openings on these segments. In *P. heideri* these glands lie in chaetiger 8 and open close to the female pores (so-called adanal glands of Reisinger (1960)). These glands consist of two types of cells and probably produce the egg capsules. Such capsules can now be demonstrated for *S.*

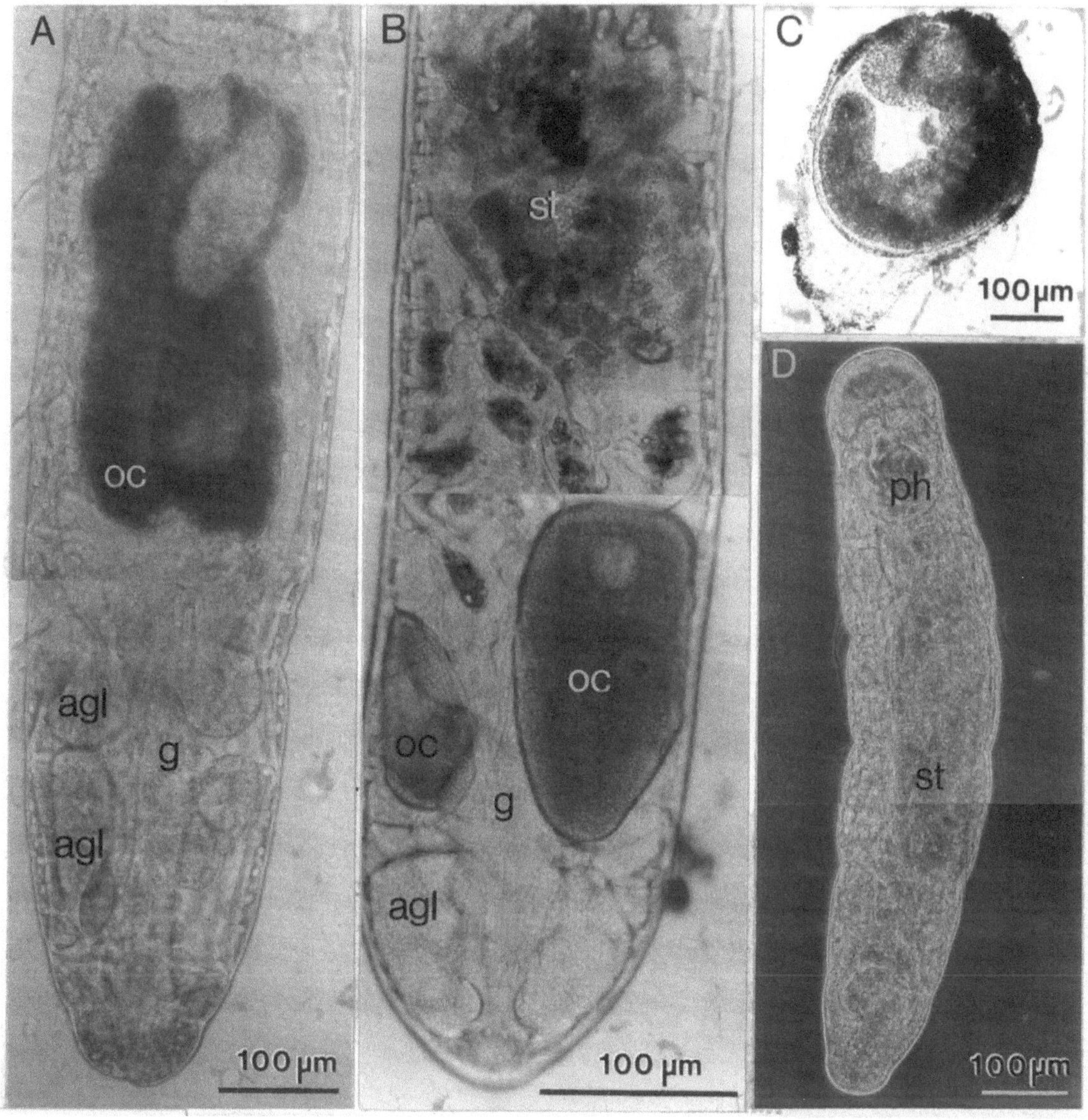

Figure 3. Female reproductive organs and development. Light micrographs. (A) *Stygocapitella subterranea*. Posterior end of female with ovaries containing vitellogenic oocytes (oc) and accessory genital glands (agl). (B) *Parergodrilus heideri*. Posterior end of female with ovaries each containing a vitellogenic oocyte of different size and accessory genital glands. (C–D) *S. subterranea*. (C) Cocoon with developing juvenile. (D) Newly hatched juvenile with 4 chaetigerous segments. g – gut; ph – pharynx; st – stomach.

subterranea as well (Figure 3C). They consist of a more or less transparent and leathery casing to which detritus and other particles may be attached. They are deposited between the sand grains and have occasionally been found in those samples with high numbers of individuals. Each egg capsule contains only one developing egg lying within a liquid secretion. Development is direct and juveniles possess 4 chaetigers upon hatching (Figure 3D). Although numerous mature individuals were cultured for long periods no egg capsules were obtained from these individuals. For this reason cleavage could not be observed in *Stygocapitella*.

The spermatozoa are thread-like in *Stygocapitella* and preliminary ultrastructural observations show that there is a single mitochondrion surrounding the proximal part of the axoneme and the distal centriole is situated immediately below the nucleus. The structure of the acrosome has not yet been clarified.

Central nervous system and sensory structures

In *S. subterranea*, the brain has a basiepidermal position and lies within the prostomium. Ventrally, it gives rise to two basiepidermal circumoesophageal connectives; the dorsal roots could not be detected. Although the ventral nerve cord lies more or less deeply within the body it still retains its basiepithelial position throughout its length. In *P. heideri*, the brain is internal, extending somewhat into the first segment. The brain is surrounded by a thin extracellular matrix and peritoneal lining. There is a single pair of circumoesophageal connectives without dorsal roots. Circumoesophageal connectives and the ventral nerve cord have basiepithelial positions.

In *H. periglandulata*, the prostomium, recognizable by the first external transverse groove, is very small and the brain is situated in the following achaetigerous segment (or peristomium), which comprises two annuli (Figure 4). With the exception of the circumoesophageal connectives the central nervous system is subepidermal. The basiepithelial circumoesophageal connectives emerge anterioventrally (Figure 4). Dorsal roots of the circumoesophageal connectives are lacking. There are two pairs of stomatogastric nerves.

In *H. periglandulata*, two pairs of sensory organs are associated with the brain: a pair of ciliary photoreceptor-like organs (pers. observ.) and a pair of nuchal organs (Figure 5A–D). The nuchal organs are externally almost invisible (Figure 5B, C). The organs are displaced into the body and communicate with the exterior through a narrow canal, the nuchal pit (Figure 5A, D), which opens in the furrow between pro- and peristomium. From this pore the organs extend posteriorly. The nuchal nerve and nuchal ganglion lie beside the brain and the nerve enters the brain posteriorly. The cilia of a few supporting cells always lie within the pit and at most their tips are discernible in the scanning electron microscope (Figure 5C). These cells are the only epidermal cells with kinocilia in *H. periglandulata*. Below the ciliated cells there is the olfactory chamber containing the sensory processes (microvilli and cilia) of about 5 primary sensory cells.

Other sensory structures have likewise been observed in all three species. These are penetrative multiciliate sensory cells, up to 3 of which may form inconspicuous sensory buds (Figure 6A–D). These sensory buds are especially abundant on the prostomium of *S. subterranea*. Each sensory cell bears 2–9 sensory cilia which extend 150–250 nm above the cuticular surface. The most common type of primary sensory cells are nonpenetrative multiciliated cells (Figure 6F–J). These are especially abundant on the prostomium and the following segment. The dendrites terminate in small bulbs extending somewhat above the surrounding epidermal cells. They give rise to numerous sensory cilia which run parallel to the epidermis in various directions. They have modified axonemal patterns such as $9\times2+0$, which are transformed into 9 singlets in *Parergodrilus* or 9 irregularly arranged doublets (Figure 6I–J). The number of microtubules progressively decreases along the cilia. In the parergodrilids a small vacuole is situated a few micrometers above the basal body (Figure 6H). In *H. periglandulata* there is another type of sensory cell with two short and straight cilia which are surrounded by epidermal cells (Figure 6E).

Pharynx

In *Hrabeiella*, the buccal cavity is unciliated with a dorsal ciliary pad (Figure 4). The buccal cavity and the oesophagus are lined by a thin cuticle (about 0.9 μm thick including the layer of epicuticular projections). In the oesophagus epicuticular projections are lacking. A complex system of pro- and retractor muscle fibres attach to this pad and there are four paired glands of comparable large size. These glands consist of the cell bodies of numerous gland cells, each of which sends a long and thin process ventrally. The gland cells open between the ciliated cells of the dorsal pad. The ciliary pad, with comparatively short cilia of equal length, is surrounded by a cuticular velum. The ciliated oesophagus starts dorsally behind a small fold following the dorsal pad. On the ventral side, the first ciliated cells lie at the posterior end of the pad (Figure 4).

Discussion

The Clitellata are characerized by an extraordinarily large number of autapomorphies (Table 1). The monophyletic nature of this group is not seriously questioned (Erséus, 1987; Brinkhurst, 1994; Rouse & Fauchald, 1995, 1997; Omodeo, 1997; Westheide et al., 1999). It is remarkable that they constitute a rather homogeneous group and many autapomorphies are of high diagnostic significance such as the complex reproductive biology and the genital organs (Westheide et al., 1999). Although the high number of apomorphies in itself speaks against a basal position within the

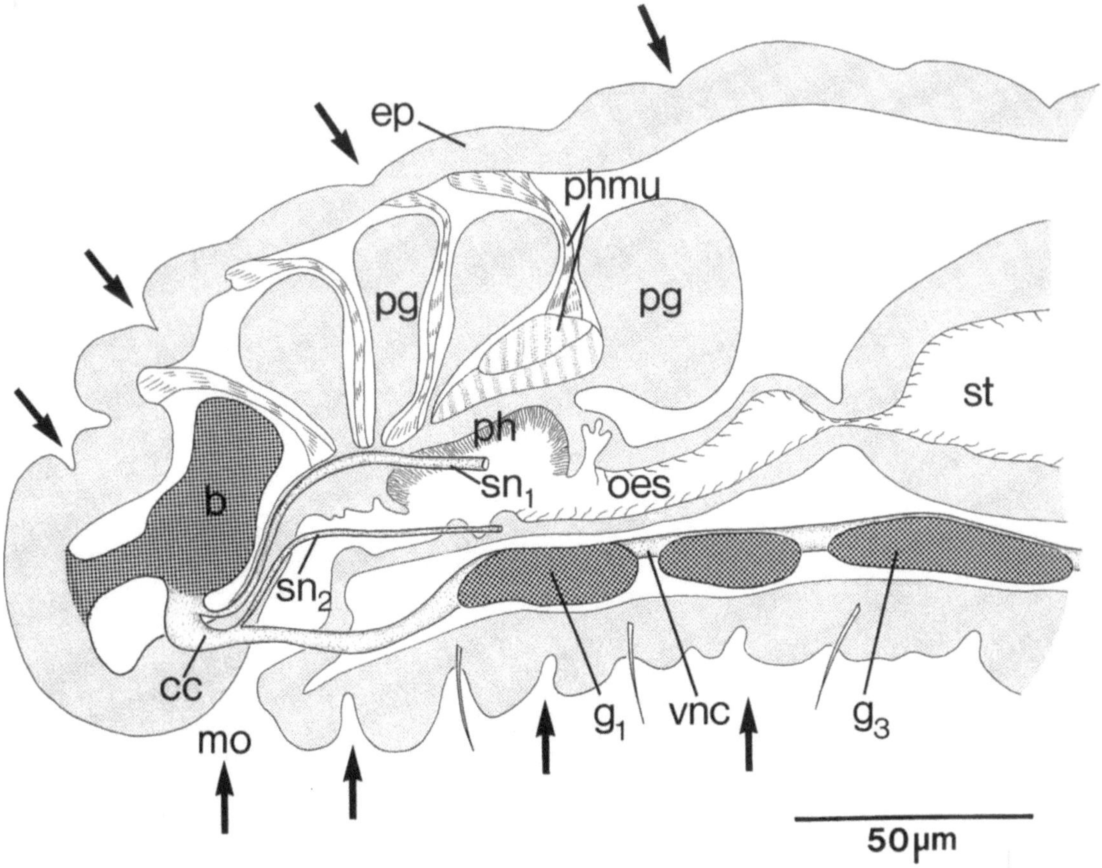

Figure 4. Hrabeiella periglandulata. Schematic reconstruction of the anterior end from longitudinal sections. Based on TEM observations. Arrows indicate boundaries of prostomium, peristomium (=1st segment), 2nd segment (=1st chaetigerous segment) etc. b – brain; cc – circumoesophageal connective; ep – epidermis; $g_{1,3}$ – 1st and 3rd ganglion of ventral nerve cord; mo – mouth; oes – oesophagus; pg pharyngeal gland; ph – pharyngeal pad; phmu – pharynx muscle; $sn_{1,2}$ – 1st and 2nd stomatogastric nerve; st – stomach; vnc – ventral nerve cord.

Annelida, the demonstration of convergently evolved characters in the highly derived polychaete taxa Parergodrilidae and *H. periglandulata* would further support the view of a highly derived position of the Clitellata (McHugh, 1997; Westheide, 1997; Kojima, 1998; Westheide et al., 1999) rather than their basal position as the sistergroup of the Polychaeta (Rouse & Fauchald, 1995, 1997). Moreover, this would likewise provide additional evidence for an evolution of the Clitellata in the terrestrial environment as suggested by Hessling et al. (1999).

The three polychaete species investigated are comparatively small and have meiofaunal sizes (between 0.7 mm (*P. heideri*) and 2.5 mm (*S. subterranea*) long and 100 μm (*H. periglandulata*) and 250 μm (*S. subterranea*) wide). Due to the lack of prostomial appendages and parapodia they have an oligochaetoid habitus (Figure 1A, C, E). The body surface is smooth and there are no epidermal kinocilia except for the ciliated supporting cells in the internal nuchal organs of *S. subterranea* and *H. periglandulata*. As a rule meiofaunal polychaetes of comparable size living in water-saturated marine or limnetic sediments possess external kinocilia. These occur either as ventral ciliary bands or segmentally arranged ciliary rings used for locomotion, ciliary fields around the mouth opening playing an important role in the feeding process, or at least in the form of external nuchal organs (Westheide, 1990; Eibye-Jacobsen & Kristensen, 1994; Purschke & Tzetlin, 1996; Purschke, 1997). Such a ciliation is present in the limnetic Aeolosomatidae and Potamodrilidae as well (Bunke, 1967).

In the Clitellata, an epidermal ciliation is generally absent. This holds also true for taxa with marine

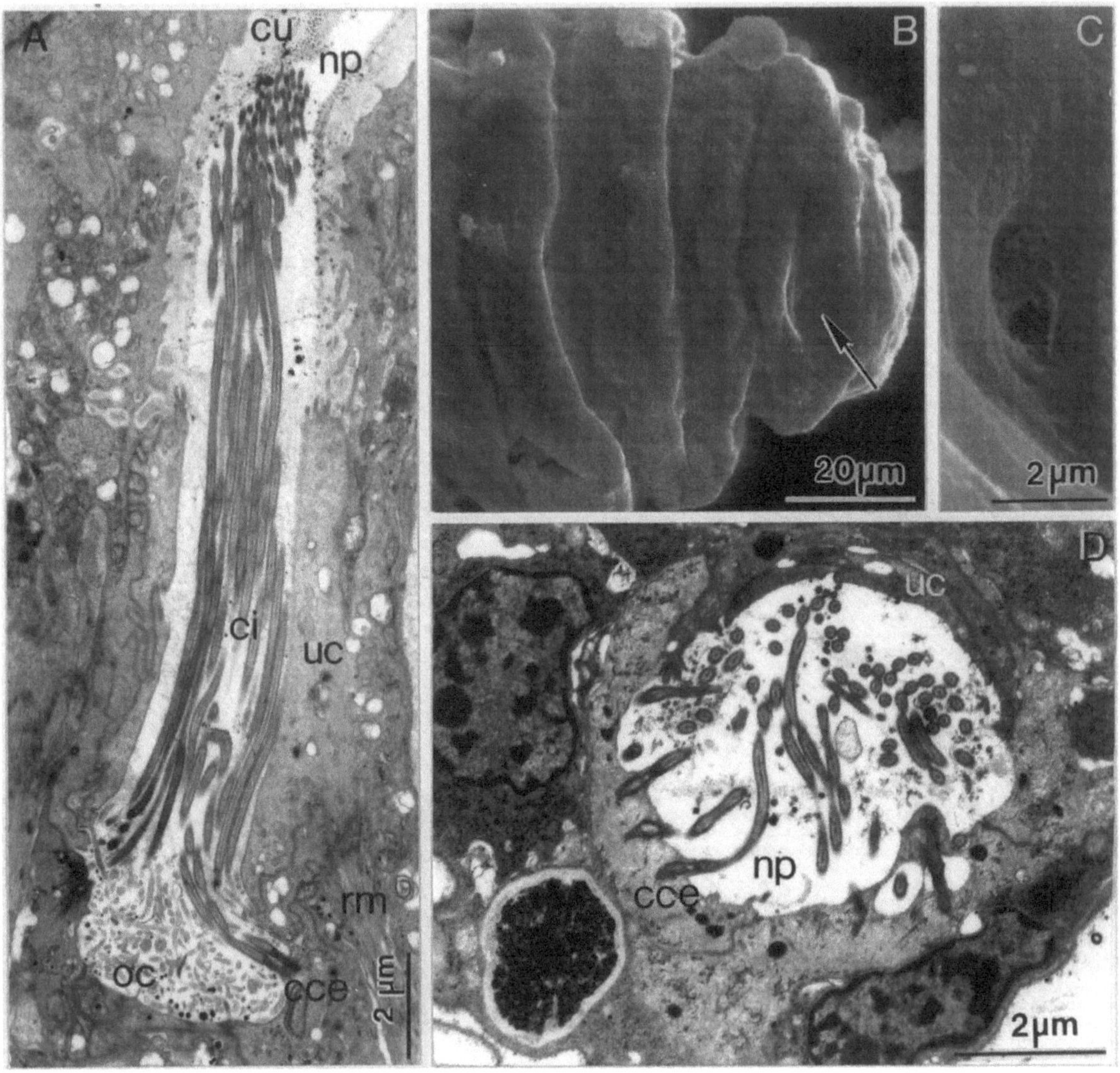

Figure 5. Hrabeiella periglandulata. Nuchal organ. TEM and SEM. (A) Nuchal organ in longitudinal section. (B) Anterior end with opening of nuchal pit (arrow). SEM. (C) Enlargement of opening. SEM. (D). Cross section of nuchal pit (np) at the level of the ciliated supporting cells (cce) with motile cilia. ci – cilium; cce – ciliated supporting cell; cu – cuticle; np – nuchal pit; oc – olfactory chamber; rm – retractor muscle; uc – unciliated supporting cell.

and limnetic species such as Tubificidae and Naididae. There are only a few aquatic taxa possessing branchiae (Brinkhurst & Jamieson, 1971; Jamieson, 1992). These gills are either ciliated diverticula of the ciliated rectum as in the naidids *Dero* and *Aulophorus* or unciliated extensions of the external body wall in *Branchiura* (Tubificidae) and *Phreodrilus* (Phreodrilidae) (Perrier, 1872; Brinkhurst & Jamieson, 1971). The only exception appears to be the naidid *Branchiodrilus* which is reported to possess segmentally arranged ciliated branchiae (see Brinkhurst & Jamieson, 1971; Coates, pers. com.).

The chaetae of the polychaetes investigated have a simple structure and are grouped in two or four bundles, respectively (Figure 1B, D, F). In *S. subterranea* there are three types of chaetae arranged in a specific pattern (see Riser, 1980); simple bilimbate chaetae and forked chaetae are present in each chaetigerous segment. In each of the other species the chaetae are of only one type. In *P. heideri*, the chaetae resemble those of enchytraeids, whereas in *H.*

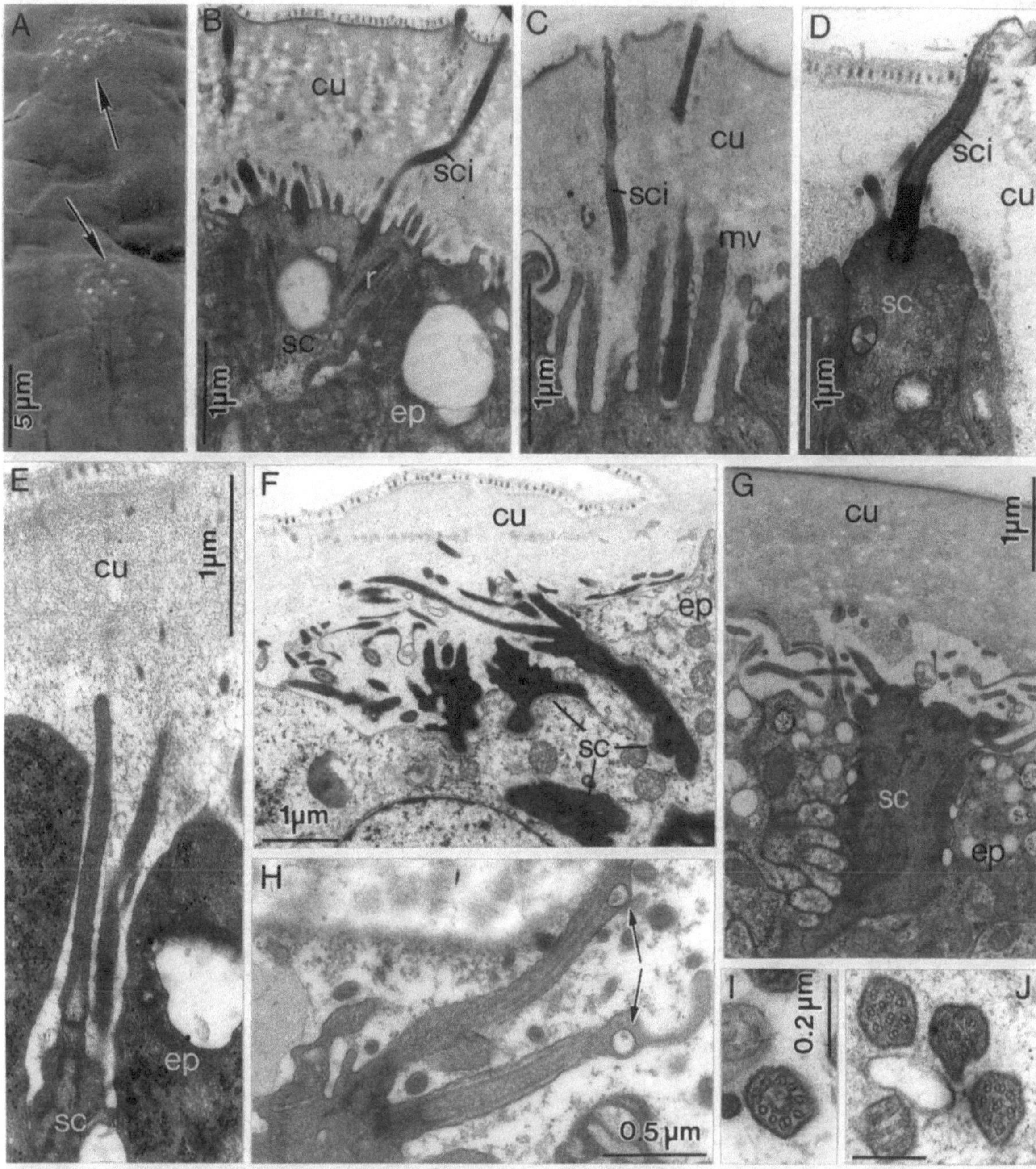

Figure 6. Sensory cells. (A) Sensory buds (arrows) with short penetrative cilia in *Stygocapitella subterranea*. SEM. (B–D). Multiciliated sensory cells with penetrative cilia. TEM. (B) *S. subterranea*. (C) *Parergodrilus heideri*. Note comparatively thick microvilli (mv) surrounding sensory cilia (sci). (D) *Hrabeiella periglandulata*. (E–J) Multiciliated sensory cells with non-penetrative cilia. TEM. (E) *H. periglandulata*. Sensory cell with two short cilia surrounded by epithelial supporting cells (ep). (F) *H. periglandulata*. Group of sensory cells with long basicuticular cilia. (G) *P. heideri*. Same cell type as in F. (H) *S. subterranea*. Same cell type as in F. Note vacuoles in ciliary shafts (arrows). (I–J) Axonemal structures. (I) *P. heideri*. (J) *H. periglandulata*. cu – cuticle; r – rootlet; sc – sensory cell.

periglandulata they have a shovel-like tip, one side of which gives rise to numerous, densely packed fibres (Rota & Lupetti, 1996; Rota, 1998). The structure of the chaetae of my specimens of *H. periglandulata* does

Table 1. Diagnostic and apomorphic characters of the Clitellata. From Jamieson (1992), Purschke et al. (1993), Westheide & Müller (1996), Omodeo (1997), Westheide (1997). Diagnostic features are marked with *, Apomorphic features not generally accepted marked with ()

– hermaphroditism	– a small prostomium without appendages ()
– gonads in specific segments	– brain situated one or a few segments behind the prostomium
– specific type of filiform spermatozoon	– absence of nuchal organs ()
– spermathecae which are not part of the female ducts	– epidermis without kinocilia ()
– eggs laid in cocoons formed by the clitellum, a girdle of at least two types of gland cells	– sensory cells with short cilia projecting only slightly beyond the cuticle*
– the cocoon is taken off by moving the anterior end backwards	– phaosomous photoreceptors (even in eyes, if present)*, ()
– external fertilization inside the cocoon	– eversible pharynx roof with conspicuous pharnygeal (salivary) glands
– specific type of development	– pharynx roof densely ciliated
– no larva	– secretion of pharyngeal glands sticks food particles to the pharynx
– no parapodia ()	
– a few and simple chaetae ()	

not significantly differ from that observed by Rota & Lupetti (1996), suggesting that fixation artifacts are the reason for the dissimilarities described by Jans & Römbke (1989) rather than genetic differences.

Parergodrilus heideri usually occurs in the leaf litter of humid beech woods, inhabiting thick, moist layers of decomposing leaves (Reisinger, 1925; Rota, 1997). Single specimens have been found in the banks and in the hyporeal interstices of freshwater streams; these findings have been explained as accidental invasions from the surrounding woods (Schwank, 1981). The species is known from various localities in Europe from Sweden to central Italy (Rota, 1997). *Hrabeiella periglandulata* has a similar distribution (Rota & Lupetti, 1996). However, it occurs in the upper mineral layers of well-drained sandy soils under various vegetation types (Rota, 1998). It is known that both species cannot withstand long periods of either desssication or submersion (Reisinger, 1925; Rota, 1998).

Stygocapitella subterranea has an amphiboreal-mediterranean distribution (for localities see above and Riser, 1980). Records from the Asian Pacific are lacking. However, it is also known from New Zealand and Australia (Riser, 1984; Hartmann-Schröder, 1996). Although this species has a worldwide distribution, no significant morphological and molecular differences between different populations have so far been found (pers. observ.; Schmidt & Westheide, pers. com.). The species inhabits the eulittoral of sandy beaches in every locality studied.

The habitat of *S. subterranea* can be described as the zone of permanently damp sand (moist sand stratum, zone of retention; see Giere [1993] for terminology) extending into the upper dry sand stratum (pers. observ. and see Schmidt, 1970; Lasserre & Renaud-Mornant, 1973; Westheide, 1990). In tidal beaches, the habitat of only part of the population becomes water saturated during high tide and above high tide level lower values of water saturation persist. In non-tidal beaches, the whole population inhabits areas far away from the water level, which are only occasionally influenced by the sea and normally never become saturated with water (Schmidt, 1970, 1972). These areas are characterized by low water saturation (at least for a part of the tidal cycle) and high fluctuations in temperature and salinity (Giere, 1993). In Sylt, *S. subterranea* occurs in various levels of water content from water saturation in the ground water (21.5% of the dry weight) down to 2.5% of the dry weight (pers. observ.). The lower levels observed are comparable to those in terrestrial sandy soils, where under dry conditions the water content reaches values of about 5% of the fresh weight (Tietze, 1968; Kache, Westheide & Wacker, pers. com.).

Relatively little is known about these animals' reproductive biology. The genital organs and spermatozoa of the parergodrilids and *Hrabeiella* have been shown to be completely different from those of clitellates indicating a high probability that other features shared with clitellates are convergent (Reisinger, 1925, 1960; Karling, 1958; Pizl & Chalupský, 1984; Rota & Lupetti, 1997; Rota, 1998). The spermatozoa of *S. subterranea* do not show the clitellate apomorphies of the midpiece and tail (Ferraguti, 1984; Jamieson, 1992). In parergodrilids the occurrence of seminal receptacles, which are part of the female genital ducts, and the structure of the male genital organs indicate a high probability of direct transfer of sperm and internal fertilization. The genital glands in the females were known to produce the egg capsules in *P. heideri* (Reisinger, 1960). This can now be verified for *S. subterranea* as well. These glands and the egg capsules are comparable to the clitellum and cocoon of oligochaetes. As in clitellates development is direct. Cleavage has only been observed in *Parergodrilus*; and it follows the pattern typical of polychaetes with large and yolky eggs (Reisinger, 1960). Transfer of sperm, egg deposition in a cocoon and direct development appear to be likely in *H. periglandulata* as well but have not yet been observed (Rota, 1998).

Important features of the central nervous system to distinguish between polychaetes and clitellates are the position of the brain and the structure of the circumoesophageal connectives. In polychaetes, the brain normally lies within the prostomium and communicates with the ventral nerve cord through the circumoesophageal connectives (Bullock, 1965; Orrhage, 1995). These originate as dorsal and ventral roots, which merge after a short distance to form the single pair of circumoesophageal connectives. Especially in small species the central nervous system may retain a basiepithelial or intraepithelial position. In oligochaetes the brain normally lies outside the prostomium in one of the following four segments (Bullock, 1965). The circumoesophageal connectives arise anteriorly and are not split into dorsal and ventral roots. The nervous system always has an internal position. In the Parergodrilidae, there are no clitellate features (Reisinger, 1925, 1960; Karling, 1958) except for the absence of dorsal roots in the connectives. In *H. periglandulata* the nervous system exhibits several similarities with oligochaetes: the position of the brain behind the prostomium, the anterior-ventral origin of the circumoesophageal connectives, the lack of dorsal roots in these connectives and the subepidermal position of the central nervous system. In clitellates these features most likely have evolved in the terrestrial environment and are related to the reduction of the prostomium and a habit of burrowing in soil (Hessling et al., 1999; Westheide et al., 1999). The backward displacement of the brain can be observed during embryogenesis (Hessling & Westheide, pers. com.) and is present in exclusively limnetic or marine taxa such as the Naididae and Tubificidae as well.

Nuchal organs are typical for polychaetes (Purschke, 1997; Rouse & Fauchald, 1997) and their absence in *P. heideri* (see Purschke, 1986) is thus interpreted as a loss. Usually these organs are more or less large ciliated areas situated posteriorly on the prostomium and mostly exposed to the exterior. In *S. subterranea* (see Purschke, 1986) and *H. periglandulata*, the organs are deeply displaced into the body and communicate with the exterior through the nuchal pit opening via small pores in the furrow between pro- and peristomium. The main function of the kinocilia in the supporting cells is generating water currents for a rapid exchange of sensory stimuli and protection of the sensory elements (Purschke, 1997). An internal position of the organs would allow to maintain these functions even if the substrate is not saturated with water.

The absence of nuchal organs in clitellates can thus be interpreted as a loss as well. Such a complete reduction might have occurred during invasion of the terrestrial environment. The reduction of the prostomium and the posterior displacement of the brain might have subserved this process. Interestingly, in one of the rare examples of polychaetes without nuchal organs, the Pisionidae, the prostomium is small and the brain is situated posteriorly (Siewing, 1953; Purschke, 1997).

The other sensory structures described in the three species are common in annelids (Jamieson, 1992; Verger-Bocquet, 1992). The cilia of the penetrative multiciliated sensory cells are comparatively short in these organisms as they are, for instance, in the Enchytraeidae, whereas they are several micrometers long in limnetic Naididae or in marine polychaetes (Purschke, 1993; pers. observ.). In contrast to what was previously thought (see Jamieson, 1992), such sensory buds cannot be considered as equivalents to nuchal organs in either these polychaetes or the Clitellata because in nuchal organs the cells with penetrative cilia are supportive cells and not sensory (Purschke, 1997). The sensory cells in nuchal organs are usually monociliated and their cilia lie inside the olfactory chamber.

The presence of a dorsal pharynx is typical for oligochaetes (Jamieson, 1992). The parergodrilids have a ventral pharynx which differs from those found in marine polychaetes in certain aspects (Purschke, 1987): the buccal cavity is unciliated, there is a conspicuous tongue-like organ with a thick cuticle and a bifurcate cuticular apophysis, and the muscle bulb – which usually is large and well-developed – is either reduced (*S. subterranea*) or absent (*P. heideri*) and replaced by a large complex of gland cells. In most other polychaetes with a ventral pharynx, the epithelium of the buccal cavity is ciliated forming so-called dorsolateral folds (Purschke & Tzetlin, 1996). The foregut structures of *H. periglandulata* are unique within the non-clitellate annelids, whereas such pharyngeal structures with a dorsal ciliated pad and prominent glands are characteristic of oligochaetes (Brinkhurst & Nemec, 1987; Jamieson, 1992; pers. observ.). Probably the pharynx structures of the Parergodrilidae on the one hand and of *Hrabeiella* and the Clitellata on the other represent different structural solutions for a similar mode of feeding in a similar habitat, i.e. ingestion of decaying plant material and detritus in environments which are not saturated with water.

In conclusion as it has been shown previously (see Rota, 1998) and in the present investigation the terrestrial polychaetes *H. periglandulata* and *P. heideri* as well as the interstitial polychaete *S. subterranea* have many features in common with the Clitellata. Although certain characters may occur in other annelids as well and some data are not yet available for every species, the density of these characters is remarkable. An invasion of the terrestrial environment via the zone of damp sand in beaches can at least be assumed for the Parergodrilidae, a view supported by Graefe (1977). Therefore, a direct invasion of the terrestrial environment from marine habitats appears to be conceivable for the Clitellata as well. This hypothesis is in conflict with many common concepts of oligochaete evolution which postulate a limnetic origin (Clark, 1969; Timm, 1981; Brinkhurst, 1984, 1994; Omodeo, 1997) but allows a conclusive explanation of all structural adaptations of this group. As shown by these polychaetes, certain clitellate-like characters may have evolved in similar marine habitats characterized by periodic water deficiency and are thus preadaptive for a successful invasion of terrestrial habitats. An origin in terrestrial habitats implies that the Clitellata are highly derived annelids and thus a basal position as the sister group of the polychaetes (Rouse & Fauchald, 1995, 1997) can be ruled out. This is supported by molecular data as well (McHugh, 1997; Kojima, 1998). The polychaetes are seen as a paraphyletic assemblage of primarily marine annelids, one subtaxon of which constitutes their sister group, though what subtaxon that could be is still an open question (Westheide, 1997).

Acknowledgements

I am indebted to Prof. Dr W. Westheide and R. Hessling for valuable discussions and suggestions. I thank Dr P. Scharnofske, Dr K. Koßmagk-Stefan and Drs B. and U. Ehlers who collected specimens at Helgoland, Mellum, Weser-Estuary and Seignosse le Penon. I express my thanks to Dipl.-Biol. R. Schmelz, who collected soil samples in Osnabrück. Thanks are also due to Anna Stein, who did the ultrathin sectioning and prepared the drawings, and to Andrea Noël for her help in preparing the manuscript.

References

Brinkhurst, R. O., 1984. The position of the Haplotaxidae in the evolution of oligochaete annelids. Hydrobiologia 115: 25–36.

Brinkhurst, R. O., 1994. Evolutionary relationships within the Clitellata: an update. Megadrilogica 5: 109–112.

Brinkhurst, R. O., & B. G. M. Jamieson, 1971. Aquatic Oligochaeta of the world. Oliver & Boyd, Edinburgh: 860pp.

Brinkhurst, R. O., & A. F. L. Nemec, 1987. A comparison of phenetic and phylogenetic methods applied to the systematics of Oligochaeta. Hydrobiologia 155: 65–74.

Bullock, T. H., 1965. Annelida. In T. H. Bullock & G. A. Horridge (eds), Structure and Function in the Nervous System of Invertebrates. Vol. I. Freeman and Company, San Francisco, U.S.A. 661–789.

Bunke, D., 1967. Zur Morphologie und Systematik der Aeolosomatidae Beddard, 1895 und Potamodrilidae nov. fam. (Oligochaeta). Zool. Jb. Syst. 94: 187–368.

Clark, R. B., 1969. Systematics and phylogeny: Annelida, Echiura, Sipuncula. Chemical Zoology 4: 1–68.

Eibye-Jacobsen, D. & R. M. Kristensen, 1994. A new genus and species of Dorvilleidae (Annelida, Polychaeta) from Bermuda, with a phylogenetic analysis of Dorvilleidae, Iphitimidae and Dinophilidae. Zool. Scr. 23: 107–131.

Eibye-Jacobsen, D. & C. Nielsen, 1996. The rearticulation of annelids. Zool. Scr. 25: 275–282.

Erséus, C., 1987. Phylogenetic analysis of the aquatic Oligochaeta under the principle of parsimony. Hydrobiologia 155: 75–89.

Ferraguti, M., 1984. The comparative ultrastructure of sperm flagella central sheath in Clitellata reveals a new autapomorphy of the group. Zool. Scr. 13: 201–207.

Giere, O., 1993. Meiobenthology. Springer, Berlin: 328 pp.

Graefe, U., 1977. *Adenodrilus* Graefe (in press) and *Parergodrilus* Reisinger two aberrant Annelida in forest soils. P. Cent. pir. Biol. exp. 9: 25.

Hartmann-Schröder, G., 1996. Annelida, Borstenwürmer, Polychaeta. In H. Schumann (ed.), Die Tierwelt Deutschlands 58, Gustav Fischer, Jena: 648pp.

Hessling, R., M.C. Müller & W. Westheide, 1999. CLSM analysis of serotonin-immunoreactive neurons in the central nervous system of *Nais variabilis*, *Slavina appendiculata* and *Stylaria lacustris* (Oligochaeta: Naididae). Hydrobiologia 406: 223–233.

Jamieson, B. G. M., 1992. Oligochaeta. In F. W. Harrison (ed.), Microscopic Anatomy of Invertebrates, Vol. 7 Annelida. Wiley-Liss, New York: 217–322.

Jans, W. & J. Römbke, 1989. Funde eines terrestrischen Polychaeten (Annelida) in Wäldern Baden-Württembergs. Carolinea 47: 158–162.

Karling, T. G., 1958. Zur Kenntnis von *Stygocapitella subterranea* Knöllner und *Parergodrilus heideri* Reisinger. Ark. Zool. Stockh. 11: 307–342.

Knöllner, F., 1934. *Stygocapitella subterranea* nov. gen. nov. spec. Schr. naturw. Ver. Schles. Holst. 20: 68–72.

Kojima, S., 1998. Paraphyletic status of Polychaeta suggested by phylogenetic analysis based on the amino acid sequences of elongation factor-1α. Mol. Phylogenet. Evol. 9: 255–261.

Lasserre, P. & J. Renaud-Mornant, 1973. Resistance and respiratory physiology of intertidal meiofauna to oxygen-deficiency. Neth. J. Sea Res. 7: 290–302.

McHugh, D., 1997. Molecular evidence that echiurans and pogonophorans are derived annelids. Proc. natn. Acad. Sci. U.S.A. 94: 8006–8009.

Omodeo, P., 1997. History of Clitellata. Ital. J. Zool. 65: 51–73.

Orrhage, L., 1995. On the innervation and homologues of the anterior end appendages of the Eunicea (Polychaeta), with a tentative outline of the fundamental constitution of the cephalic nervous system of the polychaetes. Acta Zool. (Stockh.) 76: 229–248.

Perrier, M. E., 1872. Historie naturelle du *Dero obtusa*. Arch. Zool. exp. gén. Paris 1: 65–97.

Pizl, V. & J. Chalupský, 1984. *Hrabeiella periglandulata* gen. et. sp.n. (Annelida) – a curious worm from Czechoslovakia. Vest. cs. Spolec. zool. 48: 291–295.

Purschke, G., 1986. Ultrastructure of the nuchal organ in the interstitial polychaete *Stygocapitella subterranea* (Parergodrilidae). Zool. Scr. 16: 13–20.

Purschke, G., 1987. Anatomy and ultrastructure of ventral pharyngeal organs and their phylogenetic importance in Polychaeta (Annelida). III. The pharynx of the Parergodrilidae. Zool. Jb. Anat. 115: 331–362.

Purschke, G., 1993. Structure of the prostomial appendages and the central nervous system in the Protodrilida (Polychaeta). Zoomorphology 113: 1–20.

Purschke, G., 1997. Ultrastructure of nuchal organs in polychaetes (Annelida) – New results and review. Acta Zool. (Stockh.) 78: 123–143.

Purschke, G. & A. B. Tzetlin, 1996. Dorsolateral folds in the polychaete foregut: structure, prevalence and phylogenetic significance. Acta Zool. (Stockh.) 77: 33–49.

Purschke, G., W. Westheide, D. Rohde & R. O. Brinkhurst, 1993. Morphological reinvestigation and phylogenetic relationship of *Acanthobdella peledina* (Annelida, Clitellata). Zoomorphology 113: 91–101.

Reisinger, E., 1925. Ein landbewohnender Archiannelide – zugleich ein Beitrag zur Systematik der Archianneliden. Z. Morph. Ökol. Tiere 3: 197–254.

Reisinger, E., 1960. Die Lösung des *Parergodrilus*-Problems. Z. Morph. 48: 517–544.

Riser, N. W., 1980. The aberrant Polychaete *Stygocapitella* from some American beaches. Wasman J. Biol. 38: 10–17.

Riser, N. W., 1984. General observations on the intertidal interstitial fauna of New Zealand. Tane 30: 239–250.

Rota, E., 1997. First Italian record of the terrestrial polychaete *Parergodrilus heideri* Reisinger, with anatomical and ecological notes. Ital. J. Zool. 64: 91–96.

Rota, E., 1998. Morphology and adaptations of *Parergodrilus* Reisinger and *Hrabeiella* Pizl & Chalupský, two enigmatic soil-dwelling annelids. Ital. J. Zool. 65: 75–84.

Rota, E. & P. Lupetti, 1996. An ultrastructural investigation of *Hrabeiella* Pizl & Chalupský, 1984 (Annelida). I. Chaetae and body wall organization. Hydrobiologia 334: 229–239.

Rota, E. & P. Lupetti, 1997. An ultrastructural investigation of *Hrabeiella* Pizl & Chalupský, 1984 (Annelida). II. The spermatozoon. Tiss. Cell 29: 603–609.

Rouse, G. W. & K. Fauchald, 1995. The articulation of annelids. Zool. Scr. 24: 269–301.

Rouse, G. W. & K. Fauchald, 1997. Cladistics and polychaetes. Zool. Scr. 26: 139–204.

Schmidt, P., 1970. Zonation of the interstitial polychaete *Stygocapitella subterranea* (Stygocapitellidae) in european sandy beaches. Mar. Biol. 7: 319–323.

Schmidt, P., 1972. Zonierung und jahreszeitliche Fluktuationen des Mesopsammons im Sandstrand von Schilksee (Kieler Bucht). Mikrofauna Meeresboden. 10: 6–60.

Schwank, P., 1981. Turbellarien, Oligochaeten und Archianneliden des Breitenbaches und anderer hessischer Mittelgebirgsbäche. II. Die Systematik und Autökologie einzelner Arten. Arch. Hydrobiol. Suppl. 62: 86–147.

Siewing, R., 1953. Morphologische Untersuchungen am 'Kopf' der Pisionidae (Pisione puzae nov. spec. Annelida, Polychaeta). Zool. Anz. 150: 298–313.

Storch, V. & U. Welsch, 1972. Ultrastructure and histochemistry of the integument of air-breathing polychaetes from mangrove swamps of Sumatra. Mar. Biol. 17: 137–144.

Tietze, F., 1968. Untersuchungen über die Beziehungen zwischen Bodenfeuchte und Carabidenbesiedlung in Wiesengesellschaften. Pedobiologia 8: 50–58.

Timm, T., 1981. On the origin and evolution of aquatic Oligochaeta. Eesti NSV Tead. Akad. Toim. (Biol.) 30: 174–181.

Verger-Bocquet, M., 1992. Polychaeta: Sensory structures. In F. W. Harrison (ed.), Microscopic Anatomy of Invertebrates, Vol. 7 Annelida. Wiley-Liss, New York: 181-196.

Westheide, W., 1990. Polychaetes: Interstitial families. In D. M. Kermack & R.S.K. Barnes (eds), Synopsis of the British Fauna 44. Universal book services/Dr W. Backhuys, Oegstgest: 152pp.

Westheide, W., 1997. The direction of evolution within Polychaeta. J. Nat. Hist. 31: 1–15.

Westheide, W. & M. C. Müller, 1996. Cinematographic documentation of enchytraeid morphology and reproductive biology. Hydrobiologia 334: 263–267.

Westheide, W. & G. Purschke, 1988. Organism processing. In R. P. Higgins & H. Thiel (eds), Introduction to the Study of Meiofauna. Smithsonian Institution Press, Washington. 146–160.

Westheide, W., D. McHugh, G. Purschke & G. Rouse, 1999. Systematization of the Annelida: different approaches. Hydrobiologia 402: 291–307.

Hydrobiologia **406:** 101–110, 1999.
B. M. Healy, T. B. Reynoldson & K. A. Coates (eds), Aquatic Oligochaetes.

Clitellate diversity in Nationalstadsparken, an urban national park in Stockholm, Sweden

Christer Erséus[1], Reinmar Grimm[2], Brenda Healy[3], Stefan Lundberg[4], Emilia Rota[5] & Tarmo Timm[6]
[1]*Department of Invertebrate Zoology, Swedish Museum of Natural History, Box 50007, SE-104 05 Stockholm, Sweden*
[2]*Zoologisches Institut und Zoologisches Museum, Universität Hamburg, Martin-Luther-King-Platz 3, DE-20146 Hamburg, Germany*
[3]*Department of Zoology, University College, Belfield, Dublin 4, Ireland*
[4]*Service Center for Taxonomic Zoology, Swedish Museum of Natural History, Box 50007, SE-104 05 Stockholm, Sweden*
[5]*Dipartimento di Biologia Evolutiva, Universitá di Siena, Via Mattioli 4, IT-53100 Siena, Italy*
[6]*Võrtsjärv Limnological Station, EE-61101 Rannu, Tartumaa, Estonia*

Key words: Clitellata, Oligochaeta, Hirudinea, biogeography, biodiversity, urban habitats

Abstract

To increase the overall faunistic knowledge of Swedish Clitellata, a survey was performed in Nationalstadsparken, a mosaic of fresh and brackish waters, parklands alternating with remnants of natural biota, and areas of urban development, within the cities of Stockholm and Solna on the Baltic coast of Sweden (59° N, 18° E). One hundred and twenty-one species (113 Oligochaeta, 8 Hirudinea) were collected at 31 aquatic and 27 terrestrial or littoral localities; 49 species of Oligochaeta were found at the aquatic localities, 70 species at the terrestrial/littoral sites. Twenty-four species are new records for Sweden, and this urban national park thus features 64% of all non-marine clitellates known from the country. The result is consistent with a view that diversity is enhanced when human activities increase habitat complexity, even in areas in direct contact with a large city.

Introduction

Knowledge of the composition and geographic distribution of the clitellate fauna of Sweden is still incomplete. Existing information indicates that leech species (Hirudinea) are few and, for the most part, common members of a North West European assemblage (Elliott & Mann, 1979; Degerman et al., 1994). However, for Oligochaeta (here referring to Clitellata excluding Branchiobdellida and Hirudinea), many unrecorded species are expected from the Scandinavian Peninsula.

Freshwater oligochaetes (largely Tubificidae and Naididae) of Sweden and Norway have been treated by several authors (e.g. Ekman, 1915; Alm, 1916; Piguet, 1919a,b; Sperber, 1948; Milbrink, 1969, 1970, 1973, 1978, 1994; Holmquist, 1985; Bremnes & Sloreid, 1994), yet only 62 species are known from this area to date. In 'Limnofauna Europaea', over 30 years ago, Brinkhurst (1967) listed as many as 189 species of limnic Oligochaeta for Europe as a whole. For the Enchytraeidae, most of which are terrestrial or semi-terrestrial, apart from scattered earlier reports (summarized by Lohm, 1979) and work on marine taxa (e.g. Erséus, 1977), there has only been one major account dealing with the Swedish fauna (Chalupský, 1992). The faunistics of Swedish earthworms has been poorly studied since the work by Julin (1949), who treated over 20 species (only 16 of which are valid today).

To increase the overall faunistic knowledge of Swedish Clitellata, we performed a survey of local diversity in Nationalstadsparken, a mosaic of freshwater and brackishwater bodies, and parklands alternating with natural biota, intermixed with urban developments; this area offers a wide representation of both aquatic and terrestrial habitats typical of Sweden. The park covers 27 km^2 within the city limits of Stockholm and Solna on the Baltic coast of Sweden (59° N, 18° E), i.e. in the middle of a metropolitan region with over one million inhabitants. To minimize further exploitation and to reinforce a proper management, Nationalstadsparken was declared a national park by the Swedish Government in January, 1995. It is the only urban national park in Sweden and, as far as we know, the only one in the world.

The paper is a contribution of the Swedish Worm Project (SWORM) at the Swedish Museum of Natural History (SMNH). A more comprehensive report (Erséus et al., 1998) is available at the Department of Invertebrate Zoology, SMNH. Taxonomic aspects of the oligochaete material have been published separately (Timm et al., 1997; Rota & Healy, 1999).

Nationalstadsparken

This area comprises

1. Södra Djurgården, and
2. Norra Djurgården, both parts of an old royal hunting park (partly established by King Johan III around 1580, but extended by King Carolus XI at the end of the 17th century),
3. the former military drilling ground Ladugårdsgärdet, and

4, 5. the two parks associated with the palaces Ulriksdal and Haga.

It is partly surrounded by brackish waters of the Baltic archipelago and all parts are within walking distance of the Stockholm downtown area.

Ponds and marshes, one small lake (Laduviken) and one stream (Igelbäcken) provide a range of freshwater habitats. There are extensive woodlands of spruce and pine, or oak and other deciduous trees, and mixed stands of conifers and hardwoods. Some of the woods are remnants of natural, herb-rich forests formerly widely distributed in central Fennoscandia, others are man-made habitats including ornamental parks, reflecting a complex cultural history of more than 450 years. Characteristic is the multitude of several hundred years old oak trees (*Quercus robur*) throughout the area. Swamps with partly flooded woods (largely alder, *Alnus glutinosa*) or reed beds are frequent near the water bodies.

Parts of Norra Djurgården, in particular, have been exploited during the twentieth century by the growth of Stockholm city. The northern and southern parts of Nationalstadsparken are thus connected only by narrow 'green' corridors through areas of residential and industrial development. The park offers a mixture of natural and culturally enriched biota, with high biological and recreational values. More than 100 species of birds breed there, and the oak woods harbour rich assemblages of insects and fungi. Von Proschwitz (1995) found 55% of all land and freshwater Mollusca known from Sweden in Nationalstadsparken. In general, the number of red-listed species of plants and animals is high.

There are a few records of Oligochaeta from the park. Sperber (1948) collected *Chaetogaster diaphanus, Nais barbata* (as '*N. obtusa*'), *N. communis, N. simplex, N. variabilis, Stylaria lacustris* and *Lumbriculus variegatus* (material in SMNH) in Brunnsviken bay in 1939. Chalupský (1992) reported *Enchytraeus buchholzi* s.l., *Fridericia bisetosa, F. bulboides, F. deformis, F. galba, F. ratzeli* s.s., *Henlea perpusilla, Marionina argentea* s.l. and *M. communis* from Norra Djurgården (his stations 19 and 20), and Julin (1949) collected *Allolobophora chlorotica* in the Bergian Botanical Garden (material in SMNH).

Material and methods

The field work of this study was carried out during four workshops at the SMNH, which is located in Nationalstadsparken. Material from other studies was also included: oligochaetes from grab samples, taken in the archipelago adjacent to Nationalstadsparken by Stockholm Water Co. (in 1992), and in Stora Skuggan pond and Laduviken by the World Wildlife Fund project 'Ekovatten' (in 1995; see Lundberg, 1996); additional bottom fauna samples (kick samples) from Igelbäcken collected by S. Lundberg in May, 1988, and the winter of 1994–1995.

Clitellate material from a total of 58 (31 aquatic and 27 terrestrial/littoral) localities, described in detail by Erséus et al. (1998) were used in the study.

Aquatic sampling

Samples were collected from the water bodies in October 1993, August 1994, and May and August 1995. Qualitative sediment samples were collected with a bottom scraper or a small dredge. Other samples (largely for naidids) were taken by scraping epiphyton, or by scooping water out of the submerged vegetation and sieving the material through a net with about 200 μm mesh. These samples were fixed in 4% formalin in the field to prevent the small worms from

Table 1. List of species of Clitellata collected in Nationalstadsparken, Stockholm. Frequency of occurrence at aquatic (totally 31) and terrestrial localities (totally 27), as well as habitat types (B=brackish water; BL=brackish-water littoral; F=freshwater; FL=freshwater littoral; T=terrestrial; WS=wet soil) are given. Asterisked species were not previously known from Sweden

Taxon names	Frequency of occurrence		Habitat types
	Aquatic	Terrestrial	
FAMILY NAIDIDAE:			
Chaetogaster diaphanus (Gruithuisen, 1828)	15	-	F, B
Chaetogaster diastrophus (Gruithuisen, 1828)	12	-	F, B
Chaetogaster cristallinus Vejdovský, 1883	4	-	F, B
Chaetogaster limnaei Baer, 1827	1	-	F
Dero digitata (Müller, 1773)	3	-	F, B
Dero nivea Aiyer, 1929	3	-	F
Dero obtusa Udekem, 1855	1	-	F
**Homochaeta setosa* (Moszýnski, 1933)	1	-	F
Nais barbata Müller, 1773	7	-	F, B
Nais communis Piguet, 1906	15	-	F, B
Nais elinguis Müller, 1773	7	-	F, B
Nais pardalis Piguet, 1906	3	-	F
Nais pseudobtusa Piguet, 1906	6	-	F, B
Nais simplex Piguet, 1906	1	-	F
Nais variabilis Piguet, 1906	14	-	F, B
Ophidonais serpentina (Müller, 1773)	11	-	F, B
Paranais frici Hrabe, 1941	2	-	B
Paranais litoralis (Müller, 1784)	1	-	B
Pristina longiseta Ehrenberg, 1828	4	-	F
Slavina appendiculata (Udekem, 1855)	6	-	F
Specaria josinae (Vejdovský, 1883)	2	-	F
Stylaria lacustris (Linnaeus, 1767)	15	-	F, B
FAMILY TUBIFICIDAE:			
Aulodrilus pluriseta (Piguet, 1906)	1	-	F
Limnodrilus claparedeanus Ratzel, 1868	3	-	F
Limnodrilus hoffmeisteri Claparède, 1862	14	-	F, B
Limnodrilus profundicola (Verrill, 1871)	2	-	B
Limnodrilus udekemianus Claparède, 1862	4	-	F, B
Potamothrix bavaricus (Oschmann, 1913)	1	-	B
Potamothrix hammoniensis (Michaelsen, 1901)	6	-	F, B
Potamothrix heuscheri (Bretscher, 1900)	5	-	F, B
Potamothrix vejdovskyi (Hrabe, 1941)	1	-	B
Psammoryctides albicola (Michaelsen, 1901)	3	-	F, B
Psammoryctides barbatus (Grube, 1861)	1	-	B
Rhyacodrilus coccineus (Vejdovský, 1875)	2	-	F
Rhyacodrilus falciformis Bretscher, 1901	3	-	WS
**Rhyacodrilus subterraneus* Hrabe, 1963	3	-	F, WS, B
Spirosperma ferox Eisen, 1879	1	-	F
Tubifex tubifex (Müller, 1774)	9	-	F, B
FAMILY LUMBRICULIDAE:			
Lumbriculus variegatus (Müller, 1774)	15	-	F, WS, B
Stylodrilus heringianus Claparède, 1862	5	-	F, WS
Trichodrilus sp. 1-F			

Table 1. Continued

Taxon names	Frequency of occurrence		Habitat types
	Aquatic	Terrestrial	
FAMILY ENCHYTRAEIDAE:			
Achaeta bibulba Graefe, 1989	-	3	T
Achaeta camerani (Cognetti, 1899)	-	2	T
**Achaeta vesiculata* Nielsen & Christensen, 1959	-	2	T
**Achaeta* sp. A (to be described as new)	-	2	T
**Bryodrilus ehlersi glandulosus* Dózsa-Farkas, 1990	-	1	T
Buchholzia appendiculata (Buchholz, 1862)	-	4	T
Buchholzia fallax Michaelsen, 1887	-	3	T, FL
Cernosvitoviella atrata (Bretscher, 1903)	-	2	BL, FL
Cernosvitoviella goodhui Healy, 1975	-	2	BL, FL
Cernosvitoviella immota (Knöllner, 1935)	-	1	BL
**Cernosvitoviella microtheca* Rota & Healy, 1999	-	4	BL, FL
**Cernosvitoviella minor* Dózsa-Farkas, 1990	-	3	T, BL, FL
Cognettia glandulosa Michaelsen, 1888	-	9	T, BL, FL
Cognettia sphagnetorum (Vejdovský, 1878)	-	7	T
Enchytraeus albidus Henle, 1837	-	3	BL
Enchytraeus buchholzi Vejdovský, 1878 s.lat.	-	13	T, BL, FL
**Enchytraeus bulbosus* Nielsen & Christensen, 1963	-	3	T
**Enchytraeus lacteus* Nielsen & Christensen, 1961	-	4	T
Enchytraeus christenseni christenseni Dozsa-Farkas, 1992[1]	-	8	T, FL
Enchytronia parva Nielsen & Christensen, 1959 s.lat.	-	6	T, FL
Fridericia bisetosa (Levinsen, 1884)	-	9	T, FL
Fridericia bulboides Nielsen & Christensen, 1959	-	4	T
Fridericia bulbosa (Rosa, 1887)	-	3	T
**Fridericia callosa* (Eisen, 1878)	-	1	BL
**Fridericia christeri* Rota & Healy, 1999	-	3	T
Fridericia connata Bretscher, 1902	-	6	T, BL
Fridericia deformis Möller, 1971	-	3	T
Fridericia galba (Hoffmeister, 1843)	-	9	T, FL
**Fridericia gracilis* von Bülow, 1957	-	1	T
**Fridericia isseli* Rota, 1994	-	1	T
**Fridericia leydigi* (Vejdovský, 1878)	-	1	T
Fridericia cf.*maculatiformis* Dózsa-Farkas, 1972	-	1	T
Fridericia nemoralis Nurminen, 1970	-	6	T, FL
**Fridericia* cf. *paranemoralis* Dózsa-Farkas, 1982	-	1	T
Fridericia paroniana Issel, 1904	-	6	T, FL
Fridericia perrieri (Vejdovský, 1878)	-	6	T, FL
Fridericia ratzeli (Eisen, 1872) s.lat.	1	8	F,T, FL
**Fridericia singula* Nielsen & Christensen, 1961	-	1	T
Fridericia striata (Levinsen, 1884)	-	1	T
**Fridericia ulrikae* Rota & Healy, 1999	-	1	T
**Fridericia waldenstroemi* Rota & Healy, 1999	-	2	T
**Hemifridericia parva* Nielsen & Christensen, 1959	-	5	T, FL
Henlea nasuta (Eisen, 1878)	-	1	FL
Henlea perpusilla Friend, 1911	-	4	T, BL
Henlea ventriculosa (Udekem, 1854)	-	4	T, BL
**Lumbricillus arenarius* (Michaelsen, 1889)	-	2	BL

[1]Replacement name for *Enchytraeus minutus minutus* Nielsen & Christensen, 1961 (see Dozsa-Farkas, 1992).

Table 1. Continued

Taxon names	Frequency of occurrence		Habitat types
	Aquatic	Terrestrial	
Lumbricillus fennicus Nurminen, 1964	1	3	BL, B
Lumbricillus lineatus (Müller, 1774)	1	-	B
Lumbricillus pagenstecheri (Ratzel, 1869)	-	1	BL
**Lumbricillus rivalis* Levinsen, 1883	-	5	T, BL
Marionina argentea (Michaelsen, 1889)	-	10	T, BL, FL
Marionina communis Nielsen & Christensen, 1959	-	1	T
**Marionina* sp. B (to be described as new)	-	5	T, BL
Marionina riparia Bretscher, 1899	-	6	BL, FL
Marionina southerni (Cernosvitov, 1937)	-	3	BL
Mesenchytraeus armatus (Levinsen, 1884)	1	5	F, T, BL, FL
Mesenchytraeus flavus (Levinsen, 1884)	-	3	T, FL
Mesenchytraeus pelicensis Issel, 1905	-	2	T
Oconnorella cambrensis (O'Connor, 1963)	-	1	T
**Stercutus niveus* Michaelsen, 1888	-	1	T
FAMILY LUMBRICIDAE:			
Allolobophora chlorotica (Savigny, 1826)	-	5	T
Aporrectodea caliginosa (Savigny, 1826)	1	9	F, T
Aporrectodea longa (Ude, 1895)	-	6	T
Aporrectodea rosea (Savigny, 1826)	1	6	F, T
**Dendrobaena attemsi* (Michaelsen, 1902)	-	1	T
Dendrobaena octaedra (Savigny, 1826)	-	1	T
Dendrodrilus rubidus (Savigny, 1826)	-	2	T
Eiseniella tetraedra (Savigny, 1826)	2	-	F
Lumbricus castaneus (Savigny, 1826)	-	5	T
Lumbricus rubellus Hoffmeister, 1843	1	5	F, T
Lumbricus terrestris Linnaeus, 1758	-	2	T
Octolasion cyaneum (Savigny, 1826)	-	5	T
HIRUDINEA:			
Erpobdella octoculata (Linnaeus, 1758)	8	-	F, B
Erpobdella testacea (Savigny, 1820)	2	-	F, B
Glossiphonia complanata (Linnaeus, 1758)	3	-	F
Glossiphonia heteroclita (Linnaeus, 1761)	1	-	B
Haemopis sanguisuga (Linnaeus, 1758)	1	-	F
Helobdella stagnalis (Linnaeus, 1758)	9	-	F, B
Piscicola geometra (Linnaeus, 1758)	1	-	B
Theromyzon tessulatum (Müller, 1774)	2	-	F, B

disintegrating. At most stations, pH and conductivity (later converted into salinity) were measured. Some untreated plant material was brought to the laboratory for extraction (see Laboratory procedures, following).

Terrestrial and littoral sampling

This sampling was carried out in August 1994 and May 1995. Samples for Enchytraeidae consisted of 500–1000 ml of soil, humus or decaying plant material, as well as plant debris, roots, algae or moss from the saturated zone at water's edge or just above it. Generally, several places were visited at a given locality; thus, a total of 88 subsites were sampled. Earthworms were collected by digging and handsorting litter and top soil. Preliminary sampling in Ulriksdal and the vicinity of the Bergian Botanical Garden was performed by Dr Astrid Lofs in October 1993.

Laboratory procedures

Most of the aquatic samples were sorted live with (largely naidids) or without magnification (for most of the Tubificidae, Lumbriculidae and Hirudinea), and the worms were preserved in 70% ethanol and for the most part identified as whole mounts in glycerine; leeches were identified under the dissection microscope. The formalin-fixed samples for Naididae were washed in the lab and the worms sorted by hand under a dissection microscope; they were identified either simultaneously under this microscope or, later, under a light microscope. The naidids were preserved in 70% ethanol; some of them were later stained with paracarmine and mounted in 'L25' (Zeiss) or 'Euparal' (Carl Roth GmbH, Karlsruhe, Germany).

Enchytraeidae and other terrestrial/littoral taxa were extracted by means of the wet funnel method (O'Connor, 1955; Healy & Rota, 1992). All enchytraeid worms were examined microscopically while alive, and most mature specimens could be identified to species without further treatment. Some specimens of all species were fixed in Bouin's fluid, Kahle's fluid or 70% ethanol, and permanent whole mounts were made for specific features (see Rota & Healy, 1998).

Soil characteristics recorded in the laboratory were

1. a rough, subjective assessment of texture, organic content and moisture content, and
2. pH, measured using equal proportions of soil and distilled water.

A reference collection of most of the clitellate taxa found is deposited in SMNH.

Results

Environmental factors

The water bodies of Nationalstadsparken are neutral to slightly alkaline (pH range 6.7–8.7), with salinity values between <0.1 and 3.4 ‰, covering limnetic and oligohaline ranges defined by the Venice System (1958). The bays that are in direct contact with the Baltic Sea all had salinities of at least 2.5 ‰ and contained *Enteromorpha*, indicating brackish conditions.

Water temperatures recorded in August 1995 were from 8 °C (spring in Haga Park) to 19.5 °C (effluent from Laduviken). The maximum summer temperatures of Laduviken and the various ponds, however, are likely to be around 25 °C.

The sediments of Brunnsviken and Laduviken suffered severe oxygen deficiency in August 1994, probably due to the exceptionally hot summer that year. Conditions had somewhat recovered in 1995, but the sediments of depths greater than about 4 m in Brunnsviken appear to be permanently azoic.

The woodland and grassland soils are generally well-drained and sandy. In most places the granite rock lies close to the surface and the soil reaction is acid, especially in the woodlands where the pH of the drier areas was 3.4–5.8 in August 1994, slightly higher in May 1995. Calcareous soils are absent. In the swamps, the substrate is black and peaty. The partially dry ditches, reed beds and other formations in the littoral zone of fresh and brackish waters contain leaf litter (reeds and trees), moss, algae, roots and gravelly deposits containing decaying organic matter. Neutral or slightly alkaline conditions occur in wet peaty soils where nutrients accumulate, and near roads and buildings. One of the highest pH values recorded (pH 7.5) was measured in a marshy grassland (Söderbrunn in Norra Djurgården) where groundwater issued from underground shell deposits.

Oligochaete fauna

On the basis of about 4000 specimens, a total of 113 species of Oligochaeta were identified (Table 1): 22 Naididae, 16 Tubificidae, 3 Lumbriculidae, 60 Enchytraeidae and 12 Lumbricidae. Six enchytraeid species were new to science (four described by Rota & Healy, 1999). Fifteen additional enchytraeids (asterisked in Table 1), one naidid (*Homochaeta setosa*), one tubificid (*Rhyacodrilus subterraneus*) and one lumbricid (*Dendrobaena attemsi*) were not previously known from Sweden.

In the aquatic biota, a total of 49 species was found (Table 1), but diversity of any one sample was generally low. *Chaetogaster diaphanus, Nais communis, N. variabilis, Stylaria lacustris, Limnodrilus hoffmeisteri* and *Lumbriculus variegatus* were the most frequently occurring species in the material, and all are common throughout the world (Brinkhurst & Jamieson, 1971).

With a total of 24 species collected, Igelbäcken stream in Ulriksdal is the most diverse aquatic locality, both in terms of species number and diversity of higher taxa, all oligochaete families here considered being represented in the torrential portion of the stream. Two additional species (*Chaetogaster langi* Bretscher,

1896 and *Pristina aequiseta* Bourne, 1891) were found by R. Grimm (August, 1995) in an upstream part of Igelbäcken and its source lake (Säbysjön). Although not yet found within the boundaries of Nationalstadsparken, they are thus likely to occur there. The effluent stream from Laduviken was a speciose site too, but here all but one species were naidids. The shallow profundal of Brunnsviken (at about 4 m depth) and Laduviken (at 1.5–3 m), was dominated by the small tubificid *Potamothrix heuscheri*. *Potamothrix hammoniensis* was absent from Laduviken but present in Brunnsviken.

Three aquatic species were extracted from wet soil samples. *Rhyacodrilus falciformis* and *R. subterraneus* represent a predominantly groundwater fauna (see also Timm et al., 1997), while *Lumbriculus variegatus* was widely distributed in the open water bodies too.

The 528 specimens collected in the weakly brackish straits surrounding the eastern part of Nationalstadsparken represented a total of nine species; a majority of the specimens were either *Potamothrix hammoniensis* or *Limnodrilus hoffmeisteri*.

The terrestrial and littoral material was dominated by Enchytraeidae with 15 enchytraeid genera represented, and more than one third of the species belonging to *Fridericia* (Table 1). *Cognettia glandulosa*, *Enchytraeus buchholzi* s.lat., *Marionina argentea, Enchytraeus christenseni christenseni*, *Fridericia bisetosa*, *F. galba* and *F. ratzeli* s.lat. were the most common taxa, occurring not only at most stations (Table 1), but also in most of the habitats available (coniferous as well as deciduous woods, grasslands and littoral zones); the last four species, however, were not found in any of the brackishwater littoral habitats. On the other hand, nine species (*Cernosvitoviella immota, Enchytraeus albidus, Fridericia callosa, Lumbricillus arenarius, L. fennicus, L. lineatus, L. pagenstecheri, L. rivalis* and *Marionina southerni*) were collected exclusively at littoral sites.

For taxonomic notes on several of the enchytraeid taxa in Nationalstadsparken, see Rota & Healy (1999).

Although earthworms were not extensively searched for in this study, 12 lumbricid species (Table 1), including the first Swedish record of *Dendrobaena attemsi* (Rota & Erséus, 1997), were encountered and identified.

Hirudinean fauna

The aquatic samples contained a total of eight leech species (Table 1), all members of the Euhirudinea and all previously well known from Sweden (see Degerman et al., 1994).

Table 2. Number of species of Clitellata known from Sweden (SE) and Nationalstadsparken (NSP). The 'non-marine' oligochaete species exclude 15 Tubificidae and seven Enchytraeidae that are either restricted to the Swedish west coast or, when present in the Baltic Sea, confined to sublittoral zones with higher salinity. Sources: Timm et al. (unpublished compilation of oligochaete literature); Degerman et al. (1994) (hirudineans). (Present authors are not aware of any compilation of Swedish records of marine leeches; hence the question-mark for the total number of species of Euhirudinea)

Group	SE, total	SE, non-marine	NSP	NSP/SE non-marine
Tubificidae	41	26	16	62%
Naididae	39	39	22	56%
Propappidae	1	1	0	0%
Lumbriculidae	6	6	3	50%
Enchytraeidae	90	83	60	72%
Lumbricidae	18	18	12	67%
Octochaetidae	2	2	0	0%
All Oligochaeta	197	175	113	65%
Branchiobdellida	1	1	0	0%
Acanthobdellida	1	1	0	0%
Euhirudinea	?	13	8	62%
All Clitellata	?	190	121	64%

Discussion

There is no single publication reviewing the whole oligochaete fauna of Sweden, but within the SWORM project a list of oligochaete taxa recorded from the country (compiled by Timm et al.) is regularly updated on the internet (http://www.nrm.se/ev/dok/oligochl.html.en); there are now records of 197 species (Table 2), of which 22 may be considered as exclusively marine. No less than 113 (65%) of the non-marine species were found in Nationalstadsparken; in fact, for 24 their only known Swedish occurrence is in this park. This high score is in part due to the great sampling effort, but it probably also reflects a high ecological diversity in Nationalstadsparken relative to that of Sweden in general. The park contains a mixture of natural as well as culturally impacted habitats, favouring flora and fauna development. The representation of Oligochaeta appears somewhat higher in the terrestrial biota than in the aquatic ones; 72% of all known non-marine Swedish Enchytraeidae are recorded in the park, with the corresponding figures for Tubificidae being 62%, Naididae 56% and Lumbriculidae 50% (Table 2).

In the most recent faunistic study of Swedish Enchytraeidae, Chalupský (1992) concluded that a total of 67 species (including marine taxa) had then been recorded from Sweden. We found 69 species reported in the literature prior to our study, and with the addition from Nationalstadsparken (six new species and 15 additional new records; Table 1) the Swedish Enchytraeidae now amount to 90 species.

The 12 lumbricids found hardly give a complete list of the earthworm fauna of Nationalstadsparken, as sampling was largely designed for microdriles. Noteworthy, however, is the rather dense population of *Dendrobaena attemsi* at one locality, which had never been collected north of Germany before (Rota & Erséus, 1997).

The limited representation of aquatic species may be explained by the lack of large, geologically old, freshwater bodies within Nationalstadsparken. The area is in close contact with the oligohaline parts of the Stockholm archipelago, which are physiologically unfavourable for freshwater as well as truly brackish-water or marine fauna. Laduviken, originally a bay was separated from the Baltic Sea by land uplift in the 18th century and is now the only true lake in the park. In other lakes near Stockholm, one of us (R. Grimm) found six species of Naididae in addition to those collected in Nationalstadsparken: *Nais bretscheri* Michaelsen, 1899, *Vejdovskyella comata* (Vejdovský, 1883), *Chaetogaster langi*, *Pristina aequiseta* (all in Mälaren, the latter two also in Säbysjön), *Ripistes parasita* (Schmidt, 1847), and *Uncinais uncinata* (Ørsted, 1842) (the latter two in the Tärnan lakes, Uppland).

Nevertheless, Igelbäcken, a stream of national interest due to the presence of the Stone Loach (*Barbatula barbatula* L., a fish red-listed in Sweden), has a well developed oligochaete fauna (26 species recorded including the two taxa collected upstream of Nationalstadsparken; see above). The oligochaetes of Swedish streams and rivers have been little studied, but considering that all freshwater fauna of Scandinavia is younger than 10 000 years, one should not expect as many species as in continental Europe. For instance, Lafont & Juget (1976) found 64 species in the River Rhône in France, Martinez-Ansemil (1990) found at least 68 species in the small River Tambre in northern Spain. The difference between Scandinavia and Continental Europe is particularly evident for the Lumbriculidae, of which only six are known from Sweden (two of these, *Rhynchelmis tetratheca* Michaelsen, 1920 and *Lamprodrilus isoporus* Michaelsen, 1901 were recently reported; Timm et al., 1997). In comparison, Hrabě (1981) listed 17 lumbriculid species for the former Czechoslovakia, Kasprzak (1981) 33 species for Poland and other parts of Central Europe, and Di Chiara Paoletti & Sambugar (1996) listed 14 for Italy.

Only one representative of the speciose genus *Trichodrilus* Claparède (see Rodriguez & Giani, 1994), *T. allobrogum* Claparède, 1862, has been identified from Sweden, from a drilled well in Täby north of Stockholm (Holmquist, 1975). A single, sexually immature, individual attributable to *Trichodrilus* was found in Igelbäcken, which is only a few km away from Täby.

Trichodrilus allobrogum is partly a subterranean species. So are *Rhyacodrilus falciformis* and *R. subterraneus*, which both may be common in wet soils associated with groundwater seeps. The Stockholm records demonstrate that suitable habitats for these taxa may exist even in the close vicinity of a large city. *Rhyacodrilus falciformis* is regarded as an indicator of oligotrophic conditions (Milbrink, 1978). On the other hand, despite its name, *R. subterraneus* is also a member of species-rich oligochaete associations in moderately eutrophic open watercourses (Timm et al., 1997).

The abundance of *Potamothrix heuscheri* in Brunnsviken and Laduviken is noteworthy, as this species here seems to replace *Potamothrix hammoniensis*, a species otherwise common in the profundal of eutrophic lakes (Timm, 1970; Milbrink, 1973, 1978). *Potamothrix heuscheri* is tolerant of anoxia (Gitay, 1968; Por & Masry, 1968; Milbrink, 1973) and a recent invader in Sweden (Milbrink, 1970).

In the two straits (Lilla Värtan and Saltsjön) containing most of the Stockholm Baltic harbours, the oligochaete fauna appears typical of eutrophic freshwater bodies that are slightly contaminated with brackish water, and the pollution level seems to be moderate (Laakso, 1968; Bagge & Ilus, 1973). The small difference in salinity between the connecting Lilla Värtan and Brunnsviken bays (recorded values 3.4 ‰ and 2.5–2.6 ‰ respectively) may be critical for non-marine species of Enchytraeidae. At one littoral locality in Brunnsviken, for instance, five oligohaline or euryhaline species (*C. immota, L. arenarius, L.*

fennicus, L. rivalis and *M. southerni*) were coexisting with 11 non-marine species, while at Lilla Värtan five brackishwater species were found but no non-marine species. The salinity difference corresponds to the α-oligohaline (3-5 ‰) and β-oligohaline (0.5–3 ‰) zones (Venice System, 1958), a subdivision which has shown to be of biological importance in the Baltic Sea.

The eight leeches found (Tables 1–2) represent 62% of all the freshwater euhirudinean species known from Sweden and contain no unexpected taxa (see Degerman et al., 1994).

Conclusions

The survey seems to corroborate data for other groups of organisms (plants, insects, molluscs, birds) which show that Nationalstadsparken in Stockholm has a high species diversity. The new oligochaete material includes six new species and 18 new distribution records. The result is consistent with a view that diversity is enhanced when human activities increase habitat complexity, even in areas in direct contact with a large city. Although at a different scale, the situation is similar to that of naturalized gardens, which are among the most species-rich habitats known (see, e.g. Owen, 1991; Miotk, 1996).

Acknowledgements

This is contribution No. 5 of the Swedish Worm (SWORM) Project. We are indebted to Mr Henrik Waldenström, who encouraged us to undertake this study, assisted in many ways and provided some initial financial support; to Mr Lars Lidén and Dr Björn Sohlenius, for assistance in the field; Dr Astrid Lofs for collecting part of the earthworms; Mr Anders Stehn (Stockholm Water Co.) for placing benthic samples at our disposal; Riksmusei Vänner and Ebba och Sven Schwarz Stiftelse for financial support; and Dr Ralph O. Brinkhurst, Dr J. Chalupský and Dr Kathryn A. Coates for constructive criticism of an earlier version of the paper.

References

Alm, G., 1916. Faunistische und biologische Untersuchungen im See Hjälmaren (Mittelschweden). Ark. Zool. 10: 1–47.

Bagge, P. & E. Ilus, 1973. Distribution of benthic tubificids in Finnish coastal waters in relation to hydrography and pollution. Oikos Suppl. 15: 214–225.

Bremnes, T. & S.-E. Sloreid, 1994. Fåbørstemark i ferskvann. Utbredelse i Sør-Norge. NINA Utredning 56: 1–42. NINA Norsk institut for naturforskning, Oslo.

Brinkhurst, R. O., 1967. Oligochaeta. In J. Illies (ed.), Limnofauna Europaea. Gustav Fischer Verlag, Stuttgart: 110–117.

Brinkhurst, R. O. & B. G. M. Jamieson, 1971. Aquatic Oligochaeta of the world. Oliver and Boyd, Edinburgh. 860 pp.

Chalupský, J., 1992. Terrestrial Enchytraeidae (Oligochaeta) and Parergodrilidae (Polychaeta) from Sweden, with description of a new enchytraeid species. Zool. Scr. 21: 133–150.

Degerman, E., B. Fernholm & P.-E. Lingdell, 1994. Bottenfauna och fisk i sjöar och vattendrag. Utbredning i Sverige. Rapport 4345. Naturvårdsverket, Stockholm. 201 pp.

Di Chiara Paoletti, A. & B. Sambugar, 1996. Aquatic Oligochaeta in Italy, with special reference to Naididae. Hydrobiologia 334: 37–49.

Dozsa-Farkas, K., 1992. List of enchytraeid synonyma. Newsletter on Enchytraeidae 3: 16–46.

Ekman, S., 1915. Die Bodenfauna des Vättern, qualitativ und quantitativ untersucht. Int. Rev. ges. Hydrobiol. Hydrogr. 7: 146–204, 275–425.

Elliott, J. M. & K. H. Mann, 1979. A key to the British freshwater leeches with notes on their life cycles and ecology. Freshwater Biological Association. Scientific Publ. No. 40: 1–72.

Erséus, C., 1977. Marine Oligochaeta from the Koster area, west coast of Sweden, with descriptions of two new enchytraeid species. Zool. Scr. 6: 293–298.

Erséus, C., R. Grimm, B. Healy, S. Lundberg, E. Rota & T. Timm, 1998. A survey of Clitellata in Nationalstadsparken, an urban national park in Stockholm, Sweden. The complete report. Unpublished; copies available at Department of Invertebrate Zoology, Swedish Museum of Natural History, Stockholm. 20 pp.

Gitay, A., 1968. Preliminary data on the ecology of the level-bottom fauna of Lake Tiberias. Israel J. Zool. 17: 81–96.

Healy, B. & E. Rota, 1992. Methods for collecting Enchytraeidae during expeditions. Soil Biol. Biochem. 24: 1279–1281.

Holmquist, C., 1975. Maskar i dricksvattnet – en för Sverige okänd oligochaet (fåborstmask) avslöjad. (Worms in the drinking water – one for Sweden unknown oligochaete revealed.) Fauna Flora, Stockh. 70: 193–197 (in Swedish, with English summary).

Holmquist, C., 1985. A revision of the genera Tubifex Lamarck, Ilyodrilus Eisen and Potamothrix Vejdovský & Mrázek (Oligochaeta, Tubificidae), with extensions to some connected genera. Zool. Jb. Syst. Ökol. Geogr. Tiere 112: 311–366.

Hrabě, S., 1981. Vodní málostetinatci (Oligochaeta) Ceskoslovenska. Acta Univ. Carolinae, Biol. 1979(1-2): 1–167 (in Czech, with English summary).

Julin, E., 1949. De svenska daggmaskarterna. (The Swedish earthworms.) Ark. Zool. 42A: 1–58 (in Swedish with English summary).

Kasprzak, K., 1981. Skaposzczety wodne, I. Polska Akademia Nauk, Instytut Zoologii, Warszawa. 226 pp. (in Polish).

Laakso, M., 1968. The bottom fauna in the surroundings of Helsinki. II. Records of Naididae and Tubificidae (Oligochaeta) and Spionidae (Polychaeta). Ann. zool. fenn. 5: 262–264.

Lafont, M. & J. Juget, 1976. Les Oligochètes du Rhône. I. Relevés faunistiques generaux. Ann. Limnol. 12: 253–268.

Lohm, U., 1979. The present faunistic knowledge of terrestrial Enchytraeidae in Sweden. Zoon 7: 63–66.

Lundberg, S., 1996. Bottenfaunaundersökningar i Laduviken och Solfångardammen - två sjöar/dammar inom Stockholms Nationalstadspark. Projekt Ekovatten WWF. (Study of the bottom fauna in Lake Laduviken and Solfångardammen pond in the Stockholm urban national park.) Information från Sötvattenslaboratoriet i Drottningholm 1996 (4): 1–18 (in Swedish with English summary).

Martinez-Ansemil, E., 1990. Etude biologique et écologique des Oligochètes aquatiques de la rivière Tambre et de ses milieux associés (Galice, Espagne). Ann. Limnol. 26: 131–151.

Milbrink, G., 1969. On the composition and distribution of oligochaetes in Lake Vättern 1967–1968. Inst. Freshw. Res. Drottningholm, Rep. 49: 149–156.

Milbrink, G., 1970. Records of Tubificidae (Oligochaeta) from the great lakes (L. Mälaren, L. Vättern, and L. Vänern) of Sweden. Arch. Hydrobiol. 67: 86–96.

Milbrink, G., 1973. On the use of indicator communities of Tubificidae and some Lumbriculidae in the assessment of water pollution in Swedish lakes. Zoon 1: 125–139.

Milbrink, G., 1978. Indicator communities of oligochaetes in Scandinavian lakes. Verh. int. Ver. Limnol. 20: 2406–2411.

Milbrink, G., 1994. Oligochaetes and water pollution in two deep Norwegian lakes. Hydrobiologia 278: 213–222.

Miotk, P., 1996. The naturalized garden – a refuge for animals? – First results. Zool. Anz. 235: 101–116.

O'Connor, F. B., 1955. Extraction of enchytraeid worms from coniferous forest soil. Nature 175: 815–816.

Owen, J., 1991. The ecology of a garden – the first fifteen years. Cambridge University Press, Cambridge: 415 pp.

Piguet, E., 1919a. Wasserbewohnende Oligochaeten der nordschwedischen Hochgebirge. Naturw. Unters. Sarekgebirge 4, Zool. 7: 778–801.

Piguet, E., 1919b. Oligochètes communs aux Hautes Alpes suisses et scandinaves. Rev. suisse Zool. 27: 1–17.

Por, F. D. & D. Masry, 1968. Survival of a nematode and an oligochaete species in the anaerobic benthal of Lake Tiberias. Oikos 19: 388–391.

Rodriguez, P.& N. Giani, 1994. A preliminary review of the taxonomic characters used for the systematics of the genus Trichodrilus Claparède (Oligochaeta, Lumbriculidae). Hydrobiologia 278: 35–51.

Rota, E. & C. Erséus, 1997. First record of Dendrobaena attemsi (Michaelsen) (Oligochaeta, Lumbricidae) in Scandinavia, with a critical review of its morphological variation, taxonomic relationships and geographical range. Ann. zool. Fenn. 34: 89–104.

Rota, E. & B. Healy, 1999. A taxonomic study of some Swedish Enchytraeidae (Oligochaeta), with descriptions of four new species and notes on the genus Fridericia. J. nat. Hist. 33: 29–64.

Sperber, C., 1948. A taxonomical study of the Naididae. Zool. Bidr. Upps. 28: 1–296.

Timm, T., 1970. On the fauna of the Estonian Oligochaeta. Pedobiologia 10: 52–78.

Timm, T., C. Erséus & S. Lundberg, 1997. New and unusual records of freshwater Oligochaeta from the Scandinavian Peninsula. Nordic J. Freshwat. Res. 72 (1996): 15–29.

Venice System, 1958. Symposium on the classification of brackish water. Oikos 9: 311–312.

Von Proschwitz, T., 1995. Ekoparkens land- och sötvattensmolluskfauna (The land and freshwater Mollusca of the Ekopark area (Stockholm, E. Sweden)). Unpublished report, Göteborg Natural History Museum, Göteborg, 58 pp. (in Swedish, with English summary).

Hydrobiologia **406**: 111–117, 1999.
B. M. Healy, T. B. Reynoldson & K. A. Coates (eds), Aquatic Oligochaetes.

Finding enchytraeid oligochaetes (Clitellata) in hot climates: species occurrence on the shores of Bermuda

Brenda Healy[1] & Kathryn A. Coates[2]
[1]*Department of Zoology, University College, Belfield, Dublin 4, Republic of Ireland. E-mail: bhealy@indigo.ie*
[2]*Bermuda Biological Station for Research Inc., 17 Biological Lane, St. Georges, GE 01, Bermuda. E-mail: kcoates@bbsr.edu*

Key words: enchytraeidae, seashores, Bermuda, habitats

Abstract

Sampling of a wide range of intertidal habitats increased the number of marine enchytraeid species for Bermuda from 12 to 31, of which 15 are believed to be new to science. The new species are in the genera *Enchytraeus*, *Achaeta* and *Marionina*, predominantly of the last named genus (12 of 15 species). Probable reasons for the large number of new records are the inclusion of habitats not previously sampled, such as algal cushions on rocks and seawater outfalls from aquaria, as well as the use of a wet funnel extractor. The ecological distribution of the species illustrates their dependence on moisture and a good supply of oxygen and confirms the view that enchytraeids are unable to live in stagnant, wet habitats, in hot climates. Species richness is comparable with that recorded in faunal studies of temperate shores. It is proposed that estimates of enchytraeid species richness in the tropics and subtropics could be much higher if attention were paid to the physiological requirements of these worms when designing sampling programs.

Introduction

Among the physiological constraints limiting the ecological distribution of enchytraeids in hot climates, two characteristics are particularly important: their poor resistance to desiccation and their high oxygen requirements. In marine littoral habitats, further limitations would be set by periodically high temperatures and salinities. Species distributions are controlled by combinations of these factors (Giere, 1980; Giere & Pfannkuche, 1982), as well as by availability of food of a certain type and substrates which protect against wave action and predators, and are appropriate to locomotory activities. The narrow ecological limits set by these requirements mean that enchytraeids tend to occur in smaller subhabitats in the tropics than in temperate regions and that it is necessary for collectors to search for the appropriate conditions. Observations by the authors in tropical regions (e.g. Healy & Coates, 1997) have shown that enchytraeids are most frequent in moist substrates where there is some atmospheric oxygen and are rare or absent in still water or saturated substrates where oxygen is depleted by high temperatures. Recognition of the requirements of enchytraeid species should make sampling more efficient, leading to more accurate estimates of regional diversity and the discovery of many more species. The purpose of this paper is to draw attention to some of the habitats on a subtropical seashore which enchytraeids have colonized, including some which have hitherto been ignored.

Investigations of marine enchytraeids have mainly been concerned with sediments or algal accumulations near the high water mark (wrack beds), and until recently most studies were in temperate zones. Few species have been recorded from rocky shores, or from any type of sea shore in the tropics. Published records of 12 enchytraeid species in Bermuda (*Enchytraeus albidus* (Henle, 1837), *Grania bermudensis* (Erséus & Lasserre, 1976), *Grania americana* (Kennedy, 1966), *Grania hylae* (Locke & Coates, in press), *Grania laxarta* (Locke & Coates, in press), *Stephensoniella marina* (Moore, 1902), *Stephensoniella sterreri* (Lasserre & Erséus, 1976), *Stephensoniella trevori* (Coates, 1980), *Marionina achaeta* (Lasserre, 1964), *Marionina spicula* (Leuckart, 1847), *Marionina swed-

marki (Lasserre & Erséus, 1976), and *Marionina subterranea* (Knöllner, 1934)) are from sediments and tidal debris on sandy beaches, mud flats with mangroves and from the shallow subtidal (Moore, 1902; Lasserre & Erséus, 1976; Coates, 1983a; Locke & Coates, in press). Preliminary investigations in a wider range of habitats indicated that the marine enchytraeid fauna was far richer than previous studies had suggested. This paper describes the habitats in which enchytraeids occur, and demonstrates the degree of specialization which some species have attained.

The Bermuda Islands

The Bermudas are a group of small, oceanic islands in the north-west Atlantic (32° 20′ N, 65° 50′ W) surrounded by the waters of the Sargasso sea. They lie on the Bermuda Pedestal, one of three volcanic peaks comprising the Bermuda Seamount which rises over 4000 m from the sea floor on the axis of the mid-Atlantic ridge. The peak of the extinct volcano is capped by a shallow coral reef platform which has the form of a pseudoatoll (Verrill, 1900) in which the outer reefs are submerged. The platform comprises about 130 islands and islets, surrounded by reefs, bays and lagoons. The total land area is approximately 55 km^2 and the maximum altitude is 79 m (Morris et al., 1977; Thomas & Logan, 1991). The original Bermuda Pedestal probably dates from the Cretaceous but there was a second period of eruptions 30–50 m y ago which is believed to have caused renewed emergence of the islands (Morris et al., 1977).

Owing to the proximity of the Gulf Stream, the islands have an atypically warm climate for their latitude and, consequently, Bermuda possesses the most northerly extensions of coral reefs and mangroves in the Atlantic ocean. The climate is subtropical with two seasons: summer, with a mean monthly air temperature of 29.6 °C in August, and winter, with a mean of 18.5 °C in February. The mean annual rainfall of 1500 mm is spread throughout the year and is very variable. Droughts lasting up to three months can occur in summer.

Marine habitats

Information in the following three paragraphs is based on Thomas & Logan (1991) and personal observations.

Tides are semidiurnal and the tidal range varies between 0.5 and 1.4 m with a mean of 0.75 m. The intertidal zone is therefore narrow and zonation poorly developed in comparison with mesotidal shores. The south-facing shores are exposed to greater wave action, while those to the north where the reefs lie some distance offshore are more sheltered and have somewhat finer sediments. Salinity of the seawater is 36–37‰.

Most of the coastline is rocky with only about 20% occupied by sandy beaches or muddy bays, the latter fringed by mangroves. The surface rocks of the islands are mainly limestones of aeolian origin which weather to give a karst topography. Rocks washed by the sea are generally pitted by holes, fissures and hollows, formed with the aid of organisms (phytokarst), which provide pockets of water or moist air allowing development of algal mats. Most rainwater disappears into the porous rock and there are, therefore, no natural streams running onto the shore, although groundwater runoff and seepage may be significant after heavy rains. There are, however, a large number of landlocked saltwater or brackish ponds which are a feature of the islands. Most are anchialine ponds, i.e. receiving seawater through underground fissures, and many are tidal. Littoral and sublittoral sediments are primarily biogenic carbonates.

The fauna and flora of the islands show close affinities with those of the Caribbean and the American coast of the Gulf of Mexico, and marine species probably colonized by way of currents. In spite of the geological age and isolation (approximately 600 km from the nearest point of N. America), the islands hold a relatively low proportion of endemics, and the diversity of the biota is much reduced in comparison with that of the Caribbean islands.

Materials and methods

Sampling of potential habitats was carried out at 14 localities during dry periods in March–April 1994 and April–May 1996. Preliminary sampling had shown that enchytraeids were most frequent in moist, organic substrates in six main habitat types: tidal debris on beaches, algal mats on intertidal rocks, soil, leaf debris and algae at seawater outflows from marine aquaria, leaf debris and algal mats under mangrove trees, and moist sand in certain regions of beaches (see Results for more details). Samples were collected from several localities for each of these habitat types, generally where some shade from direct sunlight was provided by rocks or trees. Samples of the substrates were removed into plastic bags and the enchytraeids extracted within 1–3 days, the wettest samples being treated first. Sampling was qualitative only and the size of samples depended largely on the availability of the substrate. Moist tidal debris was plentiful at several localities but algal turf in rock pockets at appropriate tide levels was always scarce and only small samples were taken in the interests of conservation. Worms were extracted from sediments by washing on a 0.3 mm sieve and from algae, tidal debris and root-bound sediments and soil by the wet funnel method (O'Connor, 1955; Healy & Rota, 1992). Specimens were examined live before fixing in 70% ethanol for further studies.

Results

Habitats

The habitats from which enchytraeids were frequent, i.e. at least ten specimens in a 200 ml sample, fall into six categories:

Tidal debris at and above high water mark (8 localities)

Accumulations of organic material were mainly found in sheltered bays on the north shores of the islands and around saline ponds. At these sites, the main constituents were leaves and twigs of mangroves and land-based plants, sargassum weed and seagrass remains, the dominant components varying according to season and wind direction (personal observations). They only formed a thick bed at one locality during the sampling period. Most of the accumulations sampled were on sandy beaches where debris tended to dry out. Enchytraeids here were present in the debris itself and in organic material buried in the sand beneath. Tidal debris is a temporary and seasonal habitat on most beaches with worms migrating into the sand when it becomes too dry.

This habitat contained the richest enchytraeid fauna with four genera and 18 species recorded (Table 1). Eight species were confined to tidal debris. Enchytraeids were accompanied in moist areas by talitrid amphipods, with *Pontodrilus* (Oligochaeta: Megascolecidae) and Gammaridae (Amphipoda) at some sites. Woodlice, mites and centipedes characterized drier material at higher levels.

Rocky shores (5 localities)

Enchytraeids were obtained from two distinct habitats: deep potholes on the lower shore with a sediment of limestone or coral rubble (Locke & Coates, in press), and algal mats in shaded cavities in the rock, mostly 10–15 cm across, in the middle and upper part of the intertidal zone. The algae, *Bostrychia* and similar types, mixed with green filamentous algae at the higher levels, formed cushion-like mats with a close structure which trapped some sediment and retained moisture. Sediment in shallow pools, which were heated by the sun during low tide, yielded few specimens. No enchytraeids were obtained from algal mats or mussel patches on the lower shore, which generally remained wet.

Ten species were recorded from algal mats, nine of which were *Marionina* (Table 1). The potholes were characterised by *Grania* with occasional specimens of *Marionina*.

Sandy beaches (excluding tidal debris and the sand beneath, 3 localities)

Few enchytraeids were obtained from beach sands, although most previous records for Bermuda were from beaches. Specimens examined during this survey were from near high water mark where particulate organic matter was plentiful, or from the surface of low rocks where sand was trapped in algal turf forming a stable crust.

Low densities of enchytraeids yielded only seven species belonging to three genera (Table 1).

Mangroves (5 localities)

The substrate under mangrove trees was mud, muddy sand or 'soil' (largely invertebrate frass) near the back of swamps. All had a high organic content and were generally devoid of worms. Enchytraeids were obtained chiefly from surface algal mats and from the algae growing on the pneumatophores of *Avicennia germinans* (black mangrove). None were found among algae on the prop roots of *Rhizophora mangle* (red mangrove) which form the lowest zone and are immersed for much longer.

Seven species recorded, with *Marionina* dominating (Table 1).

Seawater outflows (2 localities)

Samples were taken from two sites where seawater from aquaria was discharged continuously onto the shore, creating habitats with constant wetting and high

Table 1. Distribution of enchytraeid species in the main intertidal habitats. Asterisks denote species recorded by Lasserre & Erséus (1976) denotes species not found in these collections. Bold type indicates species confined to the habitat

Tidal debris	Mangrove	Seawater ouflows	Rocks	Beach sands
***Enchytraeus* sp. 1**	*Enchytraeus minutus*	*Enchytraeus minutus*	*Algal mats*	*Enchytraeus* sp. (not id.)
***Enchytraeus* sp. 2**	*Mar. appendiculata*	*Mar. appendiculata*	*Stephens. sterreri*	*Stephens. sterreri**
Stephensoniella marina	*M. spicula*	*M. nevisensis*	*Mar. appendiculata*	***Marionina achaeta****
S. trevori	***M. waltersi***	***Marionina* sp. 3**	*M. spicula*	*M. appendiculata*
S. sterreri	*Marionina* sp. 1	*Marionina* sp. 4	it Marionina sp. 2	*M. spicula**
***Achaeta* sp.**	*Marionina* sp. 2	*Marionina* sp.10	*Marionina* sp. 4	***M. subterranea****
Marionina coatesae	*Marionina* sp. 4		*Marionina* sp. 5	*M. swedmarki**
M. nevisensis			***Marionina* sp. 6**	
M. spicula			*Marionina* sp. 7	
M. swedmarki			*Marionina* sp. 8	
Marionina sp. 1			***Marionina* sp. 12**	
Marionina sp. 4				
Marionina sp. 5			*Potholes*	
Marionina sp. 7			***Grania americana***	
Marionina sp. 8			***Grania laxarta***	
***Marionina* sp.9**			*Marionina* spp	
Marionina sp. 10				
***Marionina* sp. 11**				
18 species	7 species	6 species	12 species	7 species

oxygen content. The effluent from the Bermuda Biological Station emptied into a narrow mangrove zone where it was shaded from the sun, and enchytraeids were plentiful among algae growing on stones, and in wet leaf litter and soil. Water from the Bermuda Aquarium discharged onto rocks in the shade of a bridge and worms were obtained from small patches of mussels and algal turf, habitats which did not harbour worms on other parts of the coast. The continuous flow of water was apparently necessary for the occurrence of the worms.

All but one of the six recorded species (five *Marionina*) were taken from other intertidal habitats and are believed to be permanent residents at these sites and not just contaminant species of the outfall water.

Subtidal sediments

Sampling at 8–10 m (Locke & Coates, in press) yielded relatively few enchytraeids, all species of *Grania*. Other oligochaetes, e.g. Tubificidae, were only abundant and diverse in potholes on the rocky shore, and a few other wet intertidal habitats or in subtidal samples. A large number of tubificid species are known from shallow subtidal habitats of Bermuda (Erséus, 1990).

Species distribution

Among 27 species recognized, eight had already been recorded from Bermuda: *S. marina, S. sterreri, S. trevori, M. spicula, M. swedmarki, E. 'albidus'*, *G. americana*, and *G. laxarta*; five were new records: *Marionina coatesae* (Erséus, 1990), *Marionina appendiculata* (Nielsen & Christensen, 1959), *Marionina nevisensis* (Righi & Kanner, 1979), *Marionina waltersi* (Healy, 1994), and *Enchytraeus minutus* (Nielsen & Christensen, 1961). Fifteen (12 *Marionina*, one *Achaeta* and two *Enchytraeus*) are new species which will be described; they include the species recorded as *E. albidus* by Lasserre & Erséus (1976), but noted by them to be different from the typical form (*Enchytraeus* sp. 1 in Table 1). Four previously recorded species: *M. achaeta, M. subterranea, G. bermudensis* and *G. hylae*, the last two being primarily subtidal, were not found during this survey.

The genus *Marionina* predominated in intertidal and supralittoral habitats, with *Stephensoniella* and *Enchytraeus* species only found at six and eight, respectively, of the 14 localities sampled. A species of *Achaeta* was present in supralittoral leaf debris at one locality. *Grania* spp were restricted to potholes on the lower shore and sublittoral sediments.

Tidal debris and the sand beneath contained the greatest number of species (Table 1), and also the greatest abundance and size range of adult specimens; other habitats had either predominantly large or small species but not a mix of these. Eight species were confined to this habitat. Two species were always associated with living algae, either on rocks or on the surface of sediments. *Marionina waltersi*, which on Bermuda shores was found in mangroves, was previously known only from *Spartina* salt-marshes (Healy, 1994).

Some of the enchytraeid species recorded from beaches by Lasserre & Erséus (1976) were found in sand around high water mark with buried sargassum and seagrasses, but in most cases, samples were taken from depths below 10 cm. As it appears likely those sample conditions resembled beach sands rather than tidal debris, as distinguished in the present survey, these species are listed under 'sandy beaches' in Table 1.

Discussion

The new records increase the number of marine enchytraeids reported for Bermuda from 12 to 31. Only four of the previously recorded species were not found during this survey and two of those are primarily subtidal species, whereas we sampled in the intertidal. This enumeration takes into account that the species originally recorded from Bermuda as *Enchytraeus albidus* was recognized, even in those records, as somehow different from the true *E. albidus* and we call it here *Enchytraeus* sp. 1. The new species and records mostly concern small worms, some not exceeding 2 mm. Most of these are species of *Marionina* which appears to be the dominant genus on tropical seashores (see also Healy & Coates, 1997). There are two likely explanations for the number of new finds: the sampling of habitats not previously investigated with emphasis on the littoral fringe, and the use of the wet funnel extractor which is particularly useful for obtaining very small worms from organic substrates. Previous workers have relied on methods most suitable for loose sediments, i.e. washing and decanting through sieves in the field or elutriation of small samples in the laboratory.

Although the tidal zone is narrow (mean tidal range <1 m, i.e. microtidal) and only a fraction of the area between low water and the upper reach of the waves was shown to support enchytraeids, species richness is comparable with that recorded during faunistic surveys in mesotidal (range 2–4 m) regions experiencing temperate climates. Tynen (1972) recorded 27 species on 30 beaches on the Welsh island of Anglesey but did not sample rocks or the sublittoral. In British Columbia, 27 species were recorded from intertidal and shallow sublittoral habitats (Coates & Erséus, 1980; Coates & Ellis, 1981; Coates, 1983b), and a survey of 123 localities on European North Sea and Baltic coasts (meso-micro-non-tidal), including the sublittoral, also yielded 27 species (Kossmagk-Stephan, 1985). Only 24 species were recorded from intertidal beaches, salt-marshes and rocks in south-east Ireland (Healy, 1979, 1996). In the only equivalent survey undertaken in the tropics, only 12 species were found by the authors in a similar range of habitats to those sampled in Bermuda in Darwin Harbour, a macrotidal region (Coates & Stacey, 1997; Healy & Coates, 1997), thus the findings for Bermuda may not be applicable to all areas experiencing hot climates.

The distribution of enchytraeids on the Bermuda shores illustrates their dependence on moisture and a good supply of oxygen. Obviously, an adequate source of organic matter in an appropriate state for ingestion is another major requirement and may account for the low numbers in intertidal sands. However, wet substrates rich in organic matter quickly become deoxygenated at high temperatures and those exposed to the sun were almost devoid of worms. Wet substrates only supported enchytraeids where oxygen was continuously replenished by running or agitated water, e.g. at the aquarium outfalls.

On beaches, enchytraeids were diverse and often abundant in and under somewhat dry debris around high water mark, but were scarce or absent in the wet deposits of the mid- and lower-shore except in surface sand trapped by algae. A rich food supply, stable habitats and an aerated sediment were apparently preferred over the wet sand of the middle and lower tidal zones where the subsurface on low energy beaches tends to be anoxic (Giere, 1973; Wieser et al., 1974), or where in higher energy beaches the organic content is low and the surface wave-washed and unstable. However, the number of species obtained from the high tide zone was lower than recorded by Lasserre & Erséus (1976). Observations of other workers indicate that more attention should have been paid to the subsurface sediments, (down to 10 cm or more) around high water mark and lower down near the groundwater level (Wieser et al., 1974; Pfannkuche, 1980). In mangroves, enchytraeids were frequent in leaf debris, in algae on the surface of the substratum and on pneumatophores in the upper mangrove zone which is inundated for short periods only, but scarce in the substrate itself and on prop roots which are immersed at each high tide. This distribution again suggests that littoral enchytraeids prefer moist air to free water and that oxygen availability is a major limiting factor.

The relationship between oxygen conditions in sediments and oligochaete distribution has been demonstrated by studies in both temperate and subtropical regions (Giere, 1973, 1977; Giere & Pfannkuche, 1978). For example, Giere (1973) demonstrated a marked vertical decrease in population density which followed the oxygen diffusion rate and redox potential curves in a Baltic beach. It has been shown that the oligochaete species of Bermuda tolerate higher temperature, salinities and pH levels than they do at higher latitudes, and that their upper limit of resistance is considerably higher as long as the oxygen supply is good (Giere, 1980). On temperate shores, the larger species inhabiting wrack beds where conditions are ever changing were reported to possess a particularly wide range of tolerance, while smaller, meiobenthic species are less tolerant (Giere, 1980). On the shores of Bermuda, the larger species (*S. marina* and *Enchytraeus* sp. 1) were also found in the present study in tidal debris, while other environments contained small species only.

One Bermuda beach (Tuckerstown Cove, not included in this study) has been studied by other workers in some detail (Wieser et al., 1974; Giere, 1977; Gnaiger et al., 1978). It is sheltered, with medium, well-sorted calcareous sand. An extensive flat zone remains saturated with water during emersion, but the surface layer of the upper sloping zone around mean high water mark is aerated for long periods. Conditions in summer were found to be similar throughout the mid-littoral with salinity 35–40‰, temperature 30–35 °C in summer, Eh about +250–300 mV and pH 9.0 or over; the slope differed chiefly in having a lower pH (Giere, 1977). Further investigation of pH variation showed that daytime values could reach 9.6 at the surface but remained below 8.2 at 4 cm depth (Gnaiger et al., 1978). Wieser et al., (1974) have suggested that high pH may be a limiting factor confining meiofaunal species to deeper layers in calcareous sand, even when conditions there are oligoxic or anoxic and high Eh and oxygen potentials prevail at the surface. Distribution of the tubificid *Aktedrilus monospermathecus* was basically defined by oxygen conditions. The worms were concentrated in subsurface layers around MHW where the sand was well-oxygenated and individuals could avoid extremes of salinity, temperature and pH. In the wetter, flat zone of the beach, they were confined to surface layers by the anoxic and toxic conditions of the subsurface, and densities were lower (Giere, 1977). Enchytraeids would be even less tolerant of the subsurface conditions of the wet zone, whereas more abundant food in the form of algal and land-plant remains in various stages of decomposition would make the high tide level particularly favourable for them.

The widespread occurrence of enchytraeids in algal mats on rocks in the upper tidal zone indicates that conditions in this habitat on the upper shore meet the requirements of many species and may be relatively stable. The close-structure of the cushions probably maintains a high moisture level as well as providing food and protection from wave action and predators (see also Healy, 1996). The ability to maintain populations in these small, isolated patches would require special behavioural adaptations. The absence of enchytraeids from the wetter algal mats and mussel beds of lower zones may be due to oxygen depletion and high temperatures within the substrate during emersion rather than to competition or a greater abundance of predators. Numbers of worms in algal turf were not lower at low tide levels on rocky shores in south-east Ireland where the number of oligochaete species actually increased from high to low water (Healy, 1996).

It has been pointed out (Giere, 1980) that in temperate climates most short-term tolerance thresholds of oligochaetes exceed the range of physical and chemical factors likely to be encountered in intertidal habitats so that other factors such as food and species interactions probably limit species distributions. In the subtropics, in contrast, there is a much closer correspondence between tolerance limits (upper limits) and abiotic factors and the latter may thus operate as stress factors directly limiting survival. Enchytraeid species are thus limited to microhabitats which are moist, but not wet, experiencing relatively narrow ranges of variation in temperature, oxygen supply, pH and other chemical and physical factors. Such conditions exist there in smaller areas than on temperate shores. They are most likely to exist in the transitional zone between terrestrial and intertidal habitats which is where enchytraeids were most diverse and abundant. The subtidal appears to be dominated in numbers by tubificid oligochaetes but is poorly studied for enchytraeids and species richness may be underestimated.

Acknowledgements

This research was supported by a grant-in aid from the Bermuda Biological Station for Research to B. Healy. This is contribution No. 1499 of the Bermuda Biological Station for Research, Inc., and contribution No. 22 of the Bermuda Biodiversity Project (BBP), Bermuda Aquarium, Natural History Museum and Zoo. Funding for videomicroscope facilities at BBSR were provided by NSF instrumentation grant DBI-9724335.

References

Coates, K. A., 1983a. A contribution to the taxonomy of the Enchytraeidae (Oligochaeta). Review of *Stephensoniella*, with new species records. Proc. biol. Soc. Wash. 96: 411–419.

Coates, K. A., 1983b. New records of marine *Marionina* (Oligochaeta, Enchytraeidae) from the pacific Northeast, with a description of *Marionina klaskisharum* sp. nov. Can. J. Zool. 61: 822–831.

Coates, K. A. & D. V. Ellis, 1981. Taxonomy and distribution of marine Enchytraeidae (Oligochaeta) in British Columbia. Can. J. Zool. 59: 2129–2150.

Coates, K.A & C. Erséus, 1980. Two species of *Grania* from the Pacific coast of North America. Can. J. Zool. 58: 1037–1041.

Coates, K. A. & D. F. Stacey, 1997. Enchytraeids (Oligochaeta: Annelida) of the lower shore and shallow subtidal of Darwin Harbour, Northern Territory, Australia. In J. R. Hanley, G. Caswell, D. Megirian & H. K. Larson (eds), Proceedings of the Sixth International Marine Biological Workshop. The marine flora and fauna of Darwin Harbour, Northern Territory, Australia. Museums and Art Galleries of the Northern Territory and the Australian Marine Sciences Association: Darwin, Australia 1997: 67–79.

Erséus, C., 1990. The marine Tubificidae of the barrier reef system of Carrie Bow Cay, Belize, and other areas of the Caribbean Sea, with descriptions of 27 new species and revisions of *Heterodrilus*, *Thalassodrilides* and *Smithsonidrilus*. Zool. Scr. 19: 243–303.

Giere, O., 1973. Oxygen in the marine hygropsammal and the vertical microdistribution of oligochaetes. Mar. Biol. 21: 180–189.

Giere, O., 1977. An ecophysiological approach to the microdistribution of meiobenthic Oligochaeta. I. *Phallodrilus monospermathecus* (Knöllner) (Tubificidae) from a subtropical beach at Bermuda. In B. F. Keegan, P. Ó'Ceidigh & P. J. S. Boaden (eds), Biology of Benthic Organisms. Pergamon Press and New York: 285–296.

Giere, O., 1980. Tolerance and preference reactions of marine Oligochaeta in relation to their distribution. In R. O. Brinkhurst & D. G. Cook (eds), Aquatic Oligochaete Biology. Plenum Publishing Corporation: 385–409.

Giere, O. & O. Pfannkuche, 1978. An ecophysiological approach to the microdistribution of meiobenthic Oligochaeta. II. *Phallodrilus monospermathecus* (Tubificidae) from boreal brackish-water shores in comparison to populations from subtropical beaches. Kieler Meeresforsch 4: 289–301.

Giere, O. & O. Pfannkuche, 1982. Biology and ecology of marine Oligochaeta, a review. Oceanogr. mar. Biol. ann. Rev. 20: 173–308.

Gnaiger, E., G. Gluth & W. Wieser, 1978. pH fluctuations in an intertidal beach in Bermuda. Limnol. Oceanogr. 23: 851–857.

Healy, B., 1979. Marine fauna of County Wexford. 1-Littoral and brackishwater Oligochaeta. Ir. Nat. J. 19: 418–422.

Healy, B., 1994. New species of *Marionina* (Annelida:Oligochaeta: Enchytraeidae) from *Spartina* salt marshes on Sapelo Island, Georgia, U.S.A. Proc. biol. Soc. Wash. 107: 164–173.

Healy, B., 1996. The distribution of Oligochaeta on an exposed rocky shore in south east Ireland. Hydrobiologia 334: 51–62.

Healy, B. & K. A. Coates, 1997. Enchytraeids (Oligochaeta: Annelida) of the mid and upper intertidal of Darwin Harbour, Northern Territory, Australia. In J. R. Hanley, G. Caswell, D. Megirian & H. K. Larson (eds), Proceedings of the Sixth International Marine Biological Workshop. The marine flora and fauna of Darwin Harbour, Northern Territory, Australia. Museums and Art Galleries of the Northern Territory and the Australian Marine Sciences Association: Darwin, Australia 1997: 81–97.

Healy, B. & E. Rota, 1992. Methods for collecting Enchytraeidae during expeditions. Soil Biol. Biochem. 24: 1279–1281.

Kossmagk-Stephan, K-L., 1985. Systematik, Faunistik und Lebenzyklus mariner Oligochaeta der Nord und Ostseeküste. Dissertation, Georg-August Universität zu Göttingen.

Lasserre, P. & C. Erséus, 1976. Oligochètes marins des Bermudes. Nouvelles espèces et remarques sur la distribution géographique de quelques Tubificidae et Enchytraeidae. Cah. Biol. mar. 17: 447–462.

Locke, J. M. & K. A. Coates, (in press) Redescriptions of *Grania americana, G. bermudensis* and descriptions of two new species of *Grania* (Annelida: Clitellata: Enchytraeidae) from Bermuda. Proc. biol. Soc. Wash.

Moore, J. P., 1902. Some Bermuda Oligochaeta, with description of a new species. Proc. natn. Acad. Sci. Philadelphia 50: 80–84.

Morris, B., J. Barnes, F. Brown & J. Markham, 1977. The Bermuda marine environment. A report of the Bermuda inshore waters investigation 1976–77. Bermuda Biological Station Special Publication No. 15, 120 pp.

O'Connor, F. B., 1955. Extraction of enchytraeid worms from coniferous forest soil. Nature 175: 815–816.

Pfannkuche, O., 1980. Distribution, abundance and life cycles of Oligochaeta from the marine mesopsammal with special reference to the Phallodrilinae (Tubificidae). Int. Rev. ges. Hydrobiol. 65: 835–848.

Thomas, M. L. H. & A. Logan, 1991. A guide to the ecology of shoreline and shallow-water marine communities of Bermuda. Bermuda Biological Station Special Publication No. 30. 345 pp.

Tynen, M. J., 1972. The littoral Enchytraeidae (Oligochaeta) of Anglesey and the Menai Strait with notes on habitats. J. nat. Hist. 6: 21–19.

Verrill, A. E., 1900. Notes on the geology of the Bermudas. Amer. J. Sci. Ser. 4, 158: 313–340.

Wieser, W., J. Ott, F. Scheimer & E. Gnaiger, 1974. An ecophysiological study of some meiofauna species inhabiting a sandy beach at Bermuda. Mar. Biol. 26: 235–248.

Hydrobiologia **406**: 119–132, 1999.
B. M. Healy, T. B. Reynoldson & K. A. Coates (eds), Aquatic Oligochaetes.

Spatial distribution of oligochaetes (Clitellata) in the tidal freshwater and brackish parts of the Schelde estuary (Belgium)

J. Seys[1,2], M. Vincx[1] & P. Meire[2,3]
[1]*University of Gent, Department of Morphology, Systematics and Ecology, Marine Biology Section, K.L. Ledeganckstraat 35, B9000 Gent, Belgium*
[2]*Ministry of the Flemish Community, Institute of Nature Conservation, Kliniekstraat 25, B1070 Brussel, Belgium*
[3]*University of Antwerp, Department of Biology, Universiteitsplein 1, B2610 Wilrijk, Belgium.*

Key words: macrozoobenthos, oligochaetes, tidal fresh- and brackish water, spatial patterns

Abstract

The benthic fauna of the Belgian part of the polluted Schelde estuary, called the Zeeschelde, was studied in September–October 1996 at 143 stations along 35 transects. This study is part of the OMES research program, funded by the Flemish Government, to build an ecosystem model of the Zeeschelde in order to help policy makers to decide upon the future of this unique estuarine system. Of all organisms retained on a 250 μm mesh-size, oligochaetes were co-dominant in the mesohaline part and the only dominant group in the tidal freshwater zone. The oligochaete fauna of the ß-mesohaline zone consisted of two brackish water tubificids, *Heterochaeta costata* and *Tubificoides heterochaetus*, and the euryhaline naidid *Paranais litoralis*. The freshwater species *Limnodrilus hoffmeisteri* penetrated as far as Boerenschans. Abundance in this zone is rather low with numbers increasing with exposure. The part of the oligohaline zone that coincides with the maximum turbidity zone is extremely poor in benthos, due to high physical, chemical and biological stress, with very low numbers of *L. hoffmeisteri*, *Tubifex tubifex* and *P. litoralis*. In the tidal freshwater section, mass populations of the tubificids *L. hoffmeisteri* and *T. tubifex* occur (peak densities of almost 3.10^6 ind m^{-2}, maximum biomass: 25.7 g ADW m^{-2}) with lower numbers of *Limnodrilus claparedeianus*, *Limnodrilus udekemianus* and *Limnodrilus profundicola* mainly in the part of the Zeeschelde close to Gent where better oxygen conditions are found. Oligochaetes are more abundant in finer sediments, resulting in a clear vertical gradient with low mean values in the deeper subtidal coarse sediments and highest numbers in the upper intertidal where fluid muds consolidate. In comparison with *L. hoffmeisteri*, *T. tubifex* was more abundant in the finest sediments. Vorticellidae are more common on tails of oligochaetes in the oxygen-enriched part (4–9%) than in the oxygen-deficient zone ($< 0.5\%$). Regeneration rates, being considered as a suitable measure of predation pressure, increase with salinity and size of the worms and are highly specific. In comparison with other temperate estuarine systems, the Zeeschelde has impoverished communities with oligochaetes as the dominant taxa. Though a future improvement of the water quality will definitely result in a less productive but more varied benthic community in the freshwater tidal zone, oligochaetes will probably continue to play an important role due to sustained harsh conditions.

Introduction

In recent years, the Schelde estuary has become one of the best studied estuarine ecosystems in the world (several papers in: Meire & Vincx, 1993; Heip & Herman, 1995). However, most research was done in the Dutch part of the estuary, the Westerschelde, which stretches out over about 55 km from the α-mesohaline zone to the mouth of the estuary. The Belgian part which extends about 105 km, the Zeeschelde, received little scientific attention, and moreover, most studies were limited to the mesohaline reach near the Belgian-Dutch border. In the early 1990s, renewed scientific interest in the Zeeschelde coincided with a slight improvement in the water quality (Meire et al., 1992; Van Damme et al., 1995). The OMES (Onderzoek Milieu Effecten Sigmaplan = Study Environmental Impact Sigma plan) research program, initiated by the Flemish Government, should result in an ecosystem model that will help policy makers to decide upon the future of this unique estuarine system (Meire et al.,1997). Within this research program, data on the distribution and the role of the benthic component are being collected.

From earlier studies on distribution patterns of macrobenthic invertebrates in the Zeeschelde, it appeared that a typical mesohaline estuarine community occurred which is dominated by polychaetes (e.g. *Nereis diversicolor* O. F. Müller, *Heteromastus filiformis* Claparède), bivalves (e.g. *Macoma balthica* L), crustaceans (e.g. *Corophium volutator* Pallas) and oligochaetes. This community is similar to those of other NW-European estuaries (Ysebaert et al., 1993). From the freshwater tidal part of the Zeeschelde, only scattered data are available. They indicate the occurrence of mass populations of oligochaetes. In general, little is known about the ecology and the response to disturbance of communities in tidal freshwater systems around the world (Odum et al., 1984).

On the other hand, the role of oligochaetes within food chains and their impact on the physics and biochemistry of the sediment-water interface are well-studied and known to be considerable. Their mass populations result in substantial reworking of the sediments they inhabit. Appleby & Brinkhurst (1970) showed that oligochaetes in Toronto harbour reworked the top 4–6 cm, 12 times a year, displacing quantities of mud eight times their own body weight within 24 h. The oligochaetes control the geochemical cycle within the top 4 cm and have a profound effect on Eh-discontinuities (Diaz, 1980). In a detailed study on the effects of tubificids on physical and chemical properties of Lake Erie sediments, McCall & Fisher (1980) demonstrated the huge impact tubificids have on grain size distribution, erosion, water content, diffusion, permeability and oxygen demand of the sediments, mainly by creating a pelletized layer of sediment at the sediment–water interface. Chatarpaul et al. (1978) describe the impact of tubificids on the denitrification process and several studies indicate the role they can play in depleting pollutants from sediments by their bioturbating activity (Giere & Pfannkuche, 1982).

This study aims to provide data on the distribution and role of oligochaetes in the brackish and freshwater tidal part of the Zeeschelde and its tidal tributaries. The spatial distribution of oligochaete assemblages over gradients in salinity, depth, oxygen saturation and sediment composition is discussed.

Study area

The Zeeschelde is part of the well-mixed Schelde estuary, and is characterised by strong currents, a large tidal amplitude and an extensive freshwater tidal zone (Table 1). This zone extends from Gent – where the tidal influence is levelled by a weir – to the mouth of the Rupel tributary (Figure 1). The main freshwater input comes from the upper Schelde (25%), Antwerp harbour and backwaters (25%), and the Rupel-Zenne tributaries (50%) (Lefèvre & Stronkhorst, 1988). The latter causes the major input of organic pollution, including all the untreated sewage water of the city of Brussels. Bacterial production rates in the Zeeschelde are among the highest values reported (Goosen et al., 1995).

Oxygen conditions are poor throughout the system with a significant dissolved oxygen 'sag curve' in the maximum turbidity zone between Antwerp and the mouth of the Rupel (Van Eck & de Rooij, 1990). Compared to the conditions in the 1970s, a slight improvement in oxygen concentrations and a shift of the oxygen deficient zone from Antwerpen upstream towards the Rupel has taken place (Van Damme et al., 1995). The Schelde estuary is highly contaminated with heavy metals and organic micropollutants. The concentrations of PCBs, PAHs and cadmium are high in the freshwater and brackish part of the estuary (Van Zoest & Van Eck, 1993; Zwolsman & Van Eck, 1993).

In order to preserve the shipping lanes to the port of Antwerp, a large amount of sediment is removed from the channels and relocated (Belmans, 1988). Tidal flats consist mainly of very fine sands or silts. Subtidal areas on average have coarser sediments and display a wider spectrum of substrate types including peat-, clay- and stony-bottoms.

Material and methods

Oligochaetes

Benthic fauna was sampled from September 2nd to October 8th 1996 at 143 sampling stations distributed over 35 transects, in 10 sections numbered 9 through 19, along the whole reach of the Zeeschelde (Figure 1). Intertidal stations were sampled with a PVC-corer (ø 3.5 cm) to a depth of 10 cm; subtidal samples were collected using Van Veen grab samplers, a Reineck box corer or an aluminium extendable corer, from which samples were taken with the PVC-corer. In

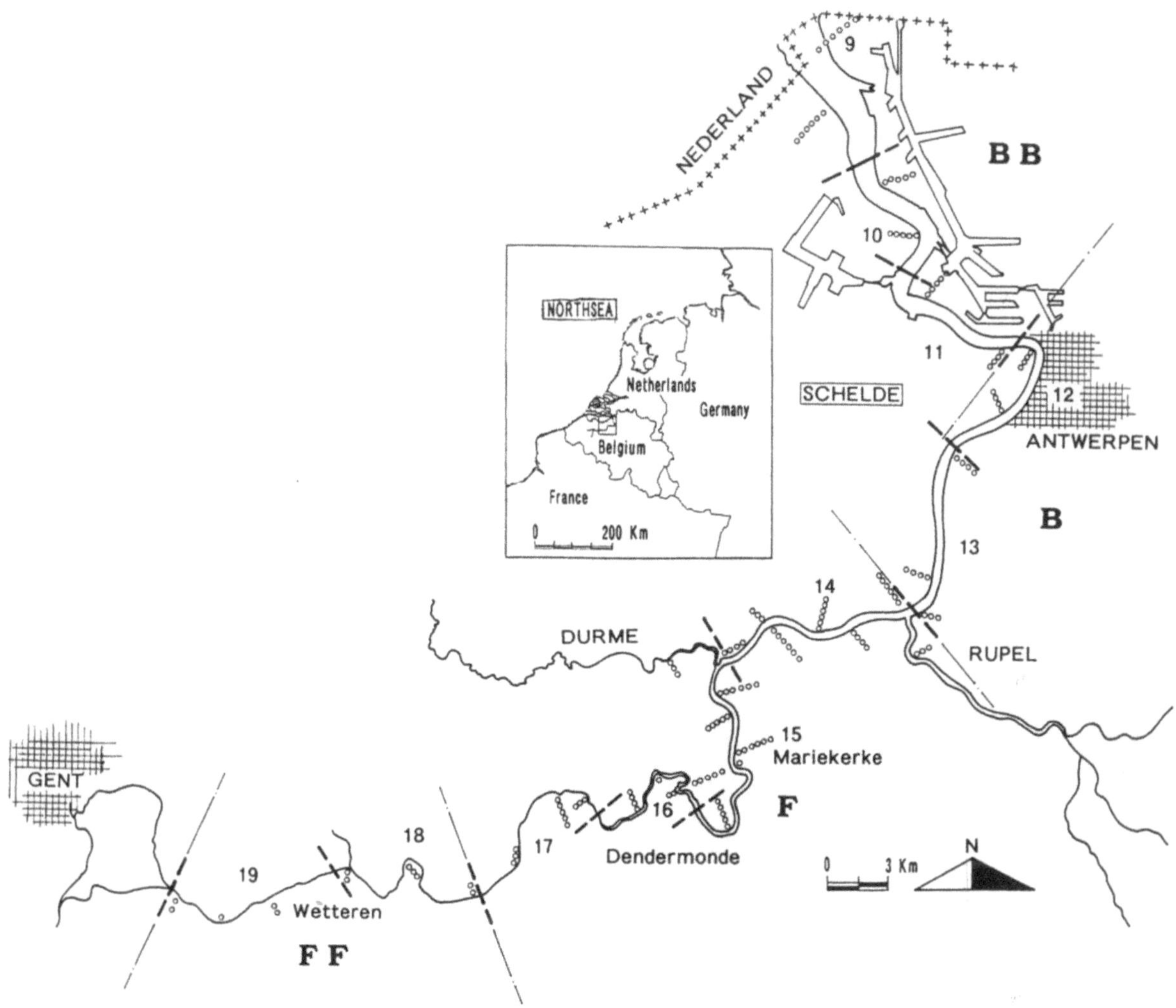

Figure 1. The Zeeschelde study area showing the ten sections (9–19), the sampling transects (oooooo) and the four main zones FF, F, B, BB.

the laboratory, the samples were preserved with a 4% formaldehyde solution after sieving over a 1mm and a 250 μm mesh-size sieve. Both fractions were sorted under a dissecting microscope and the animals were counted.

A maximum of 100 and 50 oligochaetes out of the 1 mm and the 250 μm fractions, respectively, were mounted on slides according to a standard method that was adapted for our purposes. The animals were transferred from formalin to 70% ethanol, stained with a paracarmine solution and washed with hydrochloric acid. To avoid shrinkage a gradual transfer to glycerine was carried out through a 95 parts 96% ethanol/ 5 parts glycerine medium (Seinhorst, 1959). Oligochaetes in pure glycerine were cleared by storing them overnight at 70 °C. Semi-permanent, reversible slides are obtained by sealing with paraffin rings. This method is suitable for macrobenthic as well as meiobenthic oligochaetes and provides stained and cleared material that can be stored for long periods. All oligochaetes – including immature specimens – were identified to species level, measured (width at V), and classified as one of the four age/reproductive groups:

1. juvenile= newly hatched,
2. immature,
3. adult= + penis sheath, or
4. reproductive= + eggs or spermatozeugmata in spermatheca.

They were also checked for the presence of epizoic Vorticellidae and regenerated tails. For this purpose,

Table 1. Characteristics of the Zeeschelde study area as sampled over ten sections in September/October 1996 (section 9 at the Belgian-Dutch border, section 19 near Gent). Data on salinity, O_2 saturation and BOD are derived from monthly monitoring campaigns on the Zeeschelde during the period December 1995 – November 1996 (data Institute Nature Conservation, pers.comm. Van Damme). Maxima and minima are highest and lowest measured monthly values over the one year period. For general characteristics of the Zeeschelde, references are added. (* TAW = Belgian ordnance level = MLLWS – 0.47 m)

Section	Salinity (psu)	O_2 saturation (%)	BOD (mg l^{-1})	Sampling depth (m TAW	Mean tidal range (m)	Median grain size (μm)		Silt fraction <38μm (%)	
	min–max	min–max	min–max	min–maxmin–max	min–max				
9	9.8 – 15.2	27 – 89	3 – 14	−21.0 + 4.5	4.8	19	250	2	65
10	7.2 – 14.3	7 – 82	2 – 13	−15.0 + 3.8	4.9	38	342	2	50
11	2.5 – 11.9	5 – 63	3 – 17	−12.2 + 3.2	5.0	1	289	4	73
12	0.7 – 10.3	4 – 48	4 – 24	−15.0 + 2.4	5.0	10	750	0	81
13	0.2 – 6.1	2 – 22	5 – 25	−10.2 + 2.4	5.1	9	550	0	85
14	0.2 – 2.0	1 – 38	6 – 30	−7.1 + 4.5	5.1	9	550	0	79
15	0.2 – 0.5	5 – 51	9 – 24	−10.1 + 4.3	4.9	10	800	0	81
16	≤ 0.2	3 – 54	13 – 32	−6.8 + 2.3	4.3	18	800	0	71
17	≤ 0.2	2 – 55	8 – 37	−4.5 + 4.7	3.4	13	274	3	67
18	≤ 0.2	5 – 54	11 – 36	−3.8 + 2.5	2.7	18	487	4	70
19	≤ 0.2	6 – 54	9 – 35	−2.6 – 0.8	2.3	1	260	2	80

Zeeschelde	length (km):		105	Meire et al., 1995
	width (m):		50 – 1350	–
	tidal flats (ha):		655	–
	intertidal marshes (ha):		507	–
	mesohaline &oligohaline		180	–
	freshwater tidal		322	–
	river water discharge (m^3 s^{-1})	mean	100	Lefèvre &Stronkhorst, 1988
		winter	180 (avg)	–
			500 – 600 (max)	–
		summer	50 (avg)	–
			<10 (min)	–
	maximal depth (m TAW*)		−5 à – 25	own observations

only 'complete' specimens were considered, excluding all oligochaetes with recently broken tails. The proportion of the total population identified as 3. adult plus 4. reproductive, was used as a measure of the reproductive state of the population.

Biomass was calculated based on a regression between the width of V and the ash-free dry weight of specimens of *Tubifex tubifex* and *Limnodrilus* sp. For this purpose, about 150 specimens of *Limnodrilus* sp. and 20 specimens of *T. tubifex* were measured. Eight width classes were distinguished and separately weighed on a microbalance. The ash-free dry weight (ADW) biomass was obtained by drying at 60 °C to constant weight and ashing at 550°C. Since we found no significant difference between *T. tubifex* and *Limnodrilus* sp. in the relationship width-ADW, data of both species were lumped.

The regression equation is:

$$\text{ADW}=\text{antiln}((10.84256*\text{width})-7.62636) \quad (R^2= 0.96;\ N=10;\ p < 0.01). \qquad (1)$$

Species diversity was expressed as the total number of species per sample, i.e. Hill number N_0 (Hill, 1973).

Environmental variables

Data on salinity, O_2 -saturation and BOD per section (see Figure 1) were obtained from monthly monitoring within the framework of the OMES project (data Institute for Nature Conservation, S. Van Damme, pers. com.).

Tidal levels of the sampling stations were either calculated by comparing data on tidal heights with reference points and/or depth measurements on board the research vessels, or directly measured with a theodolite. For further analysis six depth classes were established:

Depth class			
	1:	subtidally	< -6 m TAW
	2:	subtidally	−6 to −3 m TAW
	3:	subtidally	> −3 m to 0 m TAW
	4:	intertidally	on average < 33% of time exposed
	5:	intertidally	33–67% exposed
	6:	intertidally	> 67% exposed

At each sampling station, one additional sediment sample was taken to obtain data on median grain size, silt% (< 38 μm) and total organic content. The top 2 cm of the sediment was used for standard granulometry measurements. Most of the samples were processed with laser-diffraction devices, 40 samples (mostly deep subtidal stations) were analyzed with a wet-sieving technique, supplemented with an X-ray diffraction method on the fine sediment fraction. Total organic content was determined as the ADW (ash free dry weight) of a fixed amount of sediment, expressed as the weight percentage of the ADW to the DW (dry weight). Each station was fitted into one of eight sediment classes partly based on the Wentworth scale:

Sediment class			
	1	Clay	< 4 μm
	2:	Silt	4–62 μm
	3:	Very fine sand	62–125 μm
	4:	Fine sand	125–250 μm
	5:	Medium sand	250–500 μm
	6:	Coarse sand	500–1000 μm
	7:	Peat	
	8:	Stones and /or shells	

Data processing

The study area was divided into four zones based on salinity and oxygen conditions:

Zone			
	FF:	freshwater sections 19 & 18	improved oxygen conditions
	F:	freshwater sections 17 to 14	low oxygen levels
	B:	oligohaline sections 13 & 12	very low oxygen levels
	BB:	mesohaline sections 9 to 11	improved oxygen conditions

Density and biomass data for oligochaetes are expressed as units.m^{-2} and include both the 250 μm and 1 mm fraction. All statistical analyses were done with the STATISTICA-package (StatSoft, 1995). A multiple regression on the density data was carried out after carefully selecting a set of environmental variables that were not closely interrelated (R^2_{adj} <0.250).

Results

Benthos

Oligochaetes constitute more than 95% of the total density in the freshwater part of the Zeeschelde. Other taxa are rare (Nematoda: up to 50 000 ind m^{-2}), very rare (only single specimens of larval Chironomidae & Ceratopogonidae) and/or basically not benthic (Copepoda, Acari) (Table 2). The same holds for the oligohaline zone where low numbers of oligochaetes are accompanied by single specimens of the polychaete *Boccardia redeki* Horst and a few copepods.

The β-mesohaline zone is much more diverse in terms of species composition with several species of polychaetes (*B. redeki*, *Heteromastus filiformis*, *Polydora ligni* Webster, *Pygospio elegans* Claparède, *Nereis diversicolor*, *N. succinea* Leuckart, *Manayunkia aestuarina* Bourne and *Eteone longa* Fabricius), amphipods (*Corophium volutator*, *Bathyporeia* sp.), molluscs (*Macoma balthica*, *Mysella bidentata* Montagu, *Littorina littorea* L. and *Hydrobia ulvae* Pennant) and unidentified ostracods, plathelminths and barnacles. In the β-mesohaline subtidal parts, mean densities are much lower (16 000 ind m^{-2}) than in the intertidal zone (85 000 ind m^{-2}) and dominated by *B. redeki* (12 000 ind m^{-2}). The most abundant taxa intertidally are oligochaetes (52 000 ind m^{-2}), nematodes (14 000 ind m^{-2}), the polychaetes *H. filiformis* (6500 ind m^{-2}) and *M. aestuarina* (2770 ind m^{-2}), and the amphipod *C. volutator* (3850 ind m^{-2}). Benthic invertebrates were found in 80% of all the samples (143) with more samples without benthos in the brackish zone (45%) compared to the freshwater area (only 8%).

Oligochaetes

Two distinct oligochaete assemblages exist in the Zeeschelde with a β-mesohaline association of *Tubificoides heterochaetus* Michaelsen/*Heterochaeta costata* Claparède and a freshwater group of the four *Limnodrilus* species (*L. hoffmeisteri* Claparède, *L. claparedeianus* Ratzel, *L. udekemianus* Claparède and *L. profundicola* Verrill) and *Tubifex tubifex* Müller (Tables 2, 3). The naidid *Paranais litoralis* Müller is found in the freshwater section as well as in the brackish part of the Zeeschelde, with peak numbers

Table 2. Total density of benthic invertebrates, the relative abundance of different taxa and the composition of the oligochaete fauna in four sections and six depth zones in the Zeeschelde during a sampling survey in September/October 1996. For explanation of sections FF/F/B/BB and depth zones 1–6 see 'Materials and methods'

Zone	FF			F						B					BB					
depth class	2	3	4	1	2	3	4	5	6	1	2	3	4	5	1	2	3	4	5	6
BENTHOS	1	216	27	81	12	162	224	526	669	0	0	12	7	2	17	1	22	35	53	143
density (N.10 cm^{-2})																				
Gastropoda (%)	0	0	0	0	0	0	0	0	0			0	0	0	0	0	0	5.7	0	0.2
Bivalvia (%)	0	0	0	0	0	0	0	0	0			0	0	0	0	0	0	1.9	0.5	0.7
Amphipoda (%)	0	0	0	0	0	0	0	0	0			0	0	0	0	25.0	0	0	0	5.8
Acari (%)	0	0.1	0	0	0	0	0	0	0.1			0	0	0	0	0	0	0	0	0
Copepoda (%)	100	0.7	0	0.2	3.8	0.2	2.2	0.4	0.7			0	28.6	0	0	0	0	0.9	1.0	3.8
Nematoda (%)	0	0.3	7.4	0.5	1.6	0.3	0.2	0.2	1.9			0	0	0	2.9	25.0	0	5.7	0	21.1
Diptera larvae (%)	0	0	0	0	0	0	0	0	0.1			0	0	0	0	0	0	0	0	0
Polychaeta (%)	0	0	0	0	0	0	0	0	0			0	0	11.1	57.3	0	94.8	52.8	5.2	8.4
Oligochaeta (%)	0	98.9	92.6	99.4	94.6	99.5	97.6	99.4	97.1			100	71.4	88.9	23.3	25.0	5 .2	33.0	93.3	59.8
rest (%)	0	0	0	0	0	0	0	0	0.2			0	0	0	16.5	25.0	0	0	0	0.1
OLIGOCHAETA																				
density (N. 10 cm^{-2})	0	213	25	81	11	161	219	522	650	0	0	12	5	2	4	0	1	12	49	86
biomass (gADW m^{-2})	0	3.18	0.16	2.51	0.23	1.82	2.08	4.85	7.08	0	0	0.33	0.04	0.04	0.02	0.00	0.05	0.06	0.37	1.11
density species																				
T. heterochaetus	0	0	0	0	0	0	0	0	0	0	0	0	0	0	4	0	0	10	37	16
H. costata	0	0	0	0	0	0	0	0	0	0	0	0	0	0	0	0	0	1	4	64
P. litoralis	0	0	0	0	0	0	0	0	5	0	0	0	0	1	0	0	0	0	1	6
T. tubifex	0	106	15	23	2	22	47	147	241	0	0	0	1	0	0	0	0	0	0	
L. hoffmeisteri	0	44	7	51	8	79	158	319	361	0	0	12	4	1	0	0	1	0	6	0
L. claparedeianus	0	44	1	3	1	17	10	38	26	0	0	0	0	0	0	0	0	0	0	0
L. udekemianus	0	12	0	3	0	2	2	7	8	0	0	0	0	0	0	0	0	0	0	0
L. profundicola	0	6	1	1	0	3	2	5	7	0	0	0	0	0	0	0	0	0	0	0
rest	0	2	1	0	0	0	1	6	3	0	0	0	0	0	0	0	0	0	0	0

Table 3. Species composition, mean total density and total biomass of oligochaetes per station in the different Zeeschelde sections in Sept/Oct 1996. Species: Lh=*L. hoffmeisteri*, Lc=*L. claparedeianus*; Lu=*L. udekemianus*; Lp=*L. profundicola*; Tt=*T. tubifex*; Hc=*H. costata*; Th=*T. heterochaetus*; Pl=*P. litoralis*; Dd=*D. digitata*; En=Enchytraeidae sp.; Pb=*P. bavaricus*; Sf=*S. ferox*

Section	N stations	Density Oligochaetes (ind m^{-2})	Biomass Oligochaetes (g AFDW m^{-2})	Oligochaete species present	Nr of species per station
9	12	38 833	0.46	Th Hc	1.0
10	10	9300	0.18	Th Hc	0.5
11	9	24 444	0.17	Lh Pl Th Hc	1.0
12	8	3875	0.10	Lh Pl	0.4
13	8	3750	0.09	Lh Tt	0.6
14	31	216 032	2.63	Lh Lc Lu Lp Tt Pl	2.6
15	30	243 400	2.51	Lh Lc Lu Lp Tt Pl En	3.2
16	12	271 417	2.68	Lh Lc Lu Lp Tt	2.8
17	11	103 182	1.06	Lh Lc Lu Lp Tt Dd	3.1
18	7	55 000	0.67	Lh Lc Lu Lp Tt Sf	3.0
19	5	354 800	5.45	Lh Lc Lu Lp Tt Pb	4.6

of 33 000 ind m^{-2} in section 11. Four other species were found in the freshwater zone in very low numbers: *Dero digitata* Müller, *Potamothrix bavaricus* Oschmann, *Spirosperma ferox* Eisen and one unidentified Enchytraeidae. The freshwater species *L. hoffmeisteri* penetrates the brackish area as far as Boerenschans (= section 11), other freshwater species do not occur far downstream of the mouth of the Rupel. Oligochaetes were found in 85% of the freshwater stations, with most frequently observed species: *L. hoffmeisteri* (87%), *T. tubifex* (65%), *L. claparedeianus* (62%), *L. udekemianus* (45%) and *L. profundicola* (37%). Frequency of occurrence of oligochaetes in the brackish part is much lower (43%) with dominant species present in less than one third of the samples: *T. heterochaetus* (28%), *H. costata* (19%), *L. hoffmeisteri* (15%) and *P. litoralis* (6%).

Total density and biomass show a pattern of moderate values in the mesohaline zone, low values in the oligohaline oxygen-deprived part and high values all over the freshwater section. The high mean values in section 19 (Table 3) are partly due to one very rich station at a depth of −0.8 m TAW near the bridge of Melle, with a biomass of 25.7 g ADW m^{-2}. At six freshwater stations, densities exceeded 1.10^6 ind m^{-2} and at Mariekerke (section 15) mass populations amounting to almost 3.10^6 ind m^{-2} were present in very soft muds.

Throughout the salinity gradient total density increases with increasing exposure time (Table 2), i.e. toward the higher intertidal. The most exposed intertidal reaches are richest in terms of total density and densities of the dominant species. The shallow subtidal zone, however, has densities of oligochaetes comparable to the low intertidal area, with higher numbers of *T. tubifex*, *L. claparedeianus*, *L. udekemianus* and *L. profundicola*.

The relationship between total density and exposure time is caused by the close interaction between depth and sediment structure, with sediment characteristics responsible for the structuring of oligochaete communities.

Total density and total biomass are significantly correlated with sediment and silt fraction ($p < 0.05$), but not with depth as such. A multiple regression confirms the impact of salinity on density, biomass and species diversity, and reveals the relationship between mass populations of mainly *L. hoffmeisteri* and *T. tubifex* and the more exposed fine sediments (Table 4). One deep, subtidal station near the mouth of the Durme tributary, with a very fine sandy sediment at the time of sampling, contained high densities (548 000 ind m^{-2}) and biomasses (17.73 g ADW m^{-2}) of oligochaetes indicating at least the potential of deep gullies for colonisation by oligochaetes if optimal sediments are found. This also shows that oxygen limitation in the subtidal areas – where oligochaetes cannot supplement their oxygen demands by direct uptake – is not the (only) reason for lower subtidal worm densities here. The total amount of organic carbon, as measured in the top 2 cm of the sediment, was not a good predictor of oligochaete distribution.

The patterns in total density and biomass reflect the relationships between the abiotic environment and the most abundant species in the freshwater part, *L. hoffmeisteri* and *T. tubifex*. Other abundant species are less clearly determined in their distribution by sediment and/or depth properties (Table 4). The densities of the two dominant brackish water species, *H. costata* and *T. heterochaetus*, are significantly correlated with depth and not with sediment characteristics.

Reproductive specimens of *P. bavaricus* were identified from the shallow subtidal station near Melle only. Both *S. ferox* and *D. digitata* were found in very low numbers on one and two, respectively, muddy intertidal stations between Wetteren and Dendermonde.

The four *Limnodrilus* species and *T. tubifex* form an association throughout the freshwater section of the Zeeschelde, in proportions that average 59:25:10:4:2% for *L. hoffmeisteri*:*T. tubifex*:*L. claparedeianus*:*L. udekemianus*:*L. profundicola*, respectively. *Limnodrilus claparedeianus* is most abundant between Gent and Dendermonde, where it can represent up to 35% of the total density. *Limnodrilus udekemianus* and *L. profundicola* show the same trend with relatively higher proportions from Gent to Mariekerke. Downstream Mariekerke, *Limnodrilus hoffmeisteri* and *T. tubifex*, become dominant and together comprise more than 85% of the total oligochaete population.

Although *L. hoffmeisteri* and *T. tubifex* are found together in most freshwater stations and appear to prefer fine sediments, they do show differences in their habitat preference. *Tubifex tubifex* becomes relatively more abundant in sediments with high silt percentages irrespective of depth, as shown by the results of a multiple regression of the ratio *T. tubifex*/*Limnodrilus* sp. against a set of environmental variables (Table 4). Sediments with median grain sizes $\geq 125\ \mu$m on average contain not more than 10–15% *T. tubifex*; clays and silts have much higher values (38%). The very fine sands have values around 30%, except for the

Table 4. Summary results of a multiple regression on density data, total biomass, total number of species N_0 and relative abundance of adults against a selection of environmental variables; only significant standardised regression coefficients are presented; for overall coefficient of determination (R^2_{adj}): $*p < 0.05$; $**p < 0.01$; $***p < 0.001$

	Salinity max	Silt%	OC	Depth	Nr. of observations	R^2_{adj}
Biomass total	−0.190	0.235			132	0.115***
Density total	−0.190	0.212		0.224	132	0.150***
L.hoffmeisteri	−0.200			0.215	132	0.141***
L.claparedeianus	−0.180				132	0.037
L.udekemianus					132	0.038
L.profundicola	−0.200				132	0.044*
T.tubifex	−0.190	0.235		0.176	132	0.133***
T.heterochaetus	0.272				132	0.078**
H.costata	0.367			0.267	132	0.193***
P.litoralis		0.208		0.198	132	0.083**
%*T.tubifex*/Limnodrilus	−0.270	0.490			77	0.267***
Number of species	−0.460			0.416	132	0.435***
% Adults	−0.260			−0.300	65	0.180**

two depth zones near the low water line where *T. tubifex* has a lower representation in comparison to those found in coarser sediments.

The structure of the oligochaete communities as sampled in September–October 1996, depends on species composition and hence on salinity. Neither *T. heterochaetus* nor *H. costata* reproduce in late summer, resulting in an immature oligochaete population at the time of sampling. The same holds for the spring-reproducing *T. tubifex* in the freshwater part in contrast with species of the genus *Limnodrilus*. Populations of *Limnodrilus* sp. were found to constitute on average of 85% immatures, 5% newly hatched juveniles, 5% adults and 5% reproducing individuals. Cocoons of *Limnodrilus* were abundant in many freshwater stations with densities of 100 000–200 000 m^{-2} in several intertidal locations. The relative abundance of adult individuals of oligochaetes in the Zeeschelde was found to be inversely related with depth and salinity (Table 4). In the oligohaline zone, where only small numbers of *L. hoffmeisteri* live, adults become more important as a result of a high maturation rate characteristic for such impoverished habitats.

Epizoic Vorticellidae were found on tails of oligochaetes in small numbers throughout the freshwater range. Large animals (1 mm fraction) were more affected than small ones (250 μm fraction), and incidence of vorticellids tends to increase in the Zeeschelde sections near Gent up to values of 4% (250 μm) to 9% (1 mm) in the FF-zone. The oxygen-deficient freshwater section (F) had only very few worms with Vorticellidae (< 0.5%).

Regeneration rates of oligochaetes appear to increase with salinity and size (Figure 2). This is mainly due to differences in species composition: high % of regenerated animals are found in *T. tubifex* (5–10%) and *H. costata* (10%), low values in freshwater *Limnodrilus* (1–5%) and no regenerated specimens were detected in *T. heterochaetus* and *P. litoralis*. However, the very few specimens of *L. hoffmeisteri* in the brackish part that could be checked showed remarkable results: none of the 20 specimens in the oxygen-deficient oligohaline section (B) showed any sign of regeneration, contrasting with eight out of eleven specimens in the mesohaline section.

Discussion

Benthic fauna: species composition

How do the findings of this study, in terms of the species composition and distribution of benthic invertebrates, relate to results found in other estuaries ? Is the benthic fauna of the Zeeschelde impoverished due to pollution and do we find effects in mesohaline, oligohaline, and freshwater tidal reaches ?

The benthic fauna of the mesohaline part of the Zeeschelde appears similar in species composition and abundance to benthic communities in other estuar-

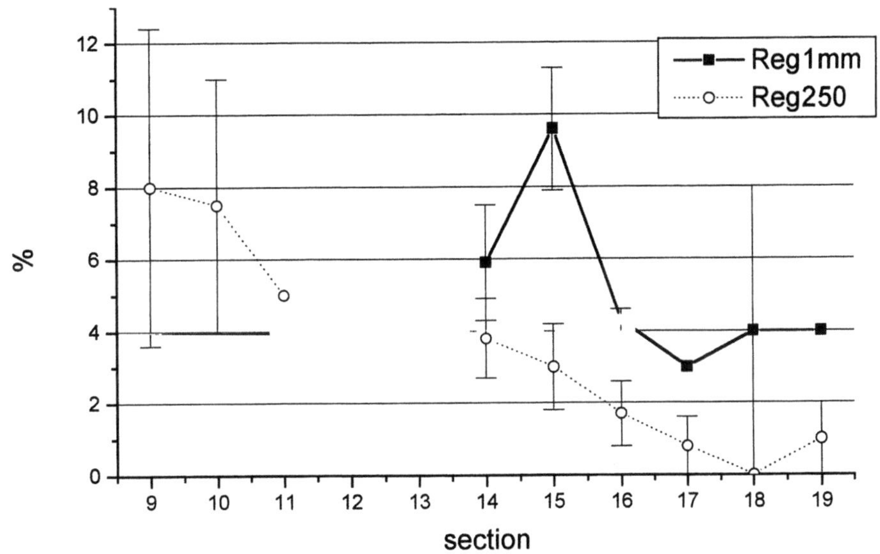

Figure 2. Regeneration rates of oligochaetes in the ten sections of the Zeeschelde for the 250 μm and the 1 mm fractions.

ies, as already observed by Ysebaert et al. (1993). We found that oligochaete populations, not identified to species level in the study of Ysebaert, are dominated by *H. costata*, *T. heterochaetus* and *P. litoralis* in the ß-mesohaline. These three species are also common inhabitants of many estuaries and brackish basins (Giere & Pfannkuche, 1982). *Heterochaeta costata* is a widespread European brackish water species, whereas *T. heterochaetus* is amphi-North Atlantic in its distribution and considered endemic to brackish water. It is reported from the oligohaline reaches of estuaries in New England (Diaz, 1980) and from Elbe (Giere & Pfannkuche, 1982) and Weser (Schuchardt et al., 1993). *Paranais litoralis* is found in many marine and brackish habitats (e.g. Schlei-Förde, Elbe, Weser, Tees, Forth estuary: Giere & Pfannkuche, 1982) but can extend into limnetic waters (e.g. Elbe). The tubificid *Monopylephorus rubroniveus,* frequently encountered in the ß-mesohaline, has not been found in the Zeeschelde. The absence of naidid species other than *P. litoralis* in the Zeeschelde is almost certainly a consequence of the sampling period. Many naidids peak in abundance in spring and become much less abundant in late summer/winter (Giere & Pfannkuche, 1982). Although oligochaetes of the α-mesohaline and polyhaline reaches of the Schelde estuary have never been studied in detail, records of *Tubificoides benedii* Udekem and *Tubificoides pseudogaster* Dahl in the Westerschelde (Verdonschot, 1981) indicate a similarity in species composition to other estuaries.The three dominant ß-mesohaline species in the Zeeschelde, *H. costatus*, *T. heterochaetus* and *P. litoralis*, reach maximum densities in the Zeeschelde comparable with peak abundances in other areas. Giere & Pfannkuche (1982) report *P. litoralis* and *T. costatus* as inhabitants of heavily polluted waters, still abundant where most animals are eliminated (also Muus, 1967; Gray, 1976), and Leppäkoski (in Giere & Pfannkuche, 1982) found *T. heterochaetus* in very polluted, oligohaline, Finnish coastal waters, increasing with the degree of pollution.

The oligohaline zone of the Zeeschelde is extremely poor in terms of species diversity and abundance of benthos in general. Only *L. hoffmeisteri* was found in densities of ca. 20 000 ind m^{-2} restricted to two shallow subtidal stations. This freshwater species is highly tolerant to salinity and thus penetrates far into brackish water (Chapman et al., 1982; Diaz, 1989). We found *L. hoffmeisteri* as far as Boerenschans (10 km downstream from Antwerp, at salinities of 4.7 psu, in September 1996), while *T. tubifex* and the other *Limnodrilus* species were absent downstream of the Rupel tributary. Since sampling took place within one month of a major freshet in the area and salinities remained reduced into September, there is a question why other freshwater species did not take advantage of the high freshwater flows. Possible explanations are:

1. High freshwater flows at the end of August did not have profound effects on interstitial salinities (Chapman & Brinkhurst, 1981);
2. *L. hoffmeisteri* is more sensitive to erosion of the bottom sediment; and

3. *L. hoffmeisteri* enters the water column sooner in an (semi-) active way under such circumstances.

An abrupt change from the brackish part towards the freshwater tidal range, with a very poor oligohaline zone, was also observed in the James River estuary (Diaz, 1980). Oligohaline zones are very often physical, chemical and biological buffers, where large fluctuations in salinity, a high turbidity and low oxygen levels prevents most species from settling. The physiography of the oligohaline part of the Zeeschelde (with a deep, clay-bottomed straight shipping lane and very few tidal flats), the oxygen-deficiency throughout the year and major concentrations of heavy industry along its course are considered the main causes for the diminished biota in this part of the estuary.

The freshwater tidal range of the Zeeschelde is very poor in species and characterised by mass populations of the eurytopic species assemblage *T. tubifex*, *L. hoffmeisteri*, *L. claparedeianus*, *L. udekemianus* and *L. profundicola*. Single specimens of *S. ferox*, *P. bavaricus*, *D. digitata* and Enchytraeidae were found upstream. Other taxa are very poorly represented in the soft sediments with only nematodes and copepods in moderately high numbers. Single specimens of Ceratopogonidae and Chironomidae larvae are the only other biota recorded in this study. Data on invertebrates from hard substrates in the Zeeschelde (De Pauw et al., 1992; own observations) confirm the general picture and reveal only very few taxa not known from the Zeeschelde soft sediments (*Erpobdella* sp., chironomids, *Gammarus* sp., *Asellus* sp., *Corixa* sp.). Intertidal vegetation has not been investigated so far for epibenthic invertebrates but is known to be populated with several small semi-terrestrial gastropods and other animals.

We can conclude that the very low species diversity reflects the heavy pollution in the Zeeschelde. Oligochaetes and particularly tubificids are being widely used as good indicators of environmental conditions in lakes and rivers, in terms of organic pollution and oxygen levels (Chapman et al., 1982; Brinkhurst & Gelder, 1991). With respect to organic pollution, *T. tubifex* and species of the genus *Limnodrilus* are considered as the most tolerant (Brinkhurst , 1980; Milbrink, 1980; Lang, 1984). Basically, tidal freshwater is dominated by pelophilic species which are opportunists and rather resistant and resilient to perturbations (Diaz, 1994), and therefore compares quite well with eutrophic large lakes, the profundal of smaller lakes or polluted harbours. High silt loads create low habitat diversity and severely restrict the composition of the biota, since suitable substrates are not available for epifaunal or crevice-dwelling organisms frequently found in fast flowing freshwater (Diaz, 1989).

Which environmental variables determine the abundance of Oligochaetes in the freshwater tidal zone of the Zeeschelde ?

The results of this study emphasised the importance of the sediment composition, in close correlation with depth, as the key factor for abundance of benthic invertebrates in the Zeeschelde. Intertidal flats are sedimentation areas for fluid mud and are inhabited by large populations of tubificids, larger than those found in the subtidal coarser sediments which are subject to strong currents. The very dense population of tubificids in at least one deep subtidal freshwater station, suggests that the oxygen conditions, as found in this part of the Zeeschelde, are not limiting the abundance of the opportunistic species *T. tubifex* and *L. hoffmeisteri*. Indications for elimination and reduced reproductive abilities due to oxygen deficiency in certain periods, as found in the Thames (Hunter, 1981), were not found.

Poor correlation of tubificid abundance with total organic carbon, as found in this study, is reported in many studies because it includes not only valuable food sources such as bacteria or recent algal remains but also substances resistant to decomposition and presumably unavailable as a food source (Pasteris et al., 1994). Wagner (1968), on the basis of experiments with freshwater tubificids, stressed the importance of finer sediment fractions ($< 63\ \mu$m) for oligochaete abundance and Birtwell & Arthur (1980) link large populations of *L. hoffmeisteri* in the Thames to substantial deposition of fine sediment ($< 63\ \mu$m). Several studies also indicate increases in populations due to an increase in sedimentation of fine particles (Pfannkuche, 1977; Lang, 1989; Sloreid, 1994) and McLusky et al. (1993) found reductions in oligochaete populations due to a reduction in the organic inflow to the Forth estuary. Juget (in Giere & Pfannkuche, 1982) discussed the relation between particle size and gut diameter of freshwater oligochaetes and formulated an index for sediment consumability. In the tidal freshwater part of the Zeeschelde, the correlation of tubificid abundance with sediment type, with highest numbers in the fluid muds, reflects the close relationship between mass populations and easily accessible organic food with associated microflora.

The impact of wave action and water currents on the abundance of tubificids is poorly studied (Birtwell & Arthur, 1980) and insufficiently understood. Sediments with particles of 180–200 μm are especially exposed to agitation by currents and waves, which could explain the avoidance reactions of both macro- and meiobenthic oligochaetes in studies on the Isle of Sylt (Giere & Pfannkuche, 1982), 200 μm also seems to be a critical level for other mechanical reasons (Wieser in Giere & Pfannkuche, 1982). This is in accordance with the sudden drop in abundance of tubificids in the Zeeschelde when one moves from the shallow subtidal freshwater part, with median grain size of 159 μm to the deeper stations which have a mean median grain size of 218 μm.

Mass populations and trophic role

Great plasticity and dependence of the life cycle of *L. hoffmeisteri* and *T. tubifex* upon local conditions, enable them to form mass populations as found in the freshwater part of the Zeeschelde. Peak densities of 3.10^6 ind m^{-2}, of which 700 000 were macrobenthic (retained on a 1 mm mesh-size), and a corresponding biomass of 25.7 g ADW m^{-2} can compete with some of the highest values ever described. Caspers (1948) reports mass populations in well-aerated, polysaprobe silts of Hamburg harbour up to 800 000 macrobenthic tubificids m^{-2} and Palmer (1968) found densities up to $5.7.10^6$ ind m^{-2} of *T. tubifex* and *L. hoffmeisteri* close to a heated effluent discharge from an electrical generating plant in the river Thames. The combination of excess of organic material with low but sufficient oxygen supply is the key for mass populations, conditions apparently also met in large parts of the Zeeschelde these days. In these populations, physical space is believed to be limiting further increase of abundance (Caspers, 1948).

It is clear that such mass populations constitute a very attractive food source for higher trophic levels. Rofritz (1977) demonstrated the dominant role of tubificids (*T. tubifex* and *L. hoffmeisteri*) in polluted, limnetic bottoms as food for ducks. Gray (1976) concluded that oligochaetes and small polychaetes form an important part of the diet of estuarine birds particularly in polluted estuaries such as the Forth. Warnes (1981) showed that *T. benedii* is a significant part of the diet of the Shelduck *Tadorna tadorna* in the same area. A study on the foraging behaviour of the Teal *Anas crecca* in the Zeeschelde (Dethier, 1997) and in the former freshwater tidal Ventjager flats in the Netherlands (Zwarts, 1976), strongly suggests that mass populations of tubificids can be exploited by the ducks. If one considers the high caloric value of freshwater tubificids (5575 cal g^{-1} DW, compared to 3503 cal g^{-1} DW for polychaetes: Cummins & Wuycheck, 1971) and the unusually high biomasses in these areas – highest biomasses in the Zeeschelde reach up to more than 1 kg wet weight m^{-2} – this is hardly a surprise. The question of whether this direct transfer of bacterial energy to top carnivores such as birds takes place is an important issue to be adressed in the near future.

Oligochaete predation by young fish has been widely documented in brackish-marine (Giere & Pfannkuche, 1982) as well as in fresh water areas (Yaroskhenko et al., 1980). Tail-cropping followed by regeneration can give an insight on predation pressure by fish on certain species of oligochaetes. Wisniewski (in Drewes & Zoran, 1989), in a study of tubificids in Polish lakes, concludes that predation on tubificids is easier to detect in terms of increasing regeneration proportions than in decreasing numbers of worms. Diaz (1980) found 1–2% of the tidal freshwater James river *Limnodrilus* sp. with cropped tails, a figure similar to the values in this study of the Zeeschelde. However, Diaz never observed any tail-cropping in brackish water oligochaetes and ascribes this to the smaller size of the species that results in their being swallowed whole, and to a different behaviour of the worms. In the Zeeschelde, higher proportions of cropped tails are found in larger specimens and when one goes from the freshwater to the brackish area. Further research will be needed to investigate whether this is due to higher predation pressure by fish, knowing that the abundance of several species of fish in the Zeeschelde increases towards the Belgian-Dutch border, with only scattered observations in the oxygen-deficient tidal freshwater part (Maes, 1997).

The Zeeschelde: in the future ?

There are no quantitative data on oligochaete distribution in the Zeeschelde from the past. The study of Konietzko (1953) in the ß-mesohaline confirms the occurrence of a similar species composition as found today (with *T. heterochaetus*, *P. litoralis*, *H. costatus*, *T. tubifex*, *Nais elinguis* and Enchytraeidae), but no quantitative information is given. Thus, it is not clear whether mass populations of tubificids existed in the tidal freshwater Zeeschelde in the past (i.e. before 1990). The situation now found is characteristic for hypertrophic water bodies with a minimal oxygen supply. With improved oxygen conditions and lower organic input we might expect to see an evolution towards higher species diversity. Pfannkuche (1977) describes 16 naidid, 1 *Aeolosoma* sp., 1 enchytraeid and 10 tubificid species from the Fährmannsand, a freshwater tidal flat in the hypertrophic Elbe with higher oxygen saturation values than in the Zeeschelde, with *Aulodrilus pluriseta*, *Psammoryctides barbatus* and *Potamothrix moldaviensis* restricted to the more sandy habitats. Mass populations of *T. tubifex* and *L. hoffmeisteri* (and *L. claparedeianus*) were found at other locations in the Elbe with higher organic pollution and hence lower oxygen levels. Hunter (1981) mentions *P. barbatus* from the freshwater part of the Thames becoming extinct where oxygen levels drop below a certain level.

Besides the possible colonisation by a limited number of oligochaete species, a substantial reduction in the organic input into the Zeeschelde will definitely affect the abundance of the worms. However, much will depend on the balance between natural organic input and sewage water discharge. Since the natural production and sedimentation of organic material will probably remain very substantial even in the case of reduction of sewage water input, we expect a rather slow and limited reduction in tubificid populations.

In the former and the small vestiges of an extensive tidal freshwater ecosystem in the lower Rhine-Meuse Delta (Amer, Biesbosch, etc.) of the Netherlands, typical assemblages included several molluscs (*Pisidium* sp., *Unio* sp., *Dreissena polymorpha*, *Sphaerium* sp., *Potamopyrgus antipodarum*), chironomids, a few amphipods and rich populations of oligochaetes (Wolff, 1973). However, in areas of the Hollandsch Diep with high sedimentation rates and soft muds, the subtidal fauna is poor and dominated by tubificids (four *Limnodrilus* species, *Quistadrilus multisetosus* and *D. digitata*), with small numbers of *Gammarus* sp., *Pisidium* sp., *Valvata* sp. and *Potamopyrgus* sp. and only two chironomid species. Tubificids were found here in populations of 320 000 ind m^{-2}, the cumulative abundance of the other species was ten times smaller (36 000 ind m^{-2}) (Smit et al., 1995). In the tidal freshwater section of the James River estuary, where good oxygen conditions prevail, tubificids are dominant (75% of density) and a rather limited number of other taxa is found (14 chironomids, 5 bivalves) (Diaz, 1989).

The tidal freshwater section of the Zeeschelde with its high tidal ranges and substantial sediment transport is a typical example of the kind of tidal freshwater system described by Diaz (1994). Interactions between physical disturbance regimes and species result in low species richness. Probably the physiography of the Zeeschelde and the high physical stress caused by the constant movement of vast amounts of fluid mud and strong water currents prevent it from becoming a system rich in benthic organisms, even though oxygen conditions improve and pollution decreases. However, at locations infrequently covered with fine muds, opportunities for the development of a richer fauna than found today certainly exist.

Acknowledgements

We are grateful for the financial assistance provided by the Ministry of the Flemish Community and for the logistical support by Ir.Taverniers, Ir.Wouters and the crews of the 'Scaldis' and the 'Veereman'. Part of the sediment samples were processed by Dr Wartel and his colleagues (KBIN). N. Deregge and T. Ysebaert (Institute Nature Conservation) analyzed more sediment samples and were very helpful during the sampling campaign. The authors thank their many colleagues at the University of Gent and at the Institute of Nature Conservation for their generous assistance with sample collection and analysis, in particular Danielle, Guy, Johan, Ranan, Annick, Koen and Ludwig. Data from the monitoring campaigns on the Zeeschelde were generously provided by S.Van Damme.

References

Appleby, A. G. & R. O. Brinkhurst, 1970. Defecation rate of three tubificid oligochaetes found in the sediments of Toronto Harbour, Ontario. J. Fish. Res. Bd Can. 27: 1971–1982.

Belmans, H., 1988. Verdiepings- en onderhoudsbaggerwerken in Wester- en Zeeschelde. Water 43: 184–194.

Birtwell, I. K. & D. R. Arthur, 1980. The ecology of tubificids in the Thames estuary with particular reference to *T. costatus* (Claparède). In R. O. Brinkhurst & D. G. Cook (eds), Aquatic Oligochaete Biology, Plenum Press, New York, London: 175–184.

Brinkhurst, R. O., 1980. Pollution biology – the North American experience. In R. O. Brinkhurst & D. G. Cook (eds), Aquatic Oligochaete Biology. Plenum Press, New York: 205–209.

Brinkhurst, R. O. & S. Gelder, 1991. Annelida: Oligochaeta and Branchiobdellida. In U. H. Thorp & A. P. Covich (eds), Ecology and Classification of North American Freshwater Invertebrates. Academic Press, San Diego: 401–435.

Caspers, H., 1948. Ökologische Untersuchungen über die Wattentierwelt im Elb-Ästuar. Verh. Dt. Zoologen Kiel 1948: 350–359.

Chapman, P. M. & R. O. Brinkhurst, 1981. Seasonal changes in interstitial salinities and seasonal movements of subtidal benthic invertebrates in the Fraser River estuary, B.C. Estuar. coast. shelf Sci. 12: 49–66.

Chapman, P. M., M. A. Farrell & R. O. Brinkhurst, 1982. Relative tolerances of selected aquatic oligochaetes to individual pollutants and environmental factors. Aquat. Tox. 2: 47–67.

Chatarpaul, L., J. B. Robinson & N. K. Kaushik, 1978. Role of oligochaete worms in nitrogen transformation in stream sediments, Abstracts, 26th Annual Meeting, N. Am. Benthological Society. Winnipeg, Manitoba.

Cummins, K. W. & J. C. Wuycheck, 1971. Caloric equivalents for investigations in ecological energetics. Mitt. int. Ver. theor. angew.Limnol. 18: 1–158.

De Pauw, N., V. Lambert & A. Van Kenhove, 1992. Kunstmatige substraten voor biologische monitoring van grote rivieren en kanalen in Vlaanderen en Nederland. Studie in opdracht van de Vlaamse Milieumaatschappij; Report University Gent: 57pp. (in Dutch).

Dethier, K., 1997. Feeding ecology of the Teal (*Anas crecca* L) on two tidal flats in the Zeeschelde: exploratory study on the foraging behaviour and food availability. MSc thesis University Gent. 79 pp. (in Dutch).

Diaz, R. J., 1980. Ecology of tidal freshwater and estuarine Tubificidae (Oligochaeta). In: R. O. Brinkhurst & D. G. Cook (eds), Aquatic Oligochaete Biology. Plenum Press, New York: 319–330.

Diaz, R. J., 1989. Pollution and tidal benthic communities of the James River Estuary, Virginia. Hydrobiologia 180: 195–211.

Diaz, R. J., 1994. Response of tidal freshwater macrobenthos to sediment disturbance. Hydrobiologia 278: 201–212.

Drewes, C. D. & M. J. Zoran, 1989. Neurobehavioral specializations for respiratory movements and rapid-escape from predators in posterior segments of the tubificid *Branchiura sowerbyi*. Hydrobiologia 180: 65–71.

Giere, O. & O. Pfannkuche, 1982. Biology and ecology of marine Oligochaeta: a review. Oceanogr. mar. Biol. Ann. Rev. 20: 173–308.

Goosen, N. K., P. van Rijswijk & U. Brockmann, 1995. Comparison of heterotrophic bacterial production rates in early spring in the turbid estuaries of the Scheldt and the Elbe. Hydrobiologia 311: 31–42.

Gray, J. S., 1976. The fauna of the polluted Tees estuary. Estuar. coast. mar. Sci 4: 652–676.

Heip, C. & P. M. J. Herman, 1995. Major biological processes in European tidal estuaries: a synthesis of the JEEP-92 Project. Hydrobiologia 311: 1–7.

Hill, M. O., 1973. Diversity and eveness: a unifying notation and its consequences. Ecology 54: 427–432.

Hunter, J., 1981. Survival strategies of tubificids in the Thames and other estuaries. In N. V. Jones & W. J. Wolff (eds), Feeding and Survival Strategies of Estuarine Organisms. Plenum Press, New York: 53–63.

Kaster, J. L., 1989. Observations of predator-prey interaction on dispersal of an oligochaete prey, *Limnodrilus hoffmeisteri*. Hydrobiologia 180: 191–193.

Konietzko, B., 1953. Notes sur les Oligochètes de Belgique. I. Eaux saumâtres du Bas-Escaut. Med. Kon. Belg. Inst. Natuurwet. 29: 1–14 (in French).

Lang, C., 1984. Eutrophication of Lakes Léman and Neuchâtel (Switzerland) indicated by oligochaete communities. Hydrobiologia 115: 131–138.

Lang, C., 1989. Effects of small-scale sedimentary patchiness on the distribution of tubificid and lumbriculid worms in Lake Geneva. Freshwat. Biol. 21: 477–481.

Lefèvre, F. & J. Stronkhorst, 1988. Chloridegehalte in de Westerschelde, zoetwaterbelasting en biotische aspekten. Rijkswaterstaat GWWS-88.514.11 pp.

Maes, J., 1997. Vis in de Zeeschelde: terug van weggeweest Schelde nieuwsbrief 12: 1–3.

McCall, P. L. & J. B. Fisher, 1980. Effects of tubificid oligochaetes on physical and chemical properties of Lake Erie sediments. In R. O. Brinkhurst & D. G. Cook (eds), Aquatic Oligochaete Biology. Plenum Press, New York: 253–317.

McLusky, D. S., S. C. Hull & M. Elliott, 1993. Variations in the intertidal and subtidal macrofauna and sediments along a salinity gradient in the upper Forth estuary. Neth. J. aquat. Ecol. 27: 101–110.

Meire, P., G. Rossaert, N. De Regge, T. Ysebaert & E. Kuijken, 1992. Het Schelde-estuarium: ecologische beschrijving en een visie op de toekomst. Gent, RUG, Rapport RUG-WWE 28/IN A 92.57 en Instituut voor Natuurbehoud: 150 pp.

Meire, P. M. & M. Vincx (eds), 1993. Marine and Estuarine Gradients. Proceedings 21st Symposium ECSA, Gent, 9–14 Sept 1991, Neth. J. aquat. Ecol. 27: 1–496.

Meire, P., M. Hoffmann & T. Ysebaert (eds), 1995. De Schelde: een stroom natuurtalent. Instituut voor Natuurbehoud, Hasselt, rapport 95.10. 32 pp.

Meire, P. M., M. Starink & M. Hoffmann, 1997. Integratie van ecologie en waterbouwkunde in de Zeeschelde: aanleiding tot en situering van het Onderzoek Milieu Effecten Sigmaplan (OMES). Water 95: 147–165.

Milbrink, G., 1980. Oligochaete communities in pollution biology: the European situation with special reference to lakes in Scandinavia. In R. O. Brinkhurst & D. G. Cook (eds), Aquatic Oligochaete Biology. Plenum Press, New York:. 433–455.

Muus, B. J., 1967. The fauna of Danish estuaries and lagoons. Distribution and ecology of dominating species in the shallow reaches of the mesohaline zone. Meddr. Danm. Fisk. -og Havunders 5: 1–316.

Odum, W. E., T. J. Smith, J. K. Hoover & C. C. McIvor, 1984. The ecology of tidal freshwater marshes of the United States east coast: a community profile. U.S. Fish Wildl. Serv. FWS/OBS-83/17: 177 pp.

Palmer, M. F., 1968. Aspects of the respiratory physiology of *Tubifex tubifex* in relation to its ecology. Proc. zool. Soc. Lond. 154: 463–473.

Pasteris, A., C. Bonacina & G. Bonomi, 1994. Observations on cohorts of *Tubifex tubifex* cultured at different food levels, using cellulose substrate. Hydrobiologia 278: 315–320.

Pfannkuche, O., 1977. Okologische und systematische Untersuchungen an naidomorphen Oligochaeten brackiger und limnischer Biotope. PhD thesis University Hamburg. 138 pp (in German).

Rofritz, D. J.,1977. Oligochaetes as a winter food source for the Old Squaw. J. Wildl. mgmt 41: 590–591.

Schuchardt, B., U. Haesloop & M. Schirmer, 1993. The tidal freshwater reach of the Weser estuary: riverine or estuarine? Neth. J. aquat. Ecol. 27: 215–226.

Seinhorst, J. W., 1959. A rapid method for the transfer of nematodes from fixative to anhydrous glycerine. Nematologica 4: 67–69.

Sloreid, S., 1994. Oligochaete response to changes in water flow in the Dokka Delta, Lake Randsfjorden (Norway), caused by hydroelectric power development. Hydrobiologia 278: 243–249.

Smit, H., H. C. Reinhold-Dudok van Heel & S. M. Wiersma, 1995. Sublittoral macrozoobenthic assemblages in the enclosed sediment polluted Rhine-Meuse Delta; their relationship to environmental conditions. Neth. J. aquat. Ecol. 29: 31–49.

StatSoft, Inc., 1995. STATISTICA for Windows , Statistics II, 2nd edn. 5405 pp.

Van Damme, S., P. Meire, H. Maeckelberghe, M. Verdievel, L. Bourgoing, E. Taverniers, T. Ysebaert & G. Wattel, 1995. De waterkwaliteit van de Zeeschelde: evolutie in de voorbije dertig jaar. Water 85: 244–256 (in Dutch with English summary).

Van Eck, G. Th. M. & N. M. de Rooij, 1990. Development of a water quality and bio-accumulation model for the Scheldt estuary. In W. Michaelis (ed.), Coastal and Estuarine Studies Springer-Verlag, Berlin: 95–104.

Van Zoest, R. & G. Th. M. Van Eck, 1993. Behaviour of selected PCBs, PAHs and ϒ-HCH in the Scheldt estuary, S. W. Netherlands. Neth. J. aquat. Ecol. 27: 288–300.

Verdonschot, P., 1981. Some notes on the ecology of aquatic oligochaetes in the Delta Region of the Netherlands. Arch. Hydrobiol. 92: 53–70.

Wagner, G., 1968. Zur Beziehung zwischen der Besiedlungsdichte von Tubificiden und dem Nahrungsangebot im Sediment. Int. Rev. ges. Hydrobiol. 53: 715–721.

Warnes, J. M., 1981. The impact of overwintering birds on the production ecology of estuarine benthic invertebrates. University of Stirling PhD thesis. 177 pp.

Wolff, W. J., 1973. The estuary as a habitat. An analysis of data on the soft-bottom macrofauna of the estuarine area of the rivers Rhine, Meuse and Scheldt. Rijksmuseum van Natuurlijke Historie, Leiden, Zoologische Verhandelingen 126: 3–242.

Yaroskhenko, M. F., Ol. Val'kovskaya & V. Kh. Chokyrian, 1980. Fresh-water Oligochaeta and their importance as food for fish. In G. M. Belyaev (ed.), Aquatic Oligochaete Worms: Taxonomy, Ecology & Faunistic studies in the USSR. Amerind Publishing Co.Pvt Ltd, New Delhi: 191–196.

Ysebaert, T., P. Meire, D. Maes & J. Buijs, 1993. The benthic macrofauna along the estuarine gradient of the Schelde estuary. Neth. J. aquat. Ecol. 27: 327–341.

Zwarts, L., 1976. Density-related processes in feeding dispersion and feeding activity of Teal (*Anas crecca*). Ardea 64: 192–209.

Zwolsman, J. J. G. & G. T. M. Van Eck, 1993. Dissolved and particulate trace metal geochemistry in the Scheldt estuary, S. W. Netherlands (water column and sediments). Neth. J. aquat. Ecol. 27: 280–287.

Hydrobiologia **406**: 133–142, 1999.
B. M. Healy, T. B. Reynoldson & K. A. Coates (eds), Aquatic Oligochaetes.

Distribution and dispersal capacity of the Ponto-Caspian tubificid oligochaete *Potamothrix heuscheri* (Bretscher, 1900) in Scandinavia

Göran Milbrink
Department of Animal Ecology, Evolutionary Biology Centre, Uppsala University, Sweden

Key words: Potamothrix heuscheri, Ponto-Caspian species, tubificid oligochaete, dispersal, ballast water

Abstract

The species composition of the Scandinavian freshwater oligochaete fauna is strongly influenced by its post-glacial history (Timm, 1980). Few species are likely to have survived in the core areas of glaciation and the present composition of species has thus been highly dependent upon the immigration rate of 'new' species (Milbrink, 1980). Active immigration into new localities has proven to be a very slow process, whereas passive transportation in running water and by rafting is much faster. Exchange of ballast water in harbours is likely to be the single most effective action by man, causing large-scale passive dispersal of the aquatic fauna and flora. Bottom water with stirred-up sediment containing oligochaetes could thereby be sucked into the tanks of ships and later exchanged in nearby and distant ports. Ponto-Caspian species among the Tubificidae, originating in the Black Sea-Caspian Sea region, are continuously dispersing to the west over Central Europe and to the north-west towards the Baltic Sea area (Timm, 1970, 1980; Milbrink, 1980) adding to the species list of the Tubificidae. Five such species of the genus *Potamothrix* have quite recently reached eastern Sweden (Milbrink, 1980). The dispersal front presently goes obliquely across Scandinavia from the north-east to the south-west. Comparisons with results from previous bottom-sampling series in Lake Mälaren from 1915 to 1916 revealed ongoing colonization, especially in those basins of the lake which were in recent contact with the Baltic Sea and thus had brackish bottom water. One of those species, *Potamothrix heuscheri*, which is particularly tolerant to anoxic conditions and high salinity and is likely to have reached Scandinavia in ballast water possibly via indirect sources from the Mediterranean Sea. Its uneven distribution in Scandinavia is well correlated with historic water-ways for heavy transport, for instance via cargoes of wood pulp from Lake Munksjön to the south of Lake Vättern and to and from other cellulose industries on the Baltic coast. It is speculated that distant occurrences of *P. heuscheri* in Africa and South America may have been reached through passive transport in the plumage of migrating wading-birds.

Introduction

The species composition of the Scandinavian freshwater oligochaete fauna is strongly influenced by its post-glacial history and by the immigration rate of 'new' species. Except for downstream dispersal via streams and rivers, active dispersal by oligochaetes is likely to be a very slow process, whereas passive dispersal, i.e. under the influence of man, is growing in importance. Very few oligochaete species are at all widespread and still fewer could be considered to be real cosmopolitans - about 4.5% according to Timm (1980). Exchange of ballast water in harbours is likely to be the single most effective action by man causing large-scale passive dispersal of the aquatic fauna and flora in 'conveyor belts' around the world (Carlton & Geller, 1993). Ballast water pumped into ships while in harbours often contains stirred bottom sediment. Oligochaetes could thereby enter ballast tanks with subsequent transport to new environments (Brinkhurst & Jamieson, 1971; Timm, 1980).

Historically, it seems unlikely that many oligochaete species would have survived in the core areas of glaciation in Scandinavia, thus explaining the depauperate aquatic oligochaete fauna when compared with that of the continent. In a summary of the Scandinavian situation, Timm (1980) proposed that melting of the ice started to take place some 18 000 years ago and over a period of about 8000 years the ice-border slowly withdrew, finally creating peripheral lakes outside present-day Scandinavia. The earth's crust in Scandinavia was much lower under the enormous weight of the thick ice-cover. Melt-water thus flowed downslope from the south and the south-east towards the rim of the ice and is the most likely origin of new species colonizing these peripheral lakes, though probably in low abundance due to low primary production. About 10 000 years ago the ice-border had reached what is now the Baltic Sea. Long periods of freshwater conditions were interrupted by periods of marine conditions creating a harsh environment for colonizing species. When the periphery of the ice had reached Scandinavia, Timm concluded that withdrawal was more or less uphill due to massive land-upheaval. Poor upstream dispersal capacity, probably made few of the species that inhabited the former peripheral lakes reach Scandinavia proper.

At the present time, however, the Ponto-Caspian species of Tubificidae, originating in the Black Sea-Caspian Sea region, are continuously dispersing towards the Baltic Sea area (Timm, 1970, 1980). Most of them belong to the genus *Potamothrix* and several species of that genus have reached eastern Sweden (Milbrink, 1980).

The present work is focused on the Scandinavian distribution and the reconstruction of trajectories of dispersal of one particular Ponto-Caspian species, *Potamothrix heuscheri*, both because of its great dispersal capacity but also because of its extremely high tolerance to anoxic conditions. Distribution records from various places in the world also show *P. heuscheri* to be a dominant species under conditions of environmental stress, such as in slightly brackish water under organic enrichment, on wood-fibre banks in fresh and brackish water and in weak soda lakes in the tropics (Milbrink, 1980). Questions to be raised concerning this species are: What do recent and historical records say about the rate of dispersal? Which vectors of dispersal are likely to be most effective with particular relevance to Scandinavia? Could migrating birds act as such vectors?

Material and methods

The material on which the present study is based covers analyses of the oligochaete fauna of hundreds of lakes distributed all over Scandinavia, mainly in Sweden, some in Norway and a few in Iceland. Littoral and profundal bottom samples obtained in the vicinity of sewage effluents from larger cities and wood processing industries (fibre banks) along the Swedish northern Baltic coast have also been examined. Most of this material has been collected with an Ekman sampler (5 replicate samples per station each time), sieved through a 0.6 gauge net and preserved in alcohol.

Furthermore, extensive sampling networks exist in the large lakes of southern Sweden - Mälaren, Vänern, Vättern and Hjälmaren. From these networks samples have been taken at regular intervals since the early 1960s by the Swedish Environment Protection Board. Much of this material has been identified by the author. In each of these large lakes, 5 Ekman samples have generally been taken each time at each locality and sieved through a mesh-size of 0.5 mm. The worms have been preserved in alcohol and mounted in glycerin for species identification.

Comparisons have been made with three earlier bottom fauna surveys carried out in Lake Mälaren. In the summers of 1915 and 1916, fishery biologists O. Nordqvist and G. Alm (Institute of Freshwater Research, Drottningholm) on board the research vessel Eystrasalt carried out bottom fauna surveys in that lake. In the first year 36 stations were sampled and 18 stations in the second year. Samples were taken with a marine Petersen sampler (aperture 0.3 m^2) and the material was sieved through a 1.0 mm mesh size. The worms had been preserved in alcohol. Vials of oligochaetes from these surveys were discovered in the 1970s at the Institute of Freshwater Research at Drottningholm and were analyzed by the author (Milbrink, unpublished).

In 1933, another small series of bottom samples from 9 stations in different basins of Lake Mälaren was collected (O. Nybelin and S. Vallin, Institute of Freshwater Research, Drottningholm). The authors used an Ekman sampler and the material was sieved through 1.0 mm gauge nets and preserved in alcohol. This series, however, is poorly defined in terms of exact positions and depths. Accordingly, the results are to be treated with some caution.

Results

Present world distribution of Ponto-Caspian Potamothrix *species*

A whole set of Ponto-Caspian species of the genus *Potamothrix* have their core areas of distribution within a belt stretching from the Caspian Sea to the

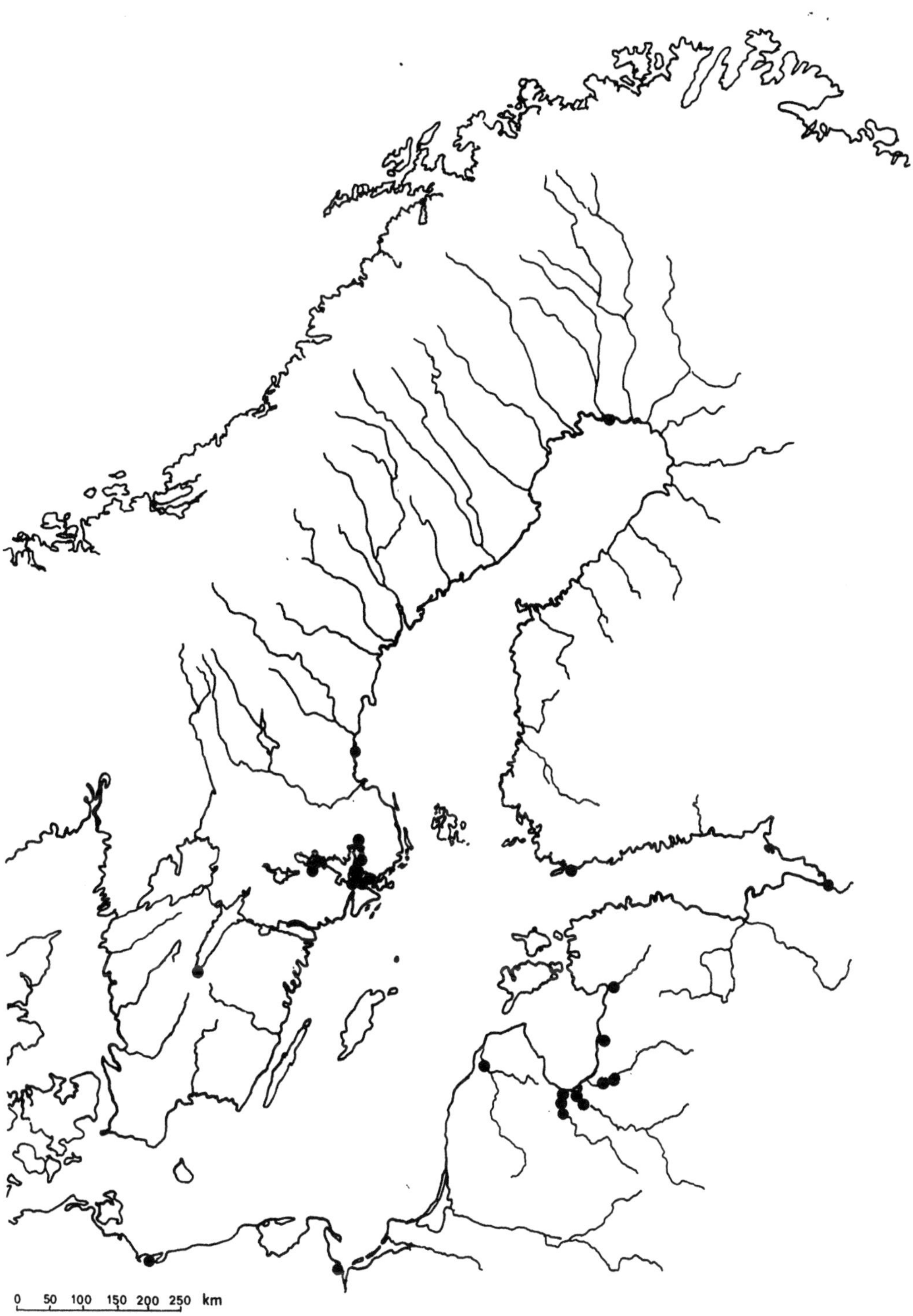

Figure 1. The present distribution of *Potamothrix heuscheri* in the states bordering the Baltic Sea (information from the Baltic States from Timm, 1980 and Timm, unpublished).

Baltic Sea. Many of these species, probably related to their origin, demonstrate a higher tolerance for salt or brackish water than most freshwater oligochaete species. This makes them very successful competitors in a wide variety of water types (Brinkhurst & Jamieson, 1971; Timm, 1980, Timm, unpublished). Both present and historical records support the ability of the Ponto-Caspian species to continuously disperse across the north-west over Europe – towards the Baltic Sea area (Timm, 1970, 1980; Milbrink, 1980), as well as to the west over Central, Southern and South-western Europe (Juget, 1967; Giani, 1979; Lang, 1984; Bonacina et al., 1992; Prat et al., 1992; Bazzanti et al., 1993; Lang & Reymond, 1996). Some *Potamothrix* species have even reached distant continents. *P. bavaricus* has been recorded from New Zealand and Australia (preferably estuarine environments) to the Great Lakes of North America (Brinkhurst & Jamieson). This species has also been found in Lake Titicaca in Bolivia in South-America (Juget & Lafont, 1994). *P. heuscheri* has been recorded in Bolivia, Kenya and Kongo in Africa (see below) and *P. vejdovskyi*, though less often associated with brackish or salt water, has likewise been recorded in the Great Lakes of North America – as have also *P. moldaviensis* and *P. bedoti* (Brinkhurst & Jamieson, 1971; Timm, 1980). As for the distribution of several of the Ponto-Caspian species in Europe there is a more or less distinct distribution limit separating the north-western parts of Europe – not yet reached by these species – from the rest (Milbrink, 1980). This line presently goes approximately obliquely through Scandinavia (Figures 1 & 3). The only Ponto-Caspian species that has been recorded beyond this line in Scandinavia is *P. bedoti* (Milbrink, unpublished) and has been recorded in southern Norway (S.-E. Sloreid, personal information).

World distribution of Potmothrix heuscheri

Potamothrix heuscheri has a particularly wide but discontinuous world-wide distribution. Its present distribution range covers a good deal of eastern Europe (Timm, unpublished), and to an increasing extent Central Europe – Switzerland and France (Juget, 1967; Giani, 1979; Lang, 1984; Lang & Reymond, 1996), Southern Europe – Italy (Bonacina et al., 1992; Bazzanti et al., 1993), as well as South-western Europe – Spain (Prat et al., 1992). As mentioned above, this species may, in fact, be found in waters of varying quality and size, often under anoxic conditions, in brackish and estuarine water, as well as in freshwater habitats.

Outside of Europe *P. heuscheri* has been found in seasonally anoxic sediments of Lake Titicaca in Bolivia (Juget & Lafont, 1994) and in Lake Naivasha (eutrophic, weak soda lake) in Kenya (Milbrink, 1977). There is also a record from eastern Congo (Brinkhurst & Jamieson, 1971). Furthermore it is considered a characteristic species of anoxic profundal sediments of the slightly saline Lake Tiberias in Israel (Gitay, 1968). Due to this extreme tolerance it often replaces two other tubificid species in fresh and brackish water, *Potamothrix hammoniensis* and *Limnodrilus hoffmesteri,* when conditions deteriorate (Milbrink, 1980; Lang, 1984). The former species is, according to Timm (1980), of Euro-Siberian origin and generally the best oligochaete indicator of eutrophic lowland-lakes in Europe (Mibrink, 1980; Timm, 1980; Lang, 1984). There are, so far, no records of *P. heuscheri* from North-America or South-East Asia with Australia and New Zealand.

Distribution pattern of Potamothrix heuscheri *in Scandinavia*

So far, eight Ponto-Caspian tubificid species have been recorded from the Baltic states (Timm, 1980) six of which have colonized parts of eastern Sweden (Milbrink, 1980), thereby increasing the known number of tubificid species in Sweden from 21 to 27.

Milbrink (1980) described the distribution of Ponto-Caspian species (*Potamothrix vejdovskyi, P. heuscheri, P. moldaviensis* and *P. bedoti*) and other invading tubificid species (*Branchiura sowerbyi,* of tropical origin and the likely *Limnodrilus cervix*, a North-American species) in Lake Mälaren, south-central Sweden (Figure 2). Records of a fifth species are few. Thus *Potamothrix bavaricus* has not been included. Although slightly differing in their distribution patterns these *Potamothrix* species are largely limited to eastern and to some extent central Lake Mälaren, i.e. largely in those areas which less than 100 years ago had brackish bottom water due to their connection with the Baltic Sea (maximum extension of brackish bottom water indicated in Figure 2). *P. heuscheri* is, as elsewhere in the world, nearly always associated with eutrophic conditions in the deep profundal, where oxygen levels are particularly low (Brinkhurst & Jamieson, 1971; Milbrink, 1980; Timm, personal information). In Lake Mälaren this species is most abundant at depths between 40 and 50 m, such as in

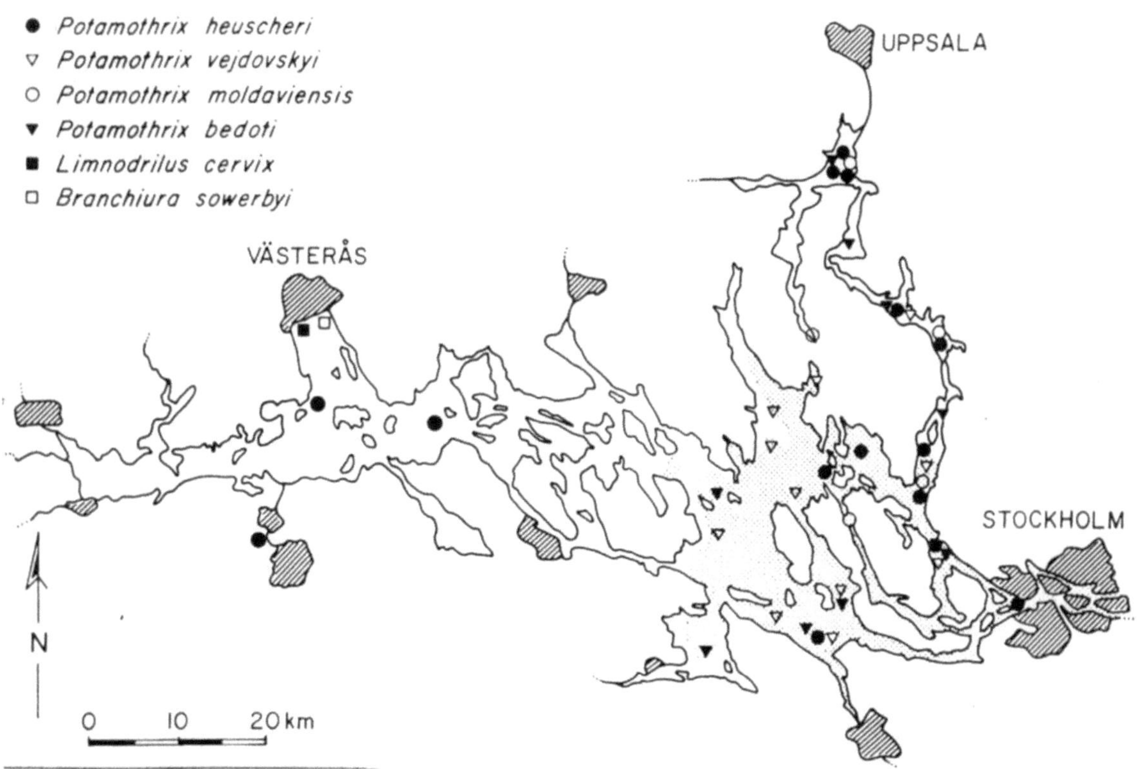

Figure 2. The present distribution of immigrating tubificid species in the Lake Mälaren drainage area (partly redrawn from Milbrink, 1980). The occurrence of a fifth *Potamothrix* – species in the lake, *P. bavaricus*, is not indicated in the figure. Maximum distribution of brackish bottom water indicated in grey.

the eutrophic basin of Ekoln near the city of Uppsala and in the more mesotrophic, central basins of the lake. In Figure 2 three new localities for *P. heuscheri* have been added. The first is from likely partly anoxic sediments (Mr. H. Torstensson, personal information) in the River Eskilstunaån outside the sewage plant of the city of Eskilstuna (about 50 000 inhabitants). There are also two 'new' localities in the basins of Blacken and Granfjärden in western Lake Mälaren. So far, however, only single specimens have been found.

The known distribution range of *P. heuscheri* in Scandinavia has thus been extended. The time-table for this extension is not entirely known, however. At the southern tip of Lake Vättern (Figure 3) is a small lake, Munksjön, connected with the latter lake by a short canal. Water levels are about the same in the two lakes. Until recently ships entered Lake Munksjön to take cargoes of paper pulp from a large wood-processing industry located there. The lake is eutrophic with anoxic conditions occurring at irregular intervals on wood-fibre banks at depths between 3 and 9 m (maximum depth, 20 m). On these banks *P. heuscheri* was recorded in monoculture in 1977 (Milbrink, unpublished). Mean levels of total-phosphorus and total-nitrogen in the surface water over the period 1990–1993, were, for instance, around 40 μg/l and 2200 μg/l, respectively (Report 11/94, the County Administration of Jönköping).

Along the northern Baltic coast several industrial point-outlets from wood-processing industries are located. Two new brackish water localities for *P. heuscheri* have thus been identified, both on fibre banks outside two such industries, at Karlsborg in the north and at Norrsundet in the middle of Sweden. Densities, however, remain low at either locality. The dominant species in both cases is *P. hammoniensis* (Milbrink, unpublished).

Finally, there are new records of *P. heuscheri* in brackish water localities in the inner archipelago of Stockholm and in a few small adjacent coastal lakes (Timm, personal information; Milbrink, unpublished).

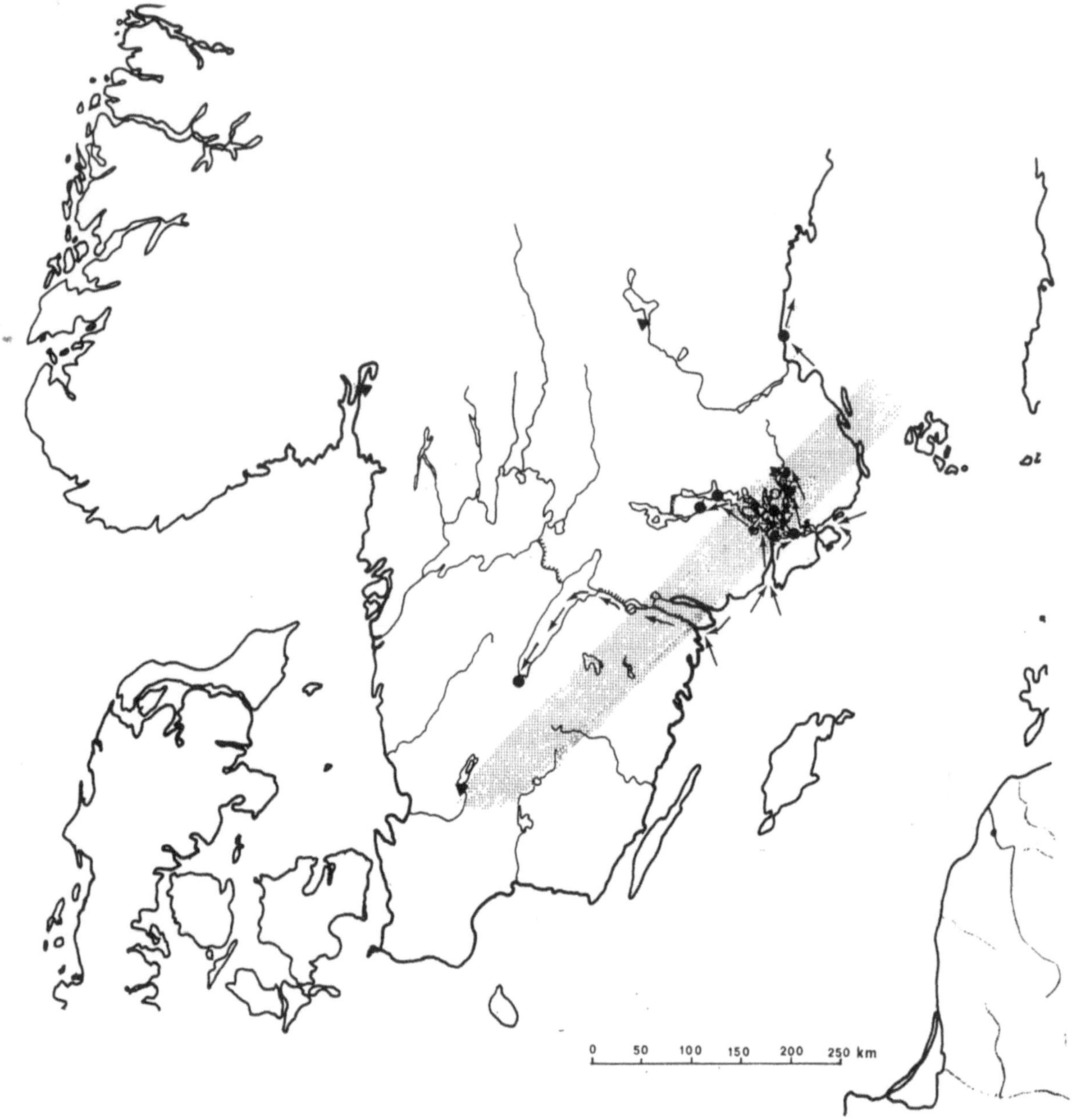

Figure 3. Distribution of species of the genus *Potamothrix* in southern Scandinavia (*Sweden*: Milbrink, 1980; Milbrink, unpublished; *Norway*: S.-E. Sloreid, personal information).Likely trajectories of dispersal – equivalent to old shipping routes – are indicated by arrows and the position of a likely 'front of dispersal' is marked in grey. The symbols are the same as in Figure 2.

Discussion

Trajectories of dispersal over Scandinavia, ballast water - mediated transport

Ballast water in ships is perhaps the overall most efficient agent of dispersal of aquatic invertebrates and planktonic algae (Carlton & Geller, 1993). Toxic dinoflagellate species may, for instance, be transported to new marine environments sometimes with disastrous environmental consequences (Hallegraeff & Bolch, 1991). As mentioned, ballast water pumped into ships while in harbours often contain stirred sediment, thereby facilitating the dispersal of oligochaetes (Brinkhurst & Jamieson, 1971; Timm, 1980).

Timm (1980) was able to reconstruct the gradual expansion of one particular Ponto-Caspian species, *Isochaetides newaensis*, transported this way over western Russia and the Baltic states. However, few existing sampling programmes or previous collections of bottom animals in the world are broad enough to allow dispersal trajectories to be reconstructed. In addition to Timm's (1980) reconstruction of dispersal trajectories of *Isochaetides newaensis*, Lang (1984) and Lang & Reymond (1996) similarly showed the recent, progressive expansion of several of the above Ponto-Caspian *Potamothrix* -species in Lake Geneva and other large Swiss lakes over the last decades. In the latter works the authors were able to demonstrate how eutrophication has generally favoured the colonization of *Potamothrix* species but also the decline of the same species group after oligotrophication of the same lakes, after extensive sewage diversion.

Experiments carried out by Timm (1980) in the profundal of Lake Võrtsjärv in Estonia clearly demonstrated the generally slow dispersal rate of lake-dwelling tubificids (i.e. about 10 m/year, on the average, for *Isochaetides newaensis*). In ideal cases, of course, in the drift fauna or in association with a variety of floating objects in running water – so-called raft dispersal – downstream dispersal may proceed much faster.

In the two bottom fauna surveys carried out in Lake Mälaren in 1915 and 1916 the same Ponto-Caspian *Potamothrix*-species, were found at roughly the same localities as they are found today, i.e. in the eastern and central basins of the lake and were thus well represented already at the start of the century. *Potamothrix heuscheri*, for instance, had nearly the same distribution pattern in the lake 80 years ago as it has now (Milbrink, unpublished). A few isolated records from western Lake Mälaren, however, suggest a further extension of its distribution limits (Figure 2). *Potamothrix vejdovskyi*, on the other hand, was not found in any samples from these early surveys and is thus an exception. Since the size of all these species (except *P. bedoti*) is fairly similar, *P. vejdovskyi* would have had about equally good chances of being retained in the sieving process as any of the other species. Accordingly, there are good reasons to believe that *P. vejdovskyi* at that time really had not yet reached Lake Mälaren and would therefore be a good example of ongoing dispersal.

Exchange of ballast water in harbours is likely to be the single most effective action by man facilitating passive dispersal (Brinkhurst & Jamieson, 1971; Timm, 1980). Ship epifauna would through gradual shedding also be a possible means of dispersal. There exist faunistic records of a diversity of invertebrate species which mainly occur in the upper rivulets of adjacent drainage areas suggesting historic anthropogenic connections between such areas, e.g. through the habit of the Vikings of hauling their ships over land (Timm, 1980).

Publications of records of *P. heuscheri* from the lower reaches of rivers in the Baltic States may provide a very coarse time-table for the arrival of this species and in some cases the rate of upstream-disperal, e.g. in the River Lielupe (E.A. Parele in a series of works in Latvian – T. Timm, personal information). Accordingly, the first publications of records of the species from the Rivers Lielupe and Daugava in Latvia, which appeared around 1970, and the River Pärnu in Estonia (Timm, 1970), and Venta, Rauna and Vaidavas in Latvia in the 1980s, give some indications. The first published records from the Rybinsk Reservoir in the River Volga are from 1974 (Samernoy – Timm, unpublished). In Finland the first records of *P. heuscheri* were published from brackish water habitats at Tvärminne (Laakso, 1969). In Poland the species was found at the mouth of the River Oder (Legezynski, 1971). All these records suggest a rather late arrival in the Baltic area.

Since *P. heuscheri* was frequently found in the bottom samples from eastern Lake Mälaren in 1915–1916, i.e. about 50 years before other records from the countries bordering the Baltic Sea, it is believed that the species arrived comparatively early in Sweden – and not necessarily directly from the south-east. Despite ample evidence that these Ponto-Caspian species are steadily expanding their distribution in Europe to the north-west and to the west (Timm, 1980; Milbrink, 1980; Lang, 1984), it is not certain that all Ponto-Caspian species show the same trajectories of dispersal as *I. newaensis*, i.e. along the inner waterways of the former Soviet Union. Timm (1980) thus gave likely trajectories of dispersal for that species from the Caspian and Black Sea areas via the upper tributaries of the Rivers Volga and Don, a network of canals and reservoirs – among them the Rybinsk Reservoir, the River Neva, Lake Ladoga and the port of St Petersburg. From there, passive transport may have taken place by ships to adjacent ports followed by upstream dispersal in rivers of the Baltic States eventually reaching inland lakes such as Lake Peipsi in Estonia. According to Timm (personal information),

I. newaensis is likely to have arrived via the Valdai heights in present-day Russia by the 18th century.

Potamothrix heuscheri would seem to be a particularly good representative of the peregrine fauna invading Scandinavia from the south-east. The question remains, however, whether this species arrived directly or indirectly from the south-east? Since *P. heuscheri* is regionally a common species in southern Europe, for instance around the Adriatic Sea, where it may have arrived long ago (cf. Kerovec & Mrsic, 1981; Vagner, 1987), it is not unreasonable to believe that this species may have reached the Baltic area from indirect sources, i.e. via ports, in the Mediterranean Sea (T. Timm, personal information). Once having reached the Baltic Sea, further dispersal would be accomplished by local ships for coastal and river traffic.

Distribution of the genus Potamothrix *in Sweden and likely routes of dispersal*

Most species of the genus *Potamothrix* are found in eutrophic situations and often brackish water. However, their ecological demands do differ (Milbrink, 1980). *Potamothrix heuscheri* is the most obvious example of brackish water tolerance. Another species showing considerable salt-tolerance (Brinkhurst & Jamieson, 1971), but remaining rare in Lake Mälaren and adjacent coastal areas of the Baltic, is *P. bavaricus*. *Potamothrix vejdovskyi* , on the other hand, is less often associated with eutrophic conditions or with brackish water than the other species of the genus. Accordingly, its main occurrence in Lake Mälaren is in the central mesotrophic basins. *P. moldaviensis* is mostly associated with flowing water or upwelling spring water in the profundal more or less irrespective of the actual trophic level (Milbrink, 1980; Timm, 1980). *Potamothrix bedoti*, finally, has a slightly more wide-spread distribution in Sweden with one record from Norway (see above) and is largely associated with eutrophic conditions.

Otherwise most Scandinavian records of the Ponto-Caspian *Potamothrix* – species are limited to the eastern part of Lake Mälaren (Figure 2), i.e. to those bottoms where brackish water inflows have frequently occurred (Milbrink, 1980), probably as late as early this century . There are actually still a few brackish water invertebrate species occurring in the eastern parts of Lake Mälaren, among them the nemertean *Prostoma obscurum* otherwise commonly occurring in the Baltic Sea (Milbrink, unpublished).

All localities where *P. heuscheri* has been found in Scandinavia thus far have been readily accessible, at least until fairly recently, by ships. Hence much of the present distribution is likely to be a delayed consequence of shipping activities decades ago. The two 'new' records in western Lake Mälaren are both located along the shipping route to the heavily industrialized city of Västerås (Figure 2). Shipping to and from ports in Lake Mälaren is still of considerable importance. Ships from all the states bordering the Baltic Sea are regular visitors in Lake Mälaren. At least until the 1950s smaller ships could ascend the Eskilstuna River to reach the harbour of the city of Eskilstuna. Thus, records of *P. heuscheri*, in the river outside the city sewage work, would seem to be easy to explain.

It seems possible that ships that visited Lake Munksjön adjacent to Lake Vättern to load paper pulp may have also visited other wood processing industries along the Baltic coasts, thereby contributing to the passive dispersal of *P. heuscheri*, for instance along the northern coast (cf. Figure 1).

Alternative theories explaining the present distribution

There is, of course, a possibility that the invasion of Scandinavia by *P. heuscheri* – as well as by other species of the genus *Potamothrix* – might not be of recent origin. The present distribution would in that case rather reflect the ecological preferences of the species (slightly brackish or soda water in combination with outflows of domestic sewage, outflows from wood-processing industries, etc.) than the routes of dispersal due to shipping activities. *Potamothrix heuscheri* is certainly the species of the genus which is most often associated with eutrophication. Other species of the genus, however, showing rather similar distribution patterns in Scandinavia as *P. heuscheri* are not necessarily bound to brackish water or heavy eutrophication, for instance *P. vejdovskyi* and *P. moldaviensis*. Localities revisited in Lake Mälaren with about 30 year-intervals rather tell a history of recent colonization – a 'front' of species dispersing to the north-west (see above). It is true that Lang (1984) and Lang & Reymond (1996) have shown that the abundance of *Potamothrix* species – as well as of some other tubificid species – first increased with eutrophication of the large Swiss lakes but later decreased with oligotrophication of the same lakes. Lang, however, in a series of papers has also been able to demonstrate the gradual colonization by *Potamothrix* species (among them *P.*

heuscheri) of 'new' bottom areas indicating active dispersal. Active dispersal by bottom-bound oligochaetes is otherwise normally a very slow process (Timm, 1980, see above). Shipping via dense networks of canals on the European continent is otherwise likely to have greatly facilitated passive dispersal to the west over central Europe and France. The most striking examples of likely passive dispersal are, of course, the 'sudden leaps' across the Atlantic made by *P. moldaviensis*, *P. vejdovskyi* and *L. cervix* (Brinkhurst & Jamieson, 1971).

Do we, however, really know the recent history of dispersal of, for instance, *P. heuscheri* in the Baltic states and Eastern Russia? Could trajectories of dispersal at all be identified? I would argue yes. Oligochaete taxonomy and zoogeography were no doubt, very early developed in Eastern Europe with many specialists involved, a fact which has been of great help in the identifiction of specific trajectories of dispersal. Through the access to published (mostly in Russian) and unpublished distributional data from the former Soviet Union, Dr T. Timm, Tartu, Estonia, has been able to follow the gradual colonization of waters by *Potamothrix* species and other Ponto-Caspian species. One of the best known examples of dispersal over vast river systems indicating passive transport by ships is, in fact, the case of *Tubifex newaensis* (Timm, 1980).

Birds as vectors of dispersal?

The few records of *P. heuscheri* that exist from the southern continents of the world are particularly difficult to explain with anthropogenic vectors of dispersal. Ballast water could certainly not be blamed. In order to explain the present distribution of the tropical tubificid species *Branchiura sowerbyi* in warm-water effluents in Europe and North-America the import of tropical plants to hot-houses and botanical gardens, from which waste-water eventually could reach the recipients, was blamed (Mann, 1958; Brinkhurst & Jamieson, 1971). In a similar way the distribution of *P. heuscheri* in Lake Naivasha in Kenya could possibly be a result of the extensive growing, for many decades, of flowering plants in the area, e.g. carnations, for export to Europe, along with exchange of nursery plants grown in hot-houses.

Alternatively, it has (Benham, 1903) been suggested that birds could be regarded as possible vectors of dispersal for oligochaetes. Other explanations have not been proposed. Cocoons could stick to the plumage and feathers of bathing birds and then be shed later, as are small landsnails which are known to be able to stick to the plumage of migratory birds (Dundee, 1967; Vagvolgyi, 1976) or even to the legs of bees and bumblebees (Rees, 1965). There are, in fact, some wading bird species commonly occurring along the Baltic coasts and along lakes and rivers in northern Europe which have wintering quarters in Kenya, Tanzania and further south, such as the Dunlin (*Calidris alpina*). The Common Sandpiper (*Tringa hypoleucos*) is another example, etc. Some populations of these migratory bird species are likely to follow an easternly flight route like that of the eastern race of the white stork and thus fly over the strait of the Bosphorus, mainland Turkey, the Middle East and along the Rift Valley System south to East Africa and even further. This could also be a logical explanation for the existence of *P. heuscheri* in Lake Tiberias, which is a well-known resting spot for wading birds on their migration to East Africa. The same explanation would apply to the dispersal of oligochaete species in South America.

It is well-known that wading birds like many other birds have a habit of 'rinsing' their plumage from external parasites and other foreign objects in quite shallow water. Cocoons of oligochaetes could be resuspended and get 'caught' in the plumage of the belly. If the fine feathers of the belly have the capacity of keeping moisture for some time as in the Sandgrouse, given the cocoons are to some extent resistant to dessication, these birds could at least theoretically function as vectors of dispersal. It is a fact that the cocoons of terrestrial lumbricid oligochaetes may withstand long drought periods – even up to months–under experimental conditions (Prof. N. de Pauw, University of Ghent, Belgium, personal information). This has yet to be demonstrated for aquatic species

Acknowledgements

The author is most indebted to Dr Tarmo Timm for much support with data and references from Eastern Europe.

References

Bazzanti, M., M. Seminara & C. Tamorri, 1993. Eutrophication in a deep lake: depth distribution of profundal benthic communities as an indicator of environmental stress. Verh. int. Ver. Limnol. 25: 784–789.

Benham, W. B., 1903. A note on the Oligochaeta of the New Zealand lakes. Trans.Proc. N.Z. Inst. 36: 192–198.

Bonacina, C., G. Bonomi & A. Pasteris, 1992. Some remarks on the macrobenthos community of the profundal zone of the large Italian lakes. Mem. Ist. ital. Idobiol. 50: 79–106.

Brinkhurst, R. O. & B. G. M. Jamieson, 1971. Aquatic Oligochaeta of the world. Edinburgh: 860 pp.

Carlton, J. T. & J. B. Geller, 1993. Ecological roulette: The global transport of nonindigenous marine organisms. Science 261: 78–82.

Dundee, D. S. 1967, Snails on migratory birds. The Nautilus 80: 89–91.

Giani, N., 1979. Les Oligochètes aquatiques du Sud-Ouest de la France. Bull. Soc. d'Hist. nat Toulouse 115: 347–358.

Gitay, A., 1968. Preliminary data on the ecology of the level-bottom fauna of Lake Tiberias. Israel J. Zool. 17: 81–96.

Hallegraeff, G. M. & C. J. Bolch, 1991. Transport of toxic dinoflagellate cysts via ships' ballast water. Mar. Poll. Bull. 22: 27–30.

Juget, J. & M. Lafont, 1994. Distribution of Oligochaeta in some lakes and pools of Bolivia. Hydrobiologia 278: 125–127.

Juget, J., 1967. Quelques données novelles sur les Oligochètes du Leman: composition et origine du peuplement. Annls Limnol. 3: 217–229.

Kerovec, M. 1980. Fauna malocekinjasa (Oligochaeta) Hrvatske. Bioloski vestnik (Ljubljana) 28: 39–48.

Kerovec, M. & Mrsic, 1981. Catalogus faunae Jugoslaviae III/1. Oligochaeta Ljubljana: 39 pp.

Laakso, M., 1969. Oligochaeta from brackish water near Tvärminne, south-west Finland. Ann. Zool. Fenn. 6: 98–111.

Lang, C., 1984. Eutrophication of Lakes Léman and Neuchatel (Switzerland) indicated by oligochaete communities. Hydrobiologia 115: 131–138.

Lang & Reymond, 1996. Reversal of eutrophication in four Swiss lakes: evidence from oligochaete communities. Hydrobiologia 334: 157–161.

Legezynski, P., 1971. Skaposzczety wodne (Oligochaeta limicola) wyspy Wolin. Badania fizjograficzne nad Polska zachodinia 24B: 83–106.

Mann, K.H., 1958. Occurrence of an exotic oligochaeta *Branchiura sowerbyi* Beddard, 1892, in the River Thames. Nature 182: 732.

Milbrink, G., 1977. On the limnology of two alkaline lakes (Naivasha and Nakuru) in the East Rift Valley System in Kenya. Int. Rev. ges. Hydrobiol. 62: 1–17.

Milbrink, G., 1980. Oligochaete communities in pollution biology. The European situation with special reference to lakes in Scandinavia. In R. O. Brinkhurst & D. G. Cook (eds), Aquatic Oligochaete Biology. Plenum Press: 433–455.

Prat, N., M. Real & M. Rieradevall, 1992. Benthos of Spanish lakes and reservoirs. Limnetica 8: 221–229.

Rees, W. J., 1965. The aerial dispersal of Mollusca. Presidential address. Proc. Malac. Soc. London 36: 269–282.

Report 11/94, the County Administration of Jönköping. Munksjön, tillstånd och miljörisker (Lake Munksjön, allowances and risks for the environment): 42 pp. (In Swedish).

Timm, T., 1970. On the fauna of the Estonian Oligochaeta. Pedobiologia 10: 52–78.

Timm, T., 1980. Distribution of aquatic oligochaetes. In R. O. Brinkhurst & D. G. Cook (eds), Aquatic Oligochaete Biology. Plenum Press: 55–77.

Vagner, D., 1987. Prilog poznavanju faune Oligochaeta (Annelida, Clitellata) rijeke Una. Biosistematika, Acta biol. iugosl. G 13: 45–61.

Vagvolgyi, J., 1976. Body size, aerial dispersal and origin of the Pacific land snail fauna. Syst. Zool. 24: 465–488.

Hydrobiologia **406:** 143–148, 1999.
B. M. Healy, T. B. Reynoldson & K. A. Coates (eds), Aquatic Oligochaetes.

Oligochaeta and Aphanoneura in two Northern German hardwater lakes of different trophic state

Rut Collado[1], Peter Kasprzak[1] & Rüdiger M. Schmelz[2,*]
[1]*Institute of Freshwater Ecology and Inland Fisheries, Berlin, Department of Limnology of Stratified Lakes, Alte Fischerhütte 2, D - 16775 Neuglobsow, Germany*
[2]*University of Osnabrück, FB5, Systematic Zoology, D-49069 Osnabrück, Germany*
*(*author for correspondence)*

Key words: Potamodrilus, species distribution, oligotrophy, eutrophy

Abstract

Lake Haus and Lake Stechlin are well-characterized and intensely studied hardwater lakes located in the Baltic Lake District of Northern Germany. The former is classified as highly eutrophic, the latter is an oligotrophic hardwater lake. The way in which Oligochaeta and Aphanoneura species assemblages reflected the different trophic states of the lakes was investigated by a programme of qualitative and quantitative sampling over one year. This paper gives an account of the species occurrence in the two lakes. There were more species in the oligotrophic Lake Stechlin (56) than in the eutrophic Lake Haus (41), the number in the latter, however, is large for eutrophic lakes. The profundal of Lake Haus was devoid of any zoobenthos, probably because of anoxia, the profundal of Lake Stechlin exhibited an annelid species assemblage not found in other comparable oligotrophic lakes. In the littoral zone, species distribution patterns differ considerably between sites. At one site in Lake Stechlin, *Potamodrilus fluviatilis* was found, this being the first record of the species in inland lakes.

Introduction

Lake Haus ('Feldberger Haussee') and Lake Stechlin ('Stechlinsee') are well-characterised and intensely studied hardwater lakes located in the Baltic Lake District of Northern Germany (Casper, 1985; Koschel et al., 1985, 1990, 1993; Kasprzak et al., 1988, 1993; Casper & Koschel, 1995; Krienitz et al., 1996). Lake Haus has been classified as eutrophic (Krienitz et al., 1996), Lake Stechlin is a rare example of an oligotrophic hardwater lake (Casper & Koschel, 1995). Both lakes have been studied for more than 20 years and considerable information is available on their hydrological and limnological characteristics and recent history. Lake Haus was subjected to a period of heavy pollution with municipal organic waste from a nearby town, beginning in 1930, and more so, because of increased growth, from 1950 (Koschel et al., 1985), however, since 1980 it has undergone major restoration, including both load reduction and biomanipulation (Koschel et al., 1985, 1990, 1993). Monitoring programmes are being performed to observe the rate and sustainability of its recovery (e.g. Ronneberger et al., 1993). Research on Lake Stechlin was initiated by the construction of a nuclear power plant in 1959–1966 which used the lake as a cooling water reservoir. Investigations continued after shutdown of the plant in 1990. Lake Stechlin now has one of the most comprehensive species lists in the world due to an all-taxa inventory carried out from 1966 to 1975 (Casper, 1985). Oligochaeta and Aphanoneura, however, were only treated cursorily (14 spp. of Naididae). There were no previous records of Oligochaeta or Aphanoneura from Lake Haus.

The large amount of hydrological, limnological and sediment data available for the two lakes, and their similar size class, water chemistry and stratification patterns, make them especially suitable for a comparative study of species occurrence in contrasting trophic conditions. For this purpose, a programme of qualitative and quantitative sampling was performed in both lakes covering all depth zones and as many habitat types as possible. This paper describes the Oligochaeta and Aphanoneura species inhabiting the sediments of

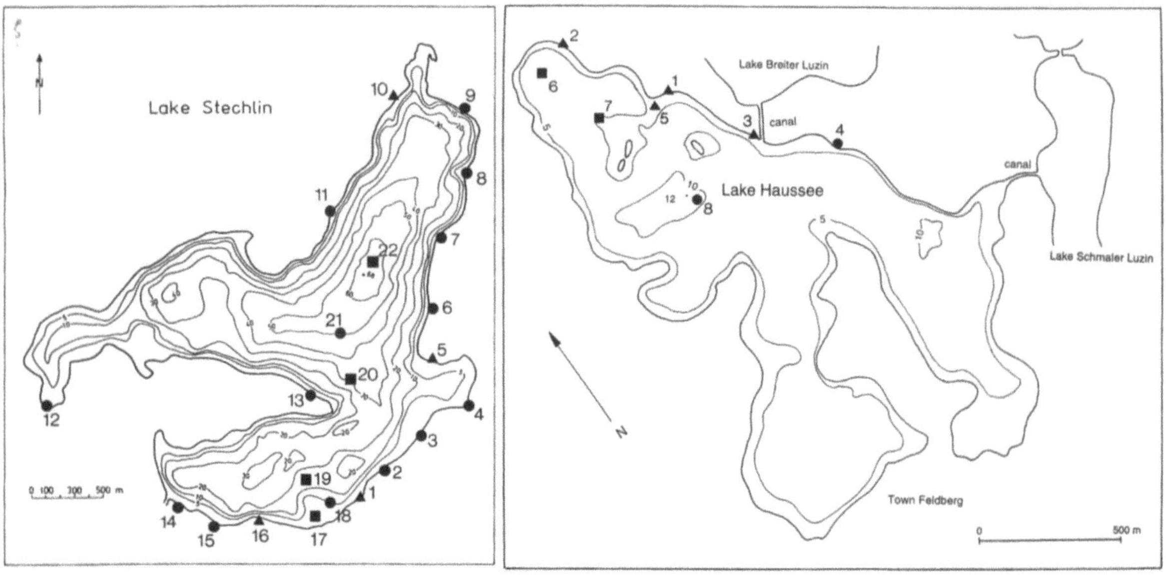

Figure 1. Lake maps with sampling stations. Symbols indicate different sampling programmes.

the two lakes. A more detailed analysis of spatial diversity and temporal changes in species distribution and abundance in relation to environmental parameters will be published elsewhere.

Descriptions of study area

Lake Haus and Lake Stechlin are part of the Mecklenburg Lake District ('Mecklenburgische Seenplatte'), a part of the Baltic Lake District, formed by melting ice after the last glaciation, about 12 000 years ago. Both are hardwater lakes and are well described in Casper (1985), Casper & Koschel (1995), Koschel et al. (1985), Mothes & Proft (1985), Proft (1995) and in Richter & Richter (1986). Their morphological and hydrological characteristics are shown in Table 1. We have divided the lake habitats into the profundal and littoral zones and in Lake Stechlin the littoral has been subdivided into eulittoral (surf shore/presence of emerged macrophytes) and sublittoral zones (presence of submerged macrophytes, mainly *Chara* species). Sampling sites were located in each of these habitats. The locations of sampling sites are shown in Figure 1, a description of the sites is given in Table 2.

Table 1. Morphological and hydrological characteristics of Lake Stechlin and Lake Haus

Parameters	unit	Lake Stechlin	Lake Haus
Surface area	km^2	4.3	1.36
Maximum depth	m	68.0	12.0
Mean depth	m	22.8	6.0
Volume	m^3 10^6	96.60	8.15
Catchment area	km^2	12.36	4.00
Altitude of water level	m (a.s.l.)	59.7	84.2

Materials and methods

A total of 22 sites in Lake Stechlin and eight in Lake Haus (Figure 1) were sampled qualitatively and quantitatively from March 1995 to August 1996. The sites were located to maximise habitat diversity. Samples were sieved (mesh size 50 μm) and worms were immediately sorted under a stereomicroscope and identified *in vivo*. Quantitative samples were fixed in formalin (4%). After washing the samples under running tap-water in a plankton-net (mesh-size 50 μm), the oligochaetes were sorted under a dissecting microscope and preserved in 70% ethanol until identification to species level. Specimens were studied as whole mounts in glycerine. Representative specimens of all species were either mounted in polyvinyl lactophenol (for chaetal characters) or stained with paracarmin, dehydrated in an alcohol through a xylene concentration series and mounted whole in Canada Balsam or

Table 2. Description of sampling sites in Lake Stechlin and Lake Haus

	Site number	Depth	Sediment	Tree litter layer	Macrophytes	Exposure
Lake Stechlin						
Eulittoral	4, 3, 16	surf shore	sand, pebbles	sparse	no	wind-exposed
	1, 2, 3, 14	20–25 cm	sand, few stones	abundant	no	wind-exposed
	6, 7, 8, 15	20–25 cm	sand, stones, pebbles, coated with algae	no	no	wind-exposed
	5, 11, 12	20–24 cm	organic mud	abundant	abundant, mainly *Phragmites australis*	sheltered
	9	10–30 cm	coarse sand	no	no	sheltered
	10	20–30 cm	sand, stones	present	submerged macrophytes	sheltered
Sublittoral	17, 18, 19	2, 5, 15 m	organic mud, rich in calcite, abundant mollusc shell debris	abundant, mainly *Chara*-species	no	
Profundal	20, 21, 22	15–68 m	planktogeneous fine-gyttja, coprogeneously transformed, black-brown in colour, oxygenated	no	no	
Lake Haus						
Littoral	1	20 cm	sand	abundant	emerged macrophytes (e.g. *Nuphar* sp.)	wind-exposed
	2	25 cm	sand, stones	very abundant	emerged, sparse	sheltered
	3 *	20–25 cm	sand, small pebbles, coated with algae	no	*Fontinalis* spp.	not exposed
	4	20 cm	fine sand	sparse	no	no
	5	2 m	fine mud, mollusc shell debris	very abundant	no	no
Profundal	6, 7, 8	5, 8, 11 m	sapropel; strong smell of H_2S	sparse	no	no

* at 25 m distance to the outlet of Lake Haus (canal), connecting it with Lake Breiter Luzin; continuous low-velocity water flow

Malinol. Altogether, 8227 specimens of Oligochaeta and Aphanoneura have been identified. The reference material is in the collection of the first author.

Results

Results of the species inventory for Lake Haus and Lake Stechlin are shown in Table 3. Altogether, 59 taxa of Oligochaeta and Aphanoneura were identified, 52 of these were determined to the species level. Eighteen species, recorded from Lake Stechlin, were not found in Lake Haus, and vice versa for three species.

In Lake Stechlin, we found 53 species, belonging to six families. The eulittoral was species-richest with 49 species. In the sublittoral we found 10, in the profundal six species. In the littoral, Naididae were the species richest family (25 species). The Enchytraeidae and Potamodrilidae were confined to the littoral. There were considerable differences as to species composition in the littoral sampling sites. Notable was Site 9 which was dominated by the meiobenthic aphanoneuran species *Potamodrilus fluviatilis*, but which was only occasionally recorded from other sampling sites. This rheophilous interstitial freshwater annelid (Bunke, 1967) is recorded here for the first time from an inland lake. It was found all year round. In the sublittoral, the Oligochaeta (Aphanoneura were not found) species composition was less diverse than in the littoral. Naididae were dominant, but only a few species occurred. *Vejdovskyella comata* and *Specaria josinae* were only found in the sublittoral. In the profundal, there were no great differences between sampling sites. Naididae, as in the sublittoral, were dominant. The only tubificid species that could be determined to species level was *Potamothrix hammoniensis*.

In Lake Haus, we found 41 species of Oligochaeta, belonging to four families. Aphanoneura were not

Table 3. Species list of Oligochaeta and Aphanoneura in Lake Stechlin and Lake Haus. The numbers indicate the sampling stations where the species were found

Species	Lake Stechlin			Lake Haus	
	(0–2 m)	(2–15 m)	(15–67 m)	(0–2 m)	(2–11)
Lumbriculidae					
Lumbriculus variegatus (Müller, 1774)	1,5,10,12,14,16			1,2,3	
Stylodrilus heringianus Claparède, 1862	1,4,5,10,15,16			1	
Rhynchelmis sp.	5			3	
Lumbriculidae sp.		19			
Naididae					
Chaetogaster diaphanus (Gruith., 1828)	1,5,7,10,11,16			1,2,3,4,5	6,7
Ch. diastrophus (Gruithuisen, 1828)	1,3,5,6,7,8,10, 11,13,14,15,16	17,18,19	22	1,2,3,4,5	
Ch. limnaei von Baer, 1827	1,5,14			1	
Nais barbata Müller, 1773	1,10,16			1,2,3,4,5	6
Nais bretscheri Michaelsen, 1899	1,3				
N. communis Piguet, 1906	1,5,10,16			1,2,3,5	
N. christinae Kasprzak, 1973	1,16			1,3,5	
N. pardalis Piguet, 1906	1,5,10,13,16			1,2,3,5	
N. pseudobtusa Piguet, 1906	1,5,10,13,15,16			1,2,3	
N. simplex Piguet, 1906	1,7	18		1,2,3	
N. variabilis Piguet, 1906	1,5,10,11			1,2,3,5	
Piguetiella blanci (Piguet, 1906)	10,16				
Amphichaeta leydigii Tauber, 1879	1		20,21,22		
Pristina aequiseta Bourne, 1891	1,4,9,10,16			1,2,3,4,5	
P. longiseta Ehrenberg, 1828	1			1,2,3,5	
Pristinella spp.	10			2	
Uncinais uncinata (Örsted, 1842)	1,16			1,2,3,4	
Slavina appendiculata (D'Udekem, 1855)	1,5	17		1,3	
Specaria josinae (Vejdovský, 1883)		18			
Stylaria lacustris (Linnaeus, 1767)	1,3,5,6,10,11,16	18,19		1,2,3,5	
Vejdovskyella comata (Vejdovský, 1883)		19			
V. intermedia (Bretscher, 1869)		17,18,19	20,21,22		
Dero dorsalis (Ferronière, 1899)	5				
D. digitata (Müller, 1773)	1,5	19	22	1,2,3,5	
D. obtusa D'Udekem, 1855	1			1,2,3,5	6
Tubificidae					
Tubifex ignotus (Stolc, 1886)	5,10,16				
T. tubifex (Müller, 1774)	5,16			1.3	
Psammoryctides albicola (Mich., 1901)	5			1.3	
Ps. barbatus (Grube, 1861)	1,13,16			1,2,3	
Ps. deserticola (Grimm, 1877)	5			3	
Spirosperma (S.) ferox (Eisen, 1879)				3,5	
Limnodrilus claparedeanus Ratzel, 1869	5,16				
L. hoffmeisteri Claparède, 1862	1,5,10,16	18		1,2,3	
L. udekemianus Claparède, 1862	1,5,16			1,2,3	
Bothrioneurum vejdovskyanum Stolc, 1888	1,2,5,7,8,10,13,16			1,2,3,4,5	
Potamothrix bavaricus (Öschmann, 1913)	10			1,2,3	
P. hammoniensis (Michaelsen, 1901)	1,5,10,14,16		20,22	1,2,3,4,5	
P. heuscheri (Bretscher, 1900)	1			1,2,3,5	6
Peloscolex sp.				3	

Table 3. Continued

Species	Lake Stechlin (0–2 m)	(2–15 m)	(15–67 m)	Lake Haus (0–2 m)	(2–11)
Enchytraeidae					
Cognettia cognettii (Issel, 1905)	1,16			1,3	
C. glandulosa (Michaelsen, 1888)	10,15			1,3	
C. sphagnetorum (Vejdovsky, 1887)	1,16			1,3	
Henlea ventriculosa (d'Udekem, 1854)	10,16			3	
Buchholzia appendiculata (Buchholz, 1862)	1,3				
Fridericia galba (Hoffmeister, 1843)	1,3				
Cernosvitoviella spp.	1,4,7			3	
Marionina argentea (Michaelsen, 1889)	3,6			3,4	
M. riparia Bretsch., 1889 augm. Cern., 1928				1,2,3	
Lumbricillus fennicus Nurminen, 1964	1,2,3,7,13,16				
Enchytraeus sp.	2			2	
Lumbricidae					
Eiseniella tetraedra (Savigny, 1826)	1				
Potamodrilidae					
Potamodrilus fluviatilis Lastockin, 1935	3,9,10,13,16				
Aeolosomatidae					
Aeolosoma cf. *olivaceum* Bunke, 1967	7,8,10,16				
A. hemprichi Ehrenberg, 1828	3				
Aeolosoma spp.	3,4,10,13		22		

found. All recorded species were present in the littoral. The profundal was nearly devoid of species, and of zoobenthos in general; four oligochaete species were recorded here, comprising, however, only five specimens in two out of 12 samples. The species richest sampling site in the littoral (Site 3, with 37 species) is located at the mouth of a canal (Luzinkanal) which connects Lake Haus with Lake Breiter Luzin.

Discussion

Species numbers of Oligochaeta and Aphanoneura in both lakes are high in comparison with other softwater lakes of similar size, e.g. 20 (Särkkä, 1982, from Lake Könnevesi) or 26/27 (Lafont, 1982, from Lakes Gérardmer and Longemer). Possible explanations for this species richness are:

1. Meiobenthic species are not generally recorded. This may be due to inadequate sampling and sorting (Särkkä, 1989, Timm et al., 1996).
2. The majority of studies are confined to the profundal, whereas the greatest species diversity is found in the littoral.
3. Most surveys do not aim at developing a comprehensive species inventory but rather look for large and easily detectable species as trophic indicators as Lang (1986, 1989) has demonstrated with tubificid and lumbriculid species in Swiss lakes.
4. Finally, Enchytraeidae, which account for 20% of the species richness in both lakes, are rarely determined to species level in other investigations on aquatic oligochaetes.

Species number is also high in Estonian eutrophic hardwater lakes. Fifty-five species are recorded for Lake Võrtsjärv (Kangur et al., 1998), 59 for Lake Peipsi-Pihkva (Timm et al., 1996). Again these numbers may rather reflect the sampling effort (over decades in Estonia) than real differences between hardwater and softwater lakes. Still it is remarkable that the species numbers in the highly eutrophic Lake Haus is not much lower compared with the oligotrophic Lake Stechlin.

The dominance of meiobenthic species in the Lake Stechlin profundal is similar to the results of Särkkä, who, in a series of papers (e.g. Särkkä, 1982, 1989), first showed that small-sized annelid worms are more or less regular inhabitants of the profundal zones of oligotrophic lakes. There are, however, differences in species composition. Some of the dominant meiobenthic species in Lake Stechlin, for example, *Vejdovskyella intermedia* and *Chaetogaster diastrophus*, were not found in the oligotrophic softwater Lake Päijänne (Särkkä, op. cit.). In Lake Haus, a more than 3 month long period of summer anoxia in the hypolimnion (Koschel, pers. comm.) is the most likely explanation for the absence of any kind of zoobenthos in the profundal.

The species distribution in the littoral of both lakes is highly diverse. 92% of all recorded oligochaete and aphanoneuran species (49 out of 53) were found at the shoreline, i.e. in sediments not deeper than 2 m below water level. A correlation between habitat diversity and species richness is probable, although not demonstrable in this paper.

Acknowledgements

This work was made possible by a postdoctoral grant to the first author, received from the Ministerio de Educación y Ciencia, Spain. We are grateful to Rainer Koschel for providing the hydrological and limnological data on the lakes studied. We thank Adelheid Scheffler, Uta Mallock, Michael Sachtleben and Roman Degebrodt for their help in the field, especially with sample taking at large depths and in frozen lakes.

References

Bunke, D., 1967. Zur Morphologie und Systematik der Aeolosomatidae Beddard, 1895 und Potamodrilidae nov. fam. (Oligochaeta). Zool. Jb. Syst. 94: 187–368.

Casper, S. J. (ed.), 1985. Lake Stechlin – A Temperate Oligotrophic Lake. Monographiae Biologicae 58, Dr. W. Junk Publishers, Dordrecht, The Netherlands: 550 pp.

Casper, P. & R. Koschel, 1995. Description of Lake Stechlin. Limnologica 25: 281–284.

Kangur, K., H. Timm, T. Timm & V. Timm, 1998. Long-term changes in the macrozoobenthos of lake Võrtsjärv. Limnologica 28: 75–83.

Kasprzak, P., J. Benndorf, R. Koschel & F. Recknagel, 1988. Applicability of the food-web manipulation in the restoration program of a hypertrophic stratified lake: Model studies for Lake Haussee (Feldberg, GDR). Limnologica 19: 87–95.

Kasprzak, P., L. Krienitz & R. Koschel, 1993. Changes in the Lake Feldberger Haussee (Germany, Mecklenburg-Vorpommern) in response to biomanipulation. Arch. Hydrobiol. 128: 149–168.

Koschel, R., S. J. Casper, W. Dolgner, O. Festersen, P. Kasprzak, H.-D. Krausch, L. Krey, M. Pächnatz, E. Penz, G. Proft & D. Ronneberger, 1985. Das Feldberger Seengebiet. Ein Beitrag zum Schutz und zur optimalen Nutzung der Feldberger Seen. Natur und Umwelt Bez. Neubrandenburg 3: 1–96.

Koschel, R., B. Giering, P. Kasprzak, G. Proft & H. Raidt, 1990. Changes of calcite precipitation and trophic conditions in two stratified hardwater lakes of the Baltic Lake District of the GDR. Verh. int. Ver. Limnol. 24: 140–145.

Koschel, R., P. Kasprzak, L. Krienitz & D. Ronneberger, 1993. Long-term effects of reduced nutrient loading and food-web manipulation on plankton in a stratified Baltic hardwater lake (Lake Hausee, Germany, Feldberg). Verh. int. Ver. Limnol. 25: 647–651.

Krienitz, L., P. Kasprzak & R. Koschel, 1996. Long term study on the influence of eutrophication, restoration and biomanipulation on the structure and development of phytoplancton communities in Feldberger Haussee (Baltic Lake District, Germany). Hydrobiologia 330: 89–110.

Lafont, M., 1982. Les Oligochètes des lacs de Gérardmer et de Longemer. Sciences de l'eau 1: 21–30.

Lang, C., 1986. Eutrophisation du Léman indiquée par les communautés d'oligochètes: campagnes 1982–1985. Schweiz. Z. Hydrol. 48: 230–239.

Lang, C., 1989. Eutrophication of Lake Neuchâtel indicated by the oligochaete communities. Hydrobiologia 174: 57–65.

Mothes, G. & G. Proft, 1985. Sediments and sedimentation. In S. J. Casper (ed.), Lake Stechlin – A Temperate Oligotrophic Lake. Monographiae Biologicae 58. Dr. W. Junk Publishers, Dordrecht, The Netherlands: 375–397.

Proft, G., 1995. Die Sedimente der Seen des Stechlinseegebietes (Deutschland): Seetyp, Verteilung im Seebecken, Trophie und Chemismus. Limnologica 25: 333–351.

Richter, W. M. & I. Richter, 1986. Zeittafel zur Entwicklung der Feldberger Seenlandschaft. Natur und Naturschutz in Mecklenburg, Greifswald-Waren 23: 12–26.

Ronneberger, D., P. Kasprzak & L. Krienitz, 1993. Long term changes in the rotifer fauna after biomanipulation in Feldberger Haussee (Germany, Mecklenburg-Vorpommern), and its relationship to the crustacean and phytoplankton communities. Hydrobiologia 255/256: 297–304.

Särkkä, J., 1982. On the ecology of littoral Oligochaeta of an oligotrophic Finnish lake. Holarct. Ecol. 5: 396–404.

Särkkä, J., 1989. Meiobenthic naidid and aelosomatid oligochaetes from the profundal zone, and relations of species to eutrophication. Hydrobiologia 180: 185–190.

Timm, T., K. Kangur & V. Timm, 1996. Macrozoobenthos of Lake Peipsi-Pihkva: taxonomical composition, abundance, and their relation to some ecological parameters. Hydrobiologia 338: 139–154.

Hydrobiologia **406:** 149–163, 1999.
B. M. Healy, T. B. Reynoldson & K. A. Coates (eds), Aquatic Oligochaetes.

Micro-distribution of oligochaetes in a soft-bottomed lowland stream (Elsbeek; The Netherlands)

Piet F. M. Verdonschot
Institute for Forestry and Nature Research, P.O. Box 23, 6700 AA Wageningen, the Netherlands

Key words: oligochaete, hydraulics, discharge, substrate, organic matter, season

Abstract

This study focuses on a low-gradient, soft-bottomed stream which shows strong fluctuations in discharge, partly natural (i.e. due to rainfall) and partly due to agricultural land-use. This dynamic discharge pattern causes silt deposition between discharge peaks which in turn cause the silt to be washed out. This process occurs independently from the height of the peak. The main substrate in the stream is quite homogeneous fine sand. Twenty-three species of oligochaetes were found. Their distribution and abundance varied with season and substrate. Mature tubificids and *Lumbriculus variegatus* occurred throughout the year. *Stylodrilus heringianus* occurred in winter and spring. Different naidids occur in different seasons, though half of the species can be collected all year. The micro-habitat of three tubificids, one lumbriculid and eight naidids were be identified. A high number of oligochaete species coexisted in a small stretch of the stream due to a separation in micro-habitat in space and time. The temporal variation in distribution was due to season. Particle size, the mineral and organic component of the substrate, and position in the stream bed influenced the spatial distribution of species. Temporal and spatial distribution of oligochaetes can not be separated.

Introduction

Among others, Hynes (1970) stated that a stream is controlled by its catchment. Similarily, the stream bed (shape and mosaic patterns) is ruled by the stream. To understand the ecology of a stream and its inhabitants, one should study the stream habitats (Southwood, 1977). The physical structure (e.g. grain size, organic particle fractions) and dynamics of a stream habitat directly determine the distribution of organisms. These physical conditions within a habitat also control food levels and modify biotic interactions (Frissell et al., 1986). The physical environment is determined by the hydraulics, current velocity and turbulence. If, as assumed, stream communities are largely adjusted to and controlled by physical factors in space and time, the question becomes one of understanding these physical factors and their mutual interactions. In addition to the overall physico-chemical conditions, these factors are the hydraulics, the substratum and the food (Cummins & Lauff, 1969).

This study focuses on the micro-distribution of oligochaetes in a lowland stream in the Netherlands. The objective is to characterise the relationship between the quantitative distribution of oligochaetes and the major constraining physical and temporal factors in a soft-bottomed lowland stream (Elsbeek; the Netherlands). A second objective was to compare substrate (habitat) characteristics established in the field as well as in the laboratory.

Study area

The study location is situated in the lowland stream 'Elsbeek' (in translation: *Alnus*-stream) in the eastern part of the Netherlands. The stream lacks a well-defined source and dries up in summer. It runs down, over a length of 5.9 km, from an ice-pushed hill-ridge towards the river Dinkel. It is a soft-bottomed lowland stream with a slope of 2.5 m/km. The stream is rain-water fed, discharge follows largely the precipitation pattern (atlantic sea-climate). The catchment (1450 hectare) is mainly in intensive agricultural use (grassland 37%, fields 17%, urbanised area 6%) though still

40% is forest. The stream is dry from half July until the beginning of August.

The surrounding agricultural land-use has a large impact on the stream because nutrients enter the system, causing hypertrophic, mesosaprobic conditions. Even more important is the impact on the hydrology. The agricultural land is drained to a depth of 0.8–1.0 m below land-level. Furthermore, the stream is canalised in parts of its upper and lower course which has resulted in an enlarged and deepened stream-bed. Canalisation and drainage have lead to stronger fluctuations in discharge. This implies higher short-term discharge peaks and longer periods of slow flow or even drought. The sampling site is situated in a semi-natural stretch in the middle course (2.4 km from the mouth). The transversal stream profile is asymmetric. The site is shaded by a wooded buffer-strip with irregular banks.

Materials and methods

Sampling

Sampling took place every two months from March 1993 to September 1993, and every two weeks from October 1993 to December 1994. Moment of sampling was defined in a nominal parameter season with dummy variables (0/1) spring, summer, autumn and winter. Three hydraulic parameters were measured at each of the habitats. Depth and current velocity were recorded at the time of sampling, whereas discharge was registered continuously downstream from the sample site.

Three measures of substrate structure were considered: substrate-class, organic matter content and grain-size fraction. The visual impression of origin (mineral or organic) and structure of the substrates were recorded according to pre-defined substrate-classes. Three mineral substrate-classes and four organic ones were distinguished:

substrate class	*description*
mineral clay	very mineral material (no grains perceptible by hand)
mineral sand	fine sand (grains perceptible by eye; < 0.5 mm)
mineral gravel	grain size of coarse sand and gravel (grains perceptible by eye; > 0.5 mm)
organic silt	very fine (no structure visible fragments) organic deposits
organic detritus	fine visible organic fragments
organic litter	coarse organic fragments such as leaves, grasses, sticks
organic leaves	leaves and large leaf fragments

At each habitat, the top two centimetres of substrate (about 200 gr. material) were sampled with a cylindric corer for quantitative substrate analysis, except for organic leaves which were sampled from leave packages. The organic matter content was analyzed separately from the substrate (ash free dry weight; 2 h at 800 °C) ate samples for substrate particles smaller and larger than 2 mm. Total organic matter content was calculated afterwards and classified:

% organic matter content	*name*
< 1%	low organic matter content (LOM)
1–5%	medium organic matter content (MOM)
> 5%	high organic matter content (HOM)

Grain-size analysis was performed in which the mineral substrate was divided into geometric particle-size fractions (Wenthworth, 1922). On the basis of these fractions the mineral substrate can be characterised according to a $Q_1M_dQ_3$ index. This $Q_1M_dQ_3$ index (Doeglas, 1968) indicates the 25%, 50% and 75% quartile of the percentages of the fraction weights in relation to the total mineral weight. The median diameter (M_d) characterises the sample. By distinguishing the two other quartiles (Q_1 and Q_3), the homogeneity of particle distribution in the sample is also taken into account. The fractions are named according to combined $Q_1M_dQ_3$ classes. The following fractions were distinguished and used in further analysis:

grain size diameter (mm)	*phi value*	$Q_1M_dQ_3$	*name*
32–64	–5	–x –5 –x	gravel
16–32	–4	–x –4 –x	gravel
8–16	–3	–x –3 –x	gravel
4–8	–2	–x –2 –x	gravel
2–4	–1	–x –1 –x, –x 1 x	gravel

1–2	0	–x x x	coarse sand
0.5–1	1	1 x x	coarse sand
0.25–0.5	2	2 2 x	fine sand
0.125–0.25	3	2 3 x, 3 3 x	fine sand
0.063–0.125	4	x 4 x	lutum
< 0.063	5–10	x 5 x	lutum

The macrofauna samples were collected with a micro-macrofauna shovel (Tolkamp, 1980) which has a sampling area of 150 cm^2 and a depth of 2 cm. The macrofauna samples were sieved (mesh sizes 1.0, 0.5 and 0.25 mm) and oligochaetes sorted by eye. The oligochaetes were preserved in formalin (4%), identified and counted.

Data processing

In total, 87 of the 93 samples contained oligochaetes. The oligochaete abundances were transformed to a logarithmic scale and the data were ordinated by Detrended Canonical Correspondence Analysis (DCCA) with the program CANOCO 3.1 (Ter Braak, 1990). DCCA assumes a unimodal model for the relationship between the response of each taxon and the ordination axes. A unimodal model is appropriate because a preliminary Detrended Correspondence Analysis (DCA) showed a relatively long ecological gradient (>3 SD: Gauch, 1982). Partial DCCA ordination can be used to control for temporal variation, and to focus on the effects of habitat variables (Ter Braak & Prentice, 1988). Partial ordination was carried out by specifying season as a covariable. This yields an ordination of the residual variation in oligochaete data after the effects of the covariables are factored out by multivariate linear regression.

The ordination results are presented as correlation biplots of oligochaetes and environmental variables (CANOCO scaling option 2). In these plots, taxa close together tend to be more highly associated than taxa far apart. The closer a taxon is projected (by a perpendicular projection of the taxon onto the arrow) towards the tip of an environmental arrow, the more it is related to this respective variable and vice versa.

The distribution of individual oligochaete species over different substrate classes and fractions was examined using the Index of Representation (IR) (Hildrew & Townsend, 1976) and statistically tested with the chi-squared test. Under the null hypothesis (H_o) an oligochaete occurs in all substrate types in equal abundance. H_o is rejected when the difference between observed and expected abundance yields chi-squared values higher than the 5% level in a chi-squared distribution table. This indicates under or over representation in one or more substrate types. Positive IR-values indicate affinity while negative ones indicate avoidance. IR-values that deviate 2 or more from zero add to an unequal species distribution (Tolkamp, 1980).

Results

Physical factors

Discharge was averaged over two week periods. Discharge patterns of the 'Elsbeek' stream show three peaks from December 1993 to the end of April 1994, and again in November 1994 (Figure 1). The discharge from late autumn until spring is about 10–20 times higher than the summer base flow. In summer, incidental peaks, 4–6 times higher than base flow, occurred irregularly in July and September 1994. The discharge follows the precipitation pattern and is irregular (unpredictable) with large fluctuations. This pattern is characteristic for low-gradient streams in agricultural areas and is due to drainage (Faber, 1972).

Current velocity was measured only at the times of sampling the benthos. No relation between velocity and discharge was observed. Still, velocity shapes the substrate and thus changes in the substrate pattern should be related to changes in discharge. The two major substrates observed in the field are mineral sand and organic silt. They account for more than 90% of the total substrate. Both sand and silt are plotted with discharge in Figure 1. In some periods no substrates could be observed due to a high discharge, turbulent water with a low transparency. Over time, sand and silt were negatively related. The changes (differences between percentage sand and silt over time) include about 25% of the stream bed surface. This pattern runs parallel with discharge peaks. After each discharge peak the percentage silt was reduced and the substrate mainly consists of sand. It is striking to see that it is not the height of the peak that is important but rather the change in discharge. The grain size classes of the substrate sample were plotted in a $Q_1M_dQ_3$ graph (Figure 2). The M_d(horizontal axis) is plotted against the three quartiles Q_1,M_d, and Q_3 all expressed in phi values. The M_d values are positioned on an imaginary line with 45 degree angle between horizontal and

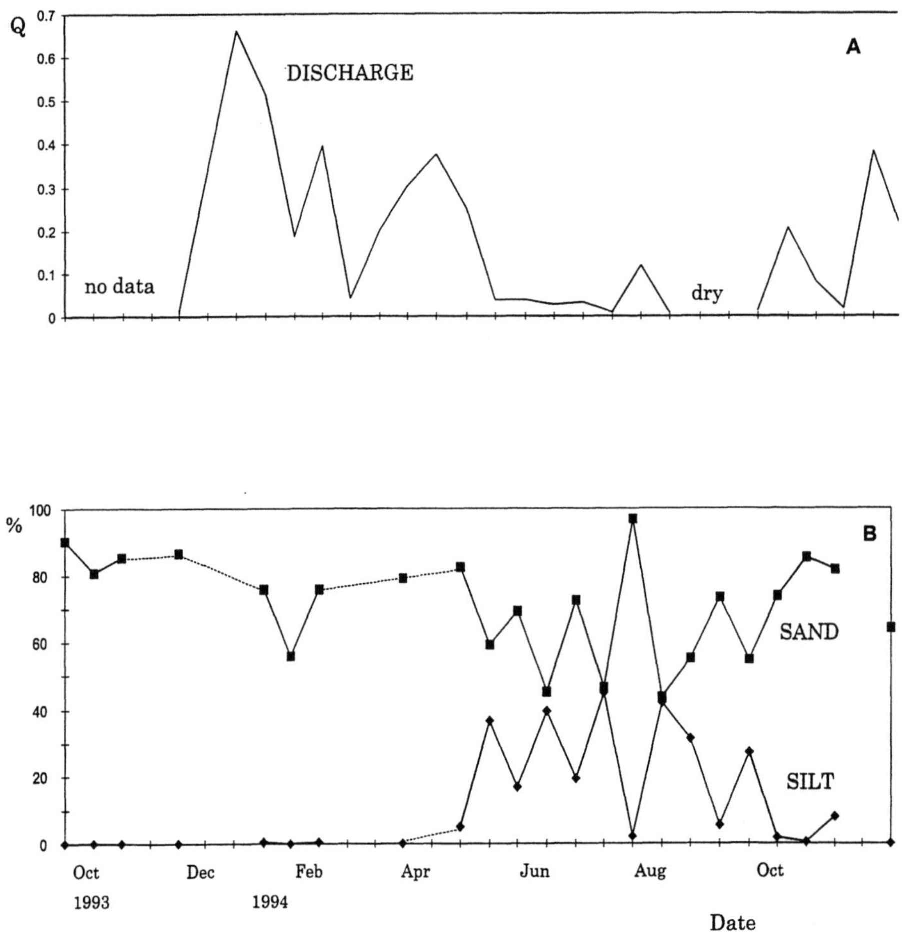

Figure 1. The discharge (above) of the 'Elsbeek' stream in 1993 and1994. In this graph the percentage of the two major substrates (sand and silt) identified by field observation are plotted below.

vertical axis. The Q_1 and Q_3 are plotted below and above the M_d value, respectively. According to names coupled with the $Q_1M_dQ_3$ index, the figure is divided into four sections (the four main substrates). They represent the substrates classified as gravel, coarse sand, fine sand and lutum. Most samples of the 'Elsbeek' are classified as homogeneous fine sand (66%). Not only the M_d but also the Q_1 and Q_3 fraction are projected in the class of fine sand. The lutum samples (5%) are also homogeneous whereas the samples of coarse sand (19%) and gravel (10%) are more heterogeneous.

In Figure 2, the open symbols represent samples without oligochaetes. In most samples oligochaetes were collected without a clear preference for only one of the four main substrate classes.

Six samples (6%) were taken in leaf packages, which implies 100% organic matter. The organic matter content was low in 46% of the sediment samples, medium in 33% and high in 16%. Thus, each of the classes contains a fair number of samples.

No relation could be established between current velocity and organic matter content nor grain size fraction, respectively. Current velocity appears to be measured on too short time scale to be useful.

A Principal Component Analysis (PCA; program CANOCO 3.1(Ter Braak 1990)) was done to compare between the field and laboratory substrate (habitat) classification systems. Hereby, also the other environmental variables were taken into account. The results are presented as a correlation biplot (Figure 3). In the right part of the diagram, the (mutual) correlation between the field classes gravel with the laboratory classes belonging to gravel (phi-values −4 to −1) is shown. Of course, current velocity correlates to these substrates. Both laboratory coarse sand classes (phi-values 0 and 1) are separated from the laboratory

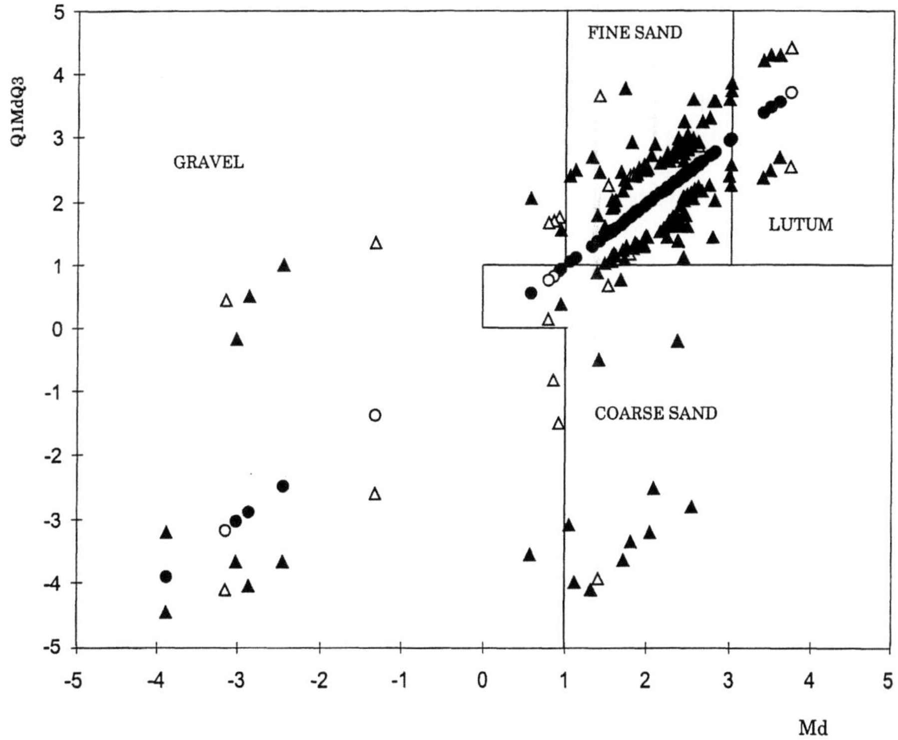

Figure 2. $Q_1M_dQ_3$ graph for mineral substrates of the 'Elsbeek' stream. The blocks indicated represent the major substrate classes. For further explanation see text.

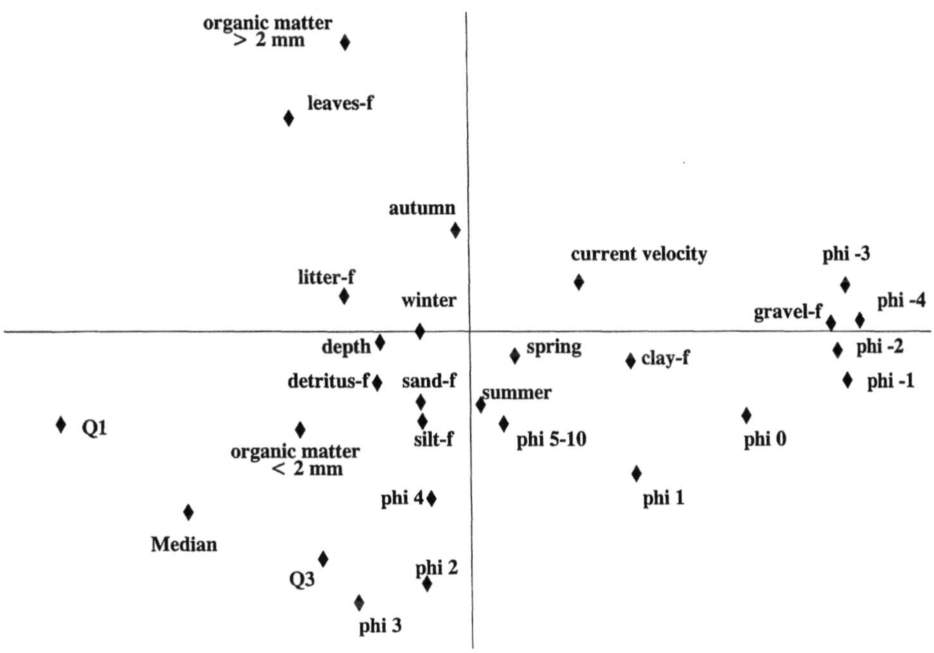

Figure 3. Ordination (PCA) diagram showing the correlation between environmental variables.

fine sand classes (phi-values 2 ad 3). They do not correspond to the field class sand in the middle of the diagram. The field class clay slightly corresponds to the laboratory lutum class (phi-value 5–10). Both are projected towards the right lower middle of the diagram. The Q_1, M_d and Q_3 characteristics do not correspond to the phi-values. Finally, the organic matter classes are projected in the middle towards the left upper corner. They show a gradient running from silty material below the middle left up to leaves and amount of organic matter (> 2 mm) in the left upper corner. The field observations correspond to the amount of organic matter established in the laboratory.

Oligochaetes

Nine species of tubificids, two lumbriculids and 12 naidids were collected (Table 1). All data collected during 1993 and 1994 were ordinated to describe the major variations in taxa distributions. Detrended Correspondence Analysis by segments was used to establish the length of gradient (the beta-diversity) in the oligochaete data. The length of gradient for the first two axes is greater than 3 meaning that the use of a unimodal model is valid (Ter Braak, 1990). Further analysis was done by using Detrended Canonical Correspondence Analysis (DCCA). The DCCA-ordination diagram (Figure 4) shows a main gradient along the first axis representing a seasonal difference between spring (at the left) and summer (at the right). Along the second axis the oligochaetes are distributed according to sites with the finest mineral substrate fraction (phi 5–10) sampled in the winter at the top, versus organic silt substrate at the bottom. The various species show clear seasonal patterns. The ordination summarised explains 40% of the variance in the oligochaete – environment data. The unrestricted permutation test demonstrated that the abiotic parameters were significantly (P=0.01) correlated with the distribution of oligochaetes among the samples.

The relation between oligochaete species and season is further examined by plotting the monthly total abundance of the species (Figure 5 & 6). Taxa identified to a coarse resolution such as family level and certain rare species (i.e. frequency of occurrence lower than 5) were excluded from the analysis. Mature specimens of all tubificids (*Limnodrilus hoffmeisteri, L. profundicola, L. udekemianus, Tubifex tubifex* and *Aulodrilus pluriseta*) occurred throughout the year but showed individual peaks in abundance at different times of the year (Figure 5). *Stylodrilus heringianus* occurred in winter and spring while *Lumbriculus variegatus* was present all year though in low abundances (Figure 5). All of these species were positioned near the centre of the ordination diagram (Figure 4), except for *Aulodrilus pluriseta* which was not found in winter. The naidids *Nais variabilis, N. communis, N. elinguis* and *Slavina appendiculata* were absent in June and July and were very abundant in May and November (Figure 6). They occur more towards the right along the first axis in the ordination diagram (Figure 4), the direction of winter and spring. In general, they can be collected throughout the year. *Nais pardalis* and *N. simplex* occurred in spring (Figure 6). *Pristina amphibiotica* was abundant in summer (located to the left in the ordination diagram) (Figure 4) and *Stylaria lacustris* occurred in spring and abundantly in autumn (Figure 6).

In the first ordination analysis, the distribution of oligochaetes is mainly due to season. As we were interested in the relationship between oligochaete species and habitat (depth, current or substrate) factors, two partial ordinations were carried out. Partial ordination yields an ordination of the residual variation in the taxa data after the covariables are factored out by multivariate linear regression (Verdonschot & Ter Braak, 1994; Ter Braak & Verdonschot, 1995). In our case, in the first run 'season' and in the second 'substrate' was considered as a covariable. The first run yielded an ordination diagram whereby the residual variation in the oligochaete data is shown after season is factored out. In Figure 7, the resulting diagram shows a primary gradient (axis 1) of shallow streams with gravel substrates to shallow streams with silt and detritus substrates. The second axis represents a gradient of streams with fine mineral substrate to sand substrate (by field observation) in the top. Somewhat to the left and below leaf packages also are important. The ordination explained 33% of the variance in oligochaete distributions (first two axes). The statistical test indicates that the ordination results approached significance (P=0.08), suggesting a weak relationship between the oligochaete community and the substrate once seasonal effects were removed. The second run yielded an ordination diagram whereby the residual variation in the oligochaete data is shown after 'substrate' (both field and laboratory parameters) is factored out. Figure 8 shows the seasonal relationship of species along the first two ordination axes. The ordination explained 85% of the variance in oligochaete distributions (first two axes). The statistical test indicates that the ordination results approached signi-

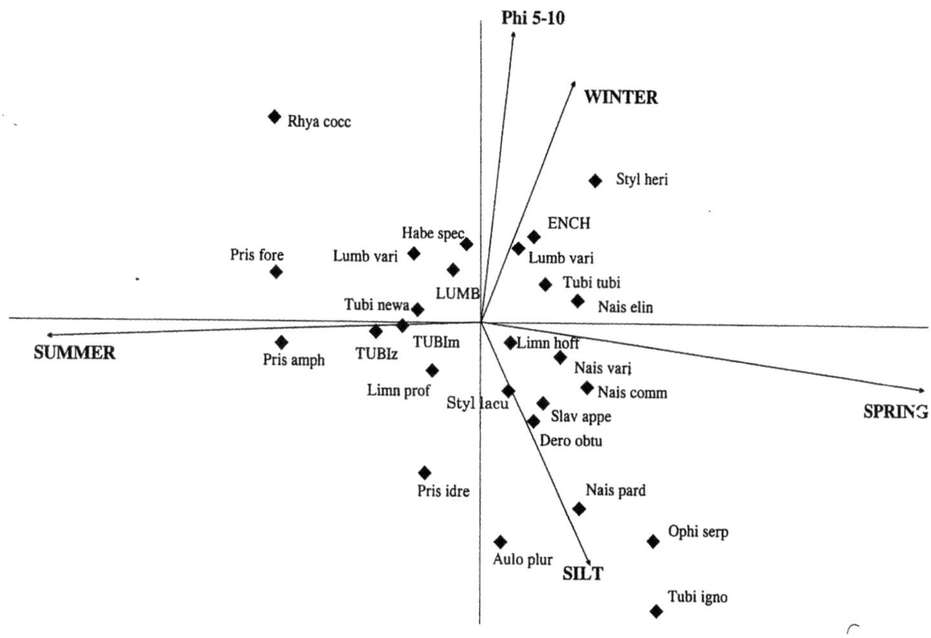

Figure 4. Ordination (DCCA) diagram showing the variation in oligochaetes (dots) among seasons and habitat related variables (arrows). See Table 1 for species codes.

ficance (P=0.01), suggesting a significant relationship between the oligochaete community and season. Comparing the eigenvalue of both partial ordinations, the differences appear small.

For the more frequently collected oligochaete species their individual relationship with substrate classes by laboratory analysis (Table 3) as well as by field observation (Table 4), and with season (Table 5) was calculated by the index of representation and tested by a chi-squared test. Clay was not included in the 'substrate' analysis because there were too few samples with clay.

The fractions of lutum and fine sand with low organic matter content were not selected by oligochaetes. The fraction of fine sand with medium organic matter content is selected by 10 of the 15 species. Only one species was most common on coarse sand with a high organic matter content, one for gravel and one for leaf packages, respectively. The other substrate classes are each preferred by three species. The substrate classes established by field observation are for the organic ones preferred by three or four (detritus) species, while the mineral substrates are preferred by one (sand) or two species (gravel).

Aulodrilus pluriseta shows a weak preference for fine sand with a medium organic matter content although no relationship with the field samples could be established. The ordination (Figure 7) indicates a strong relationship with silt and detritus. This confirms the preference indicated because a fine sandy substrate with a medium amount of fine organic material is in the field often characterised as silt. *A. pluriseta* does not shown a seasonal preference (Table 5).

Limnodrilus hoffmeisteri prefers fine sand with a medium to high organic matter content and coarse sand with a medium organic matter content and avoids all other substrates (Table 3). In the field, the preferred substrate is termed litter. This implies that the structure of the organic matter on coarse or fine sand consists of quite coarse fragments. *L. hoffmeisteri* is significantly dominant in spring (Table 5).

Limnodrilus profundicola significantly prefers coarse sand with a medium organic matter content. In the field, these substrates are classified as being gravel (this class includes the grain size of coarse sand). Thus, the organic matter is fine and mixed with the sediment. *L. profundicola* is significantly abundant in autumn (Table 5).

Limnodrilus udekemianus shows a slight preference for coarse sand with a low organic matter content. Again this substrate was in the field observed as being gravel, though this relationship is not significant. The ordination diagram indicates a relationship with fine sediments. From this, no conclusion can

Table 1. Oligochaetes collected in the 'Elsbeek' stream. Total number of samples in which species were found (87) and total number of individuals collected per taxon are indicated

	Letter code	Number of samples	Number of individuals
TUBIFICIDAE			
Tubificidae j.w.s.	TUBIz	48	260
Tubificidae j.wh.s.	TUBIm	59	801
Limnodrilus hoffmeisteri	Limn hoff	28	251
Limnodrilus profundicola	Limn prof	7	25
Limnodrilus udekemianus	Limn udek	12	18
Tubifex tubifex	Tubi tubi	21	70
Tubifex ignotus	Tubi igno	1	1
Tubifex newaensis	Tubi newa	1	1
Aulodrilus pluriseta	Aulo plur	6	12
Rhyacodrilus coccineus	Rhya cocc	2	4
Haber speciosus	Habe spec	2	4
LUMBRICULIDAE			
Stylodrilus heringianus	Styl heri	12	42
Lumbriculus variegatus	Lumb vari	7	7
LUMBRICIDAE	LUMB	38	129
ENCHYTRAEIDAE	ENCH	25	66
NAIDIDAE			
Nais variabilis	Nais vari	17	58
Nais pardalis	Nais pard	8	67
Nais elinguis	Nais elin	35	1057
Nais communis	Nais comm	23	96
Nais simplex	Nais simp	6	21
Pristina amphibiotica	Pris amph	11	58
Pristina foreli	Pris fore	4	11
Pristina idrensis	Pris idre	1	2
Ophidonais serpentina	Ophi serp	3	9
Stylaria lacustris	Styl lacu	9	84
Slavina appendiculata	Slav appe	13	47
Dero obtusa	Dero obtu	5	8

be drawn concerning its substrate preference. Apparently, other environmental factors influence its distribution. *L. udekemianus* is significantly more abundant in summer.

Tubifex tubifex prefers fine sand with a medium organic matter content and coarse sand with a medium to high organic matter content. In the field these substrates were classified as silt (organic) (Tables 3 & 4). Thus, there is a high amount of fine organic matter mixed with the mineral sediment. Also, in the ordination diagram, *T. tubifex* shows a slight preference for fine sediments. *T. tubifex* is a spring species (Table 5).

Stylodrilus heringianus prefers fine sand with a medium organic matter content. Samples classified as leaves in the field. This relationship is not shown in the ordination diagram. *S. heringianus* is significantly abundant in spring (Table 5).

Lumbriculus variegatus does not show significant relationships to specific substrate classes. This is partly due to its low abundance in the data set.

Nais communis prefers fine sand with a medium organic matter content. In the field, these samples are classified as detritus (Table 4). The structure of the organic matter is finely fragmented. This relationship

Table 2. Summary results from the DCCA ordination analysis

	axis 1	axis 2	axis 3	axis 4	trace	*P*-value
DCA-eigenvalue	0.43	0.25	0.15	0.12		
DCA-length of gradient	3.3	3.4	2.6	2.3		
DCCA-eigenvalue	0.32	0.15	0.10	0.08	1.2	0.01
DCCA-% variance explained (cumulative)	28	40	49	56		
partial DCCA-eigenvalue *season*	0.15	0.11	0.07	0.06	0.8	0.08
partial DCCA-% variance explained (cumulative) *season*	19	33	43	50		
partial DCCA-eigenvalue *substrate*	0.16	0.04	–	–	0.2	0.01
partial DCCA-% variance explained (cumulative) *substrate*	69	85	–	–		

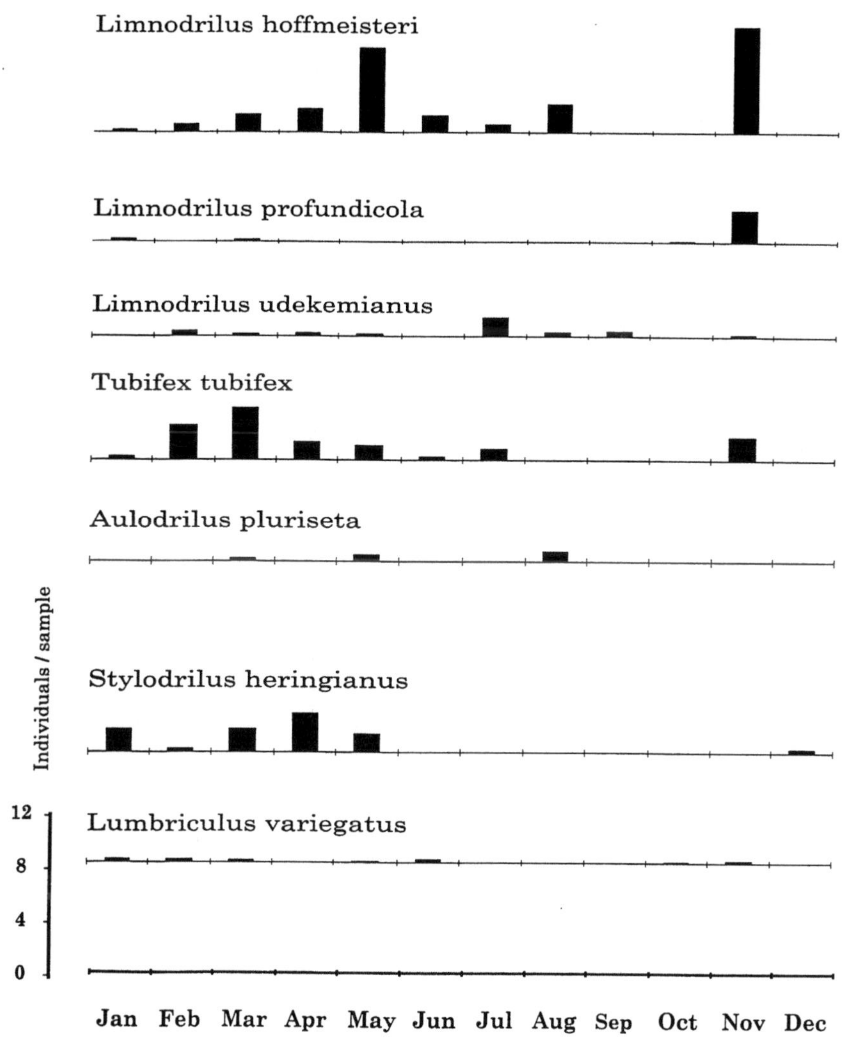

Figure 5. The monthly distribution and average abundance per sample of more common tubificids and lumbriculids.

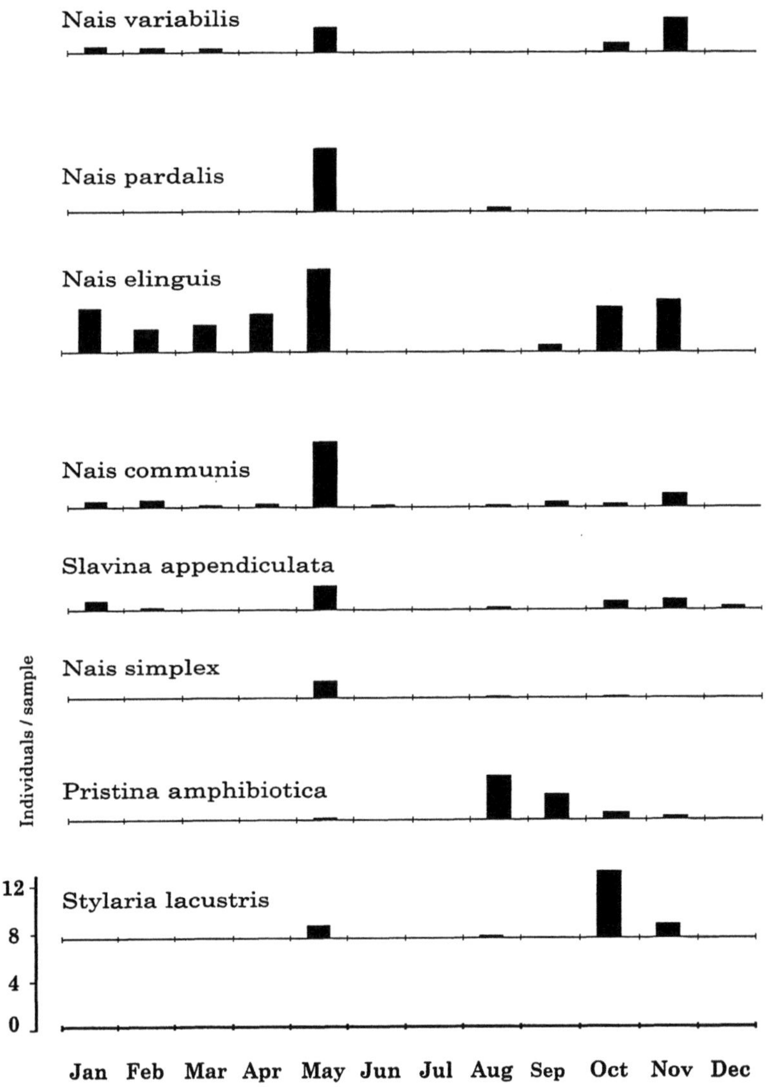

Figure 6. The monthly distribution and average abundance per sample of more common naidids.

corresponds to the ordination results. *N. communis* is significantly more abundant in spring (Table 5).

Nais elinguis strongly prefers fine sand with a medium organic matter content and a less strong preference for gravel. In the field, its habitat was termed litter and leaves. *N. elinguis* is most likely independent from the mineral fraction of the substrate but related to coarse fragments of organic material. Its centred position in the ordination diagram supports this independence from mineral fractions, though it is impossible to distinguish between variation due to substrate and to season if both run parallel. Partial ordination then removes information because it relates it to one or the other. The very signifincant relationship with spring is shown in Table 5.

Nais pardalis prefers fine sand with a medium organic matter content and coarse sand with a low organic matter content. In the field, these samples were mainly characterised as detritus and small amounts of silt. This relationship to detritus and silt is supported by the ordination diagram. *N. pardalis* is a spring species.

Nais simplex prefers fine sand with a medium organic matter content, by field observation termed detritus, a relationship which is supported by the ordination results. *N. simplex* is also a spring species.

Nais variabilis prefers fine sand with a medium organic matter content and coarse sand with a low organic matter content. In the field these samples were termed silt. Thus, there is mainly fine organic matter

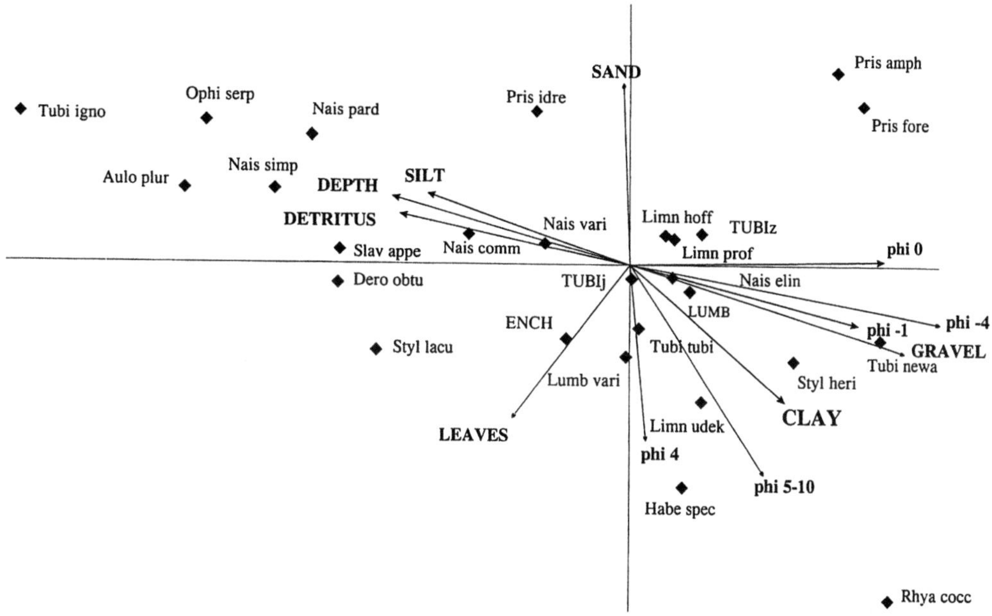

Figure 7. Partial ordination (DCCA) diagram showing the variation in oligochaetes (dots) among habitat related variables (arrows). See Table 1 for species codes.

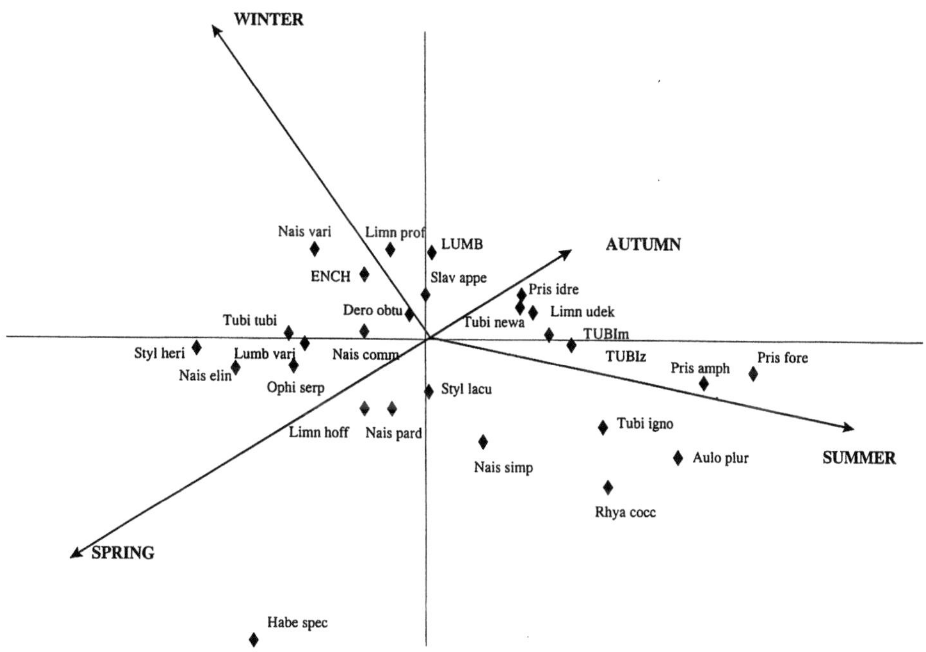

Figure 8. Partial ordination (DCCA) diagram showing the variation in oligochaetes (dots) among seasons (arrows). See Table 1 for species codes.

mixed with the mineral sediment. In the ordination its relation to silt is less pronounced. *N. variabilis* is more abundant in spring and autumn.

Pristina amphibiotica shows a slight preference for fine sand with a high organic matter content. In the field, these samples were termed litter or sand. The latter does not correspond to a high amount of organic matter, though it can be mixed in the sand. In the ordination *P. amphibiotica* is positioned in positive relation to sand and negative to any organic material

Table 3. Index of representation values per oligochaete species per substrate class established from substrate analysis. Number of samples, number of individuals, chi squared value and significance (*= P>0.05, **= P>0.01, ***= P>0.001) are indicated. Italic values indicate significant avoidance and bold italics indicate affinity. (L – low; M – medium; H – high; OM – organic material)

Mineral substrate	Lutum	Fine sand	Fine sand	Fine sand	Coarse sand	Coarse sand	Coarse sand	Gravel	Leaves	Total number of individuals	Chi squared	
organic matter	MOM	LOM	MOM	HOM	LOM	MOM	HOM	LMOM				
number of samples	4	27	18	12	7	6	3	9	6			
Aulodrilus pluriseta	−0.72	−1.34	**4.34**	−1.25	0.09	0.25	−0.63	−1.08	−0.88	12	25.16	**
Limnodrilus hoffmeisteri	*−3.30*	*−6.14*	**3.69**	**9.13**	*−3.23*	**8.31**	*−2.16*	*−2.94*	*−3.30*	251	249.36	***
Limnodrilus profundicola	−1.04	−1.60	0.05	−1.25	−1.38	**11.25**	−0.90	−1.56	−1.28	25	138.66	***
Limnodrilus udekemianus	−0.88	−0.99	0.79	−1.53	**3.96**	−0.16	0.54	−0.57	−0.16	18	21.06	**
Tubifex tubifex	−0.60	*−3.43*	**3.32**	−0.37	−0.14	**2.54**	**2.46**	−1.09	−0.26	70	37.11	***
Stylodrilus heringianus	0.87	*−3.51*	**6.90**	*−2.34*	−1.79	1.97	0.54	−1.04	−1.05	42	75.74	***
Lumbriculus variegatus	−0.55	−1.43	0.54	1.14	−0.73	−0.68	−0.48	0.38	**2.28**	7	10.52	ns
Nais communis	*−2.04*	*−2.86*	**9.97**	*−2.97*	−0.85	*−2.10*	−1.20	−0.45	−0.90	96	128.25	***
Nais elinguis	*−6.19*	*−10.12*	**23.24**	*−4.08*	0.62	*−4.57*	*−5.19*	**5.27**	*−6.86*	1057	820.36	***
Nais pardalis	−1.71	0.75	**5.22**	*−2.62*	**2.17**	−1.61	−1.48	*−2.56*	*−2.09*	67	57.98	***
Nais simplex	−0.96	−1.27	**5.87**	−1.05	−0.47	−1.17	−0.83	−1.43	−1.17	21	43.76	***
Nais variabilis	−0.96	−2.67	**5.24**	−0.57	**3.61**	−1.43	0.08	*−2.38*	−1.43	58	58.66	***
Pristina amphibiotica	−1.59	1.21	0.19	**3.07**	−1.62	−0.40	0.81	−1.96	−1.94	58	24.50	**
Slavina appendiculata	−1.43	*−3.18*	**7.85**	0.76	−1.89	−1.75	−1.24	*−2.14*	0.53	47	87.38	***
Stylaria lacustris	−0.86	*−4.97*	0.39	**13.61**	−1.34	*−2.34*	−1.66	*−2.87*	−0.20	84	229.00	***

Table 4. Index of representation values per oligochaete species per substrate class established by field observation. Number of samples, number of individuals, chi squared value and significance (*= P>0.05, **= P>0.01, ***= P>0.001) are indicated. Italic values indicate significant avoidance and bold italics indicate affinity

Substrate class	Silt (organic)	Detritus (organic)	Litter (organic)	Leaves (organic)	Sand (mineral)	Gravel (mineral)	Total number of individuals	Chi squared	
number of samples	15	7	17	10	24	18			
Aulodrilus pluriseta	−1.41	1.12	1.84	−1.15	1.03	−1.54	12	11.38	*
Limnodrilus hoffmeisteri	0.87	*−3.26*	**4.11**	−0.68	−1.99	0.05	251	32.65	***
Limnodrilus profundicola	0.43	−1.39	*−2.16*	−1.05	−1.79	**5.42**	25	40.48	***
Limnodrilus udekemianus	−0.56	−1.18	−0.20	0.73	−1.26	**2.35**	18	9.39	ns
Tubifex tubifex	**2.49**	−1.46	1.64	0.11	*−3.37*	0.85	70	23.07	***
Stylodrilus heringianus	−0.73	0.43	1.13	**3.90**	*−3.03*	−0.11	42	26.40	***
Lumbriculus variegatus	−0.14	1.99	−0.27	1.40	−1.36	−0.33	7	7.98	ns
Nais communis	1.30	**5.75**	0.96	−0.48	−0.86	*−4.36*	96	55.59	***
Nais elinguis	*−7.51*	−0.47	**10.09**	**21.90**	*−7.28*	*−10.58*	1057	803.10	***
Nais pardalis	**2.69**	**9.18**	*−3.26*	*−2.71*	0.55	*−3.64*	67	123.10	***
Nais simplex	−1.86	**8.96**	−0.97	−1.52	0.20	*−2.04*	21	91.14	***
Nais variabilis	**5.32**	−1.64	0.35	−0.94	−0.33	*−3.09*	58	41.63	***
Pristina amphibiotica	*−2.45*	*−2.11*	**5.52**	*−2.52*	**2.48**	*−2.80*	58	61.24	***
Slavina appendiculata	−0.63	7.57	−1.61	1.69	−0.96	*−2.72*	47	71.41	***
Stylaria lacustris	*−3.68*	−1.72	*−3.91*	**17.99**	−0.99	*−4.03*	84	372.46	***

(leaves, silt, detritus). It significantly prefers summer (Table 5).

Slavina appendiculata prefers fine sand with a medium organic matter content, by field observation termed detritus (Tables 3 & 4), a relationship which is supported by the ordination results. It occurs more abundant in spring.

Stylaria lacustris strongly prefers fine sand with a medium organic matter content, by field observation termed leaves. A relationship which is supported by

Table 5. Index of representation values per oligochaete species per season. Number of samples, number of individuals, chi squared value and significance (*= P>0.05, **= P>0.01, ***= P>0.001) are indicated. Italic values indicate significant avoidance and bold italics indicate affinity

Season	Spring	Summer	Autumn	Winter			
number of samples	23	19	22	13	Total number of individuals	Chi squared	
Aulodrilus pluriseta	1.28	1.77	−1.85	−1.42	12	10.20	*
Limnodrilus hoffmeisteri	**5.78**	*−2.91*	1.21	*−5.74*	251	76.31	***
Limnodrilus profundicola	*−2.37*	*−2.48*	5.93	−1.57	25	49.43	***
Limnodrilus udekemianus	−0.59	2.16	−0.94	−0.60	18	6.28	ns
Tubifex tubifex	**4.18**	*−3.19*	−1.57	0.34	70	30.20	***
Stylodrilus heringianus	**6.06**	*−3.22*	*−3.46*	0.34	42	59.17	***
Lumbriculus variegatus	−0.06	−0.55	0.00	0.75	7	0.88	ns
Nais communis	**8.65**	*−4.25*	*−2.76*	*−2.78*	96	108.25	***
Nais elinguis	**36.09**	*−16.09*	*−13.12*	*−11.49*	1057	1865.40	***
Nais pardalis	**9.83**	*−3.33*	*−4.38*	*−3.36*	67	138.21	***
Nais simplex	**5.08**	−1.84	*−2.04*	−1.88	21	36.91	***
Nais variabilis	**2.32**	*−3.78*	**2.56**	−1.85	58	29.70	***
Pristina amphibiotica	*−3.68*	**6.26**	0.35	*−3.13*	58	62.68	***
Slavina appendiculata	**3.46**	*−2.82*	−0.12	−1.04	47	21.01	***
Stylaria lacustris	−1.81	*−4.11*	**8.57**	*−3.77*	84	107.90	***

the ordination results. *S. lacustris* is significantly more abundant in autumn (Table 5).

Discussion

Oligochaete distribution and abundance is, among other factors, often related to substrate composition (Wachs, 1967; Zahner, 1967; Stimpson et al., 1975), organic matter content (Lazim & Learner, 1987), food quality of the organic component (Brinkhurst, 1967; McMurtry et al., 1983) and current velocity (Korn, 1963). Berg (1938) previously concluded that sediment composition affects oligochaetes distribution. More recently, factors like sediment chemistry (Sarkka & Paasivirta, 1972), fish predation (Kajak & Dusoge, 1971) and organic load (Aston, 1973) were added to the list of important factors. Lazim & Learner (1987) concluded that the distribution of tubificids was determined by the physical nature of the substratum and by the presence (especially amount) and kind of leaf material. Tubificids showed preferences for particular, either complete or degraded, leaf species. This preferential selection was due to differences in microbiological floras (see also Brinkhurst & Chua, 1969; Brinkhurst et al., 1972) of the various leaves and to different digestive capabilities. These differences probably also account for the detritivorous, riverine naidids (Learner, 1979). Though the latter group also feeds on diatoms (Learner et al., 1978). Tubificids are known to inhabit the substratum and penetrate it. Naidids do not penetrate the substratum unless there are larger interstitial spaces with associated higher oxygen levels (Learner et al., 1978).

Seasonality (in discharge) contributes to considerable changes in stream substrates (habitats). Unfortunately, season and substrate appear highly confounded. To distinguish between both, partial ordination was used. By using partial ordination, either season or substrate variation would be factored out. The season-abundance relationship was significant while the substrate-abundance relationship was only weak. This indicates the higher importance of season, though the low eigenvalues do not support a major role. I conclude that partial ordination removes substrate as well as seasonal information; partially, because substrate changes are seasonal in nature. Therefore, the index of representation and a chi-square analysis were used to look 'separately' at substrate and season. Most oligochaetes vary seasonally. In conclusion, all oligochaetes respond to both factors, except for *L. udekemianus* which only significantly responds to substrate.

According to these results, beside season, the structure of both organic (diameter of fragments) and mineral (grain size diameter) material adds to the micro-distribution of oligochaetes, together with the behavioural habit of each. The oligochaete species can be grouped according to their behavioural habit (Sperber, 1950). Burrowers inhabit the sediment and often are tube building, bottom dwellers (sprawlers) inhabit the top of the substrate, and swimmers cling to submersed objects between short bursts of swimming (Merrit & Cummins, 1984). In fact, these three groups inhabit different vertical layers in the stream bed. Burrowers occur in the lower layer of the sediment. Bottom dwellers occur mainly on the sediment, between the organic deposits, while swimmers also occur above the substrate in the water column. They are most vulnerable to water movement. In analyzing the tubificids and lumbriculids it appeared that three species did not show a pronounced habitat preference (*Aulodrilus pluriseta*, *Limnodrilus udekemianus* and *Lumbriculus variegatus*), although only a small number of individuals (<20) were collected for each of the three above species. All other species show a separation in micro-habitat distribution. The ecological habitat in space and time per species can be described as:

food quality (Learner et al., 1978). The larger the fragment the smaller the outer surface area. Thus, smaller fragments are associated with a richer microbiological flora.

Finally, a remark on both substrate classification systems used. Substrates were classified based on a visual as well as on a laboratory classification system. Comparison of both can only be indirect. The field classification is qualitative, subjective to the person taking the samples and only separates visually interpretable structures. The laboratory system is quantitative and accurate but strongly depends on the representativity of the sample, the sampling depth and the classes predefined. From the analysis can be concluded that both add to the interpretation of the results. Differences in taxa preferences (Tables 3 and 4) between field and laboratory are due to the definition of the organic matter content (laboratory classes) and the fact that field observations only provide information on the upper most layer of the substrate.

oligochaete species	*substrate habitat*	*time period*
burrowers:		
Limnodrilus profundicola	gravel to coarse sand (mixed with silt)	summer (all seasons)
Tubifex tubifex	coarse to fine sand mixed with silt	spring (all seasons)
Limnodrilus hoffmeisteri	coarse to fine sand covered with litter	spring (all seasons)
Stylodrilus heringianus	fine sand covered with leaves	spring (winter)
bottom dwellers:		
Nais communis	fine sand covered with detritus	spring (all seasons)
Pristina amphibiotica	mineral sand and fine sand covered with litter	summer
Slavina appendiculata	fine sand covered with detritus	spring (all seasons)
swimmers:		
Nais elinguis	litter and leaves	spring (all seasons)
Nais pardalis	coarse to fine sand covered with detritus (to silt)	spring
Nais simplex	fine sand covered with detritus	spring
Nais variabilis	coarse to fine sand mixed with silt	all seasons
Stylaria lacustris	fine sand covered with leaves	autumn

Only two species with the same behavioural habit occur in the same micro-habitat at the same time (i.e. *Nais communis* and *Slavina appendiculata*). All others coexist in different habitats. The diameter and structure of the organic material is not independent from the

Conclusions

Low-gradient, soft-bottomed streams in agricultural areas show high peaks in discharge which cause a swift shift in surface substrate composition. Discharge peaks cause silt to be washed out, while silt deposition takes place during periods of low flow. The micro-distribution of oligochaetes is mainly related to season. The approached significance in partial ordination suggests also a weak relationship with substrate. The latter is determined by the particle size and distribution (structure) of mineral and organic material. Perhaps micro-habitat is also related to food quality.

In future research, attention should be focused on the stability and food quality of the substratum as well as to the effect of turbulence on micro-habitats. A second topic should be the role of other species on the micro-distribution of oligochaetes.

The comparison of field and laboratory parameters on substrate (habitat) characteristics resulted in the conclusion that both are needed and that the observation of sand classes in the field should be refined.

Acknowledgements

Assistance with sampling and environmental data collection was provided by Joke Schot and Martin van der Hoorn. The latter also assisted in data processing. Oligochaetes were identified by Tjeerd-Harmen van den Hoek. The waterboard 'Regge & Dinkel' provided the discharge data. Patty Gillis corrected the manuscript. I am most grateful to all these individuals and organisation for their help.

References

Aston, R. J., 1973. Tubificids and water quality: a review. Envir. Pollut. 5: 1–10.

Berg, K., 1938. Studies on the bottom animals of Esrom Lake. K.Danske Vidensk. Selsk. Naturv. Math. Afd. 8: 1–255.

Brinkhurst, R. O., 1967. The distribution of aquatic oligochaetes in Saginaw Bay, Lake Huron. Limnol. Oceanogr. 12: 137–143.

Brinkhurst, R. O., K. E. Chua & N. K. Kaushik, 1972. Interspecific interactions and selective feeding by tubificid oligochaetes. Limnol. Oceanogr. 17: 122–133.

Cummins, K. W. & G. H. Lauff, 1969. The influence of substrate particle size on the microdistribution of stream benthos. Hydrobiologie 34: 145–181.

Doeglas, D. J., 1968. Grain-size indices, classification and environment. Sedimentology 10: 83–100.

Faber, Th., 1972. Regimes and regime-related basin properties of some Dutch small rivers. Thesis, University of Amsterdam, Amsterdam.

Frissell, C. A., W. J. Liss, C. E. Warren & M. D. Hurley, 1986. A Hierarchical Framework for stream Habitat Classification: Viewing Streams in a Watershed Context. Envir. Mgmt 10: 199–214.

Gauch, H. G., 1982. Multivariate analysis in community ecology. Cambridge University Press, 1–298, Cambridge.

Hildrew, A. G. & C. R. Townsend, 1976. The distribution of two predators and their prey in an iron rich stream. J. anim. Ecol. 45: 41–57.

Hynes, H. B. N., 1970. The ecology of running waters. Liverpool Univ. Press, Liverpool: 1–202.

Kajak, Z. & K. Dusoge, 1971. The regularities of vertical distribution of benthos in bottom sediments of three Masurian lakes. Ecol. Pol. 19: 485–499.

Korn, H., 1963. Studien zur Ökologie der Oligochaeten in der oberen Donau unter Berücksichtigung der Abwassereinwirkungen. Arch. Hydrobiol. Suppl. 27: 131–182.

Lazim, M. N. & M. A. Learner, 1987. The influence of sediment composition and leaf litter on the distribution of tubificid worms (Oligochaeta). A field and laboratory study. Oecologia 72: 131–136.

Learner, M. A., 1979. The distribution and ecology of the Naididae (Oligochaeta) which inhabit the filter-beds of sewage-works in Britain. Wat. Res. 13: 1291–1299.

Learner, M. A., G. Lochhead & B. D. Hughes, 1978. A review of the biology of British Naididae (Oligochaeta) with emphasis on the lotic environment. Freshwat. Biol. 8: 357–375.

McMurtry, M. J., D. J. Rapport & K. E. Chua, 1983. Substrate selection by tubificid oligochaetes. Can. J. Fish. aquat. Sci. 40: 1639–1646.

Sarkka, J. & L. Paasivirta, 1972. Vertical distribution and abundance of macro- and micro-fauna in the profundal sediments of Lake Paijanne, Finland. Ann. Zool. Fenn. 9: 1–9.

Southwood, T. R. E., 1977. Ecological methods with particular reference to the study of insect populations. Chapman and Hall, London.

Sperber, C., 1950. A guide for the determination of European Naididae. Zool. Bidr. Upps. 29: 45–78.

Stimpson, K. S., J. R. Brice, M. T. Barbour & P. Howe, 1975. Distribution and abundance of inshore oligochaetes in Lake Michigan. Trans. am. microsc. Soc. 94: 384–394.

Ter Braak, C. J. F., 1990. Update notes: Canoco version 3.10. Agricultural Mathematics Group, Wageningen, The Netherlands.

Ter Braak, C. J. F. & I. C. Prentice, 1988. A theory of gradient analysis. Adv. Ecol. Res. 18: 271–317.

Ter Braak, C. J. F. & P. F. M. Verdonschot, 1995. Canonical correspondence analysis and related multivariate methods in aquatic ecology. Aquat. Sci. 57: 255–289.

Tolkamp, H. H., 1980. Organism-substrate relationships in lowland streams. Thesis, Agricultural University Wageningen. 1–211.

Verdonschot, P. F. M. & C. J. F. Ter Braak, 1994. An experimental manipulation of oligochaete communities in mesocosms treated with chlorpyrifos or nutrient additions: multivariate analyses with Monte Carlo permutation tests. Hydrobiologia 278: 251–266.

Wachs, B., 1967. Die Oligochaetenfauna der Fliessgewässer unter besonderer Berücksichtigung der Beziehungen zwischen der Tubificiden-besiedlung und dem Substrat. Arch. Hydrobiol. 63: 310–386.

Wenthworth, C. K., 1922. A scale of grade and class terms for clastic sediments. J. Geol. 30: 377–392.

Zahner, R., 1967. Experimente zur Analyse biologischer, chemischer und physikalischer Vorgänge in der Wasser-Sediment-Grenzschicht stehender und langzam strömender Gewässer. I. Beschreibung der Versuchsanlage mit vorläufigen Ergebnissen über das Verhalten der Tubificiden in Wahlversuchen. Int. Rev. ges. Hydrobiol. 52: 627–645.

Hydrobiologia **406:** 165–174, 1999.
B. M. Healy, T. B. Reynoldson & K. A. Coates (eds), Aquatic Oligochaetes.

Oligochaeta from the abyssal zone of Lake Baikal (Siberia, Russia)

Patrick Martin, Koen Martens & Boudewijn Goddeeris
Institut royal des Sciences naturelles de Belgique (I.R.Sc.N.B.), Section Biologie des Eaux douces (Royal Belgian Institute of Natural Sciences, Freshwater Biology Section), 29 rue Vautier, B-1000 Brussels, Belgium

Key words: aquatic Oligochaeta, Lake Baikal, abyssal zone, sediment, abundance, distribution, diversity, deep-water fauna

Abstract

Lake Baikal is unique because the water circulation carries oxygen to its deepest point (1637 m), which makes it the only freshwater lake in the world with an inhabitable abyssal area. The sampling of the abyssal of the Lake was recently made possible, allowing a study of the bathymetric and vertical distribution in the sediment of Oligochaeta. Samples were taken with a Reineck box corer and subsamples were extracted and subsequently divided into slices. Factors likely to affect oligochaete abundance with depth and in the sediment were then evaluated. Identification to the species-level allowed discussion of the possible role of the abyss of Lake Baikal in the origin of oligochaete taxa and to assess if genuine deep-water taxa exist. Abundance of Oligochaeta generally follows an exponential decline with depth. An exception was one station located near a deep hot vent. In the abyssal area, all families of Oligochaeta are concentrated near the surface of the sediment. While there are generally no Naididae below 50 m, Tubificidae, Lumbriculidae, Propappidae, Enchytraeidae and Haplotaxidae are present at all depths. Evidence suggests, for the first time, that food abundance is a limiting factor of oligochaete distribution. The possibility of a genuine deep-water oligochaete fauna in Lake Baikal cannot be excluded but the low densities and the very small sizes of animals in this environment might have caused biased samples.

Introduction

Lake Baikal, situated in the Great Eastern Siberian Rift, is the deepest (1637 m) of all extant lakes (Kozhov, 1963; Martin, 1994). Its location in an active graben trough has permitted its preservation despite continuous deposition of sediments. Lake Baikal is truly 'ancient' or 'long-lived' (Gorthner, 1994) (25–30 Ma) and holds a very diverse and largely endemic fauna. A total of 152 oligochaete species was recently reported, 111 of these being endemic (Martin, 1996), today we know of 155 and 114 species, respectively (Martin & Brinkhurst, 1998; Martin et al., 1998a).

The area deeper than 250 m is commonly defined as the abyssal zone of Lake Baikal (Kozhov, 1963). Although a very different depth compared to abyssal or deep sea depths (Gage & Tyler, 1991), the abyssal of Lake Baikal and the deep sea have common characteristics. They are both very stable and homogeneous in physical and chemical parameters (Kozhov, 1963; Falkner et al., 1991; Weiss et al., 1991; Liebezeit, 1992; Hohmann et al., 1997a). Major environmental characteristics of the abyssal of Lake Baikal are low temperatures (3.2–4.0 °C), high concentrations of dissolved oxygen (347 μmol kg^{-1}), elevated hydrostatic pressures and distance from the zone of primary organic production, indicative of a psychrosphere (Martens, 1997).

Due to a deep-water renewal below the dimictic 250 m upper layer (Kozhov, 1963; Weiss et al., 1991; Shimaraev et al., 1993; Killworth et al., 1996; Hohmann et al., 1997a,b), oxygen is carried to the deepest point and makes these depths inhabitable by metazoans. Other ancient and deep lakes, such as Lake Tanganyika (East Africa), are virtually anoxic below the first 100–250 m (Coulter, 1994).

Although Lake Baikal has been an active centre of research for about 100 years, studies of ecology and distribution of oligochaetes in the vast abyssal zone really started only twenty years ago. Tubificidae from depths exceeding 500 m were studied for the first time by Chekanovskaya in 1975. A detailed study of the northern basin by Snimschikova (1982, 1985a, 1987) filled in many taxonomic gaps but mainly focused on shallow areas down to 500 m. Semernoi (1983) was the first to give a comprehensive list of oligochaete species, with their respective distribution ranges, from the abyssal zone of Lake Baikal. Semernoi's list included species from the families Naididae, Tubificidae, Lumbriculidae, Enchytraeidae and Haplotaxidae. Takhteyev et al. (1993) gave a description of the general benthos of the abyssal zone but the distribution of oligochaetes in the deep lake was still imprecise.

Recent fieldwork on Lake Baikal has been possible within the framework of BICER (Baikal International Center for Ecological Research). In 1994, a joint Russian-Belgian expedition focused on sampling the abyssal zone of the lake. Data collected during this expedition enabled us to study the bathymetric and vertical distribution of Oligochaeta in the sediment. This is compared to similar data from shallower zones accumulated during previous cruises. Specimens were identified to species-level, allowing us to discuss the possible role of the abyss of Lake Baikal in the origin of extant oligochaete diversity and to assess whether or not genuine deep-water taxa exist.

Material and methods

Sampling was done at the widest part of the lake, along a transect from Barguzin Gulf to the deepest point (1680 m as measured by the echo sounder). Four sample stations were on the slopes of the underwater feature called 'Akademichesky ridge', which marks the boundary between the central and northern basins (Figure 1). Transects in previous studies were in shallower depths: three in front of Selenga delta (18–84 m, 1991 cruise) and two in the southern basin, on Murinsk bank (135–650 m, 1991) and in the vicinity of Posolskaya Bank (20–400 m, 1990; Martin et al., 1994), including a few isolated stations deeper than 250 m (Figure 1).

All samples were taken with a modified Reineck box corer (0.079 m^2). Subsamples were extracted for faunal analysis with Perspex tubes, the surface area of the cores was modified for each cruise in order to obtain adequate numbers of animals and minimize sample processing (1990: 37.7 cm^2, 1991: 84.9 cm^2; 1994: 62.2 cm^2). Each subsample was mounted on an extruder and subsequently divided into slices. Samples were fixed in 7% buffered formalin, washed in the laboratory through a sieve (250 μm mesh), and preserved in 70% alcohol.

All mature oligochaetes from the 1994 samples were identified. Larger specimens were dissected, smaller ones were mounted whole. Worms were stained with alcoholic carmine and mounted in Canada balsam. These animals are deposited in the Royal Belgian Institute of Natural Sciences (RBINSc; Brussels) under No I.G. 28051.

The bathymetric distribution of oligochaete abundance was studied using all available quantitative data. Data from the 1990 study were excluded from the analyses of vertical distribution within the sediments because of the different slice thickness.

Results

The abundance of Oligochaeta (no. individuals per m^2) follows an exponential decline with depth, although a slightly better curve fit is achieved using a power decay (Figure 2). The latter can be expressed as $Y = ax^{-b}$ where Y is density, x is depth, and a and b are constants. As the b value is very close to 1, this means that for each doubling of depth, densities decrease by a factor of 2. Thus, there is a rapid decrease in oligochaete abundance in the first 200 m but abundance is almost constant in the abyssal zone of the lake. Density decreases 300 times from 20 m to the greatest depth. Stations at 1500–1680 m have densities 3–15 times lower than those at 225 m. One station, at 420 m, stands distinctly out of the data sequence, with a density of 15 956 ind. m^{-2} (open circle, Figure 2). This station is located near the Frolikha Bay vent, a deep lake vent discovered more than 10 years ago (Golubev et al., 1993).

The vertical distribution of Oligochaeta in the sediment was studied by estimating the interaction between:

1. depth zone and sediment depth for each family that was significant in our abyssal samples (Tubificidae, Lumbriculidae and Enchytraeidae) and
2. family and sediment depth for each depth zone (Figures 3 & 4).

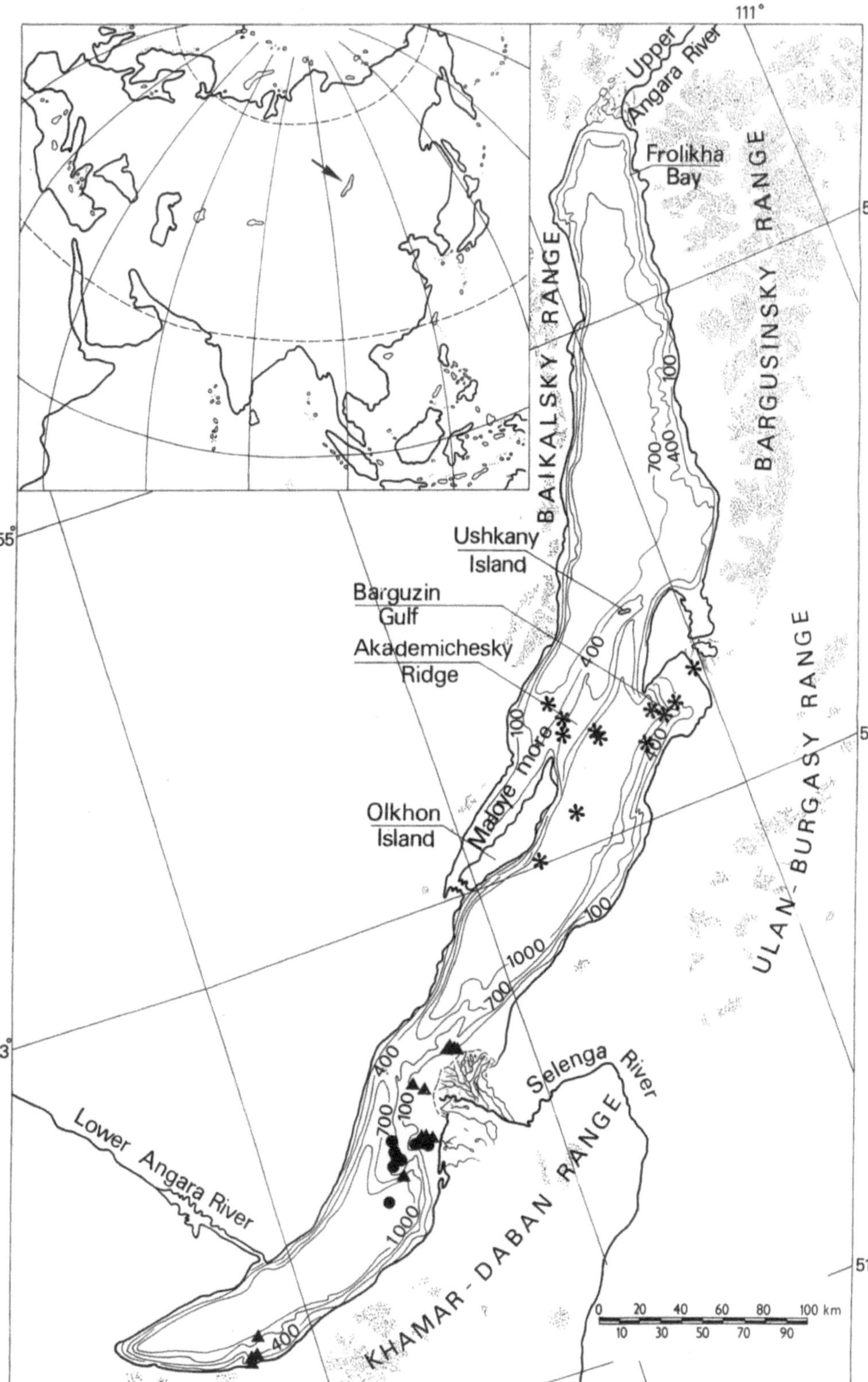

Figure 1. Location of Lake Baikal and the stations sampled in 1990 (dots), 1991 (triangles) and in 1994 (stars).

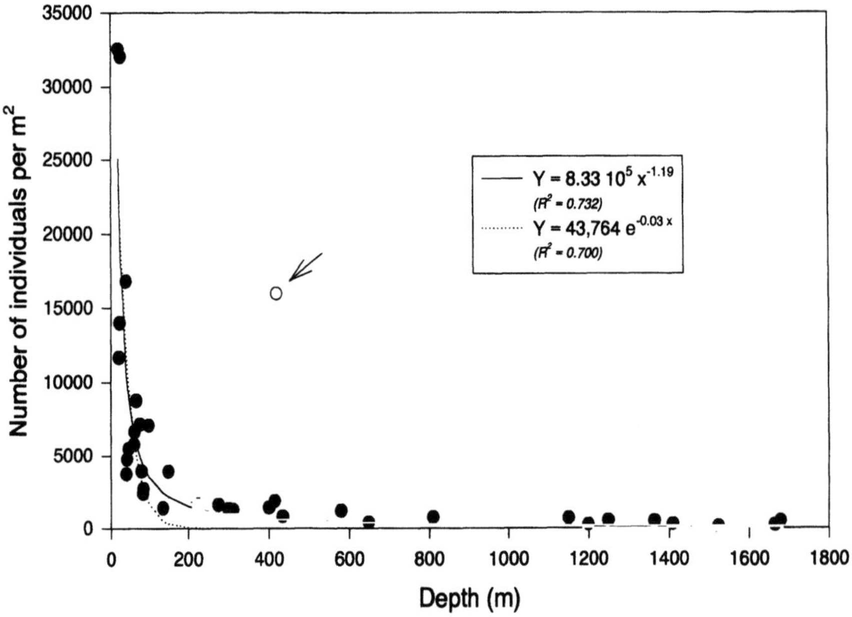

Figure 2. Bathymetric distribution of density of Oligochaeta (dashed line: exponential curve fit; solid line: power decay curve fit). The open circle (arrowed) refers to one station located near the Frolikha Bay vent in the northern basin of the lake, at 420 m depth. This value was excluded from computation of the curve fits. Data are from 1990, 1991 and 1994 collections.

Table 1. Summary of size measurements of selected Oligochaeta from the abyssal zone of Lake Baikal (from Martin & Brinkhurst, 1998)

	Breadth at widest	Length	Depth
Burchanidrilus petitbonum Martin & Brinkhurst, 1998	240 μm	fragment	1250 m
Lymphachaeta pinnigera Snimschikova, 1982	186 μm	4.50 mm	580 m
Rhyacodriloides abyssalis Chekanovskaya, 1975	260 μm	7.30 mm	275 m
Rhyacodriloides gladiiseta Martin & Brinkhurst, 1998	135 μm	3.24 mm	310 and 415 m
Tubifex bazikalovae Chekanovskaya, 1975	310 μm	4.73 mm	415 and 580 m

Tubificids burrow into the sediment of the dimictic area (Figure 3), while they are more concentrated near the surface of abyssal sediments (Figure 4). In contrast, there is no apparent difference in sediment penetration by Lumbriculidae and Enchytraeidae in either depth zone. This is fully confirmed by a two-way ANOVA test which reveals a statistically significant interaction between water depth zonation and sediment depth for Tubificidae ($F_{(7,160)} = 4.57$; $p < 0.0001$) (Figures 3 & 4), but none for Lumbriculidae and Enchytraeidae.

There is a statistically significant interaction between family and sediment depth in the dimictic zone ($F_{(14,240)} = 1.89$; p = 0.0281) and none in the abyssal zone (Figures 3 & 4). While the test isolates Tubificidae as being responsible for this significant interaction, an examination of median values suggests a different pattern of vertical distribution in the sediment between Enchytraeidae, Lumbriculidae and Tubificidae. In the abyssal zone, all families follow a similar pattern and are concentrated near the sediment surface.

Twenty-three species of oligochaetes were identified in all samples (1990, 1991 and 1994 cruises), of a total of 155 reported species from the Lake (Figure 5). The distribution of oligochaete species (Figure 5) is eurybathic for the families Haplotaxidae, Enchytraeidae, Propappidae and Lumbriculidae. All these species were already mentioned in the deep-lake studies (refs in Figure 5) and their distributions are only extended a few hundred metres deeper or even to the greatest depth (1680 m; *Haplotaxis* sp.) with data from our study. Naididae are generally restricted to the first 50 m, with a notable exception at 700 m deep (*Nais abissalis*, Semernoi 1988). One immature specimen of *Nais* ?sp. n. was found at 200 m, extending the distribution range of the family close to the dimictic-abyssal boundary. In Tubificidae, most

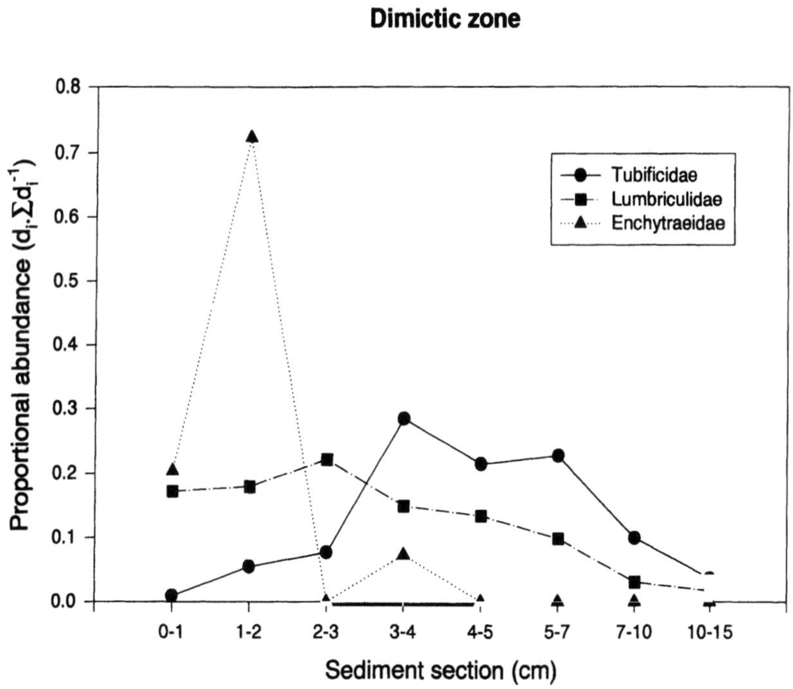

Figure 3. Vertical distribution in the sediment of Oligochaeta in the dimictic zone of Lake Baikal. Proportional abundance calculated as $d_i.\Sigma d_i^{-1}$, in which d_i = density of oligochaetes in sediment core segment i, expressed as the number of individuals per m^2. Data are from 1991 and 1994 collections.

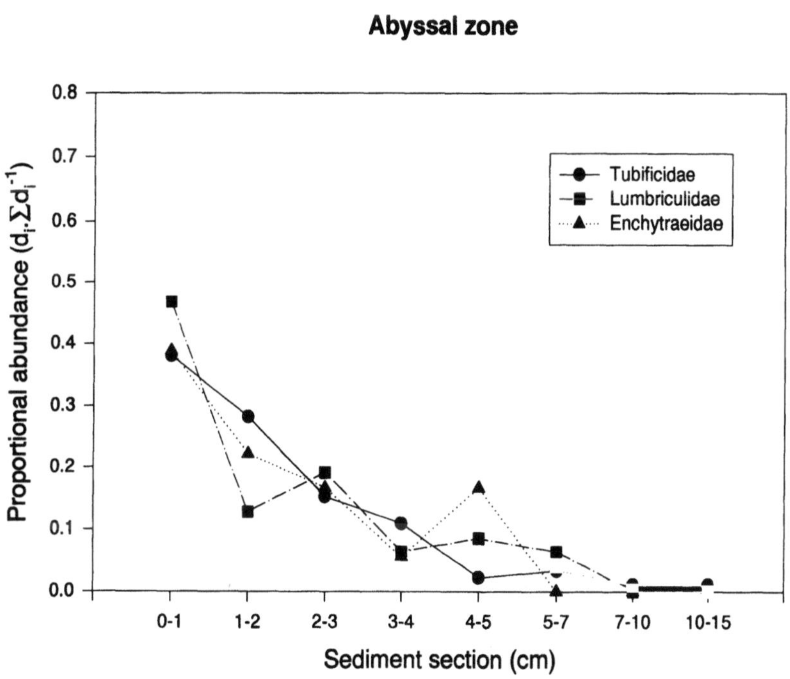

Figure 4. Vertical distribution in the sediment of Oligochaeta in the abyssal zone of Lake Baikal. Proportional abundance calculated as $d_i.\Sigma d_i^{-1}$, in which d_i = density of oligochaetes in sediment core segment i, expressed as the number of individuals per m^2. Data are from 1991 and 1994.

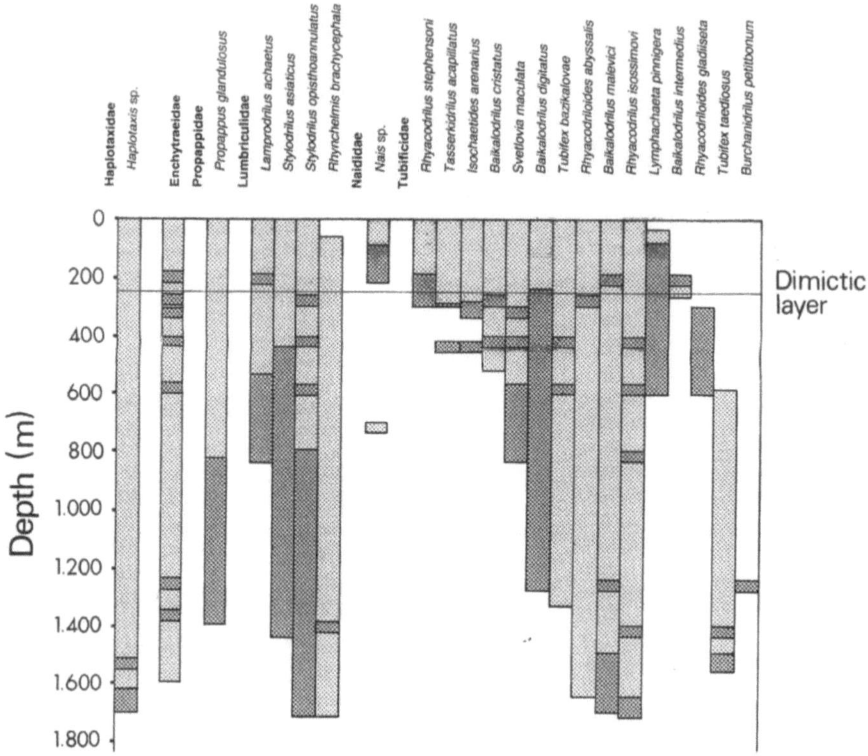

Figure 5. Bathymetric distribution of oligochaetes. Lightly shaded: range of bathymetric distribution as known from literature; darkly shade: ranges from data of the present study. When the present data fall into the former bars, only the depth of the station is indicated as a narrow band; when the data enlarge the previously known distribution, the bar extends as far as the deepest station where the species was found. Isolated stations indicate a possible artefact in Naididae distribution (at 700 m) or correspond to the deep vent station in Frolikha Bay (*T. acapillatus* and *I. arenarius*). Literature data are from Chekanovskaya (1975), Semernoi (1983, 1987), Snimschikova (1985a,b, 1987), Snimschikova & Timm (1992) and Snimschikova & Akinshina (1994).

species are found in both the dimictic and the abyssal zone of the lake, but some seem to be restricted to the shallower zone and others to the abyss.

Discussion

Distribution of abundance

Abundance of food controls density and biomass of both pelagic and benthic deep-sea animals (Sanders & Hessler, 1969; Rowe et al., 1974; Gage & Tyler, 1991; Childress, 1995). During its slow descent from the surface waters to the benthos, organic matter is progressively utilised by mid-water organisms and degraded by autolysis and bacterial decay. The refractory or non-usable proportion of the organic matter increases with depth. Other factors (oxygen concentration, sediment type, temperature) either do not show correlated changes or never reach levels regarded as limiting (Sanders & Hessler, 1969; Khripounoff et al., 1980). There is no reason to believe that such a pattern cannot also apply to Lake Baikal.

In addition to food, other factors could explain variation with depth in oligochaete abundance in Lake Baikal and their possible effects are worthy of being evaluated.

Three factors which co-vary with depth are:

1. hydrostatic pressure,
2. ambient light, and
3. temperature.

Oxygen is neglected here because the range of dissolved O_2 in the water column (0–1637 m) is very narrow in Lake Baikal (347–383 μmol; Martin, 1994) and the sediment is well-oxygenated at all depths (Martin et al., 1998b).

Although hydrostatic pressure could limit the species present, there is no apparent way in which it can regulate abundance. Suggestions that pressure might, in some way, act so as to limit metabolic rates were recently rejected (Childress, 1995).

Ambient light has an indirect influence on abundance because it determines the thickness of the euphotic zone which, in turn, affects phytoplankton production and, hence, food availability to benthic organisms. However, this effect is mainly restricted to the first 100 m, as attested by the concentration of phytoplankton in this upper layer, with subsurface maxima (Bondarenko et al., 1996).

An intuitively attractive variable that could be responsible for lower abundances at depth is temperature. This suggestion can, however, be rejected. In situations with deep-sea water mass being much warmer than elsewhere, benthic standing crop is always lower (Gage & Tyler, 1991). This is so, because pelagic consumption is enhanced, due to a high energy demand of temperature-related metabolism and much of the organic matter available at the bottom is lost for benthic production. In Lake Baikal, while temperatures can be considered to be constant from 100 m down, in August, the pattern of distribution still follows a power decay ($R^2 = 0.797$) along this depth range.

Another test of the potential importance of temperature on abundance comes from the hydrothermal vent oligochaetes. Increased temperature in the vent is not higher than 0.06 °C due to rapid dilution of vent water by ambient water. At 10 cm into the sediment, temperature is less than 5 °C, compared with 3.2 °C for bottom water at 420 m (Golubev et al., 1993). The independence of abundance at the vent from temperature indicates that much higher densities cannot be attributed to this variable. Hot vent communities apparently have much higher densities than other abyssal oligochaetes communities (15 956 *vs*. 807 ind. m^{-2} on average, respectively) because they do not depend on energy coming from the lake surface but from an *in situ* chemosynthetic primary production (Grachev et al., 1995).

Vertical distribution in the sediment

Lower food availability at depth is the most likely explanation for the limited vertical distribution of oligochaetes in the sediment of the abyssal zone. In the dimictic zone, the vertical distribution in the sediments confirms the preference of Enchytraeidae for the sediment surface and of Tubificidae for deeper sediment, while Lumbriculidae usually penetrate into intermediate depths (Healy, 1987). In the abyssal zone, the impressive concentration of all families near the sediment surface suggests that food is restrictive at such depths. In the dimictic zone, where food scarcity is probably less stressing, oligochaetes are not impeded from fully expressing their burrowing behaviour.

Many attempts were made to estimate available food. Total organic carbon in the sediment has been used but there are numerous examples showing that it is not a good index (Sanders & Hessler, 1969; Bekman & Mizandrontzev, 1971; Rowe et al., 1974; Khripounoff et al., 1980). In Lake Baikal, an increase in organic carbon content of the sediments with depth was noted, parallel to a decrease in benthic biomass (Bekman & Mizandrontzev, 1971). Estimation of available food cannot be carried out in a simple way. It requires either a careful biochemical analysis of organic matter in the laboratory (Khripounoff & Rowe, 1985) or the use of sediment traps (estimation of the flux of organic matter; Hinga et al., 1979; Khripounoff & Rowe, 1985), which are beyond the scope of the present paper. Such a work in Lake Baikal is clearly needed in the future, however, in order to bring a confirmation of the above assumptions.

Bathymetric distribution of species

The bathymetric distribution of species is one of the oldest problems in the study of lakes and is related to presence or absence of a specially evolved deepwater fauna. Due to its age and habitable abyssal zone, Lake Baikal offers a unique opportunity to study this issue. It should be noted, however, that the Baikal abyss, as a biosphere, is rather recent, relative to the age of the lake itself. Siberia had a warm subtropical climate in the late Oligocene-Miocene periods (30–6 Mya) and the palaeothermohaline regime of Lake Baikal was similar to that of the modern Tanganyika (Sherstyankin et al., 1995), resulting in an anoxic abyssal zone.

In spite of this, the presence of genuine deep-water species in Lake Baikal have been documented in major benthic groups such as Gammaridae (Takhteyev, 1996; Takhteyev & Mekhanikova, 1996), Ostracoda (Mazepova, 1994), and Turbellaria (Timoshkin, 1994). In Oligochaeta, only two apparently 'truly' abyssal species are known to exist, *Lamprodrilus inflatus* Michaelsen, 1905 and *Lamprodrilus bythius* Michaelsen, 1905. These species are assumed to have specific morphological adaptations to deep-lake conditions (thinning down of tegument, extension of ventral setae, transparency of the body), although their bathymetric distribution includes depths shallower than 90 m (Isossimov, 1962).

As a rule, Naididae mainly feed on algae and higher plants (Brinkhurst & Gelder, 1991). As the biomass of phytoplankton is not significant below the first 100 m (Bondarenko et al., 1996), their restriction to this depth range is expected. The presence of *Nais abissalis*, described from one immature specimen found at 700 m (Semernoi, 1988) and one immature specimen of *Nais* ?n. sp. found at 200 m in our study, are therefore enigmatic and probably artefacts linked to sinking of plant debris or transport by water circulation. Transport of metazoans via sinking of large aggregates to the deep-sea floor is not uncommon (Shanks & Edmondson, 1990; Bochdansky & Herndl, 1992) and during deep-water renewal in Lake Baikal, considerable water masses can suddenly sink to a greater depth as a result of 'thermobaric instability' (Weiss et al., 1991).

All species of Haplotaxidae, Enchytraeidae, Propappidae or Lumbriculidae found are eurybathic and give no indications of potential deep-water speciation. The family Tubificidae is currently considered to be eurybathic (Snimschikova & Akinshina, 1994). Here we report the presence of *Tubifex taediosus* Chekanovskaya, 1975, *Rhyacodriloides gladiiseta* Martin & Brinkhurst, 1998 and *Burchanidrilus petitbonum* Martin & Brinkhurst, 1998 exclusively below the dimictic layer of the lake, suggesting the existence of genuine abyssal species in this family. However, two of these species (*R. gladiiseta* and *B. petitbonum*) were recently described for the first time and their presumed restricted depth range could result from a biased sampling effort.

Some tubificid species are only found in the dimictic zone: *Baikalodrilus intermedius* Snimschikova, 1991, *Rhyacodrilus stephensoni*, Cernosvitov 1941, *Isochaetides arenarius* (Michaelsen, 1926) and *Tasserkidrilus acapillatus* (Finogenova, 1972). A restricted depth distribution due to food shortage is suggested by the latter two species, as the only place they are found deeper than 250 m is at 420 m near the Frolikha hot vent.

In the abyssal zone, oligochaetes are not only rare but also very small, probably a consequence of food shortage at depth. As a result, classical sorting-out methods, based on the use of a sieve of 250 μm mesh, are perhaps unsuitable for organisms coming from this environment. A summary of some species measurements suggests that mesh size lower than 250 μm should be used in future studies (Table 1). Sampling the abyssal region of Lake Baikal is thus hampered, not only by low densities, but also by the small sizes of organisms. As this problem is underscored here for the first time, it is impossible to evaluate its importance on estimates of abundance distributions of Baikalian oligochaetes in our samples.

In conclusion, our evidence suggests that food restricts oligochaete distribution in Lake Baikal. As such, it is an excellent candidate for being a factor at the origin of segregation in populations, a necessary prerequisite for speciation. Clinal preferences along an ecological gradient are indeed deemed to create so-called 'dumb-bell' population structures, with limited gene-flow in intermediate environments, leading to parapatric speciation as a result of ecological segregation (Martens et al., 1994).

Acknowledgements

We are especially grateful to Dr M. Grachev, Director of the Limnological Institute of Irkutsk, for continuing to support international research on Lake Baikal. We are indebted to the captain and the crew of the R/V Vereschagin (Baikal) for their active assistance during the fieldwork and Cl. Devries-Duchène who inked various figures. This work was organised under the auspices of BICER (Baikal International Center for Ecological Research) and with financial support of the Belgian Ministry of National Scientific Institutions and the Siberian Branch of the Academy of Sciences of Russia.

References

Bekman, M. Yu. & I. B. Mizandrontzev, 1971. O svyazy mezhdu raspredeleniyem bentosa i organicheskogo veshchestva v osadkakh (=On the relation between distribution of benthos and organic matter in sediments). In G. I. Galazii (ed.), Limnologija Prideltovyh Prostranstv Baikala (=Limnology of Near-Estuarine Regions of Baikal). Nauka, Leningrad, SSSR: 127–132 [in Russian].

Bochdansky, A. B. & G. J. Herndl, 1992. Ecology of amorphous aggregations (marine snow) in the Northern Adriatic Sea. III. Zooplankton interactions with marine snow. Mar. Ecol. Prog. Ser. 87: 135–146.

Bondarenko, N. A., N. E. Guselnikova, N. F. Logacheva & G. V. Pomazkina, 1996. Spatial distribution of phytoplankton in Lake Baikal, Spring 1991. Freshwat. Biol. 35: 517–523.

Brinkhurst, R. O. & S. R. Gelder, 1991. Annelida: Oligochaeta and Branchiobdellida. In V. H. Thorp & A. P. Covith (eds), Ecology and Classification of North American Freshwater Invertebrates. Academic Press, San Diego, USA, Chap. 12: 401–435.

Chekanovskaya, O. V., 1975. Novye tubifitsidy (Oligochaeta, Tubificidae) iz abissali ozera Baikal (= New tubificids (Oligochaeta, Tubificidae) from the abyssal zone of Lake Baikal). Trudy

Sibirsk. otdel. limnol. inst. Akad. nauk SSSR (Proc. Limnol. Inst. Siber. Dep. USSR Acad. Sciences), 18: 112–130 [in Russian].

Childress, J. J., 1995. Are there physiological and biochemical adaptations of metabolism in deep-sea animals ? TREE 10: 30–36.

Coulter, G. W., 1994. Lake Tanganyika. Arch. Hydrobiol. Beih. Ergebn. Limnol. 44: 13–18.

Falkner, K. K., C. I. Measures, S. E. Herbelin & J. M. Edmond, 1991. The major and minor element geochemistry of Lake Baikal. Limnol. Oceanogr. 36: 413–423.

Gage, J. D. & P. A. Tyler, 1991. Deep-sea biology. A natural history of organisms at the deep-sea floor. Cambridge University Press, Cambridge: 504 pp.

Golubev, V. A., J. Klerkx & R. Kipfer, 1993. Heat flow, hydrothermal vents and static stability of discharging thermal water in Lake Baikal (south-eastern Siberia). BCREDP 17: 53–65.

Gorthner, A., 1994. What is an ancient lake ? Arch. Hydrobiol. Beih. Ergebn. Limnol. 44: 97–100.

Grachev, M., V. Flalkov, T. Nakamura, T. Ohta & T. Kawai, 1995. Extant fauna of ancient carbon. Nature 374: 123–124.

Healy, B., 1987. The depth distribution of Oligochaeta in an Irish quaking marsh. Hydrobiologia 155: 235–247.

Hinga, K. R., J. McN. Sieburth & G. R. Heath, 1979. The supply and use of organic material at the deep-sea floor. J. mar. Res. 37: 557–579.

Hohmann, R., M. Hofer, R. Kipfer, F. Peeters, D. M. Imboden & M. N. Shimaraev, 1997b. Distribution of Helium and Tritium in Lake Baikal. J. Geophys. Res. 103: 12823–12838.

Hohmann, R., R. Kipfer, F. Peeters, G. Piepke, D. M. Imboden & M. N. Shimaraev, 1997a. Processes of deep-water renewal in Lake Baikal. Limnol. Oceanogr. 42: 841–855.

Isossimov, V. V., 1962. Maloshchetinkovye chervy semeystva Lumbriculidae (=Oligochaete worms of the family Lumbriculidae of Lake Baikal). Proc. Limnol. Instit. Acad. Sci. USSR, Siberian Sect., 1: 3-126 [in Russian].

Khripounoff, A. & G. T. Rowe, 1985. Les apports organiques et leur transformation en milieu abyssal à l'interface eau-sédiment dans l'Océan Atlantique tropical. Oceanol. Acta 8: 293–301.

Khripounoff, A., D. Desbruyères & P. Chardy, 1980. Les peuplements benthiques de la faille Vema: données quantitatives et bilan d'énergie en milieu abyssal. Oceanol. Acta 3: 187–198.

Killworth, P. D., E. C. Carmack, R. F. Weiss & R. Matear, 1996. Modeling deep-water renewal in Lake Baikal. Limnol. Oceanogr. 41: 1521–1538.

Kozhov, M., 1963. Lake Baikal and its life. In W. W. Weisbach & P. Van Oye (eds), Monogr. Biol. 11, Junk.

Liebezeit, G., 1992. Water and porewater chemistry of a Lake Baikal central basin station. Mitt. Geol.-Palaont. Inst. Univ. Hamburg 72: 41–50.

Martens, K., 1997. Speciation in ancient lakes. TREE 12: 177–182.

Martens, K., G. Coulter & B. Goddeeris, 1994. Speciation in ancient lakes – 40 years after Brooks. Arch. Hydrobiol. Beih. Ergebn. Limnol. 44: 75–96.

Martin, P., 1994. Lake Baikal. Arch. Hydrobiol. Beih. Ergebn. Limnol. 44: 3–11.

Martin, P., 1996. Oligochaeta and Aphanoneura in ancient lakes: a review. Hydrobiologia 334: 63–72.

Martin, P. & R. O. Brinkhurst, 1998. A new genus and two new species of Tubificidae (Oligochaeta) from the abyssal zone of Lake Baikal with redescriptions of *Lymphachaeta pinnigera*, *Rhyacodriloides abyssalis* and *Tubifex bazikalovae*. Zool. Scr. 27: 197–207.

Martin, P., M. Ferraguti & I. Kaygorodova, 1998a. Description of two new *Rhynchelmis* species from Lake Baikal (Russia), using classical morphological and ultrastructural spermatozoal characters. Ann. Limnol. 34: 283–293.

Martin, P., B. Goddeeris & K. Martens, 1994. Depth distribution of oligochaetes in Lake Baikal (Siberia - Russia). Hydrobiologia 278: 151–156.

Martin, P., L. Granina, K. Martens & B. Goddeeris, 1998b. Oxygen concentration profiles in sediments of two ancient lakes: Lake Baikal (Siberia, Russia) and Lake Malawi (East Africa). Hydrobiologia 367: 163–174.

Mazepova, G., 1994. On comparative aspects of ostracod diversity in the Baikalian fauna. Arch. Hydrobiol. Beih. Ergebn. Limnol. 44: 197–202.

Rowe, G. T., P. T. Polloni & S. G. Horner, 1974. Benthic biomass estimates from the northwestern Atlantic Ocean and the northern Gulf of Mexico. Deep Sea Res. 21: 641–650.

Sanders, H. L. & R. R. Hessler, 1969. Ecology of the deep-sea benthos. Science 163: 1419–1424.

Semernoi, V. P., 1983. Species composition and distribution of oligochaetes in the abyssal zone of Lake Baikal. In B. E. Kurashvili (ed.) Vodnye Maloshchetinkovye Chervi (=Aquatic Oligochaeta). Proceedings of the 4th All-Union Symposium, Tbilisi, 5–7 October, Metsniereba Publ. House, Tbilisi, USSR: 99–104 [in Russian].

Semernoi, V. P., 1987. Oligochaeta of the Maloye More of Lake Baikal. In Aquatic Oligochaeta, Proceedings of the 6th All-Union Symposium, Salaspils, 27-30 April, Riga: 123–127.

Semernoi, V. P., 1988. K morfologii i sistematike roda *Nais* Mueller (Oligochaeta, Naididae) v ozere Baikal (= The morphology and systematics of the genus *Nais* Mueller (Oligochaeta, Naididae) in Lake Baikal). Bulletin of the Siberian Division of the Academy of Sciences of USSR, Biological Sciences, 1: 69–75.

Shanks, A. L. & E. W. Edmondson, 1990. The vertical flux of metazoans (holoplankton, meiofauna and larval invertebrates) due to their association with marine snow. Limnol. Oceanogr., 35: 455–463.

Sherstyankin, P. P., L. N. Kuimova & M. N. Shimaraev, 1995. On paleothermohaline regime of Lake Baikal. Abstracts of the second Vereshchagin Baikal conference, 5–10 October 1995, Irkutsk, 230–231.

Shimaraev, M. N., N. G. Granin & A. A. Zhdanov, 1993. Deep ventilation of Lake Baikal waters due to spring thermal bars. Limnol. Oceanogr. 38: 1068–1072.

Snimschikova, L. N., 1982. Novye tubifitsidy (Oligochaeta, Tubificidae) iz severnoi okonechnosti Baikala (= New tubificids (Oligochaeta, Tubificidae) from the Northern Extremity of Lake Baikal). In G. I. Galazii (ed), Novoe o Faune Baikala (=New Information on the Fauna of Lake Baikal). Academy of Sciences USSR, Sib. Branch, Novosibirsk, USSR: 86–99 [in Russian].

Snimschikova, L. N., 1985a. Tubifitsidy (Oligochaeta, Tubificidae) severnogo Baikala (= Tubificids (Oligochaeta, Tubificidae) of northern Lake Baikal). In Bespozvonochnye i ryby (Invertebrates and Fishes). Nauka, Novosibirsk: 39–61 [in Russian].

Snimschikova, L. N., 1985b. K sistematike roda *Svetlovia* (Oligochaeta, Tubificidae) iz ozera Baikal (= A contribution to the systematics of the genus *Svetlovia* (Oligochaeta, Tubificidae) from Lake Baikal). Zoologicheskii zhurnal 64: 1148–1153.

Snimschikova, L. N., 1987. Oligokhety Severnogo Baikala (= Oligochaeta of northern Baikal). In A. A. Linevich (ed), The Fauna of Baikal. Academy of Sciences USSR, Sib. Branch, Nauka, Novosibirsk, USSR: 105 pp. [in Russian].

Snimschikova, L. N. & Akinshina, T. W., 1994. Oligochaete fauna of Lake Baikal. Hydrobiologia, 278: 27–34.

Snimschikova, L. N. & Timm, T., 1992. Review of the genus *Baikalodrilus* Holmquist, 1978 (Oligochaeta, Tubificidae). Bull. Inst. r. Sc. nat. Belg., Biologie, 62: 53–85.

Takhteyev, V. V., 1996. On the ecology of a rare species of amphipods, *Polyacanthisca calceolata* (Crustacea, Amphipoda) of Lake Baikal, with reference to parallelism in the evolution of the Baikal and Oceanic Deepwater Faunas. Hydrobiol. J. 32: 107–109.

Takhteyev, V. V. & I. V. Mekhanikova, 1996. Raspredeleniye endemichnikh nektobenthicheskikh bokoplavov v ozere Baikal (=Distribution of the endemic nectobenthic amphipods in Lake Baikal). Bull. of the Moscow Soc. of Naturalists, Sec. Biology, 101: 39–48 [in Russian].

Takhteyev, V. V., L. N. Snimschikova, G. L. Okuneva, O. A. Timoshkin, L. A. Obolkina & A. I. Tanitchev, 1993. Kharakteristika donnovo nasseleniya glubinnoï zoni Baikala (=Description of the benthic population of the abyssal zone of Baikal), Ecology 6: 60–68.

Timoshkin, O. A., 1994. Free-living Plathyhelminthes – a model group for the evolution of invertebrates in Lake Baikal. Arch. Hydrobiol. Beih. Ergebn. Limnol. 44: 183–196.

Weiss, R. F., E. C. Carmack & V. M. Koropalov, 1991. Deep-water renewal and biological production in Lake Baikal. Nature 349: 665–669.

Hydrobiologia **406:** 175–182, 1999.
B. M. Healy, T. B. Reynoldson & K. A. Coates (eds), Aquatic Oligochaetes.

Oligochaetes of Karavasta Lagoon (Albania). Preliminary results

S. Casellato
Dipartimento di Biologia, Università di Padova, Via Colombo 3, I-35121, Padova, Italy

Key words: Oligochaeta, Tubificidae, lagoon, benthos, Albania

Abstract

In a study of the composition and structure of the benthic community of the Lagoon of Karavasta (Albania), carried out in June 1996, a peculiar distribution of oligochaetes was observed. Four species of tubificids were collected, together with 32 species of other benthic invertebrates (crustaceans, polychaetes, molluscs, nemerteans and chironomids). The oligochaete species found are well-known for the Adriatic coast. Their distribution in the lagoon was mostly restricted to a longitudinal zone near the sea and their densities decreased landwards, to be completely replaced by other faunal groups (particularly chironomids and molluscs). Zonation was distinct according to a salinity gradient and variable hydrodynamic and sediment characteristics.

Introduction

This work was part of a wider investigation of the ecology of the Karavasta Lagoon, which included studies of benthos and plankton. Its aim was to acquire knowledge of a natural environment which might be exploited for sustainable aquaculture. A first summer field campaign was to have been followed by a second during winter, but this proved to be impossible due to the political and social situation in Albania. The data presented here must, therefore, be considered preliminary to further studies which we hope will be continued in the near future.

Little is known of the biological populations of the Karavasta Lagoon. The only existing information is a feasibility study for a shrimp and clam farm, carried out by Bioprogramm (1995). There is no particular information in the literature on meiobenthic oligochaetes, either for the Karavasta Lagoon or for other lagoons in Albania.

Study area

The lagoon complex of Karavasta (Albania) (40° 56′ N, 19° 30′ E) is composed of the Lagoon of Karavasta itself, the Lagoon of Godulla and the estuaries of the rivers Shkumbin (north) and Seman (south) (Figure 1). The Lagoon of Karavasta covers an area of about 43.4 km^2, is at most 10.6 km long and 4.1 km wide, and has a maximum depth near the centre of 1.4 m.

Topographic surveys by the Meteorological Service of the Academy of Sciences of Tirana (Pano et al., 1975) provided information on the hydrogeomorphological evolution of this coastal complex, which has only formed over the last 150 years and which has undergone rapid changes due to the beach-building action of the two rivers (6.3 $\times$ 10^6 tonnes/year) and opposing action by the sea. Early in its formation, which goes back to about 1860, the Lagoon of Karavasta covered double the area it now occupies. The Lagoon of Godulla only began to form about 50 years ago. Of the three mouths communicating with the sea until the late1970s, only the central one is now open and, due to the poor tidal incursion, water exchange is limited. It has been estimated (Bioprogramm, 1995) that water exchange does not involve more than one-third of the entire volume of the Karavasta Lagoon.

The area surrounding the lagoon complex is not built up and there are only a few small villages. The only industrial complex nearby is the paper mill at Lushnje. Fishing in the lagoon is mainly for mullet, sea bass and eel. Fishing intensified between 1986 and 1990 from 153 to 246 tonnes/year (Bioprogramm, 1995). In the last few years, however, the fishermen of the local cooperative have complained of a marked

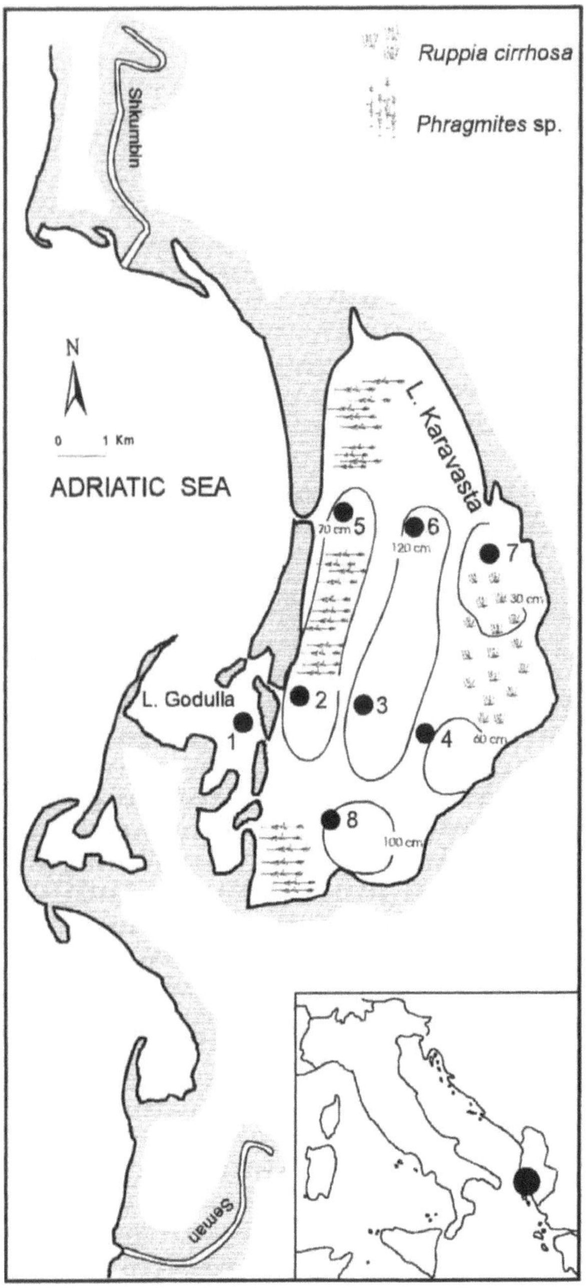

Figure 1. The Karavasta lagoon complex: sampling stations and depths (cm) are indicated.

decrease in the quantity of fish, which may be due both to reduced productivity of the lagoon, in which water exchange is progressively falling and a decline in State authority control for political reasons.

Water temperature, salinity and oxygen data from the literature (Pano et al., 1975) (Figures 2 & 3) show clearly opposing trends for the two main hydrological seasons. In summer, high evaporation, absence of rain and poor circulation explain the high salinity (up to 59 ‰ and low oxygenation (3–4 mg/l), mainly in the extreme north and south (i.e. the areas least affected by water mass dynamics). In winter, increased rainfall and lower evaporation, together with low influx and poor circulation of seawater within the lagoon, produce the opposite situation, with salinity falling to 11 ‰ in the inner part of the lagoon, and higher oxygenation (7.5 mg/l).

Materials and methods

In June 1996, I studied the benthic populations of the lagoon complex of Karavasta by means of a network of 8 stations, located along three transects parallel to the coast (Figure 1). At each station, I collected three replicate samples of benthos using a van Veen grab of 347 cm^2 sampling area and four sediment cores, 3.2 cm × 23 cm, for grain-size analysis (Buchanan, 1984). Sediment organic matter content was analyzed according to Gaudette et al. (1974). Water temperature and salinity were measured *in loco* with a YSI model 33 thermosalinometer and water oxygenation with an OXI 320 Crison water oxymeter. Water samples were collected at all stations and frozen for analysis of nutrients (P-PO_4, N-NH_4, N-NO_2, N-NO_3) and chlorophyll *a* (for further details, see Casellato et al., 1997).

Results

Environmental parameters

Grain-size analysis of sediment (Figure 4A) highlights the great difference between station 1, located in the main opening to the sea, where sediment was almost entirely sandy, and all other stations. The silt fraction predominated at stations 3, 5 and 6 and clay at stations 4, 7 and 8 in the inner part of the lagoon. The percentage of organic matter was highest at stations 6 and 7 (Figure 4B). Water salinity varied from a minimum of 32.5 ‰ to a maximum of 36.5 ‰ (Table 1). These values are well below the summer extremes reported for the lagoon in the 1970s (Pano et al., 1975) but it should be noted that our samples were taken during a period intermediate between winter minima and summer maxima. Oxygenation (Table 1) varied from 5.2 mg/l at station 8 to 10.7 mg/l at station 7, the latter near an area with abundant *Ruppia cirrhosa.*

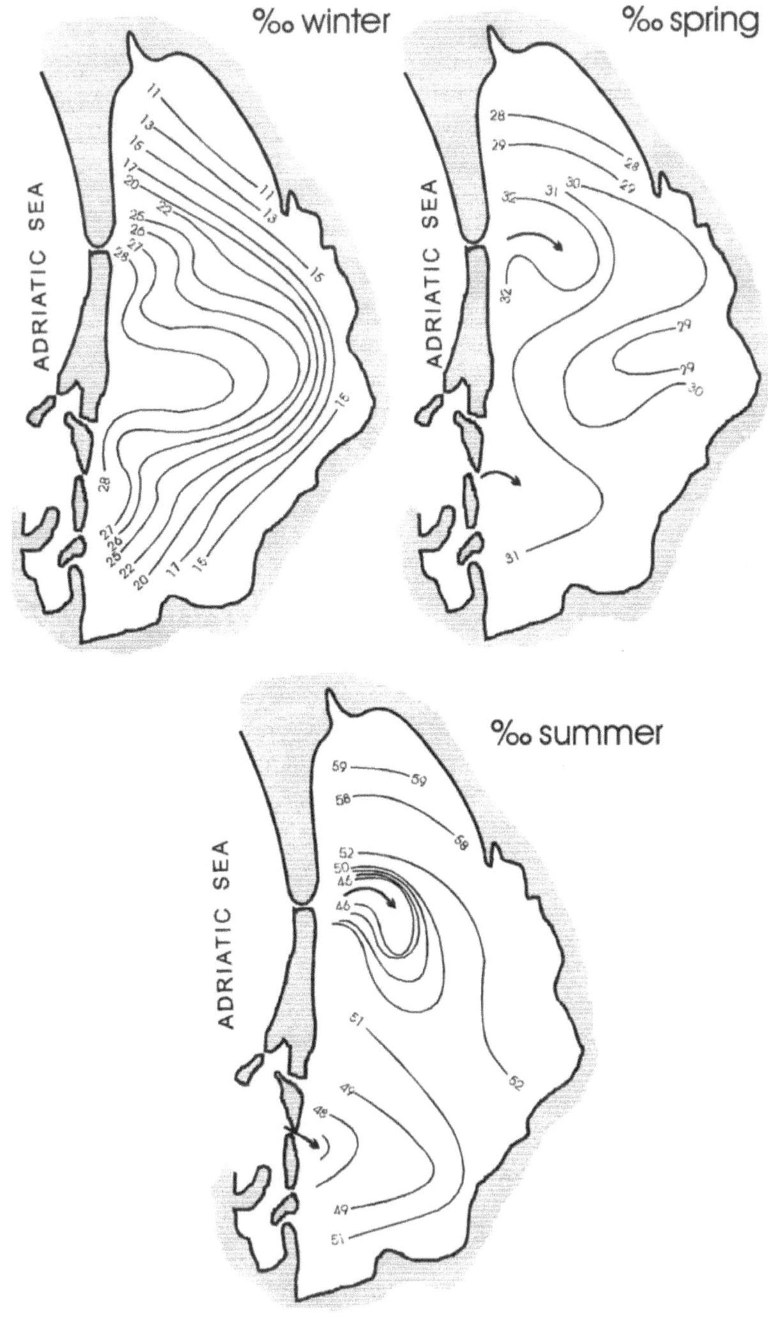

Figure 2. Isohalines (Ğ salinity) in different seasons in the Lagoon of Karavasta in 1974 (from Pano et al., 1975) .

Chlorophyll *a* and nitrate concentrations (Table 1) were highest at stations 3, 6 and 8, all located along the longitudinal axis of the lagoon, where depth was greatest. Ammonia concentration was high only at station 4, located in the south-eastern part of the lagoon. Phosphate and nitrite concentrations were high only at station 8 and at stations 3, 4, and 8, respectively.

Oligochaete density and distribution

Oligochaeta were represented by four species: *Aktedrilus monospermathecus* Knöller, 1935 (2% of individuals), *A. cuneus* Erséus, 1984 (4%), *Limnodriloides maslinicensis* (Hrabě, 1971) (2%), and *Tubificoides swirencowi* Jaroshenko, 1948 (92%), representing

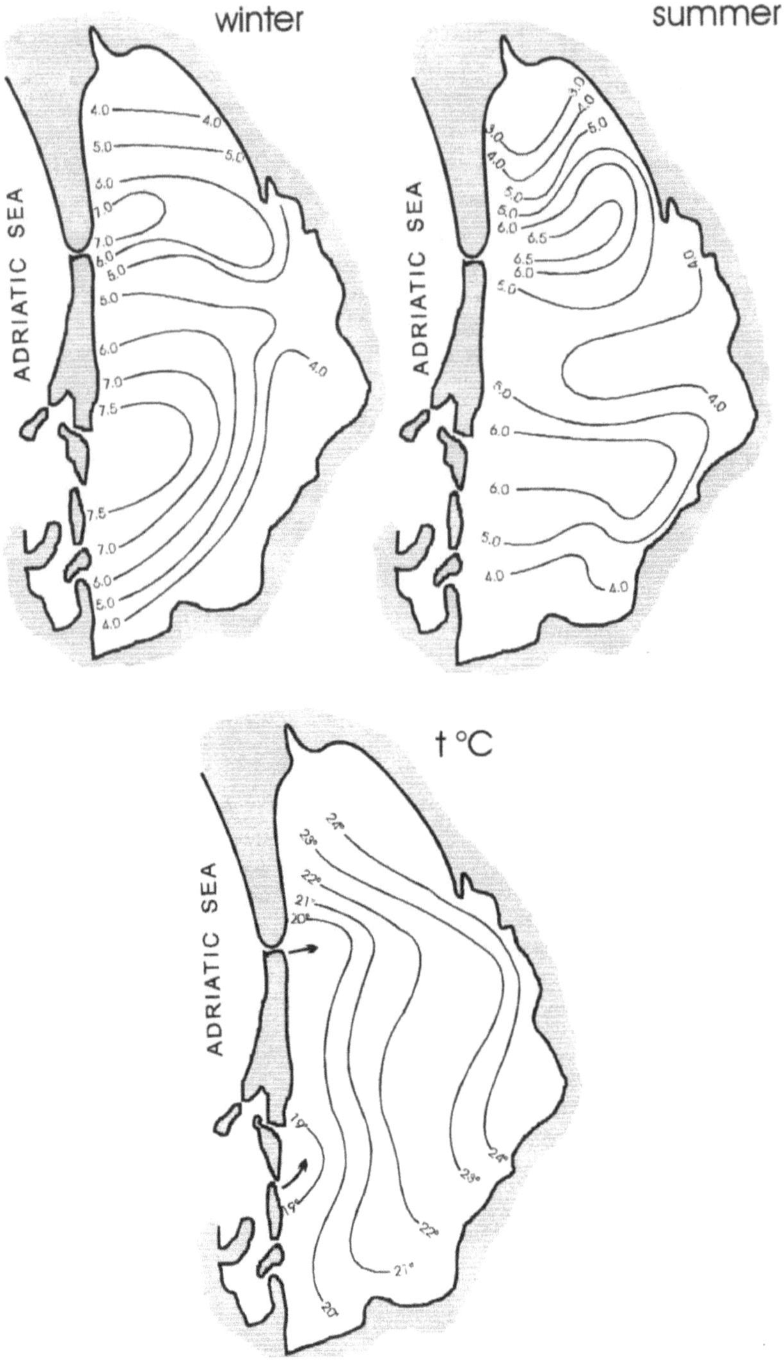

Figure 3. Oxygen isopleths (mg/l) (A) in winter, (B) in summer. (C) Mean annual water temperature (°C) (from Pano et al., 1975).

11% of species in the benthic community. The other 32 taxa reported were nemerteans, molluscs, polychaetes, chironomids and crustaceans (see Casellato et al., 1997). Oligochaetes were restricted to the seaward area of the lagoon. The highest densities (1000–2500 ind/m^{2}) were found in a belt parallel to the coastline at stations 2 and 5 (Figure 5) with a mainly silt-clay bottom (Figure 4A), a depth of less than 1 m and dense reed-beds (Figure 1). Far lower densities (20–200 ind/m^{2}) were recorded at station 1, off the centrally located lagoon mouth, with an almost completely sandy bottom and the lowest organic matter

Table 1. Physico-chemical parameters, nutrients and chlorophyll *a* concentrations at the different lagoonal stations

	T(C°)	Salin. ‰	D.O. (mg/l)	Sat.% O_2	N-NH_4 μmol/l	N-NO_2 μmol/l	N-NO_3 μmol/l	P-PO_4 μmol/l	Chlorophyll *a* mg/m^3
St.1	25.0	36.0	8.0	99.6	0.85	0.11	1.17	0.0	0.12
St.2	26.5	36.5	7.4	112.0	0.89	0.10	1.80	0.08	0.95
St.3	28.0	32.5	5.6	84.5	0.74	0.75	10.98	0.0	3.54
St.4	28.5	35.0	6.6	103.0	2.32	0.64	4.23	0.1	1.77
St.5	29.0	33.0	8.5	138.0	0.80	0.12	4.24	0.0	1.77
St.6	28.0	33.0	5.6	84.5	0.82	0.35	16.47	0.0	6.12
St.7	28.5	34.0	10.7	163.0	0.73	0.11	0.96	0.07	0.95
St.8	27.5	33.5	5.2	79.5	0.93	0.61	12.85	0.29	3.41

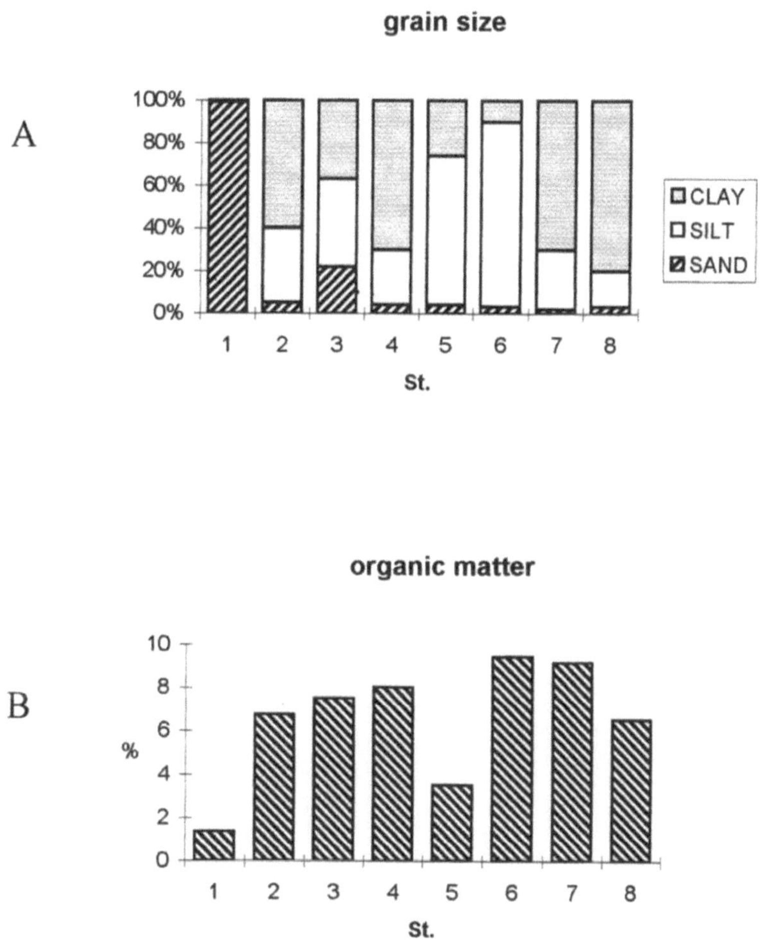

Figure 4. Grain-size composition (A) and organic matter content (B) at the different stations.

content (Figure 4B), and at stations 3 and 8, with a silt-clay bottom and a depth of 1 m (Figures 4A & 5). In the remaining areas examined, not a single specimen of oligochaete was found.

Species composition and abundance varied between stations (Figure 6). *Aktedrilus cuneus* was most abundant (144 ind/m^2) at station 1, where the sediment was almost completely sandy (Figure 4A) and the organic

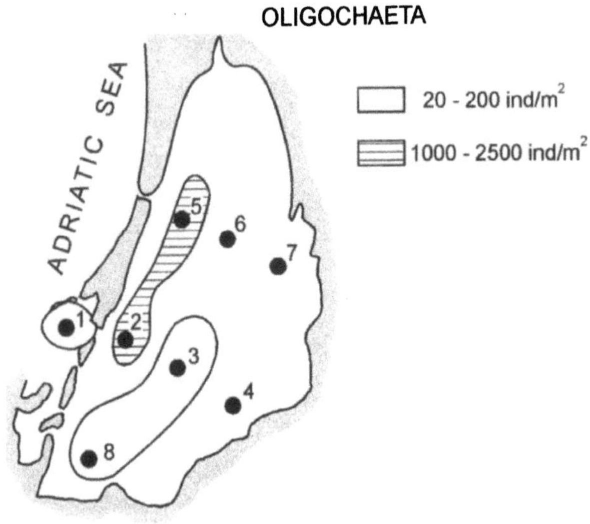

Figure 5. Oligochaete density (ind/m^2) in the different lagoonal areas.

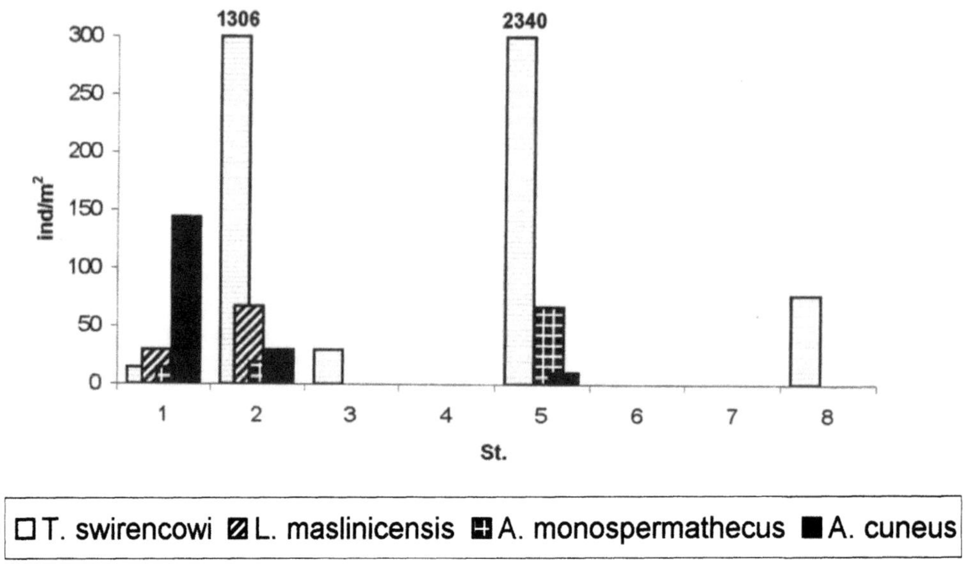

Figure 6. Total density (ind/m^2) of each oligochaete species in the different stations of the Karavasta Lagoon.

matter content low (Figure 4B). This is the preferred habitat for this intertidal species which seems to have had some difficulty in colonizing the muddy-clayey bottom of the innermost part of the lagoon, although a few specimens were also collected at stations 2 and 5 where the sediment was mostly silty. *Aktedrilus monospermathecus* was less abundant and was most numerous at station 5 (Figure 6). *Limnodriloides maslinicensis* was collected only at stations 1 and 2 where the sediment was sand and silt-clay respectively. *Tubificoides swirencowi* was the most abundant and common species in the lagoon, particularly at stations 2 and 5 (Figure 6). Moreover, it was the only species present at stations 3 and 8, where its density was lower than at stations 2 and 5. No oligochaete specimens were collected at stations 4, 6 or 7.

Discussion

The four recorded species have all been reported for the Northern Adriatic coast by Casellato (1994,

1996). *Aktedrilus monospermathecus* is a cosmopolitan species found on the south coast of Italy (Fano) in intertidal sands (Bonomi & Erséus, 1984, 1985; Erséus, 1987a), together with *A. cuneus* which is also a widespread species. *Limnodriloides maslinicensis* is reported only from the Adriatic coast of Yugoslavia (Hrabě, 1971) and Italy (Erséus, 1987b; Casellato, 1996) where it was frequent in muddy fine sands with abundant seagrass. *Tubificoides swirencowi*, first reported by Jaroshenko (1948) from the Black Sea and later by Hrabě (1964, 1965) and Moroz (1994) from the same area, was later found in the Lagoon of Venice, where it was abundant in sandy-silt sediment (Casellato, 1996), and in other brackish environments of the North Adriatic coast (Casellato, 1994). In the Lagoon of Karavasta it was confined to only two stations near sea.

Although sediment grain size probably influences settlement and colonization of this recent coastal complex, it does not completely explain the observed distribution pattern. Stations 2 and 5, where highest densities were observed, have grain-size values which are similar to those of other stations in the lagoon, e.g. 4, 7 and 8 (Figure 4A). The depth of the lagoon near the sea is not very different from that of the more landward stations. Thus, other factors have to be considered. Many authors consider that salinity is the main environmental factor governing the distribution of marine oligochaetes, especially in brackish-water areas and estuaries where it fluctuates due to tides, storms and evaporation (Giere, 1977; Diaz, 1980; Giere & Pfannkuche, 1982). Salinities in June showed no gradient but ranged between 32.5 and 36.5 ‰. However, the isohaline data reported by Pano et al. (1975) (Figure 2) show an accentuated salinity gradient appearing in winter, with salinity decreasing from the outer to the inner regions, a trend which was reversed in summer. In the present study, the salinity range in the area nearest the coast (28–48 ‰) was that in which the highest oligochaetes population densities were found. The wider range (11–52 ‰) at the inner stations may have been too great for oligochaetes compared with other more euryhaline species such as *Chironomus salinarius* Kiefer, which was abundant at stations 4 and 7 (Casellato et al., 1997).

The total absence of oligochaetes from stations 4, 6 and 7 is difficult to explain. The bed was mainly silty with abundant *Ruppia cirrhosa* at 4 and 7, conditions which do not appear unfavourable for the recorded species. However, organic matter content was highest at these stations (which were near *Ruppia* beds) and the area is probably isolated from the hydrodynamic flow as a consequence of the progressive closure of the northern and the southern lagoon mouths. In late summer, a state of anoxia in the sediment and the highest recorded values for ammonia (station 4) (Table 1) were observed and could have adversely affected the oligochaete species which are sensitive to oxygen depletion. These factors might also explain the low numbers of other faunal groups (Casellato et al., 1997). For example, polychaetes were completely absent from station 6 and of the molluscs, only small numbers of *Cerastoderma edule* (L.) and *Loripes lacteus* (L.) were found. *Chironomus salinarius*, which in other stations was very abundant, was also infrequent here. More detailed and recent data on the distribution of benthivorous fish species in the lagoon may help to explain the observed oligochaete distribution.

Acknowledgements

This study was supported by a financial grant from the Regione Veneto, within the ambit of the program: 'Institution of Regional funds for operations of national solidarity'. I would like to thank Dr Zamir Dedej of the Academy of Sciences of Tirana for his technical help during field sampling in Albania and Prof. R. Brunetti and Dr L. Masiero for analytical data on sediments and water.

References

Bioprogramm, 1995. Laguna di Karavasta: studio di fattibilità. Padova.

Bonomi, G. & C. Erséus, 1984. A taxonomic and faunistic survey of the marine Tubificidae and Enchytraeidae (Oligochaeta) of Italy. Introduction and preliminary results. Hydrobiologia 115: 207–210.

Bonomi, G. & C. Erséus, 1985. Marine Oligochaeta of Italy: a preliminary survey. Nova Thalassia 6 (suppl.): 713–714.

Buchanan, J. B., 1984. Sediment analysis. In N. Holme & A. D. Mc Intyre (eds), Methods for the Study of Marine Benthos. Blackwell Sc. Publ Oxford 2nd edn.: 41–65.

Casellato, S., 1994. Oligochaete fauna of estuarine areas and lagoons on the northern Adriatic coast (Italy). Boll. zool. 61: 261–269.

Casellato, S., 1996. Oligochaetes in the southern basin of the Venetian Lagoon: community composition, species abundance and biomass. Hydrobiologia 334: 103–114.

Casellato, S., S. Santi & L. Masiero, 1997. Ricerche ecologiche nella Laguna di Karavasta (Albania). Primi risultati sul macrobenthos. S.It.E. Atti.18: 229–232.

Diaz, R. J., 1980. Ecology of tidal freshwater and estuarine Tubificidae (Oligochaeta) In R. O. Brinkhurst & D. G. Cook (eds), Aquatic Oligochaete Biology. Plenum Press, New York: 310–330.

Erséus, C., 1987a. Taxonomic revision of the marine interstitial genus *Aktedrilus* (Oligochaeta, Tubificidae) with descriptions of three new species. Stygologia 3: 107–124.

Erséus, C., 1987b. Marine Limnodriloidinae (Oligochaeta, Tubificidae) from Italy, with description of two new species. Boll. zool. 54: 159–164.

Gaudette, H. E., R. F. Wilson, L.Toner & D. V. Folger, 1974. An inexpensive titration method for determination of organic carbon in recent sediment. J. Sed. Petrol. 44: 149–253.

Giere, O., 1977. An ecophysiological approach to the microdistribution of meiobenthic Oligochaeta. I. *Phallodrilus monospermathecus* (Knöllner) (Tubificidae) from a subtropical beach at Bermuda. In B. F. Keegan, P. O. Céidigh & P. J. Boaden(eds), Biology of benthic organisms. Pergamon Press New York: 285–296.

Giere, O. & O. Pfannkuche, 1982. Biology and ecology of marine Oligochaeta. A Review. Oceanogr. mar. Biol. ann. Rev. 20: 173–308.

Hrabě, S., 1964. On *Peloscolex svirenkoi* (Jaroshenko) and some other species of the genus *Peloscolex*. Publ. Fac. Sci. Univ. Purkyne, Brno 450: 101–112.

Hrabě, S., 1965. New or insufficiently known species of the family Tubificidae. Spisy prir. Fak. Univ. Brno: 57–77.

Hrabě, S., 1971. On new marine Tubificidae of the Adriatic Sea. Scr. Fac. Sci. Nat. Univ. Brno 1: 215–226.

Jaroshenko, M. F., 1948. Oligochaetes of the Dnjepr-Bug Liman. Nauc. Zap. Modlav. nauc. issl. Bazy Acad. Nauk. SSR 1: 57–68 (in Russian).

Moroz, T. G., 1994. Aquatic Oligochaeta of the Dnjeper-Bug estuary system. Hydrobiologia 279: 133–138.

Pano, N., Z. Skendaj, N.Rokij & M. Xheli, 1975. Kompleksi ujor i Karavastase. Akadamia E Shkencave. Instituti Hidrometeorologjik, Tirane.

Hydrobiologia **406:** 183–189, 1999.
B. M. Healy, T. B. Reynoldson & K. A. Coates (eds), Aquatic Oligochaetes.

A comparison among different population models for *Limnodrilus hoffmeisteri* Claparède (Oligochaeta, Tubificidae)

Andrea Pasteris, Martina Vecchi & Giuliano Bonomi
Dipartimento di Biologia Evoluzionistica Sperimentale, Università degli Studi di Bologna, Via Selmi 3, I-40126 Bologna, Italy

Key words: population dynamics, population structure, age, size, stage, *Limnodrilus hoffmeisteri*

Abstract

The lack of any reliable method for assessing the age of individuals collected in the field has often been considered a major obstacle for population studies in aquatic oligochaetes. One possible solution could be the adoption of other variables, such as weight or stage, for the definition of the population structure; this approach would be useful if allowed good predictions about population growth. We measured, under laboratory conditions, the effect of age, size and life stage on survival, growth and fecundity of *Limnodrilus hoffmeisteri* Claparède. The results are used to establish four matrix population models, based respectively on the classification of the individuals by age, weight and stage and on a mixed classification. Matrix population models make the assumptions that the individuals in a population can be arranged in a number of discrete classes and that time is a discrete variable. In these models, the population is represented by a vector (each element in the vector is the number of individuals in a class) and the demographic coefficients (survival, growth and fecundity) are collected in a square matrix. The estimate of λ, the potential long term population growth rate and its confidence interval were taken from the four models using the *jackknife* method. The width of the confidence interval is a measure of the effectiveness of the models and thus of the classification of the individuals. The results suggest that weight is inefficient as a criterion for the definition of population structure of *L. hoffmeisteri* in comparison to age and stage.

Introduction

One major problem found in the study of population dynamics of tubificid worms is that the traditional demography, which is based on the classification of individuals by age, is not applicable to these organisms. The first reason is the lack of any reliable method for assessing the age of individuals collected in the field. However, even if it were possible to determine the age structure of natural populations, this might be of little use. Actually, the individual traits relevant to population dynamics, i.e. survival, growth and fecundity, depend on the physiological state of the individuals, which is affected by the present and past environmental conditions. Age, by itself, does not take account of the history of the individual and, therefore, is often a poor predictor of physiological state as well as demographic traits.

For similar reasons, age based demography is useless or insufficient for many other organisms. According to several authors (Lefkovitch, 1965; Law, 1983; Hughes, 1984; Ebenman & Persson, 1988; De Leo & Gatto, 1995), the solution for this problem is the adoption of a different variable for the definition of the population structure, i.e. the use of different criteria for the classification of the individuals of a population. Suitable criteria depend on the kind of organism considered and could include size, stage, and number of instar.

The effectiveness of such an approach for the study of tubificid populations, in particular *Tubifex tubifex* (Müller), was assessed by Bonomi & di Cola (1980) and Pasteris et al. (1996); this approach is here applied to *Limnodrilus hoffmeisteri* Claparède. In particular, the question whether it is useful to define *L. hoffmeisteri* population structure by using weight or life stage, as a good prediction of population growth, is tackled.

Matrix population models are probably the simplest approach to the formal description of structured populations, whatever criterion is used to define structure (Lewis, 1942; Leslie 1945, 1948; Lefkovitch, 1965; Usher, 1972; Caswell, 1989). Two main assumptions characterise these models:

1. the individuals in a population can be arranged in a number of discrete classes; and
2. time is a discrete variable, i.e. assumes only values that are integer multiples of a basic unit called the *projection interval.*

A matrix population model can be written as a system of finite difference equations:

$$\begin{aligned}
n_1(t+1) &= a_{1,1}\cdot n_1(t) + a_{1,2}\cdot n_2(t) = \cdots \\
&\quad \cdots + a_{1,j}\cdot n_j(t) + \cdots + a_{1,m}\cdot n_m(t) \\
n_2(t+1) &= a_{2,1}\cdot n_1(t) + a_{2,2}\cdot n_2(t) + \cdots \\
&\quad \cdots + a_{2,j}\cdot n_j(t) + \cdots + a_{2,m}\cdot n_m(t) \\
&\vdots \\
n_i(t+1) &= a_{i,1}\cdot n_1(t) + a_{i,2}\cdot n_2(t) + \cdots \\
&\quad \cdots + a_{i,j}\cdot n_j(t) + \ldots + a_i, m\cdot n_m(t) \\
&\vdots \\
n_m(t+1) &= a_{m,1}\cdot n_1(t) + a_{m,2}\cdot n_2(t) + \cdots \\
&\quad \cdots + a_{m,j}\cdot n_j(t) + \ldots + a_{m,m}\cdot n_m(t),
\end{aligned}$$

Where: m is the number of classes; $n_j(t)$ is the number of individuals present in the j-th class at a generic time t; $n_i(t+1)$ is the number of individuals present in the i-th class at time $t+1$ (i.e. one time unit after t); $a_{i,j}$ is a coefficient accounting for the average contribution of one individual in the class j at time t, to the class i at time $t+1$ (i.e. how many individuals will be present at time $t+1$ in the i-th class for one individual present at time t in the j-th class).

Depending on what the biological process responsible for the contribution of one class to an other is, $a_{i,j}$ represents a fecundity rate or a transition rate between classes (i.e. the result of survival and growth combined).

Matrix population models are usually written as equations of matrices and vectors (Caswell, 1989):

$$\mathbf{n}(t+1) = \mathbf{A}\cdot\mathbf{n}(t),$$

where: $\mathbf{n}(t)$ is a vector of m elements $n_j(t)$, that describes the state of the population at time t; $\mathbf{n}(t+1)$ is a vector of m elements $n_i(t+1)$, that describes the state of the population at time $t+1$; $\mathbf{A}$ is a $m \times m$ matrix; $a_{i,j}$ is the element on the i-th row and on the j-th column.

One advantage of the matrix notation is that $\mathbf{A}$, the *population projection matrix*, displays all the parameters of the model in a concise and ordered form. The elements along the i-th row represent all the input that the i-th class receives from the other classes; the elements along the j-th column represent all the output from the j-th class to the other classes.

Because of their basic assumptions, matrix population models are an obvious choice for the description of populations with clearly separate stages and seasonal breeding. However, because of their simplicity, they have also been used for populations without these characteristics.

A second goal of this paper is to establish a matrix population model, as well as to find the most suitable classification of the individuals of the oligochaete species *L. hoffmeisteri*.

Therefore, the effect of age, weight and life stage on survival, growth and fecundity of *L. hoffmeisteri* was measured in a laboratory experiment. Since the adopted experimental design matched the assumptions of population matrix models, the results could be used as estimates of the coefficients present in these models. Four distinct matrix models, based on different definitions of the population structure, were established and their *asymptotic behaviour*, i.e. their prediction on the long term population growth (Caswell, 1989), was compared.

Materials and methods

The experiment was designed to match the assumptions of matrix population models. At the beginning of the experiment age classes, life stages and weight classes were defined.

Age classes were defined as follows:

class 1: 0 – 35 days　class 2: 35 – 70 days
class 3: 70 – 105 days　class 4: 105 – 140 days
class 5: 140 – 175 days

The range of 35 days was chosen for age classes, since no worms hatched from cocoons older than 35 days, under the chosen experimental conditions. Hence, no viable eggs nor embryos older than 35 days were present and embryonic stages all were contained in the first age class.

Weight classes were defined as follows (wet weight):

class 1: 0 – 1 mg　class 2: 1 – 2 mg
class 3: 2 – 4 mg　class 4: 4 – 8 mg
class 5: 8 – 16 mg　class 6: > 16 mg

The upper limit of 1 mg was selected for the first class because it was not possible to weigh, without damage, alive individuals much lighter than 1 mg. This was the lowest threshold that could be used both safely and reliably. The other classes were defined by doubling each time the upper limit value.

Life stages are defined as follows:

class 1: eggs and embryos (individuals inside the cocoon)

class 2: immature (individuals without a distinct clitellum)

class 3: mature (individuals with distinct clitellum)

class 4: ovigerous (individuals with clitellum and visible oocytes)

During the experiment, all measurements were performed with a periodicity corresponding to a projection interval of 35 days; the width of the age classes and the projection interval are identical, like in the classic Leslie (1945) model.

The experimental procedure, outlined below, is rather complex, yet the basic idea is quite simple (Figure 1). Small groups of individuals of the same initial age, weight and stage were assembled. Each group was reared for 35 days (one projection interval). At the end of this period, newborns and survivors were counted and survivors were classified according to their final age, weight and stage. Thus, for each specific combination of age, weight and stage, it was possible to estimate fecundity, survival and transition rates to other classes. These groups of individuals will here be referred to as *experimental units*, since they are the basic elements from which data were extracted.

A laboratory stock culture of *L. hoffmeisteri* was established using worms collected from storage ponds at the electric power station of Castel S. Giovanni (Piacenza, N. Italy) near the river Po. Worms were reared in groups of 40–50 individuals in glass containers (diameter: 11 cm, height: 6 cm), half filled with sand and half filled with previously aerated tap water. Unfrozen lettuce was put under the sand as food (depth: ca. 2 cm). Every week, each container was checked, the sand washed, the residual food removed and fresh lettuce added; also cocoons where removed and either discarded or placed in a separate container to start new stock cultures. Worms, from all containers, were mixed and then re-distributed among the containers. Both the stock culture and the experimental cultures were kept in the dark, at 15 ± 1°C.

The experiment started three months after the stock culture was established. A total of 120 cocoons was randomly collected from the stock culture and placed in the culture containers; 20 per container. The experiment was done in glass containers (internal diameter: 7.6 cm; height: 4.5 cm), half filled with sand and half filled with aerated tap water. Cocoons were buried in the sand with a pipette; no food was added at this time. The rest of the weekly cocoon production was eliminated.

After one week, these containers where checked, the water changed and the sand washed. Cocoons were returned to the containers and buried in the sand while hatched young were eliminated. Another 120 cocoons were collected from the stock culture and placed in containers for rearing. This procedure was repeated for the three following weeks: every 7 days one new batch of 120 cocoons was reared while hatched worms were eliminated from older batches.

The preliminary phase, as outlined above, produced a sample of eggs and embryos with an age between 0 and 35 days, and an age distribution which approaches the embryos age distribution of a population laying eggs at a constant rate. The next step of the experiment required an estimation of the number of eggs and embryos. However, sediment particles which stick to the wall of *Limnodrilus* cocoons make it impossible to examine their content without dissection. Therefore, one half of the number of cocoons of the whole sample was used to estimate the mean number of embryos per cocoon. The other half was used in the first experimental units. Cocoons were placed in the culture containers, 10–12 per container. The number of embryos in each container was estimated as the mean number of embryos/cocoon multiplied by the number of cocoons. Cocoons were buried in the sand with a pipette.

Unfrozen lettuce was put in the culture containers, under the sand, as food. The amount of lettuce could not be entirely consumed by the worms in one week, but also caused no severe oxygen depletion nor poor culturing conditions. Every week, each container was checked, the sand washed, the residual food removed and fresh lettuce added. Also, both cocoons and hatched worms were counted and replaced in the culture containers. The cocoons were buried in the sand. After 5 weeks (i.e. one projection interval), the survivors of each experimental unit were counted and put in clean water for a few hours, without sand or food to allow the gut content to clear. Afterwards, the maturation stage was checked under a dissecting microscope. Each worm was dried with filter paper and weighed at the nearest 0.1 mg. Only immature individuals were present at this time, from two weight classes (0 – 1 mg and 1 – 2 mg). All the worms belong-

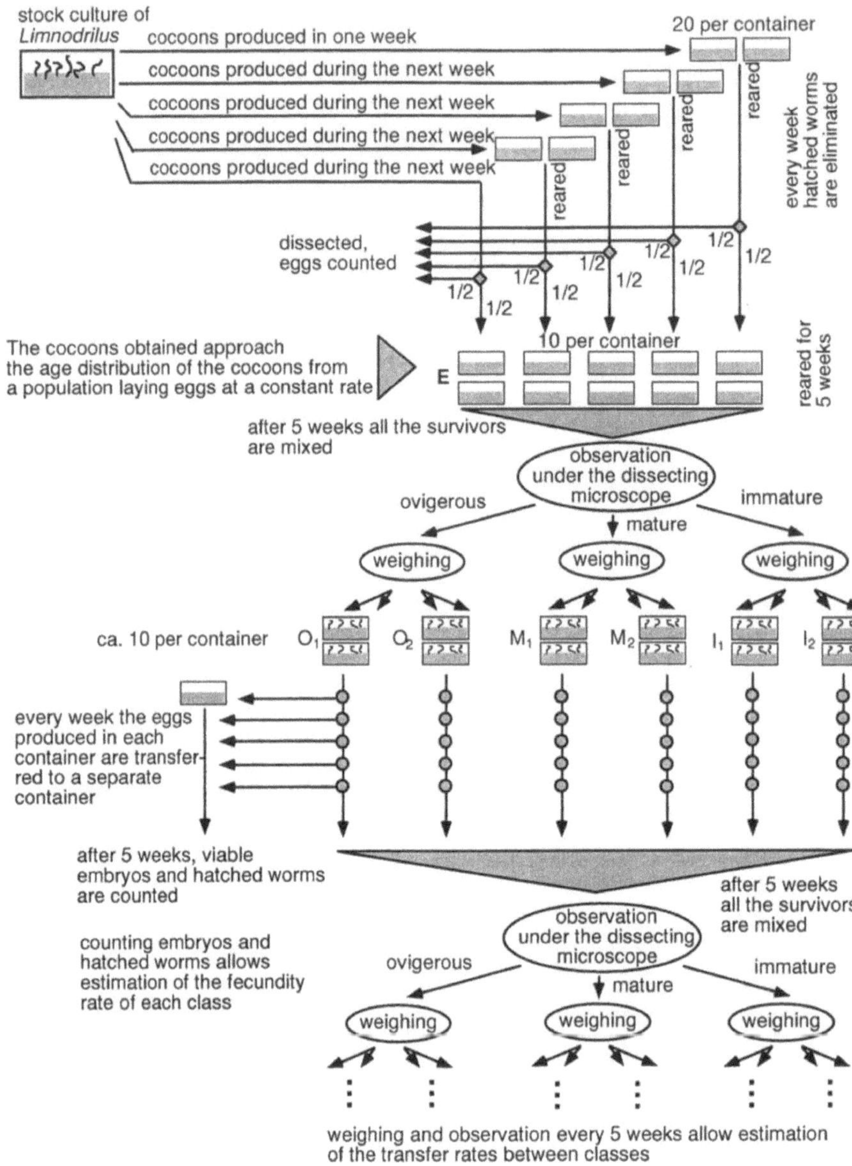

Figure 1. Diagrammatic representation of the experiment. The figure gives only a qualitative description of the procedure; the number of containers and classes represented do not correspond to the real experiment.

ing to the same weight class were mixed, regardless of the original experimental unit and the new experimental units were started by placing 5 – 10 individuals of the same class in each culture container.

These new experimental units were reared for 5 weeks. Every week newborns produced in each experimental unit were transferred to another container; also, all the containers were checked, the sand washed, the residual food removed and fresh lettuce added. After 5 weeks, newborns produced by each experimental unit, both embryos and hatched worms, were counted. Also, survivors of each experimental unit were counted and classified according to their maturation stage (both mature and ovigerous worms were present at this time) and then weighed. All the worms of the same life stage and weight class were mixed and new experimental units were assembled. Ten individuals of the same life stage and weight were assembled in each culture container. This procedure was repeated three more times.

The number of experimental units observed in each 35 day period decreased from 21 to 8 over the time span of the experiment, corresponding to a number of individuals decreasing from 418 to 75 because of mortality. Over the whole experiment, 67 experimental units were observed.

The raw data were used to estimate the elements $a_{i,j}$ of the projection matrices of three models, respectively, based on the classifications of individuals by age, weight and stage. One additional model, based on a mixed classification, was established by splitting the class 0 – 1 mg of the weight structured model in two classes: immature worms and embryos.

For the estimation of the coefficients of each model, data were pooled among experimental units belonging to the same class for the relevant criterion (e.g. for the age structured model, data were pooled among units of the same age class regardless of weight and stage). The estimate of each element in the projection matrix was:

$$\hat{a}_{i,j} = \frac{n_{j\to i}(\text{end})}{n_j(\text{start})},$$

where: n_j(start) is the total number of individuals belonging to the j-th class at the beginning of any projection interval, i.e. the total number of individual in all the observed experimental unit of class j; $n_{j\to i}$(end) is the number of individual belonging to class i found, at the end of the projection interval, in all the experimental units belonging to class j, whether they are newborns or individuals passed from class j to class i during the projection interval.

Results and discussion

The estimate of the projection matrix for the age structured model is:

$$\mathbf{A}_{\text{age}} = \begin{pmatrix} 0 & 1.26 & 28.3 & 53.5 & 27.1 \\ 0.268 & 0 & 0 & 0 & 0 \\ 0 & 0.870 & 0 & 0 & 0 \\ 0 & 0 & 0.986 & 0 & 0 \\ 0 & 0 & 0 & 1.00 & 1.00 \end{pmatrix}$$

The elements on the first row are fecundities (inputs from the other classes to class 1). The individuals younger than 35 days have no fecundity, in accordance the first element of the row is null. Individuals in class 4 have the highest fecundity ($a_{1,4} = 53.5$).

As the width of the age classes is identical to the duration of the projection interval, all the individuals that belong to class j at any time t and survive until time t+1 enter class j+1. Accordingly, only one non null transition rate is present in the column corresponding to each class, which also represents the survival rate of the class. In absence of data for individuals older than 140 days, it is assumed that these individuals have the same demographic traits of individuals of age between 105 and 140 days, and class 5 was defined as containing all the worms older than 105 days. Thus, survivors of this class remain in the same class and their survival rate is on the diagonal ($i = j = 5$). Survival is rather low for individuals in the first age class, but strongly increases in class 2. For classes 4 and 5, based on 75 individuals, no mortality has been recorded.

Only five classes were included in the weight structured model, instead of the six considered in the experiment, because only one experimental unit was observed for class 6 (> 16 mg). In the model, class 5 was defined as containing all the individuals > 8 mg. The estimate of the projection matrix for the weight structured model is:

$$\mathbf{A}_{\text{weight}} = \begin{pmatrix} 0.495 & 17.1 & 29.1 & 50.2 & 35.8 \\ 0.042 & 0.056 & 0.028 & 0 & 0 \\ 0.073 & 0.111 & 0.056 & 0 & 0 \\ 0.038 & 0.556 & 0.667 & 0.392 & 0.012 \\ 0.004 & 0.278 & 0.250 & 0.608 & 0.988 \end{pmatrix}$$

The weight structured model differs from the age based one. In the first, during any projection interval, a surviving individual may stay in the same class, enter the next class or even pass through several classes, depending on the individual growth rate. An individual may also lose weight and regress to a previous class. Thus, in each column of the projection matrix, several non null transfer rates may be present. The total survival rate for any class can be computed as the sum of all the transition rates along the corresponding column.

Total survival for individuals in class 1, which contains both embryos and worms, is 0.652. Most survivors stay in the same class (see first element on the first column, on the diagonal of the matrix), while the proportion of individuals found in other classes at the end of the projection interval is 0.157. Individuals under 1 mg can grow and pass several weight classes during the 35 day period and also reach over 8 mg, though the probability is very low ($a_{5,1} = 0.004$). No mortality was recorded for individuals in class 2 – 5; some regressions were observed from class 3 to 2 ($a_{2,3} = 0.028$) and from class 5 to 4 ($a_{4,5} = 0.012$).

The estimate of the projection matrix for the stage structured model is:

$$\mathbf{A}_{\text{stage}} = \begin{pmatrix} 0 & 2.29 & 21.9 & 37.0 \\ 0.256 & 0.226 & 2.22 & 6.38 \\ 0 & 0.322 & 0.300 & 0.024 \\ 0 & 0.339 & 0.700 & 0.976 \end{pmatrix}$$

For an immature individual at the beginning of any projection interval, it is possible to develop reproductive organs and breed before the end of the interval. However, the class transition is not recorded until the end of the interval. This explains why some fecundity is ascribed to class 2 ($a_{1,2} = 2.29$). It is also possible that, before the end of the projection interval, some worms hatch from eggs laid at the beginning of the interval. These newborns are recorded as immature (class 2), thus some elements on the second row of the stage structured matrix ($a_{2,3}$ and $a_{2,4}$) are, indeed, fecundities and not transition rates.

The estimate of the projection matrix for the model based on the mixed classification is:

$$\mathbf{A}_{\text{WS}} = \begin{pmatrix} 0 & 1.35 & 16.3 & 26.3 & 42.9 & 30.4 \\ 0.249 & 0.124 & 0.833 & 2.92 & 7.28 & 5.48 \\ 0.007 & 0.181 & 0.056 & 0.028 & 0 & 0 \\ 0 & 0.362 & 0.111 & 0.056 & 0 & 0 \\ 0 & 0.191 & 0.556 & 0.667 & 0.392 & 0.012 \\ 0 & 0.019 & 0.278 & 0.250 & 0.608 & 0.988 \end{pmatrix}$$

In general, provided that the elements of the projection matrix are constant, simulations of matrix population models lead to qualitatively similar results (there are exceptions, but they are not relevant here). The population growth rate may fluctuate initially, but eventually the population grows exponentially at a constant rate λ, which is independent from the initial population structure. Thus λ represents the potential long term population growth rate, provided environmental conditions do not change and density effects do not take place. Since λ is a function of all the elements of the projection matrix, it is a useful overall index of the performance of the model.

To compare the models and the effectiveness of the different definitions of the population structure, the precision of the estimates of λ supplied by each model was used.

Each element in the matrices was estimated by pooling the data from several experimental units, as explained above. However, elements estimated on single units can be rather different from each other. Egg production exemplifies this variability (Figure 2). As could be expected, the effects of age, weight and stage on egg production are all significant ($P < 0.01$) by Kruskal-Wallis non-parametric ANOVA (the egg production of embryos is null and thus excluded from the comparison). However, there is a wide variation within each group. For experimental units of the same class, the maximum value can be even three times higher than the minimum.

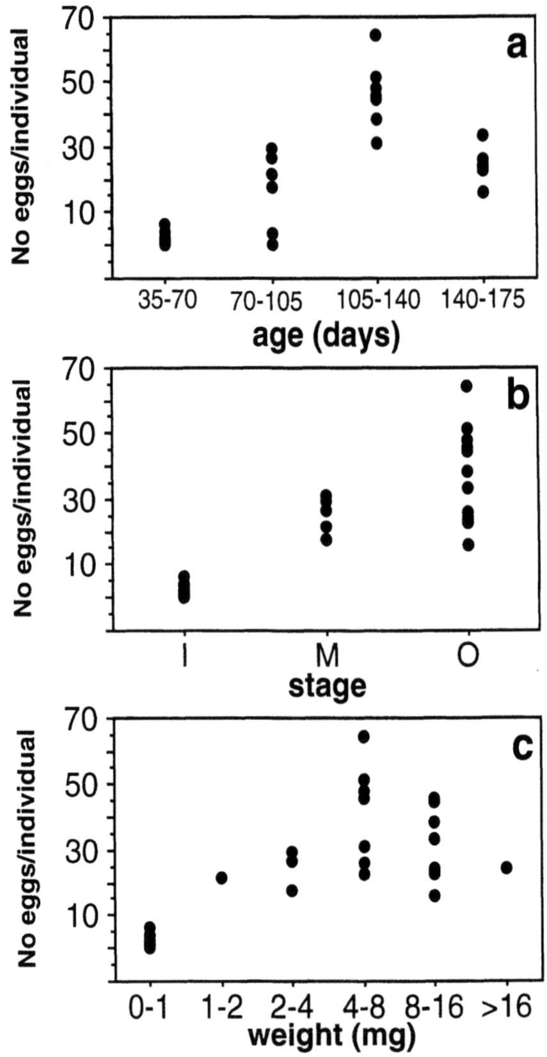

Figure 2. Effect of age (a), stage (b) and weight (c) on the mean number of eggs produced per individual over 35 days. Each point represents the number per individual in one experimental unit.

As a consequence of this variability, an estimate of λ is also affected by a sampling error. An effective definition of the population structure should minimise the differences among individuals that belong to the same class and, therefore, minimise the variance of the estimates of the elements of the matrix and of λ.

The estimate of λ and its confidence interval were taken from the four models using the *jackknife* method (Efron, 1986; Caswell, 1989). The width of the confidence interval is a measure of the effectiveness of the models and thus of the classification of the individuals. Since the same set of experimental units was used for the estimation, the observed differences in the estimates depend only on the classification criteria. A limitation of this approach is that, in absence of a 'true' value of λ, it is only possible to compare the precision of the estimates, not the bias from the actual population value.

The estimates of λ (mean $\pm$ confidence interval), for the models based on age, weight, stage and on the mixed classification are respectively:

$\lambda_{age} = 2.43 \pm 0.28$ $\quad\lambda_{weight} = 2.65 \pm 0.79$

$\lambda_{stage} = 2.97 \pm 0.33$ $\quad\lambda_{ws} = 2.56 \pm 0.40$

The mean values are similar, even though there are differences. However, the confidence intervals largely overlap. The age structured model provided the narrowest confidence interval and the stage structured model a slightly wider one. This order is reversed, if the width of the confidence interval is measured as the percentage of the mean value (23% for λ_{age} and 22% for λ_{weight}). The confidence interval for λ_{weight} is much wider, either if the absolute or the relative (60%) value is considered. When splitting the 0–1 mg class, the precision of the estimate considerably improves, but the confidence interval for λ_{ws} is still wider than for λ_{age} and λ_{stage}.

Conclusion

The results suggest that weight is inefficient as a criterion for the definition of population structure of *L. hoffmeisteri*, in comparison to age and stage. However, this outcome depends on the particular classes used and does not exclude the possibility that a different classification would have been more effective. Weight differences among individuals grouped in the 0–1 mg class can range several orders of magnitude. Moreover, clear differences in fecundity were not noted among individuals ranging from 1 to 16 mg and thereabove (Figure 1c). Therefore, a more detailed classification of the smaller individuals and a merging of the larger classes might improve the predictive capacity of weight structure. Also, a shorter projection interval could probably increase the predictive capacity, but this would also be true for the age and stage structured model.

The age and the stage structured models provided estimates of λ basically with the same precision; this suggests that stage is a suitable criterion for the definition of population structure. Moreover, the stable environmental conditions maintained throughout the experiment probably increased the predictive capacity of age, relative to stage. Individuals of the same age which are exposed to different environments are likely to be in different physiological conditions. On the contrary, if an individual is classified in a particular stage, it has reached a particular degree of development of the reproductive organs and hence a particular physiological state, regardless of the past and present environmental conditions.

References

Bonomi, G & G. Di Cola, 1980. Population dynamics of *Tubifex tubifex* studied by means of a new model. In Brinkhurst, R. O. & D. G. Cook (eds), Aquatic Oligochaete Biology. Plenum Press, New York: 185–203.

Caswell, H., 1989. Matrix population models: construction, analysis and interpretation. Sinauer Associates, Sunderland: 328 pp.

De Leo, G. A. & M. Gatto, 1995. A size and age-structured model of the European eel (*Anguilla anguilla* L.). Can. J. Fish. aquat. Sci. 52: 1351–1367.

Ebenman, B. & L. Persson (eds), 1988. Size Structured Populations. Springer-Verlag, Berlin: 284 pp.

Efron, B., 1986. The jackknife, the bootstrap and other resampling plans. CBMS-NSF Regional Conference Series in Applied Mathematics, 38. Society for Industrial and Applied Mathematics, Philadelphia: 92 pp.

Hughes, T. P., 1984. Population dynamics based on individual size rather than age: a general model with a reef coral example. Am. Nat. 123: 778–795.

Law, R., 1983. A model for the dynamics of a plant population containing individuals classified by age and size. Ecology 64: 224–230.

Lefkovitch, L. P., 1965. The study of population growth in organisms grouped by stages. Biometrics 21: 1–18.

Leslie, P. H., 1945. On the use of matrices in certain population mathematics. Biometrika 33: 183–212.

Leslie, P. H., 1948. Some further notes on the use of matrices in certain population mathematics. Biometrika 35: 213–245.

Lewis, E. G., 1942. On the generation and growth of a population. Sankhya 6: 93–96.

Pasteris, A., G. Bonomi & C. Bonacina, 1996. Age, stage and size structure as population state variables for *Tubifex tubifex* (Oligochaeta, Tubificidae). Hydrobiologia 334: 125–132.

Usher, M. B., 1972. Development in the Leslie matrix model. In: Jeffers, N. R. (ed.), Mathematical Models in Ecology. Blackwell, Oxford: 29–60.

Hydrobiologia **406:** 191–198, 1999.
B. M. Healy, T. B. Reynoldson & K. A. Coates (eds), Aquatic Oligochaetes.

Assessment of secondary production of Naididae (Oligochaeta): an example from a North German lake

Boris Löhlein
Ecosystem Research Centre, University of Kiel, Schauenburgerstr. 112, D-24118 Kiel, Germany

Key words: Naididae, P/B, secondary production, limitation of growth rates, generation times

Abstract

The annual production of the Naididae *Chaetogaster diastrophus*, *Nais* spp., and *Stylaria lacustris* inhabiting the aufwuchs on reed stems in eutrophic Lake Belau (northern Germany) was estimated. For this purpose, field biomass data taken at biweekly intervals were combined with data on growth potentials of Naididae. From biomass doubling times of laboratory cultures (own and literature) at different temperatures, relationships between temperature and instantaneous growth rate *r* were developed. Based on these functions, the temperature conditions in Lake Belau over 1993 were assessed to allow for the development of 66 generations of *C. diastrophus*, 42 of *Nais* spp. and 30 of *S. lacustris*, respectively. Corresponding annual turnover rates (P/B) are 108 for *C. diastrophus*, 96 for *Nais* spp., and 78 for *S. lacustris*, respectively. However, field data from Lake Belau provide evidence for a food limitation of naidid populations after their abundance peaks. Thus, Naididae can probably not realize their growth potentials over the whole season. Different scenarios of naidid production for various assumed food-limited growth rates are calculated. Under the assumption that no further production occurs once peak densities are reached, annual production on the reed stems in 1993 amounted to 287 μg dry mass cm^{-2} a^{-1}. This value represents only 21% of the potential production of 1342 μg dry mass cm^{-2} a^{-1}. The true production of aufwuchs-associated Naididae in Lake Belau lies between these values.

Introduction

Production of benthic invertebrates is often calculated from annual means of population biomass and literature data on annual turnover rates (P/B). Provided an appropriate taxonomic resolution is achieved (species or genus level), such procedure can be expected to yield reasonable estimations for taxa with life cycles of one year or more. However, estimations become inaccurate when dealing with multivoltine taxa, particularly if they reproduce asynchronously (Benke et al., 1984). Authors studying the production of benthic assemblages frequently use annual P/B-values $\leq$ 10 for oligochaetes (e.g. Lindegaard, 1992; Smock et al., 1992), often referring to Waters (1977) who suggested a P/B of 10 for aquatic Oligochaeta. Usually, no information is provided on the share of Naididae in total oligochaete abundance, biomass and production.

Detailed studies have shown that some Naididae are capable of annual turnover rates (P/B) of 38 (Schönborn, 1985), or even exceeding 200 (Schönborn, 1984) and may play an important role in the invertebrate production of different systems such as polluted rivers (Schönborn, 1987), wetlands (Duffy & LaBar, 1994) and lake littoral aufwuchs (Kairesalo & Koskimies, 1987). Thus, the prevailing impression of low productivity in Naididae may be wrong, resulting from the application of inappropriate methods.

It is well known that losses of small animals during sampling and sample processing may lead to severe underestimation of abundances. For example, Timm (1994) found significantly more individuals of *Uncinais uncinata* when using a 100 μm instead of a 300 μm mesh screen for sieving. Furthermore, coarse sampling intervals, with sampling occasions sometimes being several months apart, may result in the underestimation of annual means of naidid biomass as they are likely to miss the peaks of naidid populations. Production may also be underestimated when methods are used that do not pay attention to the specific reproductive mode of Naididae. Unlike the situation observed in most aquatic insect larvae, no cohorts can be distinguished in Naididae. Instead, non-synchronized asexual reproduction with short generation times dominates in naidids (e.g. Lochhead & Learner, 1983; Schönborn, 1984, 1985), and sexual reproduction does not contribute significantly to population growth (Lochhead & Learner, 1984). Therefore, the instantaneous growth method is the only method appropriate for the assessment of naidid secondary production (Benke, 1993). This method was first used by Johnson & Brinkhurst (1971) to calculate production of Tubificidae and similar approaches were made to assess naidid production (Schönborn, 1984, 1985; Juget et al., 1989).

In this paper, I estimate annual production of Naididae based on growth potentials (instantaneous growth rates) from laboratory experiments, field biomass data collected at high temporal resolution, and ambient water temperature. However, as there was evidence for biotic limitation of naidid production, the method had to be modified.

Materials and methods

Field data on abundance and biomass of aufwuchs-associated Naididae on reed (*Phragmites australis*) stems from January, 1993 to January, 1994 are those previously reported by Löhlein (1996) for the north basin of eutrophic Lake Belau (northern Germany). Four reed stems (= parallel samples) were collected at biweekly intervals and processed in the laboratory (for details see Löhlein, 1996). As oligochaete data were not normally distributed the median value of the four parallels was used instead of the arithmetical mean. At the time of sampling, lake water temperature, conductivity, pH, alkalinity, oxygen content, total phosphorus, NH_4^+- and NO_3^--nitrogen were recorded (data in Löhlein, 1996), and aufwuchs chlorophyll *a* on the reed stems was measured (2 parallels) as described by Müller (1995). Both naidid and chlorophyll data were recorded for every 10 cm section of the approximately 70 cm long submerged portions of the reed stems. For each sampling situation, vertical distribution along the reed stems (water surface to sediment) of naidid densities and chlorophyll *a* concentrations were tested for correlations by means of Spearman rank-correlation (Sachs, 1992).

Laboratory cultures of *Nais* spp. and *Stylaria lacustris* were started using animals collected in Lake Belau during their respective periods of high population growth rates in spring and early summer, 1994. Individuals of *Nais* used in cultures were not determined to species level, but *Nais barbata*, *N. pseudobtusa*, *N. variabilis* were identified in subsamples of the 1993 field samples (Löhlein, 1996). Animals were cultivated in small groups (*Nais*: 3, *Stylaria*: 2 individuals) in polystyrene petri dishes (diameter = 90 mm) containing aufwuchs and lake water. Experiments were conducted in climatic chambers at temperatures of 5, 10, 15, and 20 °C and a 12:12 h photoperiod (12–20 parallels at each temperature). The animals were counted under a stereo microscope every 5 days. After counting they were fed with aufwuchs from the lake (animals sorted out) and water losses were restored. Abundance data were converted to biomass using individual biomass values given by Löhlein (1996).

Growth of a naidid population at a given temperature can be described by the exponential growth formula $N_{(t)} = N_0\ e^{rt}$, with N = number of individuals, r = specific (instantaneous) growth rate [days^{-1}], and t = time [days]. From this, r can be expressed as $r = (\ln N_{(t)} - \ln N_0)\ t^{-1}$. For the population doubling time (t_D), $N_{(t)}$ equals 2 N_0. Thus, doubling time is $t_D = \ln 2\ r^{-1}$ [days] (Odum, 1971; Lochhead & Learner, 1983; Smith, 1986). In the same way, population biomass doubling times were calculated from field and laboratory data. As I found no seasonal shifts in individual biomass (Löhlein, 1996), it was assumed that during asexual reproduction the mean individual biomass in a population does not change. This results in population and biomass growth rates being equal. Following Juget et al. (1989), the doubling time is defined as the generation time of the respective taxon.

In Naididae, the instantaneous growth rate r rises with temperature, following a relationship of the type $r = a\ e^{b\ x}$ [d^{-1}], with x = temperature [°C] (Juget et al., 1989). Such functions were calculated for all taxa studied using specific (instantaneous) growth rates at different temperatures. The established functions were then used to assess population growth potentials from the temperature development recorded in Lake Belau over 1993 (range: 0.4–18.4 °C). Production was calculated using a modified instantaneous growth method. The growth formula of each taxon was applied to each 2–wk time interval between sampling occasions using the respective means of field biomass and water temperature of this period. For periods when growth appeared to be limited by factors other than temperature, these calculations were also carried out for growth rates being lower (10–50% of those under temperature limitation).

Table 1. Doubling times t_D [days] of Naididae (genera *Chaetogaster*, *Nais* and *Stylaria*) under laboratory conditions

Species	Temperature [°C]										Remarks	Source
	4	5	8	10	12	15	19	20	21	25		
Chaetogaster diastrophus	8.0			4.5				4.8			food high	Schönborn, 1984
	15.0							10.0			food low	Schönborn, 1984
C. langi								1.7				Taylor, 1980
Nais barbata	15.2			8.4				3.1				Schönborn, 1985
N. elinguis	13.6			10.1				2.5				Schönborn, 1985
			9.1		16.2			6.3				Lochhead & Learner, 1983
									2.4			Ratsak et al., 1993
N. pseudobtusa								4.2				Mc Elhone, 1978
N. variabilis							6.7				field data	Juget, 1976
			15.8		7.0–9.8			3.1–4.9				Lochhead & Learner, 1983
Nais spp.						14.7		8.6		4.8	*in situ* enclosures	Juget et al., 1989 [(a)]
		15.5		13.1		7.1		5.7				this study
Stylaria lacustris							3.6					Streit, 1978 [(b)]
								4.2			*in situ* enclosures	Finogenova, 1984 [(c)]
		died		16.6		11.7		8.3				this study

[a]Calculated from formula given in the paper.
[b]Calculated carbon doubling times.
[c]Time between two divisions of chains.

Results and discussion

Growth potentials of Naididae

In my experiments 'population' growth rates of both *Nais* and *Stylaria* cultures rose with temperature. Corresponding population biomass doubling times ranged from no asexual reproduction of *S. lacustris* at 5 °C to doubling times of 5.7 days for *Nais* spp. at 20 °C (Table 1). Increasing growth rates with rising temperatures are generally observed in invertebrates and have been reported for *Chaetogaster diastrophus* (Schönborn, 1984), for different species of *Nais* (Lochhead & Learner, 1983; Schönborn, 1985; Juget et al., 1989), and for *Pristina aequiseta* (Lochhead & Learner, 1983). The data obtained from animals collected in Lake Belau lie within the range observed by other authors (Table 1), although growth of *S. lacustris* at high temperatures was somewhat slower than in comparable studies (Streit, 1978; Finogenova, 1984). The apparent differences between studies may either be an effect of the different methods used or reflect adaptations of the populations studied. Despite these variations, the values underline the potentially high instantaneous growth rates and productivity of many Naididae. Population doubling times down to 5.1 days in *Nais* spp. (at 15 °C), and 3.2 in *C. diastrophus* found in small *in situ* enclosures by Schönborn (1984, 1985) prove that growth rates of field populations may even exceed those recorded under laboratory conditions.

Since the reported doubling times differ considerably (Table 1), the values used to establish equations describing instantaneous growth rates as a function of water temperature (Table 2) were chosen to be well in the range observed in different studies. Most of these doubling times represent a faster growth than observed in my own experiments, but as animals in the latter started to die off after 3 – 4 weeks (especially at higher temperatures) culture conditions were obviously not optimal.

Limitations of population development

Given their growth potential at high temperatures (Table 1), the dominating aufwuchs-associated naidids *C. diastrophus*, *Nais* spp. and *S. lacustris* could be expected to persist at high densities over the whole summer. In Lake Belau, however, summer densities of aufwuchs-associated Naididae were not linked to temperature (Figure 1a & b). After the spring maximum, abundances of *C. diastrophus* and *Nais* decreased dramatically although the high percentage of individuals undergoing paratomy proved that asexual reproduction had not ceased. Since populations did not recover for the rest of the year (total naidid abundance < 0.5 individuals cm^{-2}), the period of population breakdowns is of special interest.

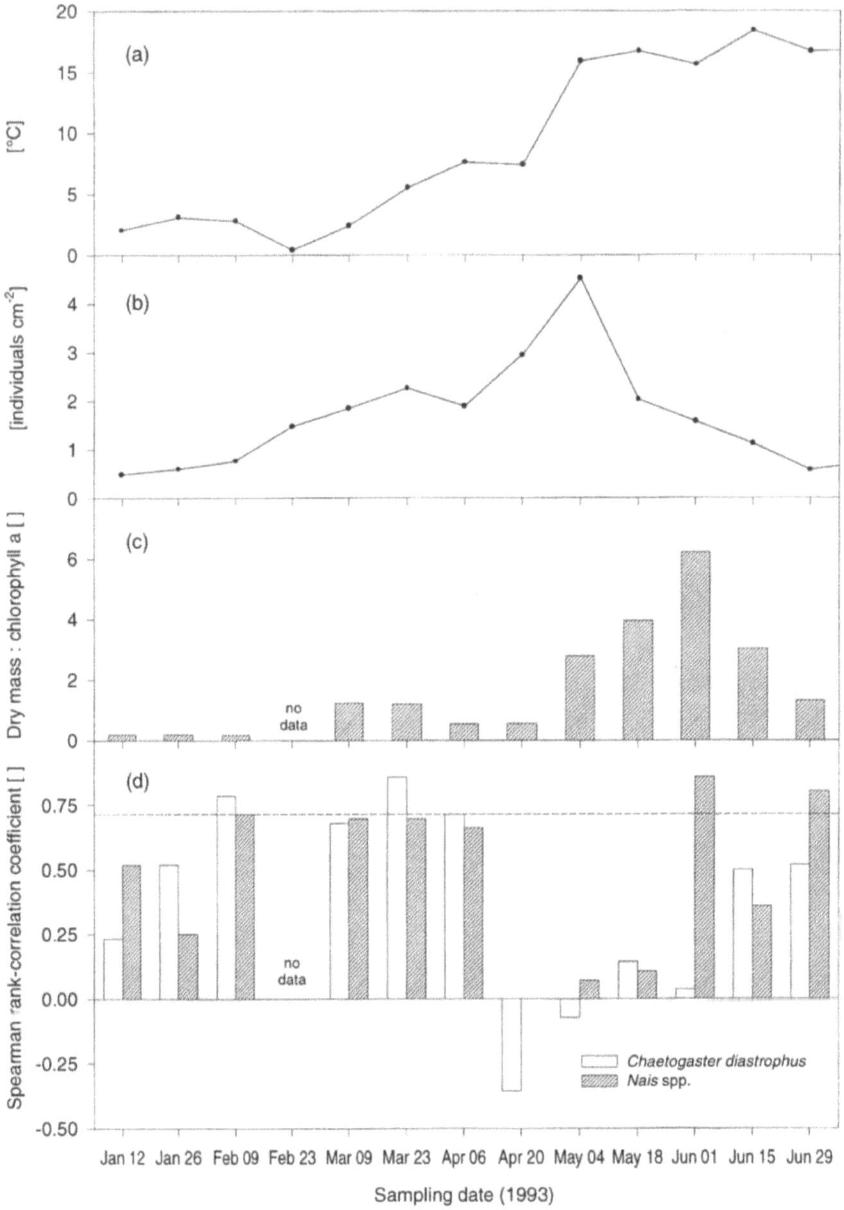

Figure 1. Seasonal development (January – June, 1993) of (a) water temperature of Lake Belau, (b) density of aufwuchs-associated Nadidae, (c) Naididae: chlorophyll *a* biomass ratio, and (d) correlation between vertical distributions of chlorophyll *a* and the abundances of *Chaetogaster diastrophus* and *Nais* spp. on the aufwuchs-carrying reed stems. Broken line: $p = 0.05$-significance level of Spearman rank-correlation coefficient.

During the early phase of the spring bloom (February – early April), Spearman rank-correlation coefficients between the vertical distribution of *C. diastrophus* and *Nais* spp. on reed stems and that of aufwuchs chlorophyll *a* were sometimes significant ($p < 0.05$), and often close to this significance level (Figure 1d). In the sampling interval from March 23 to April 6, with water temperatures around 6–7 °C, field populations had net biomass doubling times of 6.8 days for *C. diastrophus* and 12.8 for *Nais*, respectively. These values match the doubling times found in culture experiments (Table 1), indicating that at this time temperature was the factor limiting population growth in the field. From April 6 on, however, correlation coefficients were low (sometimes negative), but never significant, for three sampling occasions (= 6 weeks) in a row. During this period, populations of *C. diastrophus* and *Nais* reached their maxima (Löhlein, 1996). From mid-

Table 2. Calculation of the growth potential of the dominant aufwuchs-associated naidid taxa of Lake Belau from biomass doubling times. t_D = population doubling time [days], r = corresponding exponential growth rate (instantaneous growth rate)[days^{-1}], and $r_{(temp)}$ = resulting relationships between r and temperature [°C]

Species	Parameter	Temperature [°C] 5	10	15	20	r(temp)
Chaetogaster diastrophus	t_D	8	5	4.5	3.5	
	r	0.0866	0.1386	0.1540	0.1980	$r = 0.0725\ e^{0.0517x}$
Nais spp.	t_D	15	10	6.5	3.5	
	r	0.0462	0.0693	0.1066	0.1980	$r = 0.0273\ e^{0.0959x}$
Stylaria lacustris	t_D		15	9	4	
	r		0.0462	0.0770	0.1733	$r = 0.0117\ e^{0.1322x}$

May on, however, populations of *C. diastrophus* and *Nais* did not flourish as to be expected from lake temperature. Hence, their development must have been dominated by factors other than temperature.

Naididae appear to tolerate a wide range of most abiotic factors (e.g. Little, 1984) and physical and chemical parameters recorded in Lake Belau did not reach critical values in early summer, 1993. From April to July, lake water was well oxygenated (8.3–12.3 mg l^{-1}), poor in phosphorous and nitrogen, had a conductivity of 307–393 μS cm^{-1} and a pH of 7.7–8.4. These values are unlikely to have caused the observed population declines. Furthermore, sexually mature individuals indicating worsening of the abiotic environment (Loden, 1981) were not found before October. Hence, it can be concluded that a change took place from temperature to biotic limitation of the development of *C. diastrophus* and *Nais* populations. Biotic factors limiting naidid populations may be predation, stress induced by high population densities (McElhone, 1980), or the amount of food available (e.g. Little, 1984; Schönborn, 1987). According to Juget et al. (1989), growth of Naididae is normally limited by either temperature or food. In laboratory experiments, both food abundance (Schönborn, 1984) and quality (Irmler, 1989) have been found to influence growth rates.

In Lake Belau, the seasonal development of aufwuchs algae is characterized by a biomass maximum in spring, followed by a minimum in summer (Müller, 1994). Diatoms dominate the algal assemblage (biovolume) for most of the year, but from late April to June chlorophytes predominate (Müller, 1994). Thus, the presented results may be interpreted as a shift from temperature to food limitation. In early spring, high production of aufwuchs algae (Müller, 1995), especially diatoms, sustains growing densities of grazers. *C. diastrophus* and *Nais* spp. seek places with high algal biomass, but are unable to control the latter. Hence, vertical distributions of these naidids correlate with the distribution of algal biomass. Subsequently, as grazers reach peak densities, grazing activity outweighs primary production of aufwuchs algae, a situation as described by Hann (1991). As a result, correlations are low, or even negative. At the same time, diatoms, the preferred food of many aufwuchs grazers, are replaced by chlorophytes and the naidid: chlorophyll *a* biomass ratio (Figure 1c) indicates increasing grazing pressure on aufwuchs algae. These observations suggest that population declines were caused by a shortage of high quality food caused by overgrazing. Consequently, growth of Naididae must have been food-limited. The low abundances of *C. diastrophus* and *Nais* spp. over the rest of the year indicate a persisting high grazing pressure on the reduced algal biomass.

Stylaria densities on the reed stems were too low to allow for similar tests. However, it was apparent that the *S. lacustris* population did not decline until the end of July (Löhlein, 1996). This different development may have resulted form the ability of *S. lacustris* to exploit chlorophytes as a food source (Streit, 1978).

Assessment of naidid production in Lake Belau

A major problem associated with the calculation of production for non-cohort populations is the assessment of *in situ* growth rates (Benke, 1993). This is particularly true if growth rates are subject to shifts caused by changes in limiting factors. Due to a lack of *in situ* growth rates under food limitation, calculations were done using three different scenarios: to assess the production during phases of food limitation, food-

Table 3. Calculated number of generations per year, turnover rates and annual production (per cm^2 reed stem surface) of aufwuchs-associated Naididae assuming different ratios between temperature-limited and food-limited growth rates. For the period of food limitation three different scenarios are given assuming growth rates to be 50, 20 and 10% of those under temperature limitation

Parameter	Taxon	potential	Assumed food-limited growth rate relative to temperature-limited			before max.
			50%	20%	10%	
Number of generations [# a^{-1}]						
	Chaetogaster diastrophus	65.7	38.8	22.6	17.6	10.5
	Nais spp.	42.4	23.8	12.6	8.9	4.8
	Stylaria lacustris	29.5	19.9	14.1	12.2	10.5
Turnover (P/B) [a^{-1}]						
	Chaetogaster diastrophus	107.7	50.5	35.6	32.3	
	Nais spp.	96.3	30.0	10.6	6.0	
	Stylaria lacustris	77.8	40.8	28.5	25.5	
Production [μg dry mass cm^{-2} a^{-1}]						
	Chaetogaster diastrophus	130.2	61.1	43.1	39.1	35.8
	Nais spp.	416.8	130.1	45.8	26.0	12.0
	Stylaria lacustris	795.1	416.7	291.4	260.4	239.4
	Sum of production	1,342.1	608.4	380.4	325.5	287.2
	Proportion of potential [%]	100	45.3	28.3	24.3	21.4

limited growth rates were assumed to be (i) 50%, (ii) 20%, and (iii) 10% those under temperature limitation (Table 3). Based on the presented data, the period of temperature limitation for *C. diastrophus* and *Nais* was presumed to last until correlations with algal biomass shift. For *S. lacustris*, temperature limitation was assumed to last until two weeks prior to the population maximum similar to the situation observed in the other taxa. For all calculations, populations were treated as limited by temperature in winter and spring (from January), and as food-limited for the rest of the year (starting April 7 for *C.diastrophus* and *Nais* spp., and June 29 in the case of *S. lacustris*, respectively).

Based on the established growth functions, temperature conditions in Lake Belau over 1993 would have allowed for the development of 66 generations of *C. diastrophus*, 42 of *Nais* spp., and 30 of *S. lacustris*, respectively (Table 3). With the population biomass development recorded in the field, corresponding annual turnover rates (P/B) are 108 for *C. diastrophus*, 96 for *Nais* spp., and 78 of *S. lacustris*, respectively. If growth was assumed not to slow down under food limitation, naidid production could have amounted to as much as 1.3mg dry mass cm^{-2} a^{-1}. However, the above suggested food limitation must be expected to lower naidid production by reducing growth rates. The results of the different scenarios calculated offer a wide spectrum of values for naidid production parameters (Table 3) with turnover rates from 6.0 (*Nais*, food-limited growth rate = 10%) to 108 (*C. diastrophus*, temperature-limited all year). For lower food-limited growth rates, the parameters production, number of generations per year, and P/B-ratio decreased most remarkably in *Nais*. This was due to the high *Nais* abundance at the beginning of the phase of assumed food limitation. In contrast, corresponding values for *S. lacustris* were higher as the phase of temperature limitation was presumed to last much longer. Being the largest species, *S. lacustris* had the biggest share in annual production of Naididae (59–83%), although *C. diastrophus* and *Nais* spp. dominated total naidid abundance (Löhlein, 1996). In any case, true total naidid production values were between the 1342 μg dry mass cm^{-2} a^{-1} found for temperature limitation and the 287 μg cm^{-2} a^{-1} calculated to be produced before populations reached their peaks.

Given the reed density in the littoral of Lake Belau (60 stems m^{-2}) and the mean stem surface (approx. 120 cm^2), and assuming food-limited growth rates to be 10–50% those under temperature limitation, total annual production of the dominating taxa of aufwuchs-associated Naididae at the study site was 2.3–4.4 g dry mass m^{-2} yr^{-1}. For 46% carbon in naidid dry mass (Löhlein, unpubl.), these values correspond to a production of 1.1–2.0 g C m^{-2} yr^{-1}, which equals 2.6–4.8% of the mean annual primary production of aufwuchs algae on the reed stems (42 g C m^{-2} yr^{-1}) reported by Gessner et al. (1996).

General discussion

The rough estimations of naidid turnover rates in Lake Belau agree well with those observed by Schönborn (1985) who applied *in situ* growth rates to field biomass data and calculated the annual P/B to be 209 in *C. diastrophus*, and 38 in *Nais*, respectively. It is theoretically possible that, instead of assumed lower growth rates, summer production of aufwuchs-associated Naididae in Lake Belau was high with heavy grazing reducing abundances. As a result, high productivity would not necessarily be reflected in net biomass development. However, results from an enclosure experiment carried out in 1994 indicate that predation on aufwuchs oligochaetes is low in Lake Belau (Löhlein, unpubl.).

Both high temporal resolution in data collection and a determination to species level are necessary to properly calculate naidid production. For example, production of *Nais* spp. was probably underestimated in the present study. In Lake Belau the genus *Nais* is represented by several species (in aufwuchs: *N. barbata*, *N. pseudobtusa*, *N. variabilis*) and there is evidence for a seasonal succession within these species. If production is calculated at genus level, such a situation with overlapping rising and falling abundances of different species within the genus must lead to an underestimation of real production.

Other methods have been used to assess the production of Naididae by means of asexual reproduction. In particular, the 'stolonization rate' (Juget et al., 1989), which is the percentage of individuals in a population undergoing paratomy, has repeatedly been used to characterize the reproductive activity of naidid populations (McElhone, 1978, Schönborn, 1985; Juget et al., 1989). Comparing different methods for production assessment in *Nais*, Juget et al. (1989) found no significant differences between production calculated using instantaneous growth rate and instantaneous birth rate derived from stolonization rates of the population studied.

The high potential growth rates as well as the fact that Naididae develop extreme population densities in waters affected by organic pollution (Learner et al., 1978) are typical of opportunists. Furthermore, many Naididae show opportunistic feeding behavior (Juget & Lafont, 1994). At least the *Nais* species and *Stylaria lacustris* are able to utilize such different food sources as detritus-associated bacteria (Harper et al., 1981; Lochhead & Learner, 1983; Finogenova, 1984) and different algae (McElhone, 1979, 1980; Bowker et al., 1983). However, differences in the ability to exploit different food sources seem to play an important role for the seasonal succession of coexisting naidid species.

In systems with a predictable succession of different situations over the year, like most temperate eutrophic lakes, the temperature-limited production potential of Naididae is probably not realized. On the other hand, systems with a high frequency of fluctuations (e.g. nutrient pulses) like polluted streams may offer opportunists like naidids several occasions a year to develop population peaks resulting in high annual turnover rates as found by Schönborn (1984, 1985). Therefore, the data presented here are not necessarily applicable to other systems. Furthermore, naidids inhabiting sediments might be less productive than the aufwuchs assemblage studied, depending on food availability.

Conclusion

Assessment of Naidid production remains difficult. To date it is impossible to recommend a formula or P/B-value to be used for the estimation of naidid production. Nevertheless, it should be stressed that the P/B-values currently used to calculate naidid production are probably too low, leading to an underestimation of naidid production and their role in the respective system. For a better understanding of naidid production biology, we need more knowledge on growth rates at different food levels.

Acknowledgements

I thank Ute Müller, Hamburg, for providing me with biomass data of aufwuchs algae. This work was supported by grant no. 0339077H 'Ecosystem Research in the Bornhöved Lakes Region' from the 'Bundesminister für Bildung, Wissenschaft, Forschung und Technologie (BMBF)' of the FRG. My participation at the '7th International Symposium on Aquatic Oligochaetes' was sponsored by the 'Stifterverband für die Deutsche Wissenschaft'. Sven Petersen, Rainer Poepperl and two anonymous reviewers commented on the manuscript.

References

Benke, A. C., T. C. Van Arsdall, D. M. Gillespie & F. K. Parrish, 1984. Invertebrate productivity in a subtropical blackwater river: the importance of habitat and life history. Ecol. Monogr. 54: 25–63.

Benke, A. C., 1993. Concepts and patterns of invertebrate production in running waters. Verh. int. Ver. Limnol. 25: 15–38.

Bowker, D. W., M. T. Wareham & M. A. Learner, 1983. The selection and ingestion of epilithic algae by *Nais elinguis* (Oligochaeta: Naididae). Hydrobiologia 98: 171–178.

Duffy, W. G. & D. J. LaBar, 1994. Aquatic invertebrate production in southeastern USA wetlands during winter and spring. Wetlands 14: 88–97.

Finogenova, N. P. 1984. Growth of *Stylaria lacustris* (L.) (Oligochaeta, Naididae). Hydrobiologia 115: 105–107.

Gessner, M. O., B. Schieferstein, U. Müller, S. Barkmann & U. A. Lenfers, 1996. A partial budget of primary organic carbon flows in the littoral zone of a hardwater lake. Aquat. Bot. 55: 96–105.

Hann, B. J., 1991. Invertebrate grazer-periphyton interactions in a eutrophic marsh pond. Freshwat. Biol. 26: 87–96.

Harper, R. M., J. C. Fry & M. A. Learner, 1981. A bacteriological investigation to elucidate the feeding biology of *Nais variabilis* (Oligochaeta: Naididae). Freshwat. Biol. 11: 227–236.

Irmler, U., 1989. Population-ecologay and migration of *Dero multibranchiata* Stieren, 1892 (Naididae, Oligochaeta) in the Central Amazon inundation forest. Amazoniana 11: 31–52.

Juget, J., 1976. Les oligochètes du zoopériphyton des herbiers à *Trapa natans*: écologie et production. Ann. Hydrobiol. 7: 43–45.

Juget, J. & M. Lafont, 1994. Theoretical habitat templets, species traits and species richness: aquatic oligochaetes in the Upper Rhône River and its floodplain. Freshwat. Biol. 31: 327–340.

Juget, J., V. Goubier & D. Barthélémy, 1989. Intrinsic and extrinsic variables controlling the productivity of asexual populations of *Nais* spp. (Naididae, Oligochaeta). Hydrobiologia 180: 177–184.

Johnson, M. G. & R. O. Brinkhurst, 1971. Production of benthic macroinvertebrates of Bay of Quinte and Lake Ontario. J. Fish. Res. Bd. Can. 28: 1699–1714.

Kairesalo, T. & I. Koskimies, 1987. Grazing of oligochaetes and snails on epiphytes. Freshwat. Biol. 17: 317–324.

Learner, M. A., G. Lochhead & B. D. Hughes, 1978. A review of the biology of British Naididae (Oligochaeta) with emphysis on the lotic environment. Frehwat. Biol. 8: 357–375.

Lindegaard, C., 1992. The role of zoobenthos in energy flow in deep, oligotrophic Lake Thingvallavatn, Iceland. Hydrobiologia 243/244: 185–195.

Little, C., 1984. Ecophysiology of *Nais elinguis* (Oligochaeta) in a brackish-water lagoon. Estuar. coast. Shelf Sci. 18: 231–244.

Lochhead, G. & M. A. Learner, 1983. The effect of temperature on asexual population growth of three species of Naididae (Oligochaeta). Hydrobiologia 98: 107–112.

Lochhead, G. & M. A. Learner, 1984. The cocoon and hatchling of *Nais variabilis* (Naididae: Oligochaeta). Freshwat. Biol. 14. 189–193.

Löhlein, B., 1996. Seasonal dynamics of aufwuchs Naididae (Oligochaeta) on *Phragmites australis* in a eutrophic lake. Hydrobiologia 334: 115–123.

Loden, M. S., 1981. Reproductive ecology in Naididae (Oligochaeta). Hydrobiologia 83: 115–123.

McElhone, M. J., 1978. A population study of littoral dwelling Naididae (Oligochaeta) in a shallow mesotrophic lake in North Wales. J. Anim. Ecol. 47: 615–626.

McElhone, M. J., 1979. A comparison of the gut contents of two co-existing lake-dwelling Naididae (Oligochaeta), *Nais pseudobtusa* and *Chaetogaster diastrophus*. Freshwat. Biol. 9: 199–204.

McElhone, M. J., 1980. Some factors influencing diet of coexisting, benthic, algal grazing Naididae (Oligochaeta). Can. J. Zool. 58: 481–487.

Müller, U., 1994. Seasonal development of epiphytic algae on *Phragmites australis* (Cav.) Trin. ex. Sten. in a eutrophic lake. Arch. Hydrobiol. 129: 273–292.

Müller, U., 1995. Vertical zonation and production rate of epiphytic algae on *Phragmites australis* (Cav.) Trin. ex Steud. Freshwat. Biol. 34: 69–80.

Odum, E. P., 1971. Fundamentals of ecology, 3rd edit. W. B. Saunders Co., Philadelphia, 574pp.

Ratsak, C. H., S. A. L. M. Kooijman & B. W. Kooi, 1993. Modelling the growth of an oligochaete on activated sludge. Wat. Res. 27: 739–747.

Sachs, L., 1992. Angewandte Statistik, 7th edn. Springer, Berlin: 846pp.

Schönborn, W., 1984. The annual energy transfer from the communities of Ciliata to the population of *Chaetogaster diastrophus* (Gruithuisen) in the River Saale. Limnologica 16: 15–23.

Schönborn, W., 1985. Die ökologische Rolle der Gattung *Nais* (Oligochaeta) in der Saale. Zool. Anz. 215: 311–328.

Schönborn, W., 1987. Secondary production and energy transfer in the polluted River Saale (Thuringia, Southern GDR). Int. Revue ges. Hydrobiol. 72: 539–557.

Smith, M. E., 1986. The ecology of Naididae (Oligochaeta) from an alkaline bog stream: life history patterns and community structure. Hydrobiologia 133: 79–90.

Smock, L. A., J. E. Gladden, J. L. Riekenberg, L. C. Smith & C. R. Black, 1992. Lotic macroinvertebrate production in three dimensions: channel surface, hyporheic, and floodplain environments. Ecology 73: 876–886.

Streit, B., 1978. A note on the nutrition of *Stylaria lacustris* (Oligochaeta: Naididae). Hydrobiologia 61: 273–276.

Taylor, W. D., 1980. Observations on the feeding and growth of the predacious oligochaete *Chaetogaster langi* on ciliated protozoa. Trans. am. micros. Soc. 99: 360–367.

Timm, H., 1994. Production of some oligochaete species in large Estonian lakes. Hydrobiologia 278: 309–313.

Waters, T. F., 1977. Secondary production in inland waters. Adv. ecol. Res. 10: 91–164.

Hydrobiologia **406:** 199–211, 1999.
B. M. Healy, T. B. Reynoldson & K. A. Coates (eds), Aquatic Oligochaetes.

Ultrastructure and functional significance of intestinojunctional spermathecae in enchytraeids (Oligochaeta, Annelida)

Wilfried Westheide
Spezielle Zoologie, Fachbereich Biologie/Chemie, University of Osnabrück, D-49069 Osnabrück, Germany

Key words: Achaeta, Enchytraeus, Fridericia, Marionina, seminal receptacles, sperm phagocytosis, evolution of hermaphroditism

Abstract

The ultrastructure of paired seminal receptacles (spermathecae) communicating directly with the intestinal tract of three enchytraeid species (*Enchytraeus crypticus, Fridericia montafonensis* and *Marionina preclitellochaeta*) is described and compared with blind receptacles of an *Achaeta* species. Although the same in general configuration, comprising 1. an ectal part consisting of a single-layered epithelium with an inner lining of cuticle, 2. gland cells (rosette glands) around the outer orifice, 3. an external muscular coat, and 4. an ental part that may form a greatly expanded ampulla, the spermathecae of these three species exhibit certain marked differences in the individual parts. In *F. montafonensis*, the two ampullae are subdivided to form diverticula, where intact sperm are stored, and a common chamber that contains more or less degenerate sperm and opens into the oesophagus dorsally. In *M. preclitellochaeta*, sections revealed phagocytosis of sperm in the wall cells of the ampullae, as well as direct transfer of sperm bundles into the intestinal tract. There is evidence that the ampullae of this species are intestinal pockets and not part of the epidermal invaginations. The spermathecae of *E. crypticus* do not expand to form ampullae. Function of the spermathecae is discussed, as is the trophic significance of the allosperm, particularly in relation to the hermaphroditic organization and small body size of these animals.

Introduction

One of the basic characteristics of the Clitellata is a form of simultaneous hermaphroditism in which, although the sperm cells are transferred directly into the reproductive partner, fertilization of the female gametes occurs not while the latter are still within the body but after they have emerged into the cocoon produced by the clitellum. The partner's sperm must be stored temporarily in seminal receptacles. These are invaginations of epidermal, i.e. of ectodermal, origin (e.g. Gustavsson & Erséus, 1997), generally closed at their inner ends. They are arranged as one or more segmental pairs and are called spermathecae. Only the most highly evolved clitellates, the Hirudinea, transfer their sperm into the interior of the partner's body, by means of spermatophores or a penis, and hence, have lost the spermathecae. In these, in fact, the cocoon receives fertilized eggs.

Enchytraeids ordinarily have one pair of spermathecae opening in the intersegmental furrow between segments IV and V (3rd and the 4th chaetigerous segments) and extending into V or further backwards (Figure 1). The organs are visible by light microscopy, even in living animals or whole mounts, and because their shape varies in a genus- and species-specific manner, they play a fundamental role in enchytraeid taxonomy. Nielsen & Christensen (1959) consider them to be "perhaps the most important single criterion for the identification of Enchytraeidae". The forms of spermathecae have therefore been documented extensively and in some taxa very skilfully (e.g. Erséus, 1976, 1977; Coates & Erséus, 1985). They appear mostly saclike or bottle-shaped. Usually there is a discernible subdivision into an outer (=ectal) duct and an inner (=ental) ampulla, which may become apparent only in later stages of their morphogenetic development (Dózsa-Farkas, 1974). At the opening to the exterior, more or less large gland cells are often situated. In contrast to almost all other oligochaete taxa, the two ampullae in most enchytraeid species are connected with the

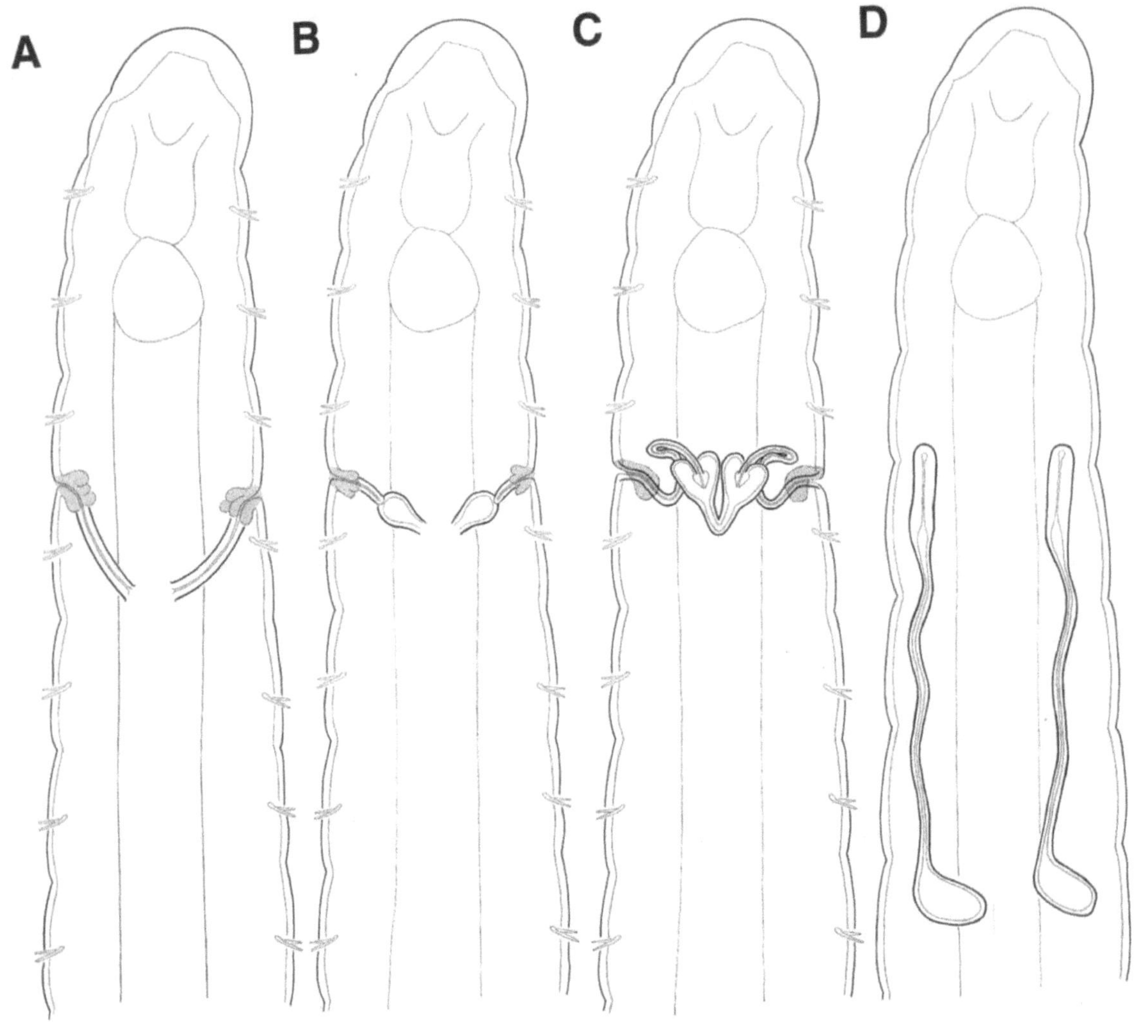

Figure 1. Overall shape and position of spermathecae in *Enchytraeus crypticus* (A) *Marionina preclitellochaeta*, (B), *Fridericia montafonensis* (C) and *Achaeta* sp. (D). Schematic presentation.

oesophagus. Spermathecae of this kind, here called 'intestinojunctional', thus have two openings: an outer opening, through which the partner's sperm cells pass to enter the receptacle, later to emerge into the cocoon fluid as the cocoon is being shed, and an inner opening, by way of which the ampullar lumen communicates with the intestinal system. Relatively few publications have given detailed histological descriptions of the spermathecae in the Enchytraeidae, and most of these are fairly old (e.g. Ratzel, 1869; Pierantoni, 1903; Eisen, 1910). So far there have been no ultrastructural studies (see Jamieson, 1981, 1988). However, ultrastructural and histochemical information is available on tubificid spermathecae (Fleming, 1981). Although Cernosvitov (1930) thoroughly documented and discussed the function of these intestinojunctional spermathecae, his work has become largely unknown. In the context of the evolution of simultaneous hermaphrodites in general, however, and specifically regarding the origin of simultaneous hermaphroditism in the Clitellata, the structure and function of these spermathecae are of great interest. Their ultrastructure, therefore, has been examined in three species possessing three considerably different types of fusion with the intestinal tract and has been compared with the 'blind' spermathecae of a fourth enchytraeid species. Here we describe the specific structures of these organs and attempt to discern the anatomical features common to them in general. Special attention is paid to the position of the sperm in the spermathecae, insofar as that can be determined, and to the alterations the sperm may undergo there. Taken together, these findings serve on the one hand as a morphological basis for a brief interpretation of these organs' function; on the other hand, the new information leads us to phylogenetic speculations about the body dimensions of the stem species of the Clitellata.

Materials and methods

1. *Enchytraeus crypticus* Westheide & Graefe, 1992: The animals (length ca. 7 mm) were taken from cultures that have been maintained in the author's laboratory for years (Westheide & Graefe, 1992). Two series of ultra-thin cross sections and one series of longitudinal sections of an anterior body fragment were examined.
2. *Fridericia montafonensis* Schmelz, 1998. The animals (10–14 mm) were collected in the Montafon, Austria by Rüdiger Schmelz, Osnabrück. Two series of ultra-thin cross sections were examined.
3. *Marionina preclitellochaeta* Nielsen & Christensen, 1963. The animals (2–3 mm) were collected from a tidal sand beach of the North Sea island of Sylt and fixed by Dr Günter Purschke, Osnabrück. One series of ultra-thin cross sections was examined.
4. *Achaeta* sp. (described as *A. dzwilloi* by Graefe (1973) in his Diploma thesis, University of Hamburg). The animals were collected by Rüdiger Schmelz in the surroundings of Osnabrück. Two incomplete series of ultra-thin cross sections were examined.

Prior to fixation, specimens were anaesthetized in sparkling (CO_2) mineral water. An osmium ferricyanide fixation modified from McDonald (1984) was used. Whole animals were fixed in a solution of 2.5% glutaraldehyde buffered with 0.05 M sodium cacodylate with 0.005 M calcium chloride for 2 h at 21 °C. After rinsing in buffer (7 changes, 1.5 h), specimens were incubated in 0.8% potassium ferricyanide in buffer for 0.5 h at room temperature, postfixed in 0.5% OsO_4 plus 0.8% K_3 [$Fe(CN)_6$] in buffer at 4 °C, rinsed once in buffer, dehydrated in a graded ethanol series and embedded in an Epon-Araldite mixture. Series of ultrathin sections (Reichert Ultracut E microtome, diamond knife) were taken from the region containing the spermathecae. Ribbons of sections were picked up on pioloform-coated single slot grids, stained in an LKB Ultrostainer with uranyl acetate and lead citrate, and examined in ZEISS EM 109 and 902A transmission electron microscopes.

Results

Enchytraeus crypticus

The spermathecae of *E. crypticus* extend diagonally backwards from their outer orifice and lead directly to the oesophagus (Figure 1A). They are elongate canals (length ca. 100–130 μm) of almost identical diameter (ca. 14–20 μm) throughout; that is, an ampullar region set off from the ectal duct does not exist, so that no border between ectal and ental parts of the organ can be distinguished by light microscopy (Westheide & Graefe, 1992). TEM investigations reveal that the ectal part, identified as the region lined with cuticle (Figure 2C), comprises only about 1/5 of the overall length of the organ. The orifices (diameter ca. 5 μm) are displaced inward by about 3 μm, so that a small but distinct vestibule is formed (Figure 2A & B). Like the whole external cuticle of the body surface, the cuticular lining of the lumen in the ectal part is penetrated by many epidermal microvilli and is covered by a dense layer of regularly arranged epicuticular projections. The orifice is surrounded by the processes of relatively small drop-shaped gland cells (rosette glands) (Figure 2A–C), which probably produce only one type of secretion: ovoid electron-dense granules of moderate size (diameter 0.15–0.32 μm).

At the vestibule outside the orifice, the basal matrix of the flat outer epidermis bends inward at nearly a right angle and merges with the tubular basal matrix of the ectal part, which continues without interruption as the basal matrix of the ental part (Figure 2D–G).

The anatomy of the ental part is uniform and relatively simple. Only the width of the lumen varies markedly, as can be seen even in the light microscope (Westheide & Graefe, 1992, Figure 2). Distended regions with a diameter of at most 9 μm alternate with almost closed sections, constricted to less than 1 μm (Figure 2D, E). The wall of the ental part is a single-layered epithelium comprising cells that in cross section taper distinctly towards the lumen wherever the latter is very narrow (Figure 2D). The most conspicuous structures in these wall cells are small mitochondria, putative glycogen granules, and spherical to ovoid nuclei, in some cases highly lobate at the periphery, with relatively large amounts of peripheral heterochromatin. In some cross sections, the nuclei are at more or less the same level in the various layers of cells, which are arranged in a regular manner, one directly above another. Although in a given cross section up to 11 outlines of nuclei can be present, the

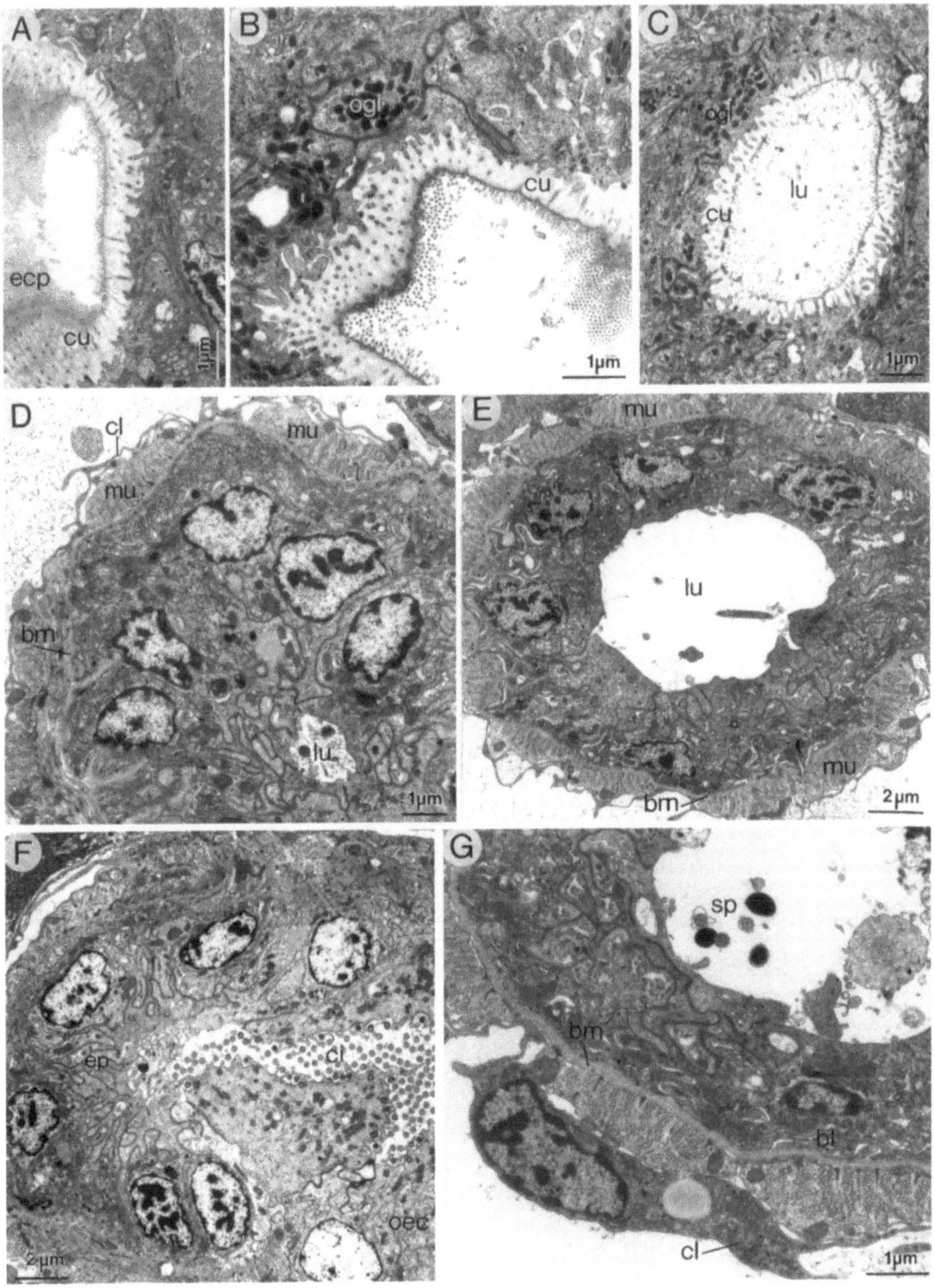

Figure 2. Enchytraeus crypticus. Spermathecae. (A), (B) Vestibule of outer orifice. (C) Cross section of ectal part. (D), (E) Cross sections of ental part with lumen of various width. (F) Communication of duct lumen with alimentary canal (oesophagus) on the right of the micrograph. (G) Transitional region of ental part with few allosperm. bl – basal labyrinth; bm – basal matrix; ci – cilia of oesophagus; cl – coelomic lining; cu – cuticle; ecp – epicuticular projections; ep – epithelial cells of ental duct; lu – lumen of spermatheca; mu – muscle cell; oec – oesophagus cell; ogl – orifice gland cell; sp – allosperm.

number of apical cell regions bounding the lumen is usually lower.

Laterally, the wall cells intermesh by way of in- and evaginations (Figure 2D–E); zonulae adhaerentes, together with septate desmosomes, are ubiquitous. The apex of the wall cell bears no cilia, but some cells exhibit a few irregularly shaped processes similar to microvilli. The base of the cell in some cases has the form of a basal labyrinth (Figure 2G). Adjacent to this labyrinth is the massive (thickness 0.13–0.25 μm) basal matrix, which in certain places sends out folds that extend between the basal regions of the wall cells (Figure 2D) forming rib-like processes. No general structural differences between the epithelium of the ectal and ental parts could be detected except for the presence of the cuticula in the ectal part.

Apposed to the outer surface of the basal matrix are muscle cells (about 12–16), oval to spindle-shaped in cross section. They are oriented more or less parallel to the long axis of the lumen; that is, they are longitudinal or spiral muscles. Most are distinctly separated from one another. External to these are the coelothelial cells, mostly very flat cells that cover the whole organ and in places push between the muscle cells to contact the basal matrix (Figure 2D–G).

The duct lumen communicates directly with that of the alimentary canal (oesophagus), through a relatively broad opening (Figure 2F); it is only here that fingerlike microvilli are relatively numerous, and a few cilia of the gut cells extend into this transitional region. Whether muscles in the transition area between spermathecae and oesophagus may function as a sphincter cannot be decided. In the specimens studied allosperm were found only in small numbers in the spermathecae, and only in the transitional region between spermathecae and oesophagus (Figure 2G).

Fridericia montafonensis

The two spermathecae are relatively large, prominent organs with a clear boundary between ectal ducts and ampullae (Figure 1B). The latter are connected to one another in the middle of the body, dorsal to the oesophagus (Schmelz, 1998). The outer orifices appear in cross sections to be lying in a skin fold, probably the intersegmental groove, containing the openings of gland cells, which together form a relatively small complex anterior to the orifice. They are relatively small, drop-shaped cells with a maximal diameter of ca. 3 μm (Figure 3A). There is probably only one type of secretory granule, irregularly shaped and moderately electron-dense, with a diameter of about 0.5 μm (Figure 3B). The relatively thick cuticle on the body surface (Figure 3A), with the pattern of crossed collagen fibrils typical of annelids and the typical oligochaete covering of rodlike, here very small, epicuticular projections, merges directly with the cuticle lining of the ectal duct. Within the duct the collagen fibrils are no longer clearly visible; the ordinary epicuticular projections also rapidly become less numerous and then vanish altogether. They are replaced by irregular cuboid projections with thick heads (height up to ca. 1 μm, diameter ca. 0.3 μm) (Figures 3A & 4). The heads are seated on thinner necks, microvilli that pass through the cuticle (Figure 4B, E). Packed close together side by side, the projections form a thick layer that can fill the duct lumen so as to leave a gap of less than 1 μm (Figure 4B). The ectal duct has a fairly uniform diameter of ca. 9 μm. Its outer surface is covered by a continuous muscular coat about 1.5 μm thick, made of longitudinal muscle cells that possibly spiral around the duct (Figure 4A).

In a given cross section, about 6–7 wall cells appear of uniform size and in a relatively regular arrangement. Characteristic of these are the deep, lateral, lobe-like projections that interlock the cells in their apical halves, i.e. the region towards the lumen (Figure 4A). In cross sections through the duct, the projections form meandering patterns. Electron-lucent spaces that open into the lumen, some containing thin processes, are presumably secretory granules, the contents of which were dissolved away during fixation (Figure 4A). The wall cell epithelium is enclosed in a cylindrical, thick (ca. 0.4 μm) basal matrix. The ectal duct (length ca. 200 μm, but see different measurements from living specimens in Schmelz, 1998) projects deeply into the ampulla, forming there an 'ectal bulb' (Figure 4C). Just outside the opening of the ectal bulb, its cuticle-lined epithelium folds outward into the lumen of the ampulla. This folding brings the angular, thick-headed epicuticular projections onto the ampullar surface of the ectal bulb (Figure 4C). In cross sections, this structure looks like a cobbled street (ca. 14 projections of this kind in 1 μm^2) (Figure 4D, E). In all the projections, filaments (actin?) are arranged in parallel, perpendicular to the duct.

Each of the two ampullae has a heart-shaped appearance, due to the presence of the ectal bulbs. The ampullae are clearly subdivided into two parts: 1. an anterior part adjacent to the ectal duct, which is separated into two diverticula (diameter ca. 10 μm) by the ectal bulb (Figures 3C & 5C), and 2. a contiguous

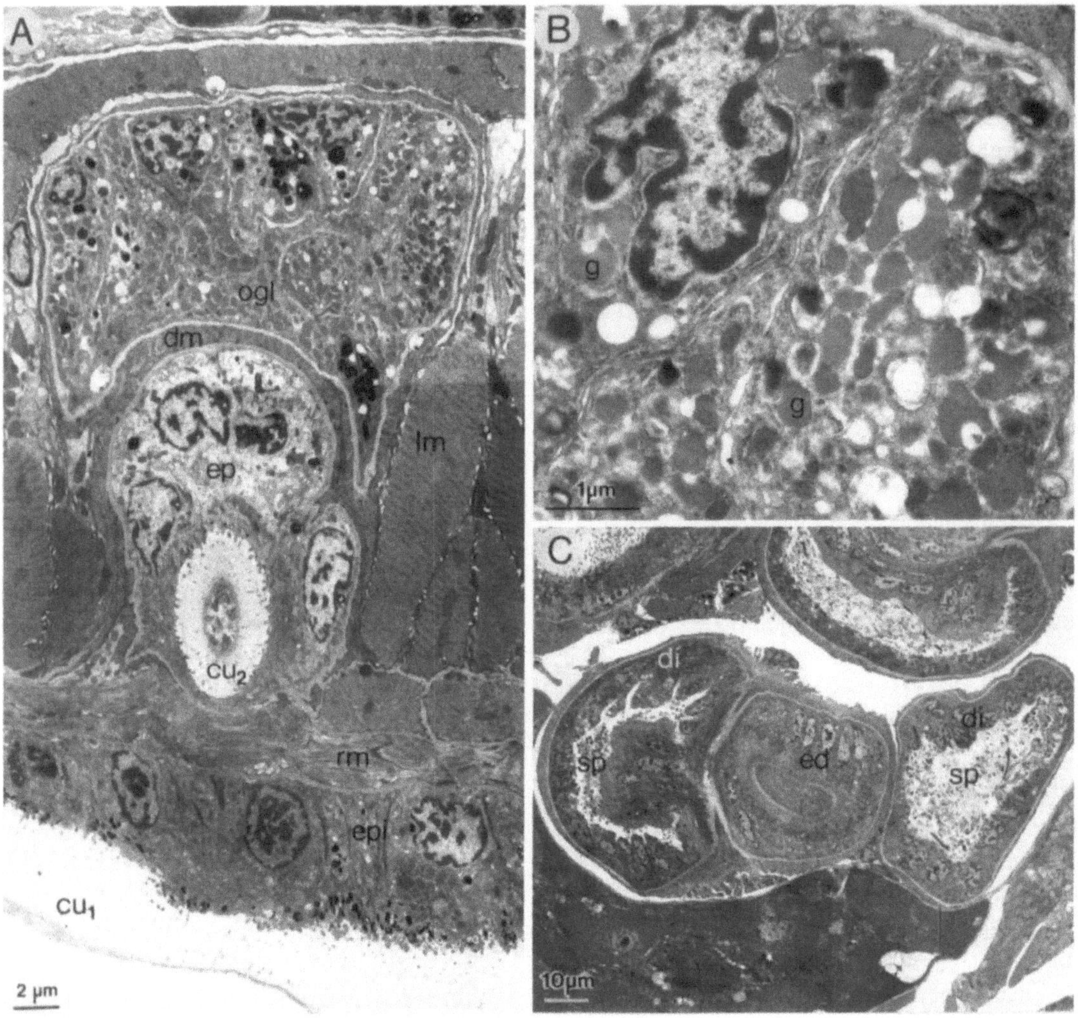

Figure 3. Fridericia montafonensis. Spermathecae. (A) Cross section of animal with ectal part of spermatheca near outer orifice; duct obliquely sectioned. (B) Gland cells near outer orifice. White spots probably extracted secretory material. (C) Cross section of ectal duct (ed) situated in between the two diverticula (di) of one of the ampullae. cu_1 – cuticle of body surface; cu_2 – cuticle lining lumen of ectal duct; dm – muscular coat of ectal duct; ep – epithelial cells of ectal duct; epi – epidermis; g – secretory granulum; lm – body longitudinal musculature; ogl – orifice gland cell; rm – body ring musculature; sp – allosperm.

posterior region with a wide lumen that communicates with that of the ampulla on the other side of the body to form one common V- or U-shaped chamber. The two parts of each ampulla are separated by a constriction, which narrows the lumen to ca. 2.5 μm. The constriction may be due to a sphincter that belongs to the relatively thick muscle coat of the entire ampullar part. Sperm are usually situated in the anterior part, forming a compact strand that stretches from one diverticulum to the other around the ectal bulb (Figure 4C) (see also Schmelz, 1998). These sperm appear to be ultrastructurally intact. The outer wall of the diverticula (as well as the wall of the other parts of the ampulla) is a single-layered epithelium about several μm thick. In the apical regions of the diverticula their luminar surface is partially very uneven. In places there are many irregularly arranged microvilli. In the epithelium, there are cells with rounded, moderately electron dense secretory granules (diameter 0.4 μm) near the apex (Figure 5D). Some of the cells are apically drop-shaped, project into the lumen, and have apically located nuclei; apical exocytotic structures can be discerned. Their basal parts reveal an extensive basal labyrinth with numerous mitochondria (Figure 5D). A number of sperm have advanced between the epithelial cells almost as far as the basal matrix. They are completely enclosed in several very thin cellular processes which are arranged in layers, that is, the sperm are not

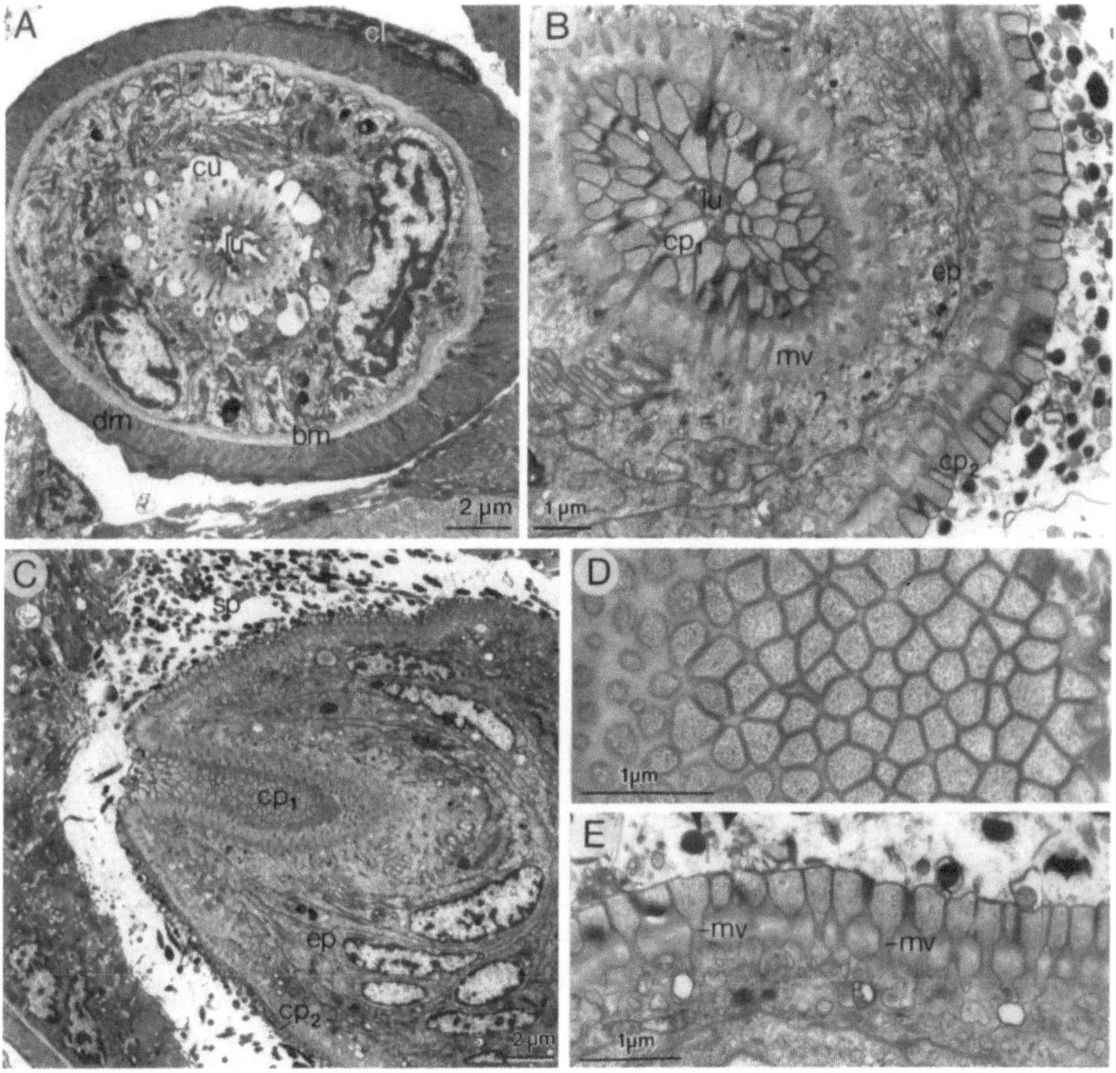

Figure 4. Fridericia montafonensis. Spermathecae. (A) Cross section of ectal duct. (B) Cross section of ectal bulb. (C) Almost longitudinally sectioned ectal bulb projecting into ampulla. (D) Cross sections of cuboid projections on the outer (ampullar) surface of the ectal bulb. (E) Longitudinal sections of microvilli with cuboid projections. bm – basal matrix; cl – coelomic lining; cp_1 – cuboid projections surrounding lumen (lu) of ectal duct; cp_2 – cuboid projections on the ampullar side of ectal bulb; cu – cuticle; dm – muscular coat of ectal duct; ep – epithelial cells of ectal duct; lu – lumen of ectal duct; mv – microvillus; sp – allosperm.

within the cells and they appear to be intact. The common U-shaped chamber (diameter ca. 50 μm) formed by the joined ental portions of the two ampullae communicates with the oesophagus through a single, broad opening (ca. 10 μm) in the dorsal oesophageal wall (Figure 5B). Its epithelium becomes remarkably thin towards this opening; no sphincter muscles could be detected here. The chamber is completely filled with sperm, all of which are more or less in degeneration (Figure 5A, B), that is, they lack at least the plasma membrane, and in many of them the pattern of microtubules in the axonema has become disordered. Many unidentifiable membranous structures are present.

Marionina preclitellochaeta

The spermathecae (Figure 6) of this relatively small marine species are about 50–70 μm long; of this length, about 35 μm is occupied by the ectal duct and the remainder by the ca. 15- to 20-μm-wide ampulla (Figure 1B). The outer orifice is funnel-shaped with a maximal lumen diameter of about 4.3 μm. The cuticle of the body surface bears a dense covering of epicuticular projections which extends uninterrupted into the lumen of the spermathecal duct, though the length of the projections decreases here by about half, to 0.1 μm. Between the epicuticular projections are

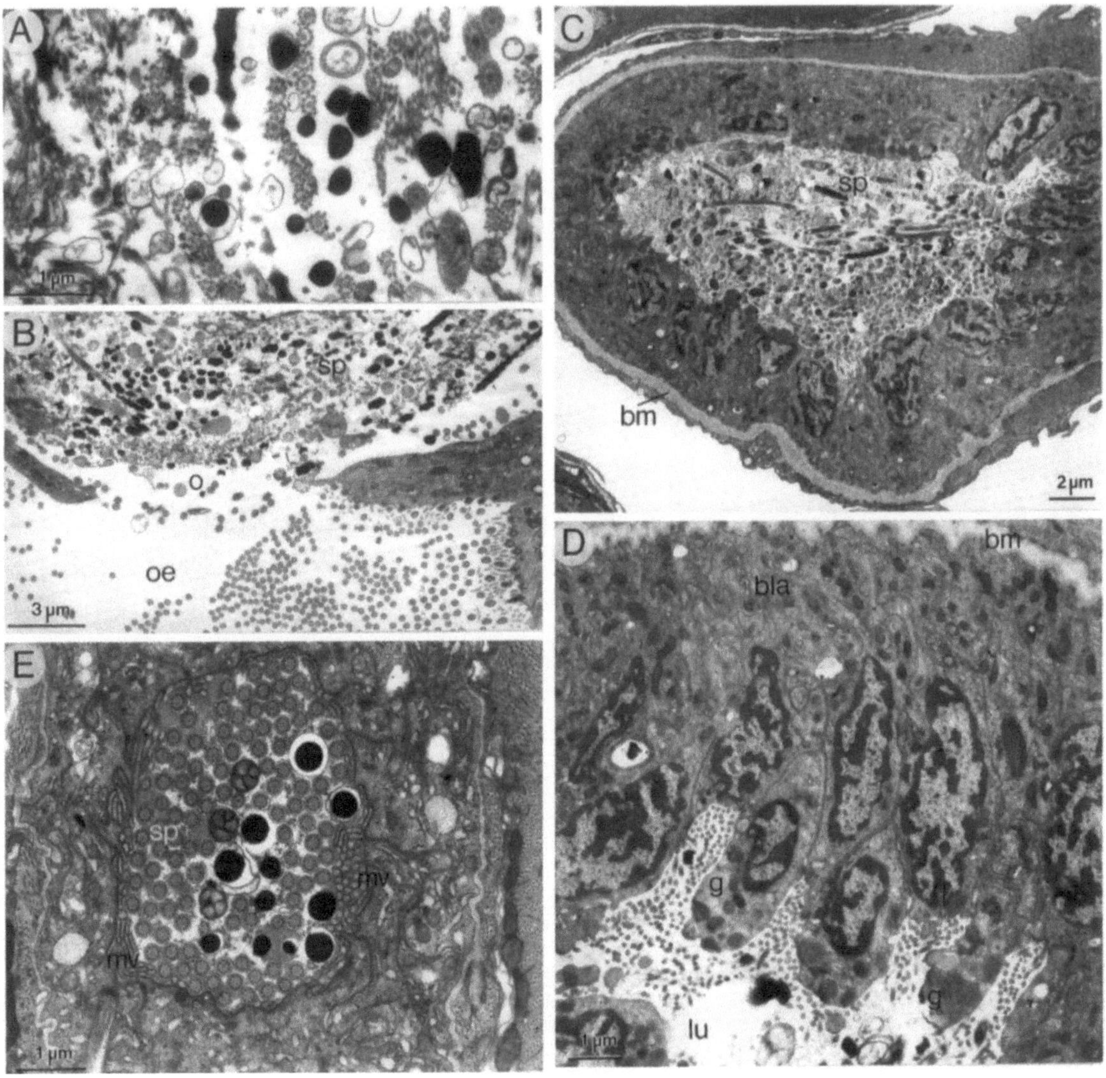

Figure 5. Fridericia montafonensis. (A) Sperm in the process of degeneration in the spermathecal chamber. (B) Single opening (o) between common chamber of the spermathecae filled with degenerate sperm (sp) (above) and ciliated oesophagus (oe) (below). (C) Diverticulum of ampullar part of spermatheca (cross section) filled with sperm. (D) Epithelium of diverticulum with drop-shaped cells with apically positioned secretory granules (g). (E) *Achaeta* sp. Cross section of the middle part of spermatheca filled with sperm. bla – basal labyrinth; bm – basal matrix surrounding diverticula; lu – lumen of diverticulum; mv – microvilli.

thicker microvilli, connected directly to the cells of the duct wall (Figure 7A). The thickness of the cuticular layer itself also decreases by half, to about 0.3 μm. The diameter of the duct and the nature of its lining are almost the same over its entire length; the duct diameter, including the peripheral basal matrix, is 2.7–3.2 μm, of which the lumen accounts for about 0.6 μm. The outer orifice is surrounded relatively uniformly by numerous drop-shaped gland cells (largest diameter over 4 μm); their spherical secretory granules (diameter 0.6–0.9 μm) are moderately electron-dense with a central electron-lucent core (Figure 7B). The outer surface of the ectal duct is covered by longitudinal muscle cells (Figure 6), either oriented parallel to the duct or probably spiralling slightly around it.

At its inner end the duct is closed and abuts against the ampulla (Figure 6), which has a very flat wall in this region; here the duct wall cells are separated by relatively large intercellular cavities (Figure 7D, E) and contact the epithelial cells of the ampullar wall directly, with no intervening basal matrix.

The ampulla consists of 1. an inflated part with large lumen and 2. a broad connecting part (ental duct) that joins the oesophagus (Figures 6 & 8). The former

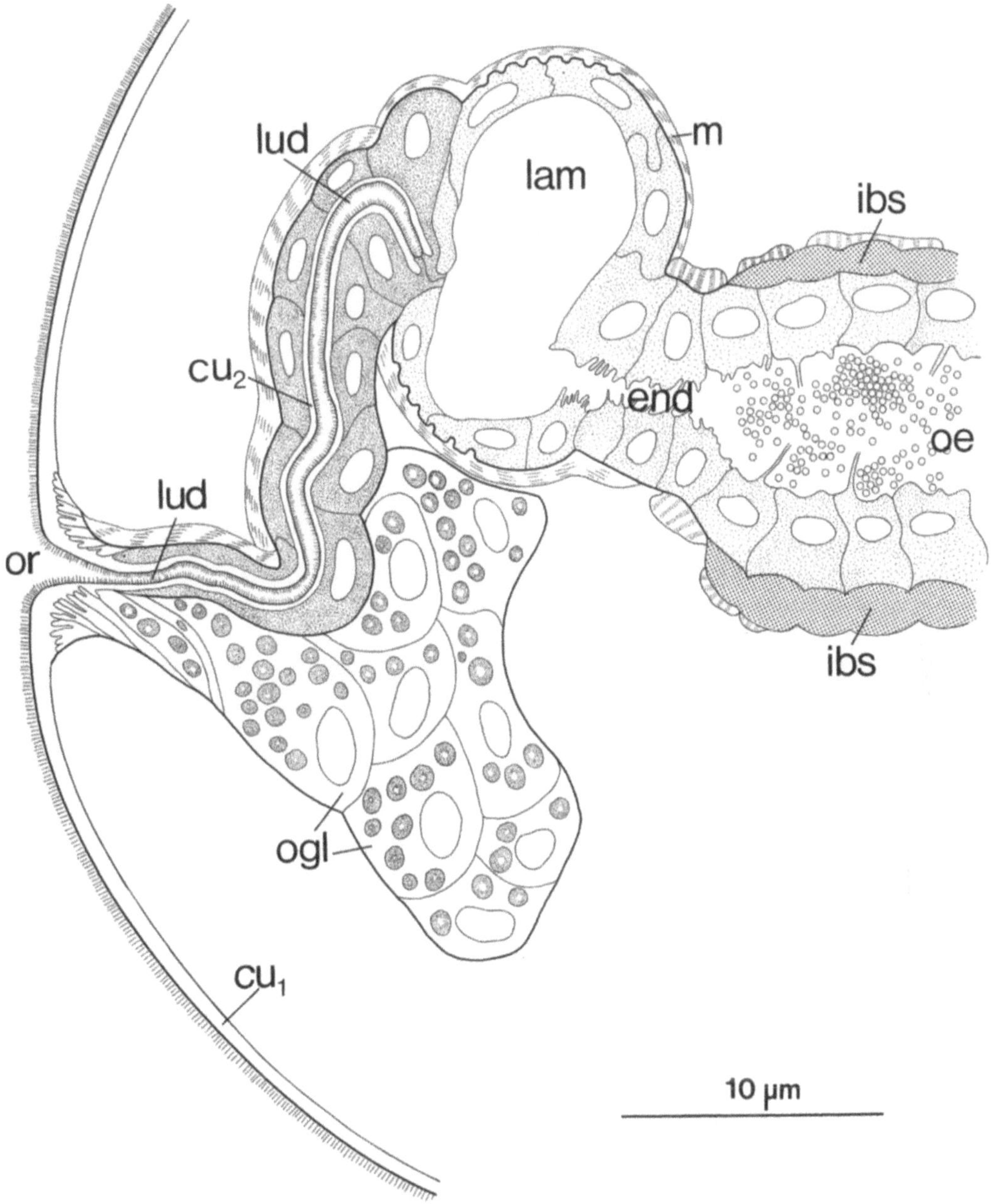

Figure 6. Marionina preclitellochaeta. Spermatheca. Reconstruction from TEM-micrographs; slightly schematic. Coelomic lining omitted. am – ampulla; cu_1 – cuticle of outer body surface; cu_2 – cuticle lining lumen of ectal duct; end – ental duct connecting ampulla and oesophagus; ibs – intestinal blood sinus; lam – lumen of ampulla; lud – lumen of ectal duct; m – muscle cells; oe – oesophagus; ogl – orifice gland cell; or – orifice.

is shaped like a laterally compressed sphere, so that its lumen has a dorsoventral diameter of about 18 μm and a lateral extent of no more than 5 μm. The flat epithelial cells, with elongate nuclei, form a wall between 1.0 and 1.7 μm thick. Nearly all these cells contain sperm probably taken up by phagocytosis (Figure 8B, C). The connecting part (ental duct) leading to the oesophagus (length ca. 3.5 μm) (Figure 8A) is constructed of considerably higher (3.3–4.3 μm), prismatic cells with nuclei oriented in the basal-apical direction. To the basal matrix of the ampulla is apposed a thin coat of muscle (Figure 8B, C). Longitudinally directed muscle cells occur near the oesophagus and may serve as a sphincter. The TEM serial sections, however, reveal the opening being widely open and bundles of sperm passing into the intestinal tract.

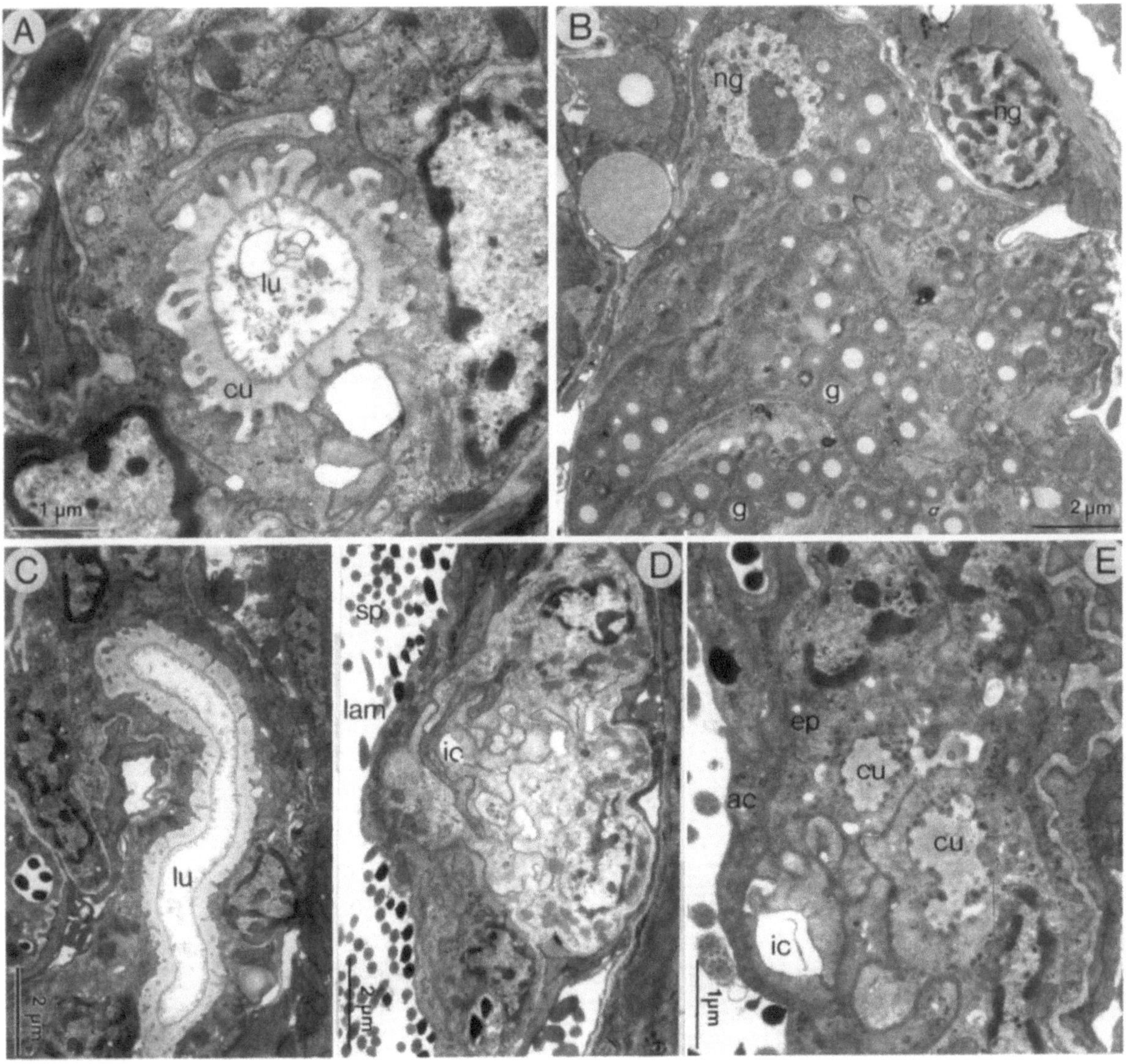

Figure 7. Marionina preclitellochaeta. Spermathecae. (A) Cross section of ectal duct. (B) Orifice gland cells (rosette glands). (C) Longitudinal section of ectal duct. (D), (E) Area where ectal duct abuts against ampulla. ac – ampulla wall cell; cu – cuticle lining lumen of ectal duct; ep – epithelial cells of ectal duct; g – secretory granulum; ic – intercellular cavities in ectal duct wall; lam – lumen of ampulla; lu – lumen of ectal duct; ng – nucleus of gland cell; sp – allosperm.

Discussion

The spermathecae of the three species studied here have much the same general structure but very different arrangements and different substructures. The points of basic structural consistency are as follows. The outer orifice of the organ leads into an epithelial, cuticle-lined passage, the 'ectal duct', which is partially or completely enclosed in longitudinally or spirally arranged musculature. Some of the epithelial cells are glandular; these are more or less greatly enlarged and surround the orifice partially or completely. They are usually called ectal glands but this term is also used for gland cells along the ectal duct. Here they are called rosette glands, although the cells are often clustered to form an asymmetrical glandular mass. Internally adjacent to the ectal duct is a section lined with simple epithelium, lacking a cuticle, which is called 'ental' and communicates directly with the oesophagus; when this ental region is distinctly enlarged, it is called the 'ampulla'.

In *E. crypticus*, no such ampulla is present; the ectal part merges directly with an ental part of identical diameter. At some places in this region, the lumen may be slightly wider. However, large numbers of sperm are evidently not stored here.

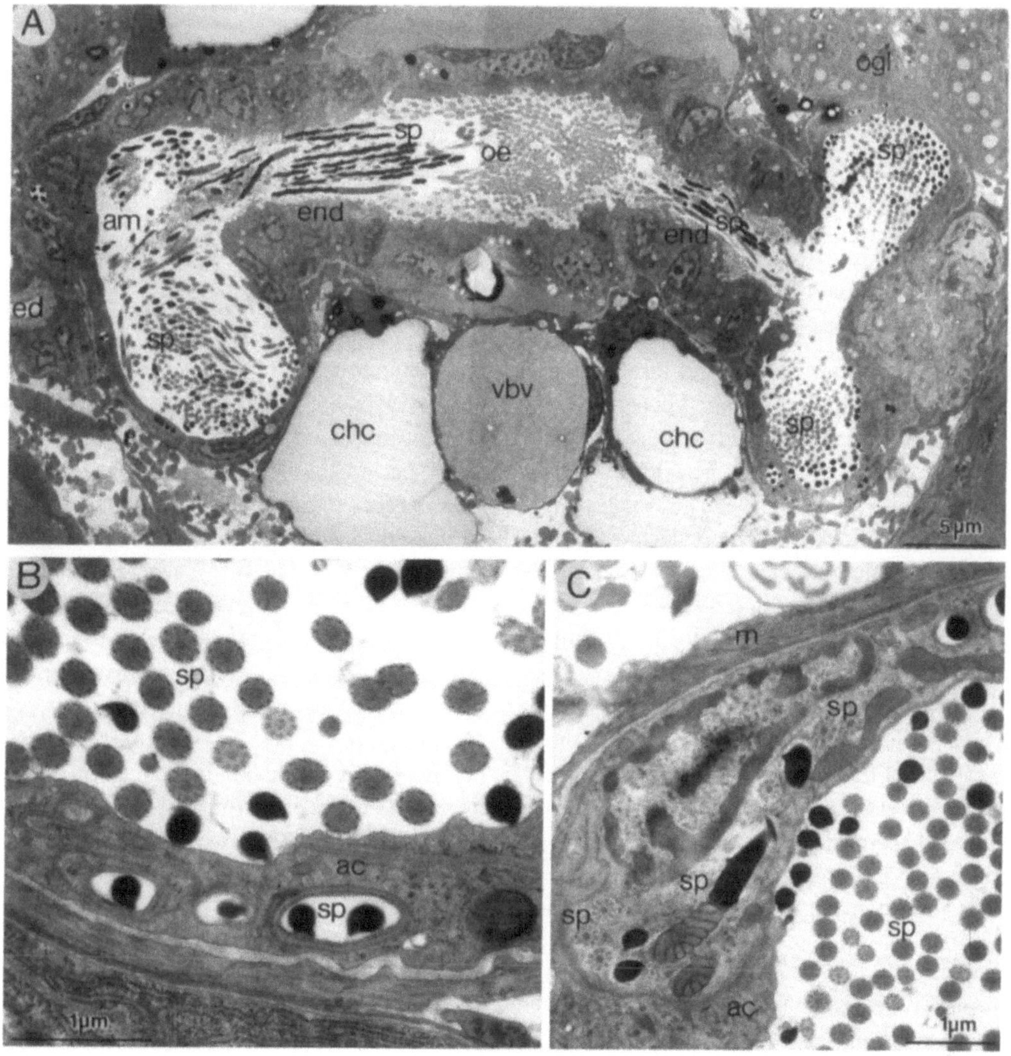

Figure 8. Marionina preclitellochaeta. Spermathecae. (A) Cross section through oesophagus (oe) with open connection to the ampullae (am) on both sides. Allosperm (sp) pass from ampullae into oesophagus. (B), (C) Peripheral areas of inflated part of ampullae. Wall cells have taken up sperm probably for digestion. ac – ampulla wall cell; chc – lipid vacuole of chloragog cell; ed – ectal duct; end – ental duct connecting ampulla and oesophagus; m – muscle coat of ampulla; ogl – orifice gland cells; vbv – ventral blood vessel.

M. preclitellochaeta differs in that a relatively long ectal duct is followed by an ampulla with distinctly expanded lumen, which communicates with the gut through a broad opening. Furthermore, there is no open connection between ectal duct and ental ampulla (Figure 6).

In *F. montafonensis*, too, the ectal ducts are distinctly separated from the ampullae. However, the latter are more complex, being subdivided into diverticula and a common chamber formed by the organs on either side, which communicates with the gut through a broad, unpaired opening (Schmelz, 1998).

It might be thought that only the part of the duct lined with cuticle, i.e. the ectal part, is produced by invagination of the epidermis and is thus of ectodermal origin, whereas the ental, ampullar part represents an evagination of the foregut. This idea can easily be refuted by a comparison with spermathecae that do not communicate with the gut: the extremely long spermathecae of *Achaeta* sp. (Figure 1D), also are largely without a cuticular lining (Figure 5E). However, there is considerable evidence that the ampullae of *M. preclitellochaeta* are intestinal pockets. The structure of the cells in the ampullae is entirely consistent with that of the gut cells (see also Pierantoni (1903) for *Grania macrochaeta*) but differs from the structure of the distinctly lighter cells in the ectal duct. Moreover, the lumen of the ectal duct is not continuous with that of the ampulla (Figures 6 & 7D, E); the epithelia of the two sections are superimposed here, so that the sperm must push between the cells in order to pass from the ectal into the ampullar part. The fact that in this species the wall cells of the ampullae are capable of absorbing the sperm (see below), a phenomenon that according to Cernosvitov (1930) is found only in species with spermathecae that communicate with the gut, is further evidence that these ampullae originate from the intestinal epithelium.

The basic function of the spermathecae must be interpreted as follows. First sperm pass through the ectal ducts into the ampullae, being pressed in by the penes of the copulatory partner and/or moving independently. Later, in order for the allosperm to emerge from the organ when the cocoon is being shed (Westheide & Müller, 1996), the muscular coat presumably contracts. The ampullae are thus primarily organs in which the sperm are stored between copulation and deposition of the cocoon. Whether, and in what manner, this occurs in *E. crypticus* is at present unclear. In *F. montafonensis* the valve-like constriction evidently produces a subdivision such that the sperm still capable of fertilization remain in the region of the diverticula, whereas in the region of the common chamber the sperm degenerate, to be eventually transferred into the gut; sperm in the diverticula are intact, those in the chamber are clearly in the process of decomposition.

In *M. preclitellochaeta*, the ampulla has no such functional compartmentation. Here, however, the wall cells take up sperm (Figure 8B, C), perhaps when they have become too old, and phagocytose them. In addition, whole bundles of sperm pass out of the ampulla and into the gut (Figure 8A). These ultrastructural investigations thus confirm older light microscopical and histological studies in which the degeneration and absorption of sperm was claimed. The first to describe absorption of sperm in spermathecae was Vejdovský (1909), and the most thorough work on the subject was done by Cernosvitov (1930). Among the Oligochaeta, he distinguished the absorption of sperm by lymphocytes (= coelomocytes) (this may not, however, concern the allosperm), in epithelial cells of the spermathecae, and in the intestinal canal. Communication between spermathecae and intestinal canal, which is also present in a few Tubificidae, Lumbriculidae and Eudrilidae, is found in most species of the Enchytraeidae and was first described in this taxon by Michaelsen (1886). Michaelsen, and later Mrázek (1900), interpreted this arrangement as providing a route for the elimination of surplus sperm. Beddard (1892) considered it possible that "spermatozoa may be conveyed a short distance by the alimentary tract and then make their way out to fertilize the ova". Cernosvitov (1930), however, was fully aware of the trophic significance of the transfer of allosperm into the gut, an idea which is adopted here without reservation.

It is assumed here that superfluous allosperm that are not used for fertilization in these oligochaetes may not only be of some nutritional value for them but inevitably contribute to the energy budget for the development of their female reproductive cells. The nutritional contribution is particularly valuable in these species, inasmuch as the energy costs of female gametes are a more critical element in the reproductive success of a species, the smaller the body dimensions of its individuals. A general crucial problem faced by all small annelid species is that the size of their gametes is not correlated with the general body size: their egg size is very similar to that of the gametes of related macrofaunal species, rarely smaller and often even larger. An inevitable result is that the number of eggs a small individual can produce is only a small fraction of that produced by a large animal. One strategy with which to confront this problem consists in ensuring the fertilization and development of all these few oocytes, as nearly as possible. It is to achieve this that the great variety of mechanisms for direct sperm transfer and brood protection has arisen, together with the equally great variety of associated genital structures in small annelids (Westheide, 1984).

The second strategy is optimally to reduce the waste of reproductive energy within a species, in order to invest energy maximally in female gametes. In principle, all sperm are wasted that do not fertilize an egg, and in general these are the great majority. In a hermaphrodite species with direct transfer of sperm, there is no loss of male gametes. In this case, each individual is capable of reproducing and all the sperm that are not transferred and remain in the body can be resorbed – that is, recycled – and contribute to the energy budget for female reproduction just like all allosperm that are transferred but fail to fertilize eggs. Special structures, such as the genitointestinal connections described in the enchytraeids presented here, may serve to introduce allosperm into the recycling process in a particularly simple and rapid manner. Finally, this idea leads to the assumption that the evolution of hermaphroditism should be favoured in small species (Clark, 1978). A conclusive phylogenetic inference that could be drawn from this reasoning for a taxon composed exclusively of hermaphrodites, as the Clitellata are, is that its stem species was small-bodied (about several millimeters) and reproduced by direct sperm transfer.

Acknowledgements

My sincerest thanks are due to Werner Mangerich for skilfully carrying out preparation and documentation of the entire electron microscopical processing. Valuable discussions with Rüdiger Schmelz and Dr Günter Purschke, who also provided part of the material, are gratefully acknowledged. I am grateful to Ulfert Graefe (Hamburg), who kindly sent me a sketch of the *Achaeta* species. Mrs Anna Stein and Mrs Andrea Noël kindly assisted in preparing the manuscript.

References

Beddard, F. E., 1892. Contribution to the anatomy of *Sutroa*. Trans. r. Soc. Edinburgh 37: 195–202.

Cernosvitov, L., 1930. Studien über die Spermaresorption. I. Teil. Die Samenresorption bei den Oligochäten. Zool. Jb., Abt. Anat. 52: 488–538.

Clark, W. C., 1978. Hermaphroditism as a reproductive strategy for metazoans; some correlated benefits. New Zeal. J. Zool. 5: 769–780.

Coates, K. A. & C. Erséus, 1985. Marine enchytraeids (Oligochaeta) of the coastal Northwest Atlantic (Northern and mid U.S.A.). Zool. Scr. 14: 103–116.

Dózsa-Farkas, K., 1974. A new *Fridericia* species (Oligochaeta: Enchytraeidae). Acta Zool. Acad. Sci. Hung. 20: 27–32.

Eisen, G., 1910. Enchytraeidae of the West coast of North America. Harriman Alaska Series XII, Smithsonian Institution, Washington: 126 pp. plates I–XX.

Erséus, C., 1976. Littoral Oligochaeta (Annelida) from Eyjafjördur, North Coast of Iceland. Zool. Scr. 5: 5–11.

Erséus, C., 1977. Marine Oligochaeta from the Koster area, west coast of Sweden, with descriptions of two new enchytraeid species. Zool. Scr. 6: 293–298.

Fleming, T. P., 1981. The ultrastructure and histochemistry of the spermathecae of *Tubifex tubifex* (Annelida: Oligochaeta). Proc. zool. Soc. Lond. 193: 129–145.

Graefe, U., 1973. Systematische und ökologische Untersuchungen an Enchytraeiden. I. Systematische Untersuchungen an der Gattung *Achaeta* (Enchytraeidae, Oligochaeta). Diploma thesis, University of Hamburg.

Gustavsson, L. M. & C. Erséus, 1997. Morphogenesis of the genital ducts and spermathecae in *Clitellio arenarius*, *Heterochaeta costata*, *Tubificoides benedii* (Tubificidae) and *Stylaria lacustris* (Naididae) (Annelida, Oligochaeta). Acta zool. (Stockholm) 78: 9–31.

Jamieson, B. G. M., 1981. The ultrastructure of the Oligochaeta. Academic Press, London, New York, Toronto, Sydney, San Francisco: 462 pp.

Jamieson, B. G. M., 1988. XXII. Oligochaete ultrastructure: Some comparisons with the Polychaeta. In W. Westheide & C. O. Hermans (eds), The Ultrastructure of Polychaeta. Microfauna Marina 4: 397–428.

McDonald, K., 1984. Osmium ferricyanide fixation improves microfilament preservation and membrane visualization in a variety of animal types. J. Ultrastruct. Res. 86: 107–108.

Michaelsen, W., 1886. Untersuchungen über *Enchytraeus Möbii* Mich. und andere Enchytraeiden. Inaug.-Diss. Kiel.

Mrázek, A., 1900. Die Samentaschen von *Rhynchelmis*. SB. K. Böhm. Ges. Wiss., Prag.

Nielsen, C. O. & B. Christensen, 1959. The Enchytraeidae. Critical revision and taxonomy of European species. Natura jutl. 8–9: 1–160.

Nielsen, C. O. & B. Christensen, 1963. The Enchytraeidae. Critical revision and taxonomy of European species. Suppl. 2 Natura jutl. 1–19.

Pierantoni, U., 1903. Studii anatomici su *Michaelsena macrochaeta* Pierant. Mitt. Zool. Stn. Neapel 16: 409–444.

Ratzel, F., 1869. Beiträge zur Anatomie von *Enchytraeus vermicularis*. Z. wiss. Zool. 18: 563–591.

Schmelz, R. M., 1998. Description of *Fridericia montafonensis* sp.n. (Enchytraeidae, Oligochaeta) from an Austrian meadow. Mitt. hamb. zool. Mus. Inst. 95: 79–88.

Vejdovský, F., 1904. O zolástnim pripadu fagocytosy. SB. K. Böhm. Ges. Wiss., Prag.

Westheide, W., 1984. The concept of reproduction in polychaetes with small body size: Adaptations in interstitial species. In A. Fischer & H. D. Pfannenstiel (eds), Polychaete Reproduction. Fortschritte Zoologie 29: 265–287.

Westheide, W. & U. Graefe, 1992. Two new terrestrial *Enchytraeus* species (Oligochaeta, Annelida). J. nat. Hist. 26: 479–488.

Westheide, W. & M. C. Müller, 1995. Organisation und Fortpflanzung von Enchytraeen (Oligochaeta) Film C 1821. Publ Wiss. Film, Biol. 22: 153–170.

Hydrobiologia **406**: 213–222, 1999.
B. M. Healy, T. B. Reynoldson & K. A. Coates (eds), Aquatic Oligochaetes.

New sperm types in Naididae and Lumbriculidae (Annelida: Oligochaeta) and their possible phylogenetic implications

Marco Ferraguti[1], Christer Erséus[2], Irina Kaygorodova[3] & Patrick Martin[4]
[1]*Dipartimento di Biologia, Università di Milano, Via Celoria 26, I-20133, Milano, Italy*
[2]*Swedish Museum of Natural History, Department of Invertebrate Zoology, Box 50007, SE-104 05, Stockholm, Sweden*
[3]*Limnological Institute, Siberian Branch of the Academy of Sciences of Russia, Ulan-Batorskaya 3, 664033 Irkutsk, Russia*
[4]*Institut royal des Sciences naturelles de Belgique, rue Vautier 29, B-1000, Bruxelles, Belgium*

Key words: Naididae, Lumbriculidae, ultrastructure, spermatozoa, phylogeny, Oligochaeta

Abstract

The sperm ultrastructure of three Naididae (*Paranais frici, P. litoralis and Stylaria lacustris*) and two Lumbriculidae (*Rhynchelmis brachycephala* and *R. alyonae*) is decribed to supply spermatological evidence towards a better understanding of the phylogeny of the two families. Naidid spermatozoa were found to be similar to one another, particularly with regard to the acrosome (with the acrosome tube bent to one side), and the shape and number of mitochondria. The various morphological characters do not militate against the previously proposed inclusion of naidids within the Tubificidae. The two lumbriculid species examined showed a characteristic conical indentation at the apex of the nucleus, a feature up to now described only in Branchiobdellida, an annelid group with possible affinities to Lumbriculidae.

Introduction

Naididae and Lumbriculidae are freshwater taxa of oligochaete annelids, the former cosmopolitan in distribution and comprising about 175 species, the latter largely confined to the Northern hemisphere and with about 145 species. Both have traditionally been treated as separate families, although they have features in common with the larger family Tubificidae (bifid setae, atria and prostate glands associated with the male ducts).

Recent phylogenetic studies using conventional morphological evidence have suggested that at least naidids are likely to represent apomorphic tubificids; they often have multiple setae of complex form as in freshwater tubificids and their genital organs are structurally of the same kind as in the tubificid subfamily Rhyacodrilinae (Erséus, 1987, 1990; Brinkhurst, 1994). Naididae differ from Tubificidae in the more forward location of these organs, a trait probably associated with the predominantly asexual mode of reproduction in the former. In oligochaetes with regular propagation by architomy, the position of genitalia is correlated with the number of segments regenerated at the anterior end (Hrabě, 1937; Sperber, 1948), and such genital shifts are prone to evolutionary shifts through cloning (Erséus, 1984, 1990).

Lumbriculidae also have atria and prostates similar to those of Rhyacodrilinae, but their basic gonadal sequence (see Brinkhurst & Nemec, 1987), semi-prosoporous arrangement of the male pores (see Brinkhurst, 1989), pairwise arrangement of somatic setae and lack of penial setae (see Brinkhurst, 1994) have been referred to as a support for the placement of lumbriculids outside the Tubificidae. On the other hand, there have been suggestions that Lumbriculidae are ancestral to Branchiobdellida, Acanthobdellida and the true leeches (Euhirudinea) (Sawyer, 1986; Siddall & Burreson, 1996), a view supported also by the phylogenetic tree presented by Brinkhurst (1994:

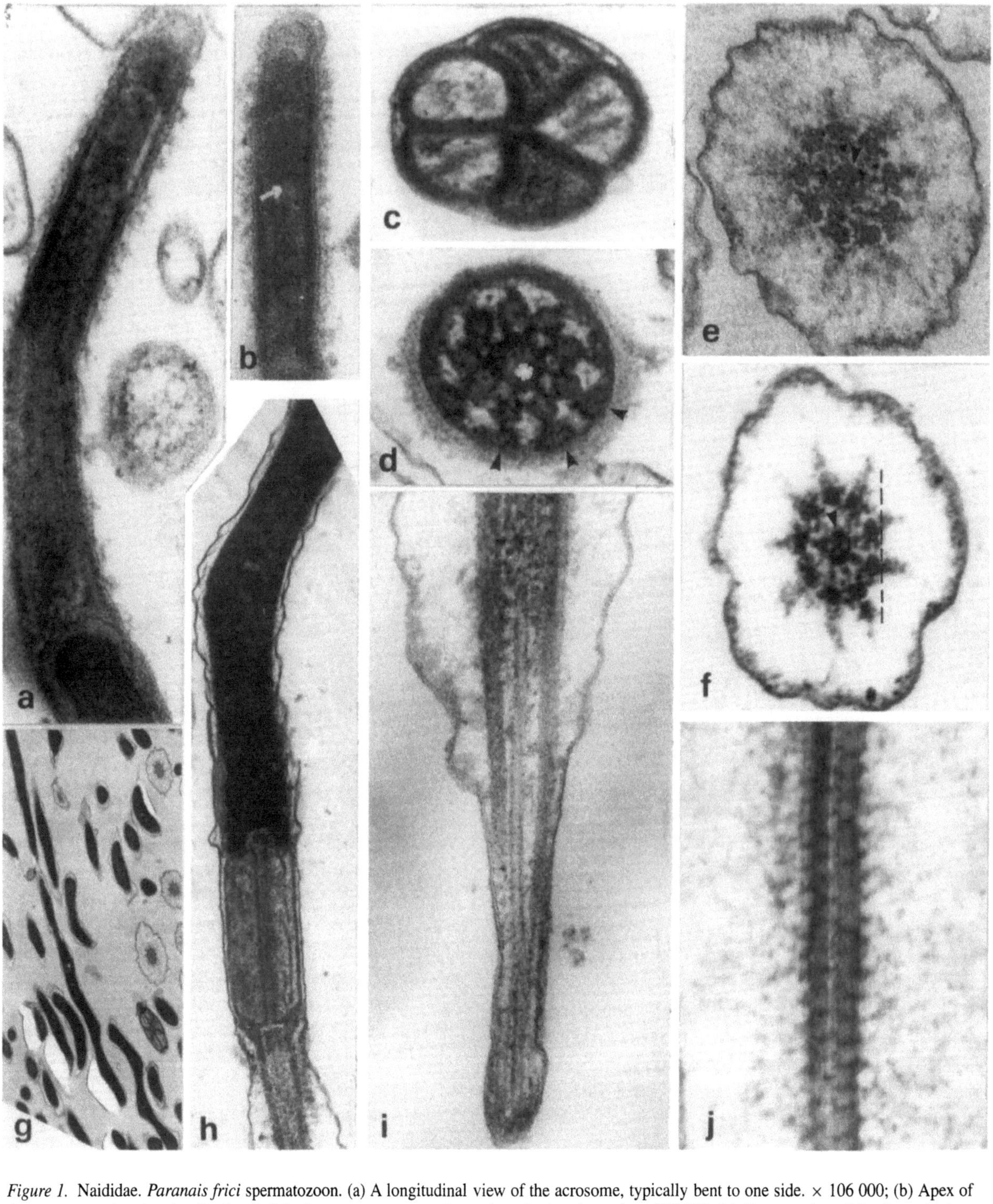

Figure 1. Naididae. *Paranais frici* spermatozoon. (a) A longitudinal view of the acrosome, typically bent to one side. × 106 000; (b) Apex of the acrosome, showing the distal end of the inner acrosome membrane invagination (arrow). × 106 000; (c) Cross section of the five midpiece mitochondria. × 90 000; (d) Cross section of the basal body area with the large basal cylinder (asterisk) and of remnants of the anchoring apparatus forming a dense ring (arrowheads) to which the doublet microtubules are somewhat connected × 90 000; (e) Cross section of the proximal tract of the flagellum. The plasma membrane is separated from the axoneme, and the central apparatus shows a prominent central sheath (arrowhead). × 90 000; (f) A more distal section of the tail with the central sheath shows the 'tetragon fibres' arrangement (arrowhead). A dotted line marks the level of section shown in Figure 1j. × 90 000; (g) The nucleus is apically straight, but basally twisted. × 7000; (h) Longitudinal section of the nuclear base and the midpiece. × 37 000; (i) The distal extremity of the tail. × 90 000; (j) Grazing, longitudinal section cut at the level of dotted line in Figure 1f. × 90 000.

Figure 1), who, however, finds this relationship improbable (op.cit.: p. 110; see also Brinkhurst & Gelder, 1989).

The ultrastructure of spermatozoa has been used for assessment of family level relationships in oligochaetes (Jamieson, 1983, 1984; 1987; Jamieson et al., 1987). However, only one lumbriculid, *Stylodrilus (=Bythonomus) lemani* Grube, 1879, has been completely studied with regard to sperm (Ferraguti & Jamieson, 1987), whereas for the Naididae, only some aspects of the spermatogenesis and sperm tail (in *Paranais* spp.) have been touched upon (Ferraguti, 1983, 1984). In the present paper, the mature spermatozoa of three species of Naididae and two Lumbriculidae are described and compared with other clitellate sperm types.

Material and methods

Specimens of *Paranais frici* Hrabě, 1941 were collected in the sediments of the River Po close to Piacenza, Italy, and fixed in a cacodylate buffered Karnovsky's (1965) mixture, diluted 1:1 with distilled water. *Paranais litoralis* Müller, 1784 was collected in the marine littoral sediments in the surroundings of Göteborg, Sweden, and fixed in a cacodylate buffered Karnovsky's mixture adjusted to the sea water molarity with the addition of sucrose. *Stylaria lacustris* (Linnaeus, 1767) was collected in littoral, low salinity water, in Husarviken Bay, in the inner archipelago of Stockholm, Baltic coast of Sweden, and fixed in the Ermak & Eakin (1976) fixative adjusted to freshwater. Both *Rhynchelmis* species were sampled in Lake Baikal (Russia), *R. brachycephala* Michaelsen, 1901 at 58 m depth in the Chivyrkuyisky Gulf (53° 48′ 59″ N, 109° 07′ 07″ E), by means of a trawl, and *R. alyonae* Martin et al. (1998) in Maloye More (Olchonskiye varota) with a dredge operating between 30 and 14 m depth (53° 00′ 25″ N, 106° 55′ 12″ E). All specimens were fixed as for *S. lacustris*. In all cases, the genital segments were cut open to facilitate penetration of the fixative. The specimens were washed overnight in 0.1 M cacodylate buffer, postfixed in cacodylate buffered osmium tetroxide, pre-stained in 2% aqueous uranyl acetate, dehydrated in a graded ethanol series, and embedded in Spurr's resin. The sections, obtained with an LKB Ultrotome V, were stained in lead citrate, carbon coated and observed under a JEOL 100 XS electron microscope operating at 80 kV.

Observations

The spermatozoa of all species have been examined at the ciliated sperm funnel, i.e. at the entrance of their deferent ducts, because of fixation of spermatozoa in the spermathecae being poor. All the sperm models here described conform to the general model of the clitellate spermatozoa (Ferraguti, in press): they are long, filiform cells, containing, in an anteroposterior sequence, the acrosome, the nucleus, the midpiece and the tail. The acrosome is characterized by the presence of an acrosome tube, a cylindrical structure containing the other acrosome structures. The midpiece is formed by tightly packed mitochondria, and the flagellar axoneme in the tail shows typical modifications of the central apparatus. This last inserts proximally into a large basal cylinder, the structure characterizing the oligochaete spermatozoa.

Naididae

The spermatozoa of *Paranais frici* (Figures 1 and 4) and *P. litoralis* are similar, except in the arrangement of the tail plasma membrane, thus they will be described together. The acrosome is slightly less than 1 μm long, and typically bent to one side (Figure 1a). It is simple in its structure: the acrosome tube contains the whole acrosome vesicle, and the latter is apically in contact with the plasma membrane at an area called the 'apical button' (*sensu* Lanzavecchia & Lora Lamia Donin, 1972) (Figure 1b). No acrosome rod is visible in the subacrosomal space. The acrosome tube terminates basally with a characteristic thickening (Figure 1a) (limen *sensu* Jamieson, 1978). The nucleus is apically straight, then slightly twisted with a decreasing pitch (Figure 1g). It terminates with a basal characteristic semispheric invagination (Figure 1h). Five elongated, parallel mitochondria, with the shape of tightly packed cylindrical sectors, form the following midpiece (Figure 1c, e, h). The tail contains a 9+2 axoneme surrounded, in *P. frici*, by a widely swollen cytoplasm for almost all its length (Figure 1f, i), whereas in *P. litoralis*, the plasma membrane is at a regular, shorter distance from the axoneme (see Figures 1 and 3 in Ferraguti, 1984). The basal body area is, as in all oligochaetes, deeply modified by the presence, at its interior, of a central basal cylinder (Figures 1d, h). The central apparatus of the axoneme is proximally of the prominent central sheath model (Figure 1e), but distally of the tetragon fibre model (Figure 1f) (for terminology, see Ferraguti, 1984). The axoneme is surrounded by a complex network (Figure 1j), generating rays connecting the doublets to the plasma membrane, a feature present also in *P. litoralis*, although in the latter the plasma membrane is closer to the axoneme than in *P. frici*.

In *Stylaria lacustris* (Figures 2 & 4), the acrosome is always bent to one side (Figure 2a), but is much shorter (0.4–0.5 μm) than in *Paranais* (Figure 2 a–c). The tube contains the whole vesicle, except for the most apical part of the latter, at the button (Figure 2b). No rod is visible inside the subacrosomal space formed by the deep invagination of the basal portion of the acrosome membrane (Figure 2c). There is a pronounced limen at the base of the acrosome tube (Figure 2c). The nucleus is twisted over its whole length, with a pitch increasing towards the posterior end (Figure 2d). The midpiece is made by five elongated mitochondria, arranged in parallel, each with the shape of a cylindrical sector (Figure 2e, h). The flagellum has a 9+2 axoneme with the typical modifications of the central apparatus (prominent central sheath, Figure 2f; tetragon fibres, Figure 2g). In the basal body area there is the basal cylinder (Figure 2h).

Lumbriculidae

The general morphology of the spermatozoa of *R. brachycephala* and *R. alyonae* is similar, but there are some important differences in the details and dimensions. The description and reconstruction of *R. alyonae* spermatozoon will be given elsewhere (Martin et al., 1998) and the results are summarized in Figure 4c. The present description will be based mainly on *R. brachycephala* (Figures 3 & 4d), with some comparisons with *R. alyonae* where necessary. The size of the different structures in the two species is reported in Table 1.

The acrosome is a long and complex structure (Figures 3a–c & 4c). At its apex there is, in *R. brachycephala*, a prominent amount of mucopolysaccharides, the apical corona *sensu* Ferraguti & Jamieson (1987) (Figure 3b). The acrosome tube is clearly divided into two portions: a stiff, basal one, and a 'softer' and more flexible, apical one (Figures 3b, c & 4c). The acrosome vesicle is mainly contained within the soft portion of the tube, and only apically is in contact with the plasma membrane of the spermatozoon, forming an enlarged button. Only its basal portion is situated inside the stiff part of the tube. The contents of the acrosome vesicle are highly electron dense in the apical portion, and assume, in *R. brachycephala*, a typical crystalline appearance (Figure 3b). In this region, the membrane delimiting the acrosome vesicle is not clearly visible. Basally, the membrane is infolded in an asymmetric way in *R. brachycephala*, more symmetrically in *R. alyonae*, delimiting a space in which the apical portion of the acrosome rod is contained. The rod continues beyond the proximal end of the acrosome vesicle, crossing the whole basal chamber, and even penetrating, in *R. brachycephala*, inside the conical invagination of the nucleus (Figure 3c). There is a secondary tube, shorter in *R. brachycephala* than in *R. alyonae*, attached to the proximal end of the acrosome vesicle (Figures 3a & 4c). The acrosome tube terminates basally with a limen. A putative nuclear pad with the shape of a circular crown separates the acrosome from the nucleus (Figure 3c). The nucleus is nearly rectilinear in *R. brachycephala* (Figure 3e), but twisted over its whole length in *R. alyonae* and is characterized, apically, by a conical invagination (Figure 4c, d). The nucleus is followed by a midpiece formed by six mitochondria with the shape of cylindrical sectors, highly twisted around their main axis (Figure 3d, f). At the centre of the midpiece, there is always an evident cavity (Figure 3f). The flagellum has the typical 9+2 appearance, with a long basal cylinder in the basal body area (Figure 3d, g, h), and a prominent central sheath of the axoneme (Figure 3i). Traces of putative glycogen granules are visible external to the axonemal doublets.

Discussion

Naidid sperm

The spermatozoa of the naidids studied in this paper present some common characters, as well as some differences. Although the study of three species only in a family comprising about 175 species does not allow a definitive identification of 'the' sperm model for the family, we can say that the two sperm models of the investigated naidids are characterized by a simple acrosome bent to one side, in which a secondary tube and an acrosome rod, structures characterizing many other oligochaete sperm models, are not visible. The limen, the basal circular projection of the acrosome tube, is in both *Paranais* and *S. lacustris* a merely progressive enlargement of the acrosome tube base. These acrosome characters are almost identical to those of the tubificid *Monopylephorus limosus* Hattai,1898 (Ferraguti et al., 1994).

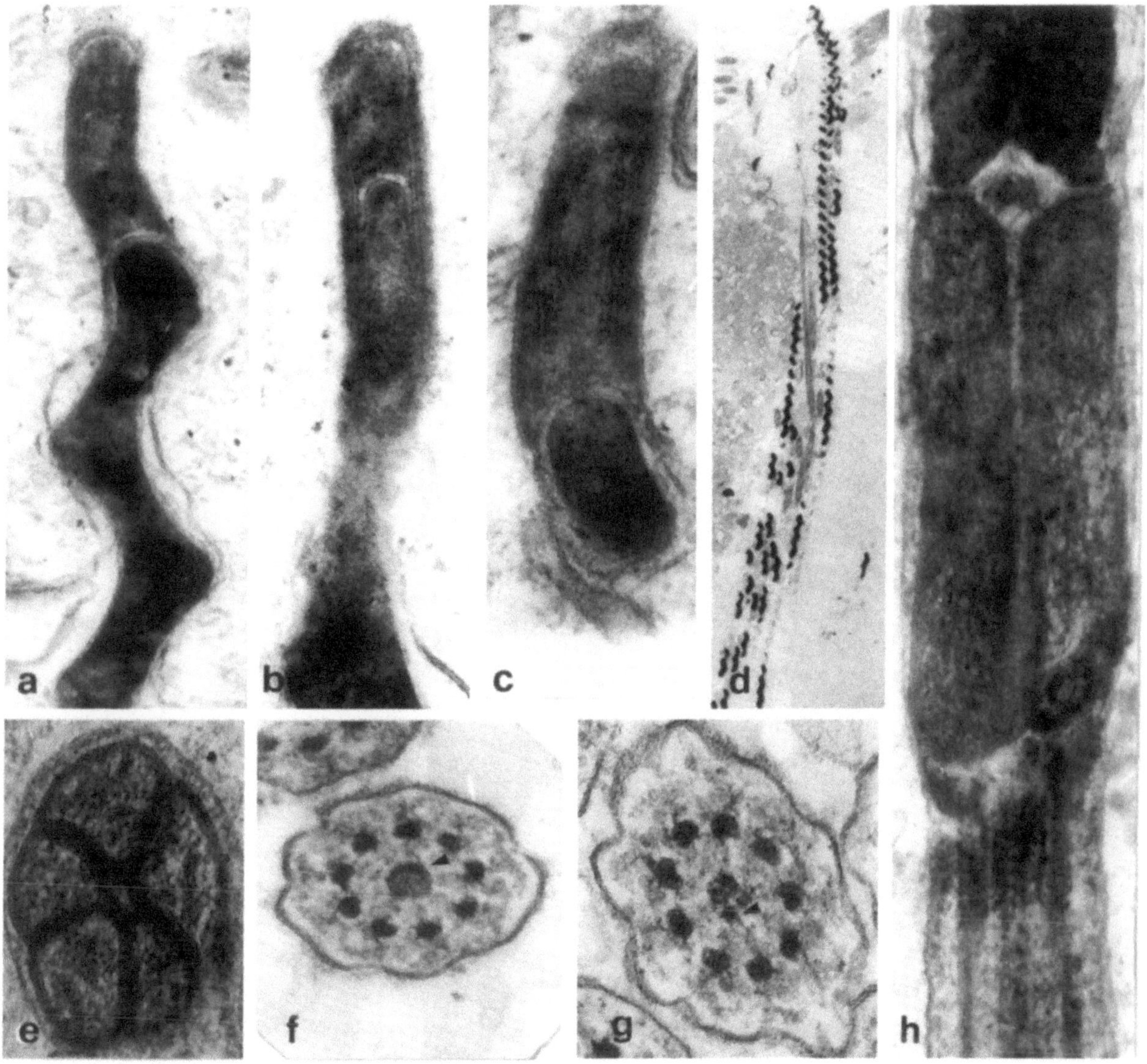

Figure 2. Naididae. *Stylaria lacustris* spermatozoon. (a) Longitudinal section of the nuclear apex and the acrosome. × 66 000; (b) Another longitudinal section of acrosome and nucleus. × 110 000; (c) An acrosome base longitudinally cut. × 148 000; (d) The sperm nucleus is twisted over most of its length. × 3200; (e) Cross section of the midpiece mitochondria. × 105 000; (f) Cross section of the flagellar basal portion. An arrowhead points to the prominent central sheath. × 98 000; (g) A more distal section. An arrowhead points to the tetragon fibres. × 98 000; (h) Longitudinal section of a midpiece. There is some dense material between the mitochondria and the nucleus. × 98 000.

Nuclear shape differs in the two genera examined, but is a highly variable character in oligochaetes, even within taxa of low taxonomic levels (see below).

The midpieces of the two *Paranais* species and *S. lacustris*, as well as of another naidid, *Chaetogaster diaphanus* (Gruithuisen, 1828) (Erséus & Ferraguti, personal observations) are very similar in shape and are formed by five mitochondria. Five is the number of mitochondria characterizing also the tubificid subfamily Phallodrilinae. However, it should be kept in mind that four–five is considered to be the plesiomorphic number of mitochondria of the 'primitive' spermatozoa (Erséus & Ferraguti, 1995).

The tails of the three naidid species differ significantly in their architecture: the tail is, in *P. frici* only, considerably swollen, leaving a large electron lucent space around the axoneme. This feature is present neither in *P. litoralis* nor in *S. lacustris*. A swollen tail identical to the one here described for *P. frici* characterizes tubificid paraspermatozoa (Ferraguti et al., 1989). Furthermore, in both *Paranais* species, the axoneme is surrounded by a complex cytoskeletal net-

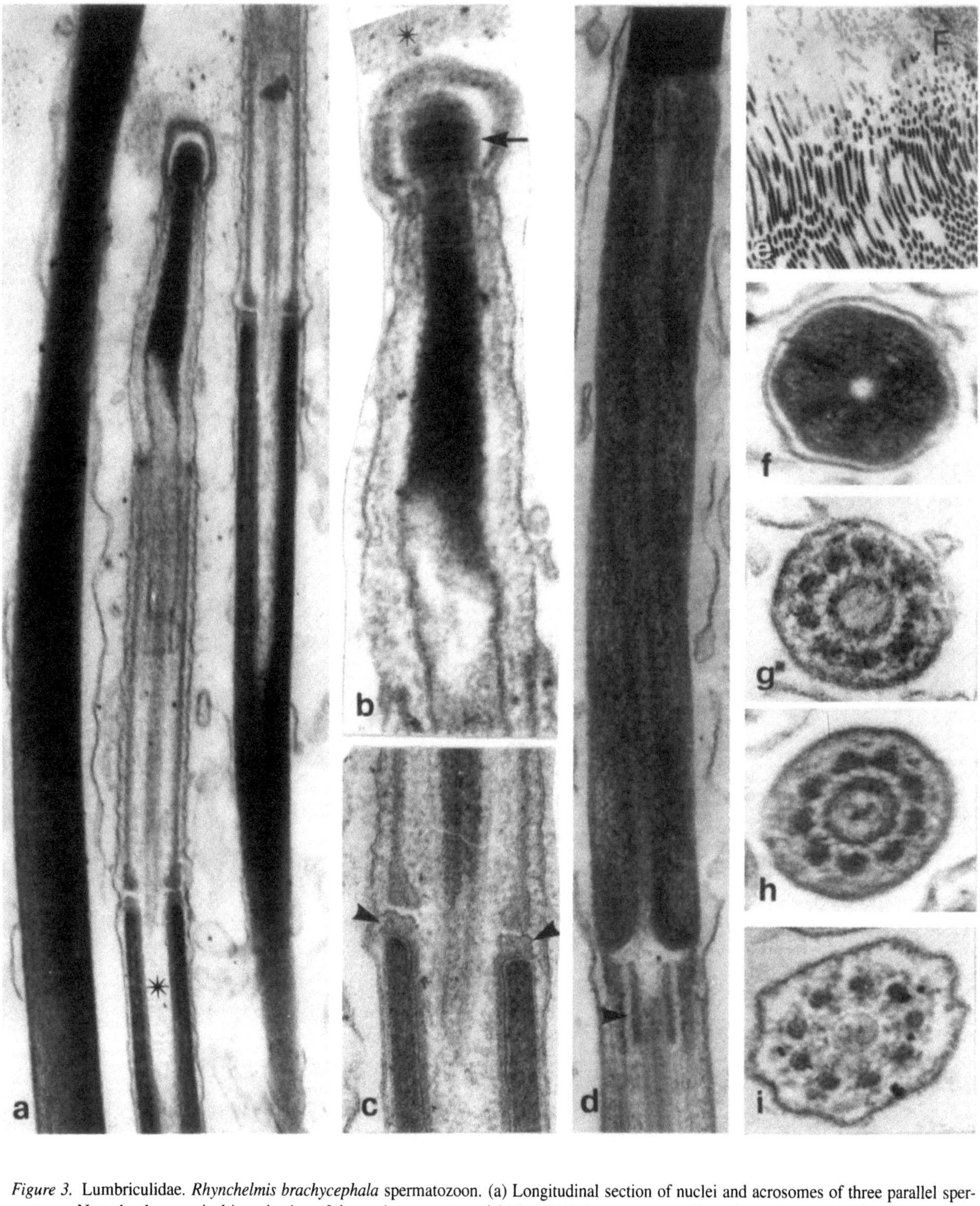

Figure 3. Lumbriculidae. *Rhynchelmis brachycephala* spermatozoon. (a) Longitudinal section of nuclei and acrosomes of three parallel spermatozoa. Note the deep conical invagination of the nuclear apex (asterisk) in which the acrosome rod terminates (compare with the scheme in Figure 4d). × 45 000; (b), (c) details of the apex (b) and base (c) of the acrosome. An apical corona (asterisk) is visible on top of the acrosome. The acrosome content is condensed in a crystalline dense matter (arrow). A putative nuclear pad is reduced to a circular crown (arrowheads). × 92 000; (d) A longitudinally sectioned midpiece. Note the twisted arrangement of the mitochondria. The basal cylinder (arrowhead) is basally penetrated by the central apparatus of the axoneme. × 63 000; (e) The nuclei at the funnels (F) are entwined in a regular way. × 1900; (f-i) Cross sections of the midpiece (f), the basal body at a progressively distal level (g, h), and the flagellum (i). All × 90 000.

Table 1. Some dimensions (in μm) of *Rhynchelmis brachycephala* and *R. alyonae* spermatozoa

	R. brachycephala	*R. alyonae*
Acrosome length	2.49 (*n*=6; *s.d.*=0.26)	1.02 (*n*=5; *s.d.*=0.08)
Acrosome tube length (soft portion)	0.87 (*n*=12; *s.d.*=0.15)	0.48 (*n*=7; *s.d.*=0.03)
Acrosome tube length (stiff portion)	1.41 (*n*=13; *s.d.*=0.16)	0.34 (*n*=7; *s.d.*=0.01)
Nuclear apical concavity: length	1.39 (*n*=10; *s.d.*=0.22)	0.23 (*n*=5; *s.d.*=0.01)
Midpiece length	2.31 (*n*=10; *s.d.*=0.19)	1.8 (*n*=7; *s.d.*=0.13)
Basal cylinder length	0.198 (*n*=5; *s.d.*=0.01)	0.09 (*n*=4; *s.d.*=0)

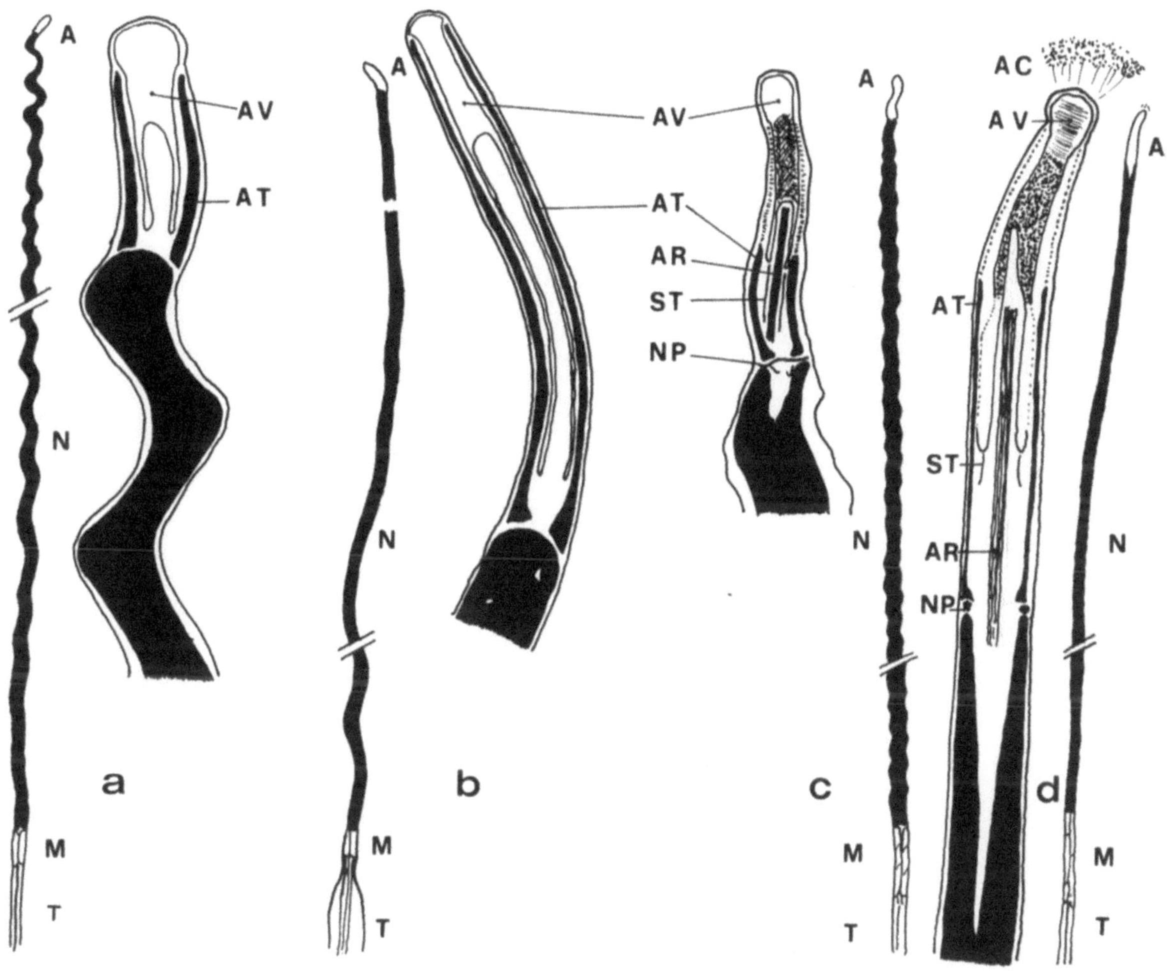

Figure 4. Schematic drawings of the whole spermatozoa and details of the acrosomes of *P. frici* (a) and *S. lacustris* (b) (Naididae), *R. alyonae* (c) and *R. brachycephala* (d) (Lumbriculidae). A = acrosome; N = nucleus; AT = acrosome tube; ST = secondary tube; AR = acrosome rod; NP = nuclear pad; AC = apical corona; M = midpiece; T = tail.

work, almost identical to that described in the tubificid *M. limosus* (Ferraguti et al., 1991). The two modifications of the central axonemal apparatus, i.e. the prominent central sheath and the tetragon fibres, also co-occur in different sections of the same tail in some tubificid species, but also in other oligochaete taxa, such as the Lumbriculidae (Ferraguti & Jamieson, 1987).

Lumbriculid sperm

Besides that of the two *Rhynchelmis* species here described, one additional sperm model is known for lumbriculids, that of *Stylodrilus lemani*, described by Ferraguti & Jamieson (1987). We also have some incomplete data on a European species of *Rhynchelmis*, *R. limosella* Hoffmeister, 1843, pointing to a marked similarity with *R. brachycephala*. The separation of the acrosome tube into two different regions, the anterior soft and the posterior stiffer, appears to be, among lumbriculids, an autapomorphy of the genus *Rhynchelmis*. A striking similarity with *S. lemani* is that the contents of the acrosome vesicle, at least apically, appear electron dense, and the membrane surrounding the acrosome vesicle in this region is not visible.

A conical cavity is present on top of the nucleus in the three *Rhynchelmis* species examined, and in another lumbriculid, *Kinkaidiana* sp. (Ferraguti, unpublished), but not in *S. lemani*. The nucleus is straight in *R. brachycephala* and *R. limosella*, twisted in *R. alyonae* and *S. lemani*, and corkscrew-shaped in *Kinkaidiana*.

In all lumbriculids examined, the midpiece has exactly the same architecture, i.e. it is made of six tightly packed, elongated and twisted mitochondria, and there are only minor dimensional differences. The axoneme shows the prominent central sheath arrangement in *Rhynchelmis* and *Kinkaidiana*, a tract with tetragon fibres being present only in *Stylodrilus* (Ferraguti & Jamieson, 1987).

While *R. brachycephala* has spermatozoa rather similar to those of *R. limosella*, there are striking differences in the dimensions of the various parts of the spermatozoa of *R. brachycephala* and *R. alyonae* (Table 1), as well as in the shape of the nucleus, supporting a closer relationship between the former two than between *R. brachycephala* and *R. alyonae*. Other conventional morphological characters give supplementary evidence of such a relationship. *R. brachycephala* and *R. limosella* have long atria, extending through many segments (to XVII in *R. brachycephala* *vs.* to XII in *R. alyonae*), and one unpaired so-called 'rudimentary atrium' in IX (absent in *R. alyonae*). Moreover, *R. alyonae* has modified penial setae[1] in X, surrounded by glandular cells (Martin et al., 1998). Such differences are remarkable in species belonging to the same genus and give confirmation of the increasingly puzzling systematics of this genus (Brinkhurst, pers. comm.).

[1] As a rule, genital setae are absent in Lumbriculidae. To date, they are only mentioned in *Thinodrilus genitosetosus* Holmquist, 1976 and *Rhynchelmis spermatachaeta* Semernoy, 1982.

Comparisons with sperm models of other taxa

The spermatozoa of naidids show characters present, although not all in the same species, in tubificids (see the review by Ferraguti et al., 1994 and Erséus & Ferraguti, 1995). In particular, as mentioned above, the acrosome of both genera studied is similar to that of *Monopylephorus limosus*. The nucleus is twisted in *Stylaria lacustris*, a feature again similar to the corresponding feature in *M. limosus* and also in other tubificids. The nucleus of *Paranais* spp. is identical to that of the tubificid *Thalassodrilides ineri* Righi & Kanner, 1979. However, the twisted column shape of the nucleus is a trait widespread in the Animal Kingdom. The midpiece, as already stated, is similar to that of phallodriline tubificids, but also to that of Glossoscolecidae (Ferraguti, in press). On the other hand, it must be noted that tubificids are the oligochaete family with the highest number of species studied, thus it may well be that the variability of sperm models within this family is an artifact caused by a bias in 'sampling' effort. However, the acrosomes, by far the most complex sperm organelles in clitellates, are, in naidids, completely different from those of the Capilloventridae (Ferraguti et al., 1996), Enchytraeidae (Westheide et al., 1991) and Lumbriculidae described so far.

The spermatozoa of the genus *Rhynchelmis* (Lumbriculidae) here examined show a conical indentation at the apex of the nucleus. This character is shared by 27 of the 30 species of branchiobdellids examined, and is considered by Ferraguti (in press) as belonging to the basal plan of branchiobdellid spermatozoa. A conical indentation on top of the nucleus is present in a single tubificid, *Rhizodrilus russus* Erséus, 1990 (Ferraguti et al., 1994), but here the indentation is the starting point of a long and thin endonuclear canal. It is also interesting to note that the apical corona, originally described by Ferraguti & Jamieson (1987) in *Stylodrilus lemani* is again a character present in all branchiobdellids (Ferraguti, 1997). However, the lack of descriptions of an apical corona may be due to technical problems (e.g. fixation), since it is clearly visible in the tubificid *Rhyacodrilus arthingtonae* Jamieson, 1978 (Figure 3 in Jamieson et al., 1978).

Some phylogenetic considerations

The present study demonstrates that spermatozoal ultrastructure at least does not refute the hypothesis that Naididae are merely a subgroup within the Tubificidae. It rather supports this relationship by the number of similarities between naidid sperm and that of individual species of Tubificidae outlined above.

Although an acrosome rod is not visible in the studied naidids, the shape of their acrosome vesicles seems to reflect the complete withdrawal of the subacrosomal meterial into the acrosome tube. Complete withdrawal, which has been considered apomorphic, characterizes a vast majority of Tubificidae (see Erséus & Ferraguti, 1995), as well as some other groups (Lumbriculidae, Phreodrilidae and all megadrile families investigated; Jamieson et al., 1987), but not the Enchytraeidae, Capilloventridae and Haplotaxidae (Jamieson et al., 1987; Ferraguti et al., 1996). However, it is not yet conclusive whether this state is synapomorphic or it has evolved independently in different families.

Spermatozoal characters give no clear clues as to whether the Lumbriculidae represent a clade outside or within the Tubificidae. The lumbriculid sperm models may be the result an evolutionary elaboration of a tubificid model, but it is equally possible to imagine the tubificid and lumbriculid sperms as independent lines from a common ancestor.

More evident is the strong indication by sperm that lumbriculid and branchiobdellids share a relatively recent common ancestor; it is likely that the apical indentation of the nucleus and, perhaps, also the apical corona, are synapomorphic features in these two groups. Furthermore, Jamieson (1988), in his cladistic analysis of the Oligochaeta, recognized that: "the relationships of leeches and branchiobdellids appear to hinge on the Lumbriculidae". In the Acanthobdellida and Euhirudinea, the nucleus is a very complex and thin multiple-start helix, and thus it is not possible to state wether the conical apical invagination of the nucleus is also present in these taxa.

Acknowledgements

The present research has been supported by a grant to M.F. from MURST (Rome), under the 40% project 'Cell Interactions', by a grant to C.E. from the Swedish Natural Sciences Research Council, by the INTAS grant 94-4465, from the Belgian Ministry of National Scientific Institutions to P.M., and by the Siberian Branch of the Academy of Sciences of Russia to I.K. I.K. and P.M. are especially grateful to Michael Grachev, Director of the Limnological Institute of Irkutsk, for supporting foreign research on Lake Baikal, and are indebted to the captain and the crew of the R/V Vereschagin for their assistance during fieldwork.

References

Brinkhurst, R. O., 1989. A phylogenetic analysis of the Lumbriculidae (Annelida, Oligochaeta). Can. J. Zool. 67: 2731–2739.

Brinkhurst, R. O., 1994. Evolutionary relationships within the Clitellata: an update. Megadrilogica 5: 109–112.

Brinkhurst, R. O. & S. R. Gelder, 1989. Did the lumbriculids provide the ancestors of the branchiobdellidans, acanthobdellidans and leeches? Hydrobiologia 180: 7–15.

Brinkhurst, R. O. & A. F. L. Nemec, 1987. A comparison of phenetic and phylogenetic methods applied to the systematics of Oligochaeta. Hydrobiologia 155: 65–74.

Ermak, T. H. & R. M. Eakin, 1976. Fine structure of the cerebral and pygidial ocelli in *Chone ecaudata* (Polychaeta: Sabellidae). J. Ultrastruct. Res. 54: 243–260.

Erséus, C., 1984. Aspects of the phylogeny of the marine Tubificidae. Hydrobiologia 115: 37–44.

Erséus, C., 1987. Phylogenetic analysis of the aquatic Oligochaeta under the principle of parsimony. Hydrobiologia 155: 75–89.

Erséus, C., 1990. Cladistic analysis of the subfamilies within the Tubificidae (Oligochaeta). Zool. Scr. 19: 57–63.

Erséus, C. & M. Ferraguti, 1995. The use of spermatozoal ultrastructure in phylogenetic studies of Tubificidae. In B. G. M Jamieson, J. Ausio & J. L. Justine (eds) Advances in Spermatozoal Phylogeny and Taxonomy. Mém. Mus. natn. Hist. nat. 166: 189–201.

Ferraguti, M., 1983. The spermiogenesis in *Paranais* (Annelida, Oligochaeta). In W Engels (ed.), Advances in Invertebrate Reproduction. Elsevier Science Publishers: 3: 578.

Ferraguti, M., 1984. The comparative ultrastructure of sperm flagella central sheath reveals a new autapomorphy of the group. Zool. Scr. 13: 201–207.

Ferraguti, M., *in press*. Euclitellata. In B. G. M. Jamieson (ed.), Progress in Male Gamete Ultrastructure and Phylogeny. Part B. John Wiley, Chichester, pp 165–222.

Ferraguti, M. & B. G. M. Jamieson, 1987. Spermiogenesis in *Bythonomus lemani* and the phylogenetic position of Lumbriculidae. Hydrobiologia 155: 123–134.

Ferraguti, M., M. Bondi & S. Lanzavecchia, 1991. Ultrastructural analysis of sperm flagella in two clitellates (Annelida). Plasma membrane and periaxonemal area. Tissue & Cell 23: 537–546.

Ferraguti, M., C. Erséus & A. Pinder, 1996. The spermatozoon of *Capilloventer australis* and the systematic position of Capilloventridae (Annelida, Oligochaeta). Aust. J. Zool. 44: 469–478.

Ferraguti, M., G. Grassi & C. Erséus, 1989. Different models of tubificid spermatozeugmata. Hydrobiologia 180: 73–82.

Ferraguti, M., D. Ruprecht, C. Erséus & O. Giere, 1994. An ultrastructural overview of tubificid spermatozoa. Hydrobiologia, 278: 165–178.

Hrabě, S., 1937. Zur Kenntnis des *Lamprodrilus mrazeki* Hr., *Aulodrilus pluriseta* Pig. und *Aulodrilus pigueti* Kow. Sbor. Kl prir. Brno 19: 1–7 (in Czech with German summary).

Jamieson, B. G. M., 1978. A comparison of spermiogenesis and spermatozoal; ultrastructure in megascolecid and lumbricid earthworms (Oligochaeta: Annelida). Aust. J. Zool. 28: 225–240.

Jamieson, B. G. M., 1983. Spermatozoal ultrastructure: evolution and congruence with a holomorphological phylogeny of the Oligochaeta (Annelida). Zool.Scr. 12: 107–114.

Jamieson, B. G. M., 1984. A phenetic and cladistic study of spermatozoal ultrastructure in the Oligochaeta (Annelida). Hydrobiologia 115: 3–13.

Jamieson, B. G. M., 1987. A biological classification of sperm types, with special reference to annelids and molluscs, and an example of spermiocladistics. In H. Mohri (ed.), New Horizons in Sperm Cell Research. Japan Science Society Press, Tokyo, & Gordon & Breach, New York: 311–332.

Jamieson, B. G. M., L. Daddow & J. D. Bennett, 1978. Ultrastructure of the tubificid acrosome (Annelida, Oligochaeta). Zool. Scr. 7: 115–118.

Jamieson, B. G. M., C. Erséus & M. Ferraguti, 1987. Parsimony analysis of the phylogeny of some Oligochaeta (Annelida) using spermatozoal ultrastructure. Cladistics 3: 145–155.

Karnovsky, M. J., 1965. A formaldehyde-glutaraldehyde fixative of high osmolality for use in electron microscopy. J. Cell Biol. 27: 137A.

Lanzavecchia, G. & C. Lora Lamia Donin, 1972. Morphogenetic effects of microtubules. II. Spermiogenesis in *Lumbricus terrestris*.J. submicr. Cytol. 4: 247–260.

Martin, P., M. Ferraguti & I. Kaygorodova, 1998. Description of two new *Rhynchelmis* species from Lake Baikal (Russia), using classical morphological and ultrastructural spermatozoal characters. Ann. Limnol. 34: 283–293.

Sawyer, R. T., 1986. Leech biology and behaviour. Clarendon Press, Oxford.

Spurr, A. R., 1969. A low-viscosity epoxy resin embedding medium for electron microscopy. J. Ultrastruct. Res. 26: 31–43.

Siddall, M. E. & E. M. Burreson, 1996. Leeches (Oligochaeta?: Euhirudinea), their phylogeny and the evolution of life-history strategies. Hydrobiologia 334: 277–285.

Sperber, C., 1948. A taxonomical study of the Naididae. Zool. Bidr. Upps. 28: 1–296.

Westheide, W., G. Purschke, G. & K. Middendorf, 1991. Spermatozoal ultrastructure of the taxon *Enchytraeus* (Annelida, Oligochaeta) and its significance for species discrimination and identification. Z. Syst. Evolut. 29: 323–342.

Hydrobiologia **406:** 223–233, 1999.
B. M. Healy, T. B. Reynoldson & K. A. Coates (eds), Aquatic Oligochaetes.

CLSM analysis of serotonin-immunoreactive neurons in the central nervous system of *Nais variabilis*, *Slavina appendiculata* and *Stylaria lacustris* (Oligochaeta: Naididae)

René Hessling, Monika C. Müller & Wilfried Westheide
Universität Osnabrück FB Biologie/Chemie, Spezielle Zoologie, Barbarastr. 11, D-49069 Osnabrück, Germany.
E-mail: hessling@biologie.uni-osnabrueck.de

Key words: 5-HT, immunohistochemistry, neuronal patterns, Annelida, Clitellata, phylogeny

Abstract

The organization of the nervous system and the distribution of serotoninergic neurons were analysed in three species of Naididae using immunohistochemical labelling methods in conjunction with confocal laser-scanning microscopy. In the posterior region of the ventral nerve cord the serotoninergic neurons are arranged in an alternating pattern in successive ganglia, clearly different from the pattern in the anterior neuromeres. This unique arrangement of 5HT-neurons is presumably an autapomorphic character of the Naididae or at least of a subtaxon including these three species. In all three species the cerebral ganglion is located in segments behind the prostomium, as has previously been found in terrestrial oligochaetes, where the relocation of the brain from the initially anterior position in the prostomium into the following segments can be explained as a functional adaptation to a life spent burrowing through comparatively firm substrates. This structural correspondence is regarded as an indication that the Naididae, being typical limnic oligochaetes, are derived from a terrestrial form.

Introduction

Serotonin (5-hydroxytryptamine, 5-HT) is an important neurotransmitter and neuromodulator, which has been detected in various metazoan taxa ranging from Cnidaria (Elofsson & Carlberg, 1989) to Vertebrata (Steinbusch et al., 1978; Steinbusch & Nieuwenhuys, 1981). Within the annelids, serotoninergic neurons have been localized with the help of classical histochemical methods; specific antibody-labelling techniques have also been used in various polychaetes (Anctil et al., 1990), in some terrestrial oligochaetes (Ehinger & Myhrberg, 1971; Csoknya et al., 1993) and also in a number of leeches (Marsden & Kerkut, 1969; Stent et al., 1992). Immunohistochemical labelling not only reveals the distribution of neuroactive substances within the ganglia but also allows the organization of the nervous system to be depicted in great detail (Hessling & Westheide, 1999; Spörhase-Eichmann et al., 1987a; Barlow & Truman, 1991).

The present investigation shows the arrangement of serotonin-immunoreactive structures in the central nervous system of *Nais variabilis* Piguet, 1906, *Slavina appendiculata* d'Udekem, 1855 and *Stylaria lacustris* Linnaeus, 1767. They provide distinct autapomorphic characters for the Naididae or at least for a subtaxon comprising at least these three genera. They also provide arguments for a terrestrial origin of these typical limnic oligochaetes.

Materials and methods

Specimens of the naidid species *Nais variabilis*, *Slavina appendiculata* and *Stylaria lacustris* were collected from various waters around Osnabrück, Germany. *Stylaria lacustris* and *Nais variabilis* were kept in culture dishes (diameter: 9 cm) half filled with 0.375% agar-medium, which was covered with an aqueous solution containing 0.09 g NaCl, 0.03 g KCl, 1.44 g $CaCl_2$, 0.06 g $MgSO_4$ per litre (Anlauf, 1989). They were cultured at 18 °C and fed with flaked oats. Approximately 25 individuals of each species, which had developed asexually by means of paratomy, were selected in order to study the distribution of serotonin-immunoreactive neurons.

For the immunohistochemical analysis the animals were fixed overnight at a temperature of 4 °C on ice, using a 4% paraformaldehyde solution with 10 mM phosphate-buffered saline (PBS, pH 7.4) (Denk, 1989; Côté et al., 1993). The specimens were preincubated for six hours in PBS containing 0.1% Triton-X-100, to increase membrane permeability (Barlow & Truman, 1991), 0.25% BSA (bovine serum albumin) as blocking reagent and 0.05% NaN_3. For an indirect immunohistochemical labelling of serotoninergic structures of the nervous system monoclonal antibodies (anti-serotonin (rabbit); Bissendorf Biochemicals GmbH, Hannover) were used in a dilution of 1:100 in preincubation solution without BSA. In order to label the primary antibodies against serotonin the specimens were incubated using secondary antibodies conjugated with the fluorophores fluorescein isothiocyanate (FITC) or indomethinecyanine (CY3) (Fine, 1993). The specimens were mounted in Citi-Fluor (UKC, Chem. Lab., Canterbury, CT27NH) on microscope slides and examined with a confocal laser-scanning microscope (Zeiss cLSM 410).

Specificity of serotonin-immunoreactivity was tested by incubating specimens in the same manner, yet omitting the primary antibody. In these controls, neuronal structures were not labelled.

Results

Nais variabilis

The immunohistochemical labelling of serotoninergic fibres with antibodies against serotonin shows the neuropile of the supraoesophageal ganglion to be located in the region of the peristomium (Figure 1A). Two pairs of serotoninergic perikarya are positioned laterally behind the neuropile of the brain in segment II (Figures 1A, B), the dorsal pair being located more medially and the second one further laterally, extending its processes almost from the side into the neuropile. All regions of the neuropile of the brain contain serotonin-immunoreactive fibres. These fibres apparently arise not only from the serotoninergic perikarya of the cerebral ganglion but also in part from neurons of the ventral nerve cord. Numerous small fibres extend from the basis of the supraoesophageal ganglion anteriorly; some isolated serotoninergic processes can be found in the region of the prostomium (Figure 1A). It was not possible to trace continuous serotonin-containing neurites contributing to the paired prostomial nerves, which can be labelled using antibodies against acetylated α-tubulin (Hessling, unpublished observations).

The circumoesophageal connectives lead from the lateral base of the brain ventrocaudally into the suboesophageal ganglion. They contain tightly packed serotoninergic fibres; individual processes are distinguishable only further posterior, in the neuropile of the ventral nerve cord.

In Naididae, a certain number of anterior segments, which are generally termed as cephalized segments (Stephenson, 1930), differ from those behind them by lacking dorsal chaetae and often possessing different ventral chaetae. In *Nais variabilis*, as well as in *Slavina appendiculata* and *Stylaria lacustris*, the cephalized region includes the prostomium, peristomium (segment I) and segments II–V.

The neuromeres of segments I–V, which form the suboesophageal ganglion, are shifted backwards and displaced longitudinally in relation to their corresponding segments. The displacement of these neuromeres can be discerned by various features, including their position posterior to the autofluorescent bundles of chaetae that mark their respective segments (Figure 1C). The more posterior neuromeres, beginning with segment VI, are situated within the corresponding segments, adjacent to the ventral bundles of chaetae.

The serotoninergic perikarya of the ventral nerve cord are distributed in a more or less regular pattern, their arrangement, however, being distinctly different in the neuromeres of the cephalized region from that in the following segments (Figures 1C, D; 4A, B). The pattern of neurons in segments I–III displays the greatest interindividual consistency. Each of these three neuromeres possesses four symmetrically arranged pairs of serotoninergic perikarya (Figures 1C; 4A). Two pairs are located medially and extend their axons into the median region of the ipsilateral longitudinal neuropile cord. The other two pairs lie laterally, the two perikarya of each side being closely adjacent to one another. Their processes pass laterally into the ipsilateral neuropile cord. At least in some cases it is evident that these processes project into the contralateral ganglionic region.

The neuromeres of segments IV and V contain fewer labelled perikarya than are found in the preceding segments (Figures 1C; 4A), but the positions of the individual somata within the neuromeres correspond to those in the anterior cell groups. Differences

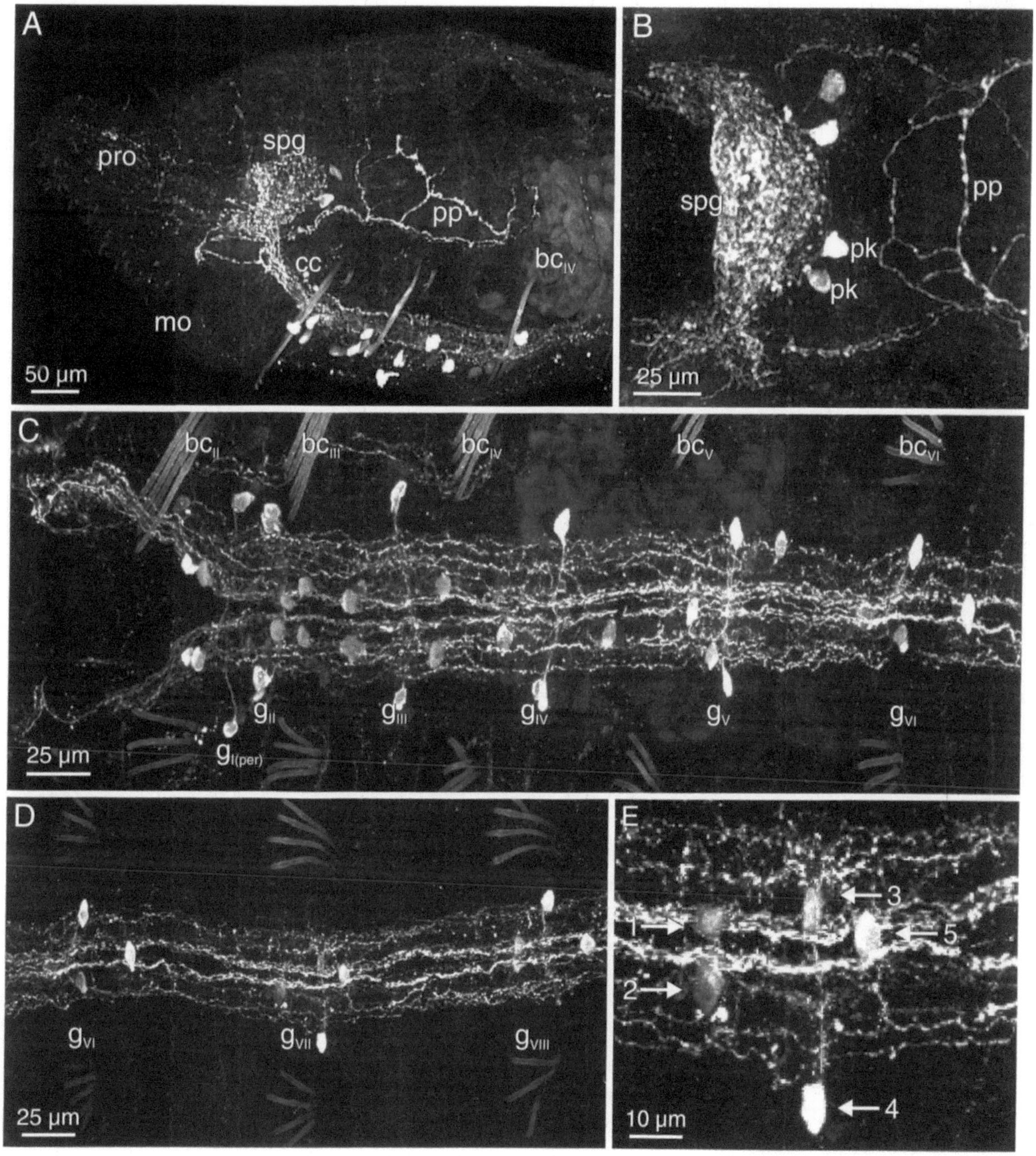

Figure 1. Nais variabilis. (**A**) Lateral view. Brain is located in segments I–II. (**B**) Dorsal view. Cerebral ganglion with two pairs of perikarya. (**C**) Ventral view. Perikarya of cephalized region. (**D**) Ventral view. Alternating pattern of serotoninergic neurons. (**E**) Ventral view. Perikarya of ganglion in non-cephalized region. **bc:** bundle of chaetae, **cc:** circumoesophageal connectives, **g:** ganglion, **mo:** mouth, **pk:** perikaryon, **pp:** pharyngeal pad, **pro:** prostomium, **spg:** supraoesophageal ganglion; roman numerals: segment numbers.

in the distribution of the perikarya also occur between corresponding neuromeres in different specimens.

In each ganglion posterior to segment V there are five labelled neurons, fewer than in the anterior neuromeres (Figures 1D, E; 4B). They are arranged in a constant pattern and are numbered **1–5**, from anterior to posterior (Figures 1E; 4B).

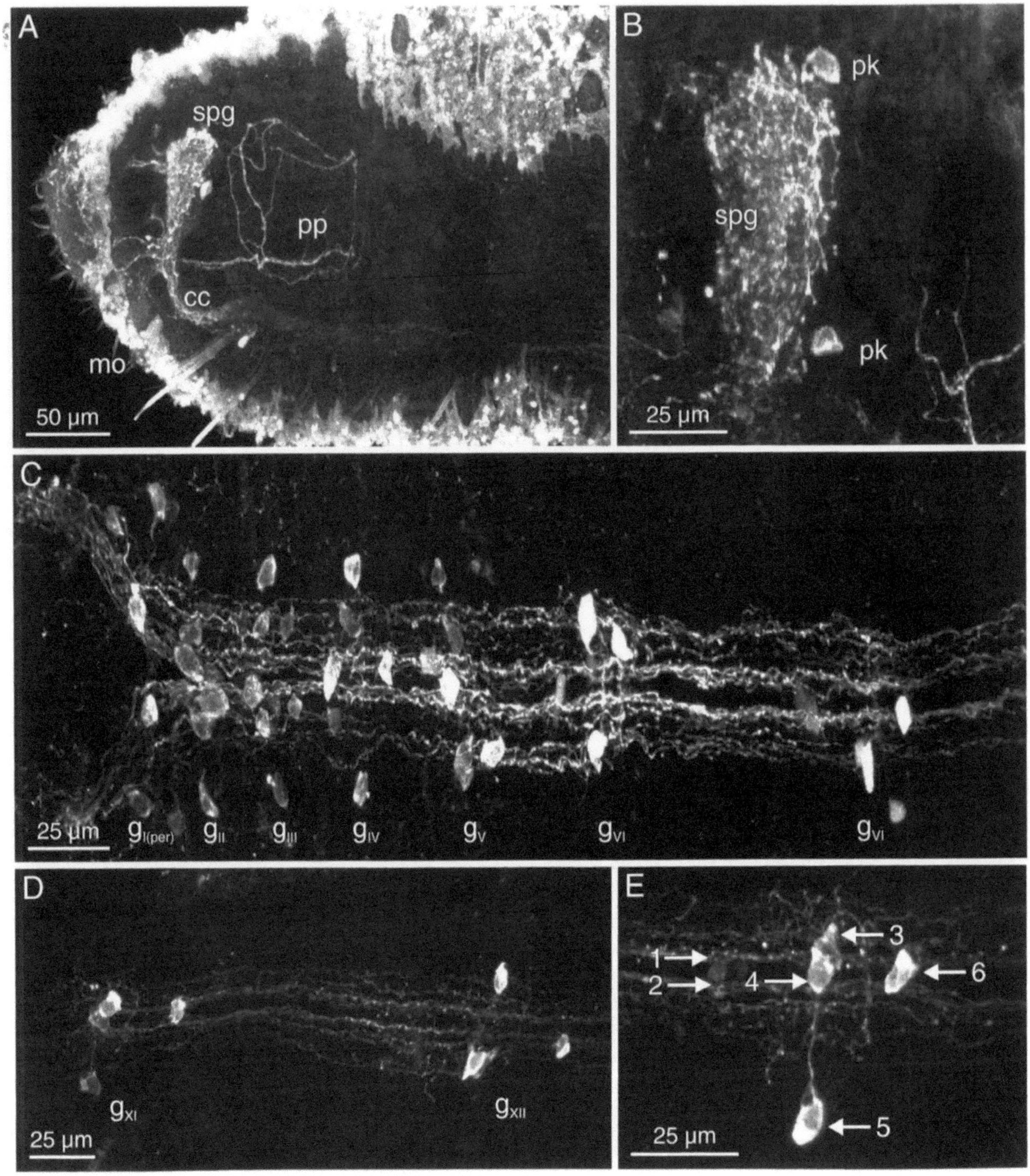

Figure 2. Slavina appendiculata. (**A**) Lateral view. Brain is located in segment I. (**B**) Dorsal view. Cerebral ganglion with one pair of perikarya. (**C**) Ventral view. Perikarya of suboesophageal ganglion. (**D**) Ventral view. Alternating pattern of serotoninergic neurons. (**E**) Ventral view. Perikarya of ganglion in non-cephalized region. **cc:** circumoesophageal connectives, **g:** ganglion, **mo:** mouth, **pk:** perikaryon, **pp:** pharyngeal pad, **spg:** supraoesophageal ganglion; roman numerals: segment numbers.

One pair of perikarya is located in the anterior region of each ganglion medially on the ventral side of the neuropile. Often these neurons are only stained weakly and in some cases, especially in posterior segments, they can not be detected at all (Figures 1E, **1** and **2**; 4B). An unpaired, usually intensely stained

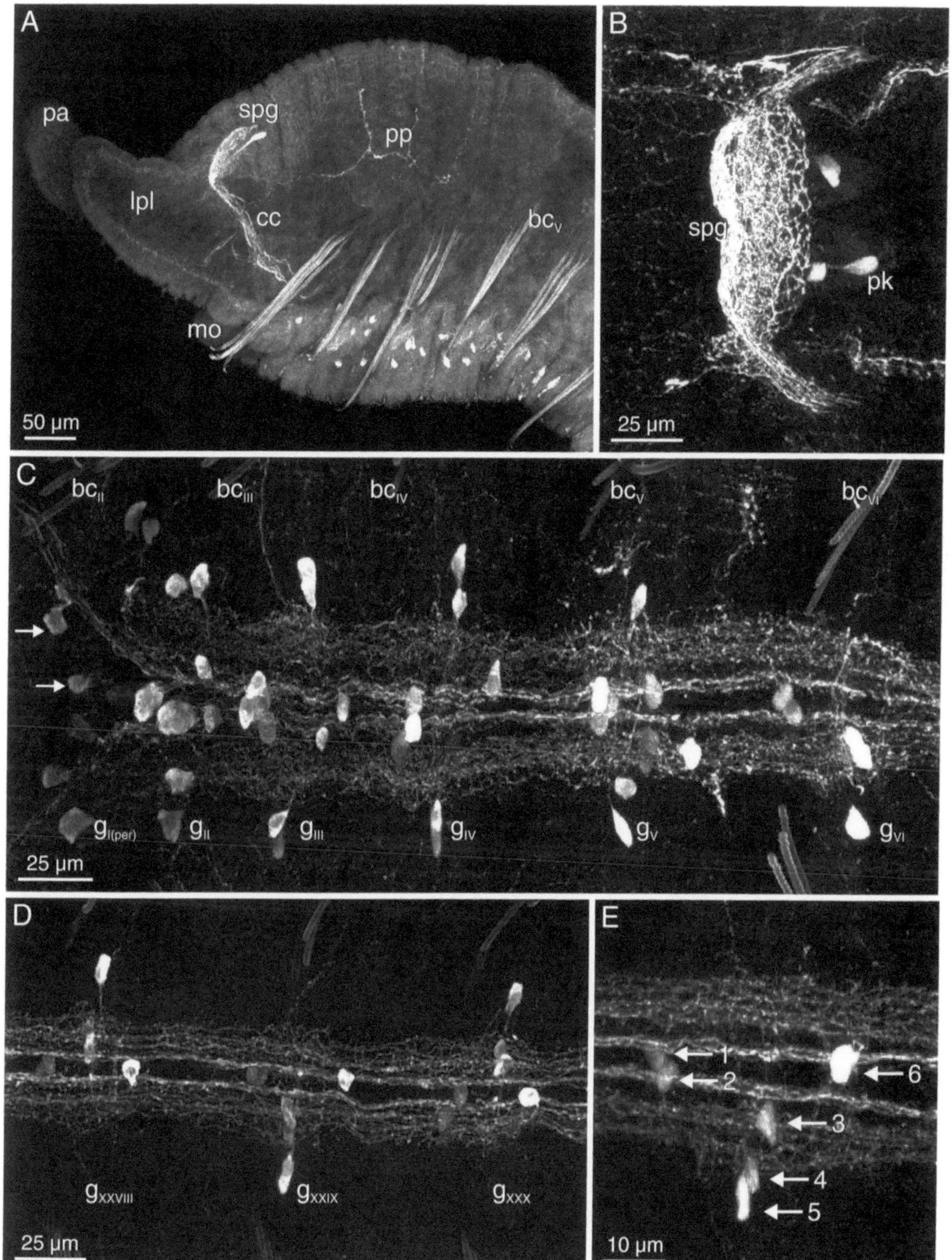

Figure 3. Stylaria lacustris. (**A**) Lateral view. Brain is located in segments I–II. (**B**) Dorsal view. Cerebral ganglion with two pairs of perikarya (only one cell visible on right side). (**C**) Ventral view. Perikarya of cephalized region. Anterior neurons of segment I on medial side of circumoesophageal connectives (arrows) (**D**) Ventral view. Alternating pattern of serotoninergic neurons. (**E**) Ventral view. Perikarya of ganglion in

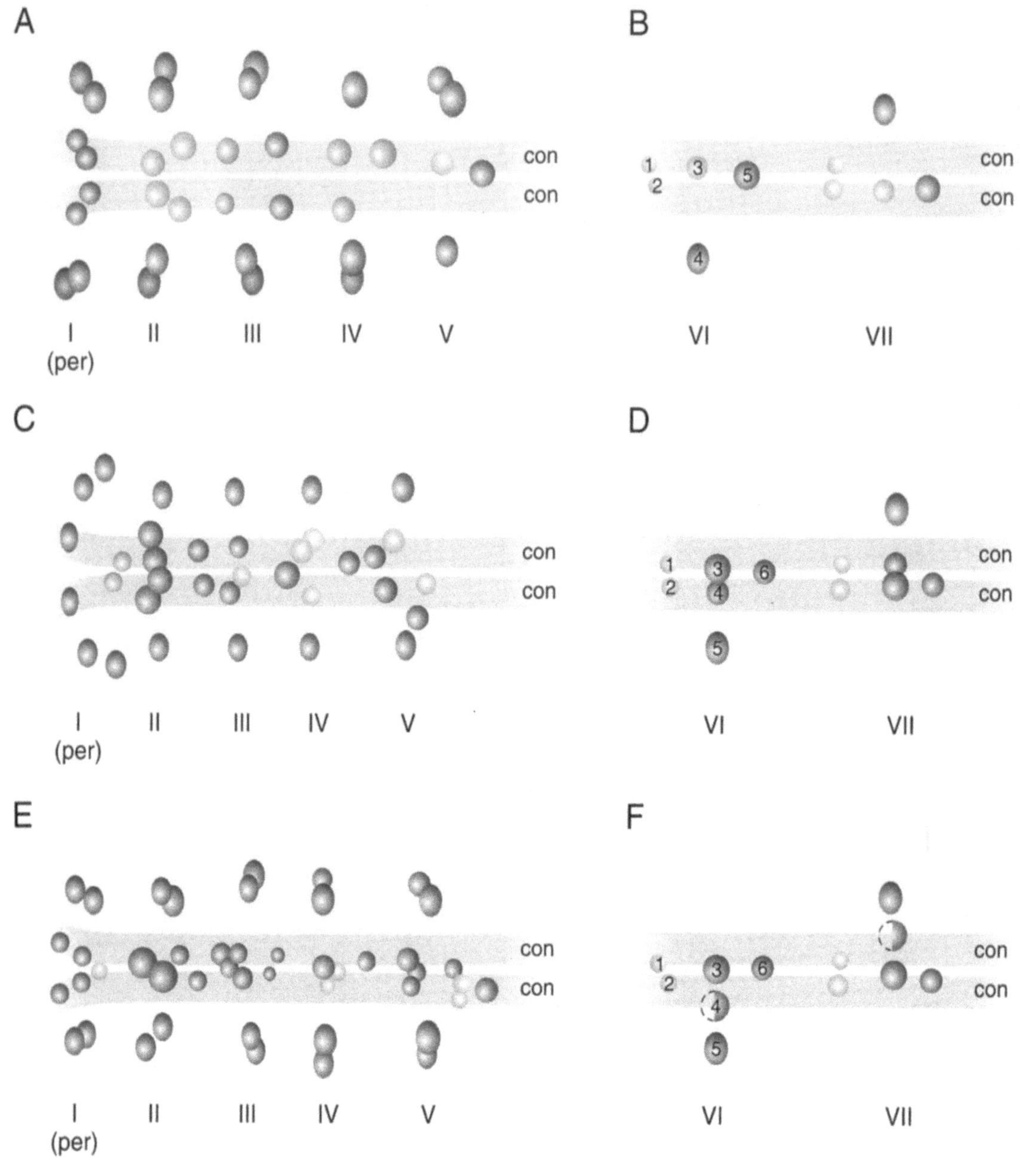

Figure 4. Patterns of serotonin-immunoreactive perikarya. Ventral view. Light gray circles correspond to neurons with weak fluorescence. **(A–B)** *Nais variabilis*, **(C–D)** *Slavina appendiculata*, **(E–F)** *Stylaria lacustris*. **(A, C, E)** Perikarya of cephalized region. **(B, D, F)** Two ganglia of non-cephalized region with alternating pattern of perikarya numbered from anterior to posterior. **con:** connective, **per:** peristomium; roman numerals: segment numbers.

neuron is located medially in the posterior section of each ganglion (Figures 1E, **5**; 4B). Two perikarya lie in the central median region of the ganglion between the neurons described above; one of them is located close to the central neuropile and often exhibits only a weak immunoreactivity (Figures 1E, **3**; 4B). The other cell is located laterally in the periphery of the ganglion and usually fluoresces intensely (Figures 1E, **4**; 4B). In adjacent ganglia, this neuron usually occurs on alternate sides, juxtaposed to the ventral midline. The neurons **3** and **5** are slightly dislocated in the same manner and thus are positioned on one or the other side of the body. This arrangement results in an alternating pattern of serotoninergic perikarya in successive segments (Figures 1D; 4B).

Apart from the neurons of the central nervous system, serotonin immunoreactive structures occur in parts of the stomatogastric nervous system. Nerve fibres extend from the anterior side of the circumoesophageal connectives, project in a curve posteriorly between these and innervate the pharyngeal pad (Figures 1A, B). These processes branch out at the ventrolateral sides of the pharyngeal pad and form two rings, which are connected by transverse fibres.

Slavina appendiculata

The neuropile of the brain is located in the peristomium (segment I) in an almost vertical position and is not, as in many other oligochaetes, slanted backwards in the dorsal region (Figure 2A). One pair of serotoninergic perikarya lies in the mid-region of the cerebral ganglion, each cell extending its process into the neuropile of the brain from the side (Figures 2A, B).

The circumoesophageal connectives lead almost vertically from the base of the brain ventrally into the suboesophageal ganglion which comprises the neuromeres of the peristomium (segment I) and segments II to V (Figures 2A, C). The connectives between these neuromeres are notably shorter than those between the following ganglia (Figure 2C). The labelled groups of neurons can be related to their corresponding segments by means of the lateral perikarya. The first neuromere evidently contains two pairs of serotonin-immunoreactive neurons, whereas each of the following neuromeres of the suboesophageal ganglion contains one pair (Figures 2C; 4C). The medial perikarya vary to a certain extent in different specimens, with the anterior neuromeres showing the greatest consistency in number and symmetrical arrangement of neurons.

The ganglia of the non-cephalized region each possess six serotonin-immunoreactive neurons, which are numbered here consecutively from anterior to posterior (**1–6** in Figures 2E; 4D). Two median cells (**1** and **2**), located in the anterior region of each ganglion, usually only exhibit weak fluorescence and cannot be detected at all in more posterior segments (Figures 2D, E). The following three perikarya are aligned in the centre of the ganglion, with the neurons **3** and **4** lying close to the ventral midline and the neuron **5** disposed laterally. The intensely labelled perikaryon **6** is located medially in the posterior region of the ganglion and sends its process into one of the longitudinal neuropile cords.

The neurons **3** to **6** are distributed on different sides of the body in adjacent ganglia, thus forming an alternating pattern of serotonin-immunoreactive perikarya (Figures 2D, E; 4D).

Some fibres of the stomatogastric nervous system display a positive serotonin-immunoreactivity as well (Figure 2A). These processes lead from the anterior side of the circumoesophageal connectives near the base of the brain posteriorly towards the pharyngeal pad, which they apparently innervate. At the anterior medial sides of the pharyngeal pad these fibres ramify and form two nerve rings, which are dorsally and caudally connected.

Stylaria lacustris

The fibre mass of the supraoesophageal ganglion is located dorsally above the mouth in the posterior region of the peristomium (segment I) (Figure 3A). The associated perikarya lie behind the brain in two more or less discrete cell packets, which project caudally into segment II. Two pairs of perikarya were labelled, the axons of which extend anteriorly and lead into the neuropile of the brain (Figure 3B). The neuropile contains numerous densely interconnected serotonin-immunoreactive fibres, especially in the anterior region of the fibre mass. Several processes extend from the lateral base of the brain into paired prostomial nerves, but can only be traced for a short distance.

The circumoesophageal connectives run posteriorly from the anterior ventral side of the brain, passing in an arc posteriorly around the foregut (Figure 3A). Especially the ventromedian regions of the circumoesophageal connectives contain numerous serotoninergic fibres.

All neuromeres of the ventral nerve cord contain serotonin-immunoreactive neurons, distributed in a more or less repetitive pattern. The arrangement of perikarya in the neuromeres of the cephalized segments is different from the distribution in the following segments (Figures 3C; 4E).

It is possible to relate the labelled groups of neurons of the suboesophageal ganglion, comprising the neuromeres of peristomium and segments II–V, to their corresponding segments by means of the lateral perikarya (Figure 3C). The anterior neuromeres are displaced longitudinally towards the posterior segments, which can be detected by the position of the segmental nerves and the chaetal bundles (Figure 3C).

From segment VI backwards the ganglia are aligned with the corresponding segments and lie adjacent to the chaetal bundles. The neuromeres of the anterior region include more serotonin-immunoreactive somata than those of the posterior region. All five neuromeres contain two pairs of lateral serotoninergic cells, but they differ in the number of median neurons (Figures 3C; 4E). Among several intensely stained median cells there are individual neurons with only weak fluorescence, making it difficult to classify them as true serotoninergic cells (Figure 3C).

The ganglia of the non-cephalized region show a pattern consisting of six serotonin-immunoreactive neurons, here numbered **1–6** from anterior to posterior (Figures 3E; 4F). Two medial cells (**1** and **2**) usually exhibiting only weak fluorescence are located in the anterior region of each ganglion. The three following neurons are aligned perpendicular to the longitudinal body axis. The cells **3** and **4** are usually located near the ventral midline, whereas the perikaryon **5** lies laterally. The neuron **4** could not be detected in all investigated specimens. The usually intensely labelled neuron **6** occurs medially in the posterior region of each ganglion. The cells **3**, **4** and **5** are located on alternate sides in adjacent ganglia. Furthermore, neuron **6** can also send its axon either into the right or the left longitudinal neuropile cord. Thus, the serotoninergic perikarya are arranged in an alternating pattern in consecutive ganglia (Figures 3D & 4F).

In this species as well, some fibres of the stomatogastric nervous system exhibit a positive serotonin-immunoreactivity (Figure 3A). Nerve processes extending from the anteriomedial sides of the circumoesophageal connectives run posteriorly above the foregut and form two interconnected rings, from which the pharyngeal pad is innervated.

Discussion

Patterns of serotonin-immunoreactive neurons

Nais variabilis and *Stylaria lacustris* each possess two pairs of serotoninergic perikarya in the supraoesophageal ganglion. In *Slavina appendiculata* and *Chaetogaster diastrophus* (Hessling, unpublished observations) on the other hand, only one pair of neurons is present. These are located dorsally behind the neuropile and extend their processes into the lateral regions of the brain.

In all investigated species, the number of serotonin-containing perikarya found in the supraoesophageal ganglion is less than in any one of the unfused ganglia of the midbody segments. There are also clearly fewer serotoninergic neurons in the brain than in the ventral-cord ganglia of other species: the polychaetes *Dinophilus gyrociliatus* and *D. taeniatus* (Windoffer, 1992) as well as the oligochaetes *Tubifex tubifex* and *Enchytraeus crypticus* (Hessling, unpublished observations). A similar situation has been found in *Lumbricus terrestris*, even though far more serotoninergic perikarya are present in the brain of this species: between 80 and 100 (Spörhase-Eichmann et al., 1987b). In Hirudinea, there are no serotonin-containing neurons at all in the supraoesophageal ganglion, in contrast to the ganglia of the ventral nerve cord (Marsden & Kerkut, 1969; Wallace, 1981). This comparatively small number of cerebral serotoninergic neurons could be attributed to the mainly sensory role of the cerebral ganglion (Stephenson, 1930; Bullock & Horridge, 1965). The coordination of locomotor behaviour, a primary function of serotoninergic neurons in various Hirudinea for example (Lent & Dickinson, 1984), is presumably concentrated in the segmental ganglia of annelids, with only a minor contribution from the brain.

In the naidid species considered here, the serotonin-immunoreactive somata of the neuromere associated with the peristomium (segment I) are located ventrally in the anterior region of the suboesophageal ganglion. Their arrangement is similar if not identical (*Nais variabilis*) to that of the perikarya in the following neuromeres. In *Lumbricus terrestris* the perikarya of the neuromere corresponding to the peristomium are located laterally along the circumoesophageal connectives (Hess, 1925; Bullock & Horridge, 1965). In leeches the cell packets of the peristomium have shifted dorsally along the circumoesophageal connectives and are considered as part of the supraoesophageal ganglionic mass (Bullock & Horridge, 1965). The neurons of these packets also differ neurochemically from those of the ventral nerve cord, in that they do not display any serotonin immunoreactivity (Sawyer, 1986). In contrast, the neuronal patterns of the peristomium in the naidids are very similar to those of the following neuromeres and differ markedly from those of the supraoesophageal ganglion.

In *Nais variabilis*, *Slavina appendiculata* and *Stylaria lacustris* the neurons in the neuromeres of the cephalized segments are arranged differently from those in further posterior segments. This anterior region comprises precisely those parts – the peristomium (segment I) and segments II–V plus the prostomium – that during the process of paratomy are formed from the posterior region of the intercalary fission zone (Stephenson, 1930). Within the neuromeres of this region, the serotoninergic neurons are arranged more or less symmetrically and are distinctly more numerous than in the neuromeres of the following segments.

This strikingly discontinuous distribution pattern of 5HT-neurons correlates with the demarcation line between these two segment regions of different origin. These results suggest that there are different developmental constraints for serotoninergic neurons that originate through the process of posterior segment proliferation, as opposed to intercalary proliferation of anterior segments. Further studies of naidid species possessing a different number of cephalized segments and a comparison with specimens having arisen through embryonic development following sexual reproduction are required to clarify whether the displayed distribution of 5HT-neurons in the cephalized region is specific of anterior segment proliferation during paratomy.

Comparison of the three species investigated here reveals some similarities in the distribution of serotonin-immunoreactive cells in the suboesophageal ganglia. Regarding the lateral neurons, two pairs are present in both *Nais variabilis* and *Stylaria lacustris*, whereas *Slavina appendiculata* has only one pair of lateral neurons in the neuromeres of segments II–V. The median cells are differently arranged in the three investigated species. The greatest inter individual consistency in number and symmetrical distribution of serotoninergic neurons is found in the anterior neuromeres of the suboesophageal ganglion.

A further point of resemblance is the alternating pattern of serotoninergic perikarya in the ventral nerve cord of the non-cephalized segments, found in all three species studied here. In *Nais variabilis* there are five immunoreactive perikarya per ganglion. In contrast, in the ganglia of *Slavina appendiculata* and *Stylaria lacustris* a repetitive pattern of six neurons is labelled. However, in *Stylaria lacustris*, one neuron, number **4**, is not visible in some of the specimens examined and often cannot be detected in all ventral ganglia of an individual. The remaining five neurons correspond in location and relative size to those of *Nais variabilis*.

Attempts to detect a similar arrangement of neurons in a *Chaetogaster* species have not yet succeeded, due to a poor immunoreactivity of serotoninergic structures. Tubificids, generally considered to be closely related to naidids (Erséus, 1990), display a clearly different and in some respects variable, distribution of serotonin-containing neurons (Hessling, unpublished results of *Tubifex tubifex*). Therefore, this alternating pattern found in *Nais variabilis*, *Slavina appendiculata* and *Stylaria lacustris* is presumably an autapomorphic character of the Naididae, or at least of a naidid subtaxon excluding the Chaetogastrinae, if such a pattern proves not to exist in the latter group.

Notable in all three species is the alternating arrangement of serotonin-immunoreactive neurons in adjacent ganglia. Several unpaired neurons containing SCP (small cardioactive peptide) have been found in *Hirudo medicinalis* displaying an alternating pattern in adjacent ganglia (Evans & Calabrese, 1989). Similar alternating arrangements of a few unpaired serotoninergic neurons are known to occur in midbody ganglia of *Lumbricus terrestris* (Spörhase-Eichmann et al., 1987a, b). In the corresponding position within anterior ganglia of this earthworm, bilaterally symmetrical neurons were found. In the same manner, the serotoninergic perikarya found singly in neuromeres posterior to segment V of the Naididae investigated here are present as more or less symmetrically arranged pairs in the anterior ganglia. It is likely that the neuronal patterns of both body regions are based upon the same, initially symmetrical distribution.

The asymmetrical, alternating arrangement of serotoninergic neurons in Lumbricidae, as well as in Naididae, could result from the existence of non-immunoreactive homologous neurons in posterior ganglia, whereas in anterior neuromeres both cells of a pair contain serotonin (Stent et al., 1992). Another explanation for the formation of asymmetrical patterns is that the neurons develop in pairs at first and that later one of these cells degenerates and dies. Leeches, for instance, possess an unpaired serotonin-containing nerve cell (pm, posteromedial neuron) in each ventral ganglion, which arises as one of a bilateral pair of neurons, of which one later dies (Stuart et al., 1987). Whereas in the case of the posteromedial neuron of leeches the member of the original pair that degenerates appears to be determined at random, this process would have to be strongly regulated in neurogenesis of Naididae in order to produce the regularly alternating pattern found here.

Position of the supraoesophageal ganglion

In the basic body plan of the Annelida, the cerebral ganglion is located in the most anterior body part, the prostomium. This is the case in polychaetes (Bullock & Horridge, 1965). In clitellate species, on the other hand, the brain lies behind the prostomium: in the Lumbriculidae it is within the peristomium (Isossimov, 1926; Stephenson, 1930), in the Lumbricidae, in segment III (Hess, 1925; Hanström, 1928) and in the genus *Pontoscolex* (Glossoscolecidae) as far back as segment IV (Hartmann-Schröder, 1993). Should the position of the brain behind the prostomium apply to all clitellates, then it could be considered an autapomorphic character of this taxon (Purschke et al., 1993), otherwise it would be an autapomorphy at least of a large clitellate subtaxon.

Functionally, the backward shift of the brain from the prostomium into following segments can be explained as a structural adaptation to a life spent burrowing through relatively firm terrestrial substrates (Westheide, 1997). A small prostomium lacking appendages was also postulated by Clark (1964) as a prerequisite for burrowing in firm substrates. In the course of this adaptation, the comparatively large prostomium of the polychaete-like ancestor of the Clitellata had to be greatly reduced in size and prostomial appendages were eliminated completely. Owing to lack of space, and in order to protect the brain, the supraoesophageal ganglion had to be shifted into the following segments (Westheide, 1997).

However, with the exception of the Chaetogastrinae (Stephenson, 1930; Sperber, 1948) the various limnic Naididae possess a comparatively large prostomium which would provide sufficient room for the brain. In *Stylaria lacustris*, the prostomium is even elongated to form a prominent unpaired appendage ('proboscis'). These species live predominantly between plants in the littoral zone of various freshwater habitats and only rarely enter loose sediments (Sperber, 1948; Wachs, 1967). They sometimes penetrate very coarse substrates and inhabit the interstitial spaces, but cannot be considered active burrowers (Learner et al., 1978). Hence, there is no obvious functional necessity for relocation of the brain from the prostomium into the following segments. Nevertheless, in *Nais variabilis* and *Stylaria lacustris*, the supraoesophageal ganglion is located in regions behind the prostomium, as in other oligochaetes, and its perikarya extend posteriorly into segment II. In *Slavina appendiculata*, with its more compressed anterior end, the brain is also clearly located behind the mouth in the region of the peristomium.

Since in many cases the nervous system is more conservative in the phylogenetic development and changes less in organization and arrangement than the structures it innervates (Lacalli, 1984), we see this concordance in the position of the brain between limnic Naididae and the terrestrial oligochaetes as a clear indication that naidids are derived from terrestrial forms and have secondarily inhabited the limnic environment.

This interpretation contradicts the view of Timm (1981), who postulated that the stem species of the Clitellata, as well as the ancestor of the Naididae and Tubificidae, was primarily limnic, constructing an evolutionary pathway from errant polychaete-like forms which proceeded from the marine to the freshwater habitat via brackish waters. The apparently polychaete-like 'tubificine' chaetation is evidently Timm's (1981) main support for this hypothesis. However, there are no extant freshwater polychaetes of this kind to corroborate such an interpretation and the homology of the tubificine forked crotchets and hair chaetae with polychaete chaetae of similar shape has not yet been proved. Omodeo (1998) assumes as well that the Clitellata have evolved in marine lagoons from where they first entered freshwater habitats and describes the Naididae in a rather basal position within the clitellate system. In particular, none of the views that postulate a limnic origin of oligochaetous clitellates considers the evolutionary origin of the cocoons produced by the clitellum. For functional reasons, these firm, tight depositories for the developing juveniles are likely adaptations to the terrestrial rather than the limnic habitat: cocoons serve as little waterproof aquaria that allow development even in dry soil. We therefore support all interpretations that consider the stem species of the Clitellata to be terrestrial and derived directly from a marine endobenthic animal or, even more plausibly, from an interstitial meiofaunal form, as Purschke's ideas (1999) may suggest.

References

Anctil, M., J.-P. De Waele & A. K. Pani, 1990. Monoamines in the nervous system of the tube-worm *Chaetopterus variopedatus* (Polychaeta): Biochemical detection and serotonin immunoreactivity. Cell. Tiss. Res. 259: 81–92.

Anlauf, A., 1989. Die Charakterisierung von Populationen des Schlammröhrenwurms *Tubifex tubifex* (Müller) mit Hilfe von enzymelektrophoretischen, populationsgenetischen und ökologischen Methoden. Doctorate thesis, Universität Köln: 127 pp.

Barlow, L. A. & J. W. Truman, 1991. Patterns of serotonin and SCP immunoreactivity during metamorphosis of the nervous system of the red abalone, *Haliotis rufescens*. J. Neurobiol. 23: 829–844.

Bullock, T. H. & G. A. Horridge, 1965. Structure and function in the nervous systems of invertebrates. Vol. 1. W. H. Freeman and Company, San Francisco. 798 pp.

Clark, R. B., 1964. Dynamics in metazoan evolution. The origin of the coelom and segments. Clarendon Press, Oxford: 313 pp.

Côté, S. L., A. Ribeiro-Da-Silva & A. C. Cuello, 1993. Current protocols for light microscopy immunocytochemistry. In A. C. Cuello (ed.), Immunohistochemistry. 2nd Ed. John Wiley & Sons Ltd, Chichester: 147–168.

Csoknya, M., I. Lengvari, L. Hiripi, K. Elekes, J. Vincze, M. Szelier & J. Hamori, 1993. Serotonin content during the regeneration of nervous system in earthworm (*Lumbricus terrestris* L., Oligochaeta). Acta Biol. Szeg. 39: 39–49.

Denk, H., 1989. Immunohistochemische Färbemethoden. In P. Böck (ed.), Romeis Mikroskopische Technik. 17th Edn. Urban und Schwarzberg, München: 252–261.

Ehinger, B. & H. E. Myhrberg, 1971. Neuronal localization of dopamine, noradrenaline and 5-HT in the central and peripheral nervous system of *Lumbricus terrestris* (L.). Histochemistry 28: 265–275.

Elofsson, R. & M. Carlberg, 1989. Gland cells in the tentacles of the jellyfish *Cyanea lamarcki* reactive with an antibody against 5-hydroxytryptamine. Cell. Tiss. Res. 255: 419–422.

Erséus, C., 1990. Cladistic analysis of the subfamilies within the Tubificidae (Oligochaeta). Zool. Scr. 19: 57–63.

Evans, B. D. & R. L. Calabrese, 1989. Small cardioactive peptid-like immunoreactivity and its colocalization with FMRFamide-like immunoreactivity in the central nervous system of the leech *Hirudo medicinalis*. Cell. Tiss. Res. 257: 187–199.

Fine, A., 1993. Immunocytochemical applications of confocal microscopy. In A. C. Cuello (ed.), Immunohistochemistry. 2nd Edn. John Wiley & Sons Ltd, Chichester: 394–407.

Hanström, B., 1928. Vergleichende Anatomie des Nervensystems der Wirbellosen Tiere. Springer Verlag, Berlin. 628 pp.

Hartmann-Schröder, G., 1993. Annelida. In H.-E. Gruner (ed.), Lehrbuch der Speziellen Zoologie, Band I: Wirbellose Tiere - 3. Teil: Mollusca, Sipunculida, Echiurida, Annelida, Onychophora, Tardigrada, Pentastomida. Gustav Fischer Verlag, Stuttgart. 276–469.

Hess, W. N., 1925. Nervous system of the earthworm, *Lumbricus terrestris*. J. Morph. 40: 235–259.

Hessling, R. & W. Westheide, 1999. CLSM analysis of development and structure of the central nervous system of *Enchytraeus crypticus* ("Oligochaeta", Enchytraeidae). Zoomorphology 119: 37–47.

Isossimov, V. V., 1926. Zur Anatomie des Nervensystems der Lumbriculiden. Zool. Jb. Anat. 48: 365–404.

Lacalli, T. C., 1984. Structure and organization of the nervous system in the trochophore larva of *Spirobranchus*. Phil. Trans. r. Soc. London 306: 79–135.

Learner, M. A., G. Lochhead & B. D. Hughes, 1978. A review of the biology of British Naididae (Oligochaeta) with emphasis on the lotic environment. Freshw. Biol. 8: 357–375.

Lent, C. M. & M. H. Dickinson, 1984. Serotonin integrates the feeding behavior of the medicinal leech. J. comp. Physiol. 154: 457–471.

Marsden, C. A. & G. A. Kerkut, 1969. Fluorescence microscopy of the 5-HT and catecholamine containing cells in the central nervous system of the leech *Hirudo medicinalis*. Comp. Biochem. Physiol. 31: 851–862.

Omodeo, P., 1998. History of Clitellata. Ital. J. Zool. 65: 51–73.

Purschke, G., 1999. Terrestrial polychaetes models for the evolution of the Clitellata (Annelida)? Hydrobiologia 406: 87–99.

Purschke, G., W. Westheide, D. Rohde & R. O. Brinkhurst, 1993. Morphological reinvestigation and phylogenetic relationship of *Acanthobdella peledina* (Annelida, Clitellata). Zoomorphology 113: 91–101.

Sawyer, R. T., 1986. Leech biology and behavior I. Anatomy, physiology and behavior. Clarendon Press, Oxford: 1–418.

Sperber, C., 1948. A taxonomical study of the Naididae. Zool. Bidr. Upps. 28: 1–292.

Spörhase-Eichmann, U., H. Gras & F.-W. Schürmann, 1987a. Patterns of serotonin-immunoreactive neurons in the central nervous system of the earthworm *Lumbricus terrestris* L. - I. Ganglia of the ventral nerve cord. Cell. Tiss. Res. 249: 601–614.

Spörhase-Eichmann, U., H. Gras & F.-W. Schürmann, 1987b. Patterns of serotonin-immunoreactive neurons in the central nervous system of the earthworm *Lumbricus terrestris* L. - II. Rostral and caudal ganglia. Cell. Tiss. Res. 249: 625–632.

Steinbusch, H. W. M. & R. Nieuwenhuys, 1981. Localization of serotonin-like immunoreactivity in the central nervous system and pituitary of the rat, with special references to the innervation of the hypothalamus. Adv. Exp. Med. Biol. 133: 7–35.

Steinbusch, H. W. M., A. A. J. Verhofstad & H. W. J. Joosten, 1978. Localisation of serotonin in the central nervous system by immunohistochemistry: a description of a specific and sensitive technique and some applications. Neuroscience 3: 811–819.

Stent, G. S., W. B. Kristan, S. A. Torrence, K. A. Freach & D. A. Weisblat, 1992. Development of the leech nervous system. Int. Rev. Neurobiol. 33: 109–193.

Stephenson, J., 1930. The Oligochaeta. Clarendon Press, Oxford: 978 pp.

Stuart, D. K., S. S. Blair & D. A. Weisblat, 1987. Cell lineage, cell death, and the developmental origin of identified serotonin- and dopamine-containing neurons in the leech. J. Neurosci. 7: 1107–1122.

Timm, T., 1981. On the origin and evolution of aquatic Oligochaeta. Eesti NSV Teaduste Akadeemia Toimetised 30: 174–181.

Wachs, B., 1967. Die Oligochaeten-Fauna der Fließgewaesser unter besonderer Berücksichtigung der Beziehungen zwischen Tubificiden-Besiedlung und dem Substrat. Arch. Hydrobiol. 63: 310–386.

Wallace, B. G., 1981. Neurotransmitter chemistry. In K. J. Muller, J. G. Nicholls & G. S. Stent (eds.), Neurobiology of the leech. Cold Spring Harbor Laboratory, Cold Spring Harbor: 147–172.

Westheide, W., 1997. The direction of evolution within the Polychaeta. J. Nat. Hist. 31: 1–15.

Windoffer, R., 1992. Immunohistochemische und elektronenmikroskopische Untersuchungen am Nervensystem zweier *Dinophilus*-Arten mit unterschiedlichem Lebenszyklus. Doctorate thesis, Universität Osnabrück. 187 pp.

Hydrobiologia **406**: 235–241, 1999.
B. M. Healy, T. B. Reynoldson & K. A. Coates (eds), Aquatic Oligochaetes.

Tubificoides benedii (Tubificidae, Oligochaeta) – a pioneer in hypoxic and sulfidic environments. An overview of adaptive pathways

O. Giere, J.-H. Preusse & N. Dubilier
Zoological Institute and Zoological Museum, University of Hamburg, Martin-Luther-King-Platz 3, D-20146 Hamburg, Germany

Key words: marine oligochaeta, hypoxia, sulfide, respiration, detoxification, tidal flat

Abstract

Eutrophic tidal flats and polluted coastal sites are the predominant habitat of the marine oligochaete *Tubificoides benedii.* The worms live in dense populations in these stressed habitats which are often characterized by high levels of hydrogen sulfide. This indicates that they have a high capacity to tolerate anoxic (and sulfidic) conditions. Respiration rates of *T. benedii* measured at various oxygen concentrations showed that aerobic respiration is maintained even at very low oxygen concentrations. This ability is combined with a high regulatory capacity of oxygen uptake. Addition of sulfide considerably reduced this capacity of maintaining aerobic metabolic pathways at low oxygen concentrations. The present work in relation to earlier physiological and structural studies (Giere et al., 1988; Dubilier et al., 1994, 1995, 1997) suggests adaptive strategies that make *T. benedii* one of the most successful inhabitants of ecologically stressed, sulfidic benthic environments. This is corroborated by comparison with other typical 'sulfide annelids' such as the polychaetes *Capitella capitata* and *Arenicola marina.*

Introduction

The ubiquitous marine oligochaete *T. benedii* is well known for its occurrence in eutrophic, often polluted, coastal sediments enriched with hydrogen sulfide (Birtwell & Arthur, 1980; Giere & Pfannkuche, 1982). In regions suffering from oxygen depletion and excessive hydrogen sulfide this worm is regularly present although it lives buried (often 10 cm deep) in the fine, anoxic sediment without contact with the surface, i.e. lacking irrigated burrows or tubes typical for limnic tubificids. The direct exposure to quite high hydrogen sulfide concentrations (based on field data from the same area by Thiermann et al. (1996)) led us to investigate the respiratory ecophysiology of the species and to relate the data to our earlier studies (Giere & Rhode, 1987; Giere et al., 1988). Possible pathways are described adapting this annelid to a life in sulfidic benthic environments.

Material and methods

The worms used for the respiration experiments were sampled in a North Sea tidal flat of the Island of Sylt (Germany). For comparison purposes, they were collected from the same population which had been used for our earlier physiological and morphological studies (Dubilier et al., 1994, 1997). The animals were maintained in small culture jars at 15 °C in artificial seawater. For each experimental run, batches of 25 worms were carefully sorted and allowed to defaecate in culture vessels. Each batch of worms was weighed on a microbalance (Sartorius) after removal of excess water with tissue paper, and then placed in the measuring chamber of a flow-through respirometer (Figure 1) with two sulfide-insensitive oxygen electrodes (Orbisphere, Geneva, Switzerland) installed before and behind the measuring chamber. The measuring unit was constructed by Cyclobios, Innsbruck, Austria. Details of the apparatus are given in Gamenick (1996).

In the incubation medium (Figure 1) different concentrations of dissolved oxygen were established by bubbling various amounts of nitrogen through oxygen-saturated seawater. After calibration of the electrodes and 1 h to allow adaptation, the experiment was started, each experimental run lasting for 1 h. Within this time period, ten readings of the oxygen content in

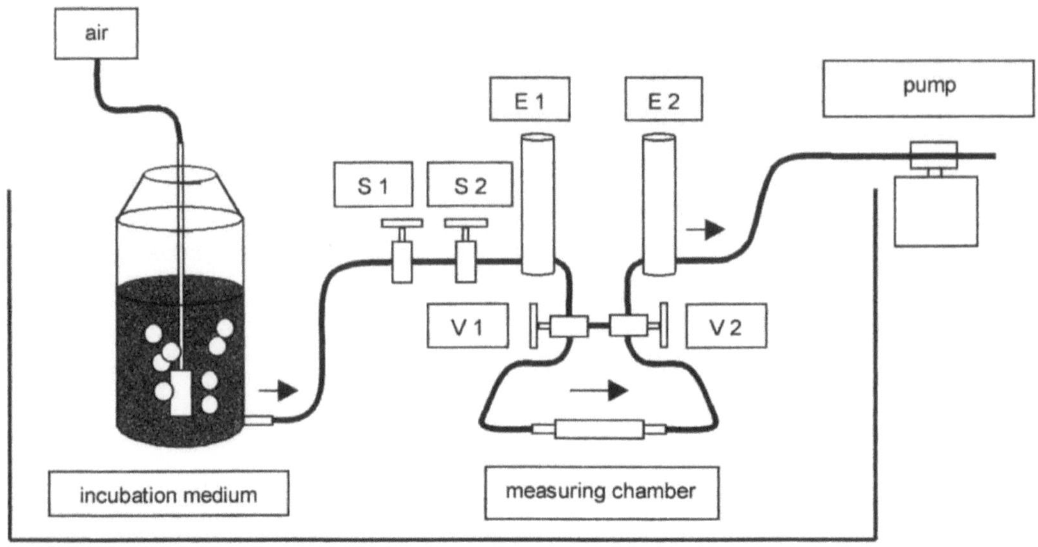

Figure 1. Schematic diagram of the flow-through respiration apparatus. E 1 and E 2: oxygen electrodes at influx and efflux side of measuring chamber respectively. S 1 and S 2, V 1 and V2: valves.

the measuring chamber were carried out at six-minute intervals. Following the measurement series, we ran a blank to account for bacterial respiration. The net respiration rate was calculated as μmol oxygen consumption per g fresh weight (Fw) per minute. Any addition of sediment was avoided since in previous test runs sediment particles had considerably increased the variances and inconsistencies in the respiration recordings. The methodology of our metabolic studies under low-oxygen-conditions with and without addition of hydrogen sulfide is described in Dubilier et al. (1994). Details of processing the specimens for ultrastructural work are given in Giere et al. (1988).

Results and discussion

The standard respiration rate of *T. benedii*, i.e. the oxygen consumption at 'normoxic' conditions (100% oxygen saturation = 21.05 kPa) was 63.1 μmol $O_2 \cdot g^{-1} Fw \cdot min^{-1}$ (n = 10, SD $\pm$7) (Figure 2, initial value).

When the oxygen concentration decreased from saturation (100%) to 80%, we observed a marked drop of the respiration rate by almost 40% (Figure 2). Although this substantial oxygen consumption at relatively high ambient oxygen concentrations is in contrast to recordings for limnetic tubificids (Berg et al., 1962), it does not seem to be caused by an experimental artifact considering the good agreement in seven runs with ten readings each. With decreasing ambient oxygen concentration, in the range between 80% and 10% saturation, an almost stable respiration level was maintained. This indicates a highly developed metabolic regulating capacity of the species. It was only at severe hypoxia, below 10% saturation, that this regulatory ability collapsed and respiration values dropped linearly. However, even at 2% oxygen a minimal, but measurable, oxygen consumption was recorded.

The curve shown in Figure 2 is that of an effective oxyregulator (Prosser & Brown, 1961; Grieshaber et al., 1994). Generally, regulatory capability is considered an important prerequisite for the ecological success of species living under the harsh, unstable conditions of tidal flats. The adaptive plateau extends to extreme hypoxia, that means, the critical PO_2 (P_c), where regulation collapses, is very low, and the range in which hypoxia is compensated for by physiological mechanisms, is wide.

Except for the initial drop (see above), the pattern of this curve is comparable with consumption curves measured by Berg et al. (1962) for *Tubifex tubifex* and other freshwater tubificids from dysoxic lake bottoms. Our data show a pronounced regulatory plateau down to a critical threshold at low oxygen concentrations beyond which a rapid drop is measured. The differences in detail between our results and those of Berg et al. (1962) are likely aggravated by the different experimental methods used for assessing oxygen

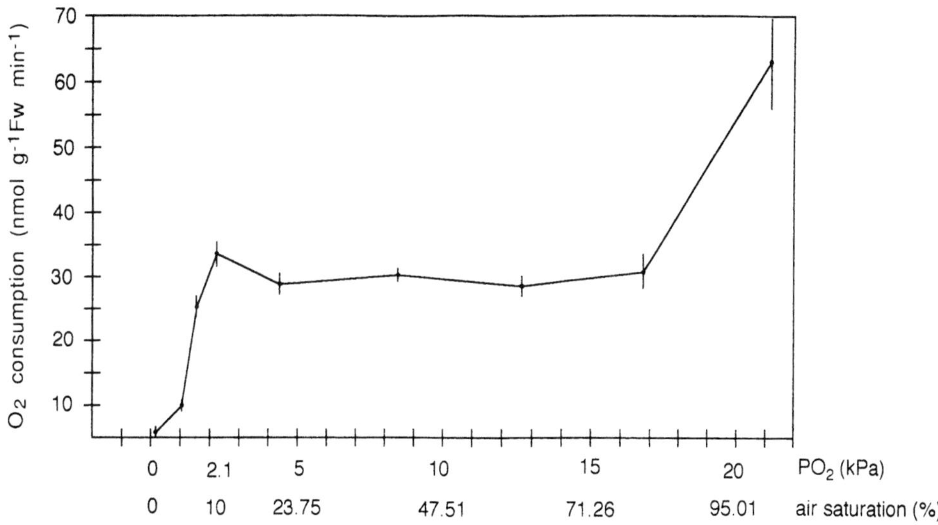

Figure 2. Curve of oxygen consumption of *Tubificoides benedii* at normoxic, oligoxic and hypoxic conditions. Each point represents the median value of 10 experimental recordings, vertical lines indicate standard deviation.

consumption. The critical threshold for *T. tubifex* of 26.7 μmol g^{-1}Fw min^{-1}, obtained after transformation of the figure given by Berg et al. (1962), is in the range of that observed for *T. benedii*, but the details of this comparison must remain questionable. The same is true for the 'critical oxygen levels' at 5–6% oxygen concentration given by Birtwell & Arthur (1980). Interestingly enough, in the material from the poikilohaline Thames estuary, the critical threshold of *Peloscolex benedeni* (= *Tubificoides benedii*) was clearly higher than that of three other common tubificid species tested. Again however, lack of information on the experimental design renders direct comparisons problematic.

On the other hand, the critical PO_2-value recorded in the present study for *T. benedii* corresponds well with physiological measurements on the same species (Dubilier et al., 1994). Metabolically, the P_c is the threshold at which the animal switches from aerobic to anaerobic pathways of energy production. In *T. benedii*, as in many annelids, the end product of this anaerobic metabolism, which takes place in the mitochondria, is succinate. Measurement of succinate is thus a valuable indicator of anaerobiosis (Grieshaber et al., 1994). In specimens of *T. benedii*, incubated under different oxygen concentrations, Dubilier et al. (1994) recorded a significant increase in succinate production at 3.8% air saturation (=10 μm oxygen = 0.75 kPa O_2), whereas at 7.6% saturation (=20 μm oxygen) the decrease was still insignificant. Thus, the onset of anaerobic metabolism occurs somewhere between 7.6% and 3.8% air saturation. This metabolic signal is close to the ecophysiological signal, the onset of the non-regulative drop of the respiratory curve, which occurred somewhere between 10% and 7.5% air saturation. This correspondance between biochemical and ecophysiological data resulting from anaerobiosis confirms the significance of these independent data sets.

The extended plateau of osmoregulation conforms also with survival experiments on *T. benedii*, kept at 13% air saturation (= 2.5 $kPaO_2$), i.e. above the critical PO_2 recorded here. Individual survival was unaffected with an LT 50 value being >1 month.

In order to compare the respiratory capacity of *Tubificoides benedii* with that of other tubificids and of co-occurring marine annelids (Table 1), the data had to be compensated for the temperature effect. For the tubificids, this temperature correction was based on the data of Birtwell (1972).

The common limnic tubificids *Tubifex tubifex* and *Limnodrilus hoffmeisteri* have been studied by Schneider (1994) with methods similar to those described here. The pertinent values in Table 1 show that these worms of comparable size had a considerably higher standard respiration rate (about 6 times higher, Table 1). In these two species, the decrease in oxygen uptake at temperatures falling from 20 °C to 15 °C is considerable, while for *T. benedii* it is only moderate (Birtwell, 1972). The limnetic worms also reached their critical PO_2 (temperature-corrected) at a threshold of about 4 kPa (Table 1). That is twice

Table 1. Comparison of standard respiration rate and critical oxygen concentration (P_c) in various annelid species. Values for *Limnodrilus* and *Tubifex* calculated from Schneider (1994), for *Capitella* from Gamenick (1996), for *Arenicola* from Hauschild & Grieshaber (1997)

Species	Habitat	Respiration type	Standard respiration (nmol O_2 g^{-1} FW min^{-1})	Critical PO_2 (kPa)
Tubificoides benedii	marine	oxyregulator	63.1 (15° C)	≈ 2
Limnodrilus hoffmeisteri	limnetic	oxyregulator	357 (15° C)*	≈ 4
Tubifex tubifex	limnetic	oxyregulator	218 (15° C)*	≈ 4
Capitella capitata, type L	marine	oxyregulator	96.3 (16° C)	≈ 17
Arenicola marina	marine	oxyconformer	32.7 (15° C)	≈ 16

*values corrected for temperature acc. to Birtwell (1972)

the oxygen concentration found critical for *T. benedii*, confirming the extraordinary metabolic capacity of this marine oligochaete. Data for limnetic tubificids derived from recordings by Berg et al. (1992) have been omitted here since they were obtained with different methods.

The common polychaete *Capitella capitata* is an annelid of comparable size and ecology that co-occurs with *T. benedii.* However, *C. capitata* is more known as a character form of hypoxic/sulfidic sediments. Its standard respiration rate is slightly higher than that of the oligochaete (Table 1; Gamenick, 1996). This larger species has been identified as an oxyregulator with a very high critical PO_2 ranging at 17 kPa.

Another species suitable for comparison is the oxyconformer *Arenicola marina* (Schöttler et al., 1984; Hauschild & Grieshaber, 1997). This marine polychaete co-occurs with *T. benedii* in the tidal flats and its irrigated burrows reach deeply into the sulfidic layer. In this large worm, the oxygen consumption at normoxic conditions is only half that of *T. benedii*, which probably refers to the considerably bigger body size (Toulmond, 1975). The critical PO_2 for *A. marina*, although difficult to assess as it is an oxyconforming species lacking a regulative capacity, has been reported to be at about 16 kPa (Hauschild & Grieshaber, 1997).

In marine sediments, severe hypoxia is mostly associated with free sulfide. What is the effect of this toxic substance on the respiratory behaviour ? At 20 μm oxygen, a concentration normally still in the range of aerobic respiration, even low additions of sulfide (30 μm) caused a significant increase of succinate production indicating the onset of anaerobic metabolism. This protracted anaerobiosis under low oxic plus sulfidic conditions is a common phenomenon of environmental anaerobiosis in benthic animals and has been termed 'sulfide dependent anaerobiosis' (Grieshaber, 1992). For comparison, if normoxic conditions were combined with sulfide (admittedly an artificial laboratory situation), the metabolism of *T. benedii* remained fully aerobic even at sulfide values of 300 μm (see below).

It was mentioned above that even under anaerobic conditions, metabolic energy can be produced in the mitochondria of the cells. In many animals incubated under hypoxic or hypoxic plus sulfidic conditions, the number and shape of mitochondria and their cristae becomes aberrant compared to tissues under oxic conditions (Maguire & Boaden, 1975; Duffy & Tyler, 1984; Menon et al., 1995). In *T. benedii* we conducted a detailed study of the mitochondrial distribution and structural variation under different incubations (Giere et al., 1988; Dubilier et al., 1997). In the epidermal layer which is most directly exposed to the ambient low-oxygen and sulfide stress, the mitochondria are characterized by their large number and aberrant shape: often very large (often 2 μm in length) and branched, they have only few cristae (see Figure 1A in Dubilier et al., 1997; Figure 13 in Giere et al., 1988). However, extensive comparative incubations of worms in normoxic, severely hypoxic and sulfidic seawater, with subsequent detailed stereological/mathematical evaluation, revealed no statistically significant changes of mitochondrial shape and fine structure. Hence, it is doubtful that dense aggregations of peripheral mitochondria and aberrant shape can be a priori considered as adaptive features for counteracting the toxic impact of sulfide (Dubilier et al., 1997).

There are some other structural peculiarities in *T. benedii* that have been related to its extreme sulfide tolerance (Giere et al., 1988). In contrast to most oligochaetes, this species lacks the characteristic epicuticular projections on the surface and, instead, has large leaf-shaped papillae in a special arrangement. Fixed on these papillae is a sticky mucus containing rich amounts of iron ions which, under sulfidic conditions, become reduced to form iron sulfide blackening the worm conspicuously. Reported also from a few other oligochaetes, these black surface areas are regularly shed. Our electron-microscopical studies on *T. benedii* (Giere et al., 1988) showed that the mucus plus the epicuticular layer of papillae become detached in this moulting process.

Regarding the extreme tolerance of *T. benedii* to anoxia and hydrogen sulfide, it had been argued that this iron-rich mucus production would serve as a means of detoxification of the highly poisonous sulfide. The sulfide would be trapped as black, insoluble iron sulfide precipitates in the mucus, which is then disposed of together with the papillae.

Similar assumptions of a mucus-mediated sulfide disposal exist for the priapulid *Halicryptus spinulosus* (Oeschger & Vetter, 1992) and for the echiurid *Urechis caupo* (Menon & Arp, 1993). However, detailed calculations considering diffusion coefficients of H_2S into the slender body of *T. benedii* (Dubilier et al., 1995) did not support the assumption of this process as an effective mode of sulfide detoxification. The rapid diffusion of the small H_2S molecules across the anterior and posteriormost body regions which are not covered by mucus-bearing papillae would suffice to flush, after a few minutes exposure, the whole body with the toxic sulfide. This would occur at very moderate ambient H_2S concentrations (25 μm) even if any influx through the papillary part of the body were blocked. Detoxification by mucus has probably only a temporary effect somewhat retarding the influx of hydrogen sulfide, but it may be one component among others.

Tubificoides benedii does not have a mutualistic symbiosis with 'sulfur bacteria' for the metabolisation and removal of sulfur compounds. There remains special metabolic pathways which are possible explanations for the insensitivity to hydrogen sulfide. The previously described reaction of *T. benedii*, switching readily to an anaerobic metabolism, was valid for dysoxic and mild sulfidic conditions (from 30 μm onwards). The situation is different, however, under experimental normoxic conditions plus sulfide (see above). Then *T. benedii* could maintain an aerobic metabolism even at fairly high sulfide concentrations (300 μm, Dubilier et al., 1994). How can the species, under highly sulfidic conditions, maintain the processes in the mitochondrial respiration chain? How can it overcome the blocking of all mitochondrial activity through hydrogen sulfide? Is it able, at least to some degree, to synthesize ATP by oxidation of sulfide into thiosulfate, as has been shown for *Arenicola marina* and some other 'sulfide animals' (Völkel & Grieshaber, 1997; Grieshaber & Völkel, 1998)? Does this phenomenon indicate presence of an alternative sulfide-insensitive cytochrome c - oxidase in the mitochondria of *T. benedii*, as has been shown for *Arenicola marina* by Völkel & Grieshaber (1997)? The study by Degn & Kristensen (1981) on a limnetic tubificid (*Tubifex* sp.) did not support this metabolic pathway. For *T. benedii*, at present, these questions must remain unanswered and await further studies.

In its normal environment, *T. benedii* inhabits the tidal flats where eutrophic fine sands and muds are covered by an oxic layer (=containing free dissolved oxygen) that is only a few mm, perhaps half a cm thick (Figure 3). Deeper down, the sediments are anoxic with an increasing amount of free hydrogen sulfide (often around 150 μm, Thiermann et al., 1996). Thus, *T. benedii* burrowing and feeding several cm down in the mud, is usually exposed to these unfavourable respiratory conditions which are compensated for by reduced predation pressure and a rich supply of detritus and bacteria.

For limnetic tubificid species common in anoxic/sulfidic sediments such as *Tubifex tubifex* and *Limnodrilus hoffmeisteri*, it has been argued that their tail waving behaviour in the water above the sediment, in combination with an intestinal respiration, provides a sufficient supply with oxygen (Alsterberg, 1922). This respiratory behaviour and intestinal respiration has also been suggested for *T. benedii*, without further investigation, to explain its occurrence in highly sulfidic marine environments (Dahl, 1960). However, in contrast to singular statements derived from experiments, our field observations could not corroborate that specimens of *T. benedii*, like their limnetic counterparts mentioned above, demonstrate such tail movement behaviour in the water above the sediment surface. Nor do their burrows have direct access to the oxic surface.

Tubificoides benedii apparently has an effective circulation of blood with a high oxygen affinity. Hence, it can probably maintain for some time its internal oxygen concentration sufficiently high to stay in the range of regulative metabolic compensation, i.e. beyond the critical P_c value, especially at low tide and in summer periods when conditions in the sediment become frequently dysoxic and mildly sulfidic. This situation corresponds to an experimental

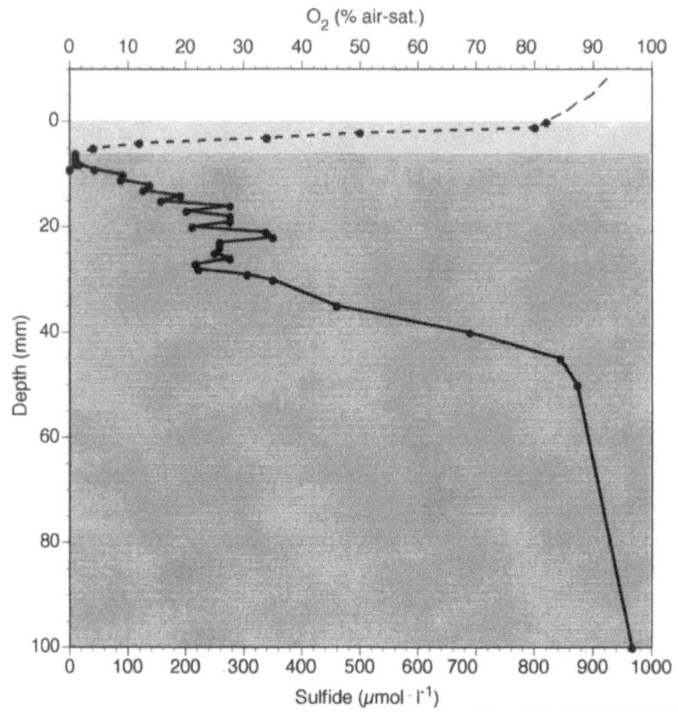

Figure 3. Typical oxic/sulfidic gradients in an experimental tidal flat sediment from the study area.

model in which the transition from an aerobic to an anaerobic metabolism was easily induced (see above). Sulfide precipitation at the external mucus layer can at best retard for a short time this switch to a less energy-yielding anaerobic pathway with accumulation of succinate (Dubilier et al., 1994).

Our tolerance experiments proved that even longer lasting sulfide stress (some days) does not jeopardize populations of *T. benedii*, since in coastal flats the length of the 'oxygen emergency periods' is limited. It will often only last for a tide cycle or, at worst, for some days. In LT 50-tests under hypoxia (= 'anoxia' produced by nitrogen bubbling), this species was superior to other tubificids tested (Birtwell & Arthur, 1980).

Tidal flats are environments with highly variable oxygen/sulfide microgradients and with numerous bioturbative oxic microniches (Fenchel, 1996). Here, the slightest intermittent oxygen supply would allow the animal to switch back to aerobic metabolism. The respiration curve at various hypoxic conditions (Figure 2) showed that in *T. benedii* this reversal could be released at oxygen concentrations far below the threshold required for most other species. Hence, under the prevailing field conditions, *T. benedii* can solve the respiratory problem by a combination of adaptive pathways. We can interpret the extreme ecological potential of this widely distributed marine oligochaete as a result of a highly specialized physiology combined with supplementing behavioural and structural adaptations. This interaction makes the worm a character species for stressed and sulfidic habitats.

References

Alsterberg, G., 1922. Die respiratorischen Mechanismen der Tubificiden. Lunds Univ. Årskr., N.F. Avd. 2, 18: 1–176.

Berg, K., P. M. Jónasson & K. W. Ockelmann, 1962. The respiration of some animals from the profundal zone of a lake. Hydrobiologia 19: 1–39.

Birtwell, I. K., 1972. Ecophysiological aspects of tubificids in the Thames Estuary. Ph.D. thesis, Univ. of London.

Birtwell, I. K. & D. R. Arthur, 1980. The ecology of tubificids in the Thames Estuary with particular reference to *Tubifex costatus* (Claparède). In R. O. Brinkhurst & D. G. Cook (eds), Aquatic Oligochaete Biology. Proceedings of the First International Symposium on Aquatic Oligochaete Biology. Sidney, B.C., Plenum Press, New York: 331–381.

Dahl, I. O., 1960. The oligochaete fauna of 3 Danish brackish water areas (taxonomic and biological observations). Meddr Kommn Danm. Fisk.- og Havunders. 2.: 1–20.

Degn, H. & B. Kristensen, 1981. Low sensitivity of *Tubifex* sp. respiration to hydrogen sulfide and other inhibitors. Comp. Biochem. Physiol. 69 B: 809–817.

Dubilier, N., O. Giere & M. K. Grieshaber, 1994. Concomitant effects of sulfide and hypoxia on the aerobic metabolism of the marine oligochaete *Tubificoides benedii*. J. exp. Zool. 269: 287–297.

Dubilier, N., O. Giere & M. K. Grieshaber, 1995. Morphological and ecophysiological adaptations of the marine oligochaete *Tubificoides benedii* to sulfidic environments. Am. Zool. 35: 163–173.

Dubilier, N., R. Windoffer, M. K. Grieshaber & O. Giere, 1997. Ultrastructure and anaerobic metabolism of mitochondria in the marine oligochaete *Tubificoides benedii*: effects of hypoxia and sulfide. Mar. Biol. 127: 637–645.

Duffy, J. E. & S. Tyler, 1984. Quantitative differences in mitochondrial ultrastructure of a thiobiotic and oxybiotic turbellarian. Mar. Biol. 83: 95–102.

Fenchel, T., 1996. Worm burrows and oxic microniches in marine sediments. 1. Spatial and temporal scales. Mar. Biol. 127: 289–295.

Gamenick, I., 1996. Ökophysiologische und enzymatische Differenzierung verschiedener Geschwisterarten des *Capitella capitata*-Komplexes (Annelida, Polychaeta). Shaker, Aachen: 112 pp.

Giere, O. & B. Rhode, 1987. Anatomy and ultrastructure of the marine oligochaete *Tubificoides benedii* (Tubificidae), with emphasis on its epidermis-cuticle complex. Hydrobiologia, 155, 159 (abstract).

Giere, O., B. Rhode & N. Dubilier, 1988. Structural peculiarities of the body wall of *Tubificoides benedii* (Oligochaeta) and possible relations to its life in sulphidic sediments. Zoomorphology 108: 29–39.

Grieshaber, M., 1992. Hypoxia and sulfide tolerance in some marine invertebrates. Hypoxie- und Sulfidtoleranz einiger mariner Invertebraten. Verh. dt. zool. Ges. 85: 55–76.

Grieshaber, M. K., I. Herdewig, U. Kreutzer & H.-O. Pörtner, 1994. Physiological and metabolic responses to hypoxia in invertebrates. Rev. Physiol. Biochem. Pharmacol. 125: 43–147.

Grieshaber, M. K. & S. Völkel, 1998. Animal adaptations for tolerance and exploitation of poisonous sulfide. Ann. Rev. Physiol. 60: 33–53.

Hauschild, K. & M. K. Grieshaber, 1997. Oxygen consumption and sulfide detoxification in the lugworm *Arenicola marina* (L.) at different ambient oxygen partial pressures and sulfide concentrations. J. comp. Physiol. B 167: 378–388.

Maguire, C. & P. J. S. Boaden, 1975. Energy and evolution in the thiobios: an extrapolation from the marine gastrotrich Thiodasys sterreri. Cah. Biol. mar. 16: 635–646.

Menon, J. G. & A. J. Arp, 1993. The integument of the marine echiuran worm *Urechis caupo*. Biol. Bull. 185: 440–454.

Menon, J. G., J. K. Willsie, A. Tauscher & A. J. Arp, 1995. Comparative ultrastructure of the epithelia of vestimentiferan tube worms from hydrocarbon seeps and deep-sea hydrothermal vents. Am. Zool. 35: 36A.

Oeschger, R. & R. D. Vetter, 1992. Sulfide detoxification and tolerance in *Halicryptus spinulosus* (Priapulida): a multiple strategy. Mar. Ecol. Prog. Ser. 86: 167–179.

Prosser, C. L. & F. A. Jr. Brown, 1961. Comparative animal physiology. (2nd edn.). Saunders, Philadelphia, 688 pp.

Schneider, A., 1994. Die Sulfidtoleranz der Schlammröhrenwürmer *Limnodrilus hoffmeisteri* (Cl.) und *Tubifex tubifex* (M.). Shaker, Aachen: 122 pp.

Schöttler, U., B. Surholt & E. Zebe, 1984. Anaerobic metabolism in *Arenicola marina* and Nereis diversicolor during low tide. Mar. Biol. 81: 69–73.

Thiermann, F., A.-S. Niemeyer & O. Giere, 1996. Variations in the sulfide regime and the distribution of macrofauna in an intertidal flat in the North Sea. Helgoländer Meeresunters. 50: 87–104.

Toulmond, A., 1975. Blood oxygen transport and metabolism of the confined lugworm *Arenicola marina* (L.). J. exp. Biol. 3: 647–660.

Völkel, S. & M. Grieshaber, 1997. Sulphide oxidation and oxidative phosphorylation in the mitochondria of the lugworm *Arenicola marina*. J. exp. Biol. 200: 83–92.

Hydrobiologia **406**: 243–251, 1999.
B. M. Healy, T. B. Reynoldson & K. A. Coates (eds), Aquatic Oligochaetes.

Decoupling of fission and regenerative capabilities in an asexual oligochaete

Alexandra E. Bely
Department of Ecology and Evolution, State University of New York at Stony Brook, Stony Brook, NY 11794-5245, U.S.A.
E-mail: abely@life.bio.sunysb.edu; Phone: 516-632-8588; Fax: 516-632-7626

Key words: anterior regeneration, evolution, fission, asexual reproduction, Naididae, Annelida

Abstract

Agametic reproduction by fission has evolved numerous times in the Annelida. In this group, as well as others, the developmental processes of fission appear to be based on those of regeneration. Thus far, all fissiparous annelids whose regenerative abilities have been investigated fully are capable of regenerating both anteriorly and posteriorly. However, in this paper, I describe a unique exception to this association between fission and regenerative abilities. Regeneration experiments demonstrate that the fissiparous annelid *Paranais litoralis* (Oligochaeta: Naididae) has lost the capacity for full anterior regeneration. The four anterior segments of *P. litoralis* are produced routinely during each round of fission, yet this species is incapable of producing these same segments by anterior regeneration. I propose that fissiparous reproduction, possibly coupled with a low susceptibility to amputation in nature, may have relaxed selection for anterior regeneration in *P. litoralis*. The decoupling of fission and regenerative abilities suggests that some steps in the developmental trajectories that lead to fission and regeneration may have diverged in this species.

Introduction

Agametic reproduction is a form of asexual reproduction that has evolved independently in numerous animal phyla (Giese & Pearse, 1975; Brusca & Brusca, 1990; Adiyodi & Adiyodi, 1993). In agametic reproduction, offspring are produced without the involvement of gametes, and develop by processes other than embryogenesis. One form of agametic reproduction, transverse fission, has evolved in a number of linear, limbless animals ('worms'), such as platyhelminths, nemerteans and annelids. In annelids alone, fission is found in nine (out of ~80) polychaete families (Schroeder & Hermans, 1975), four (out of ~14) oligochaete families (Brinkhurst & Jamieson, 1971; Christensen, 1984) and in the Aphanoneura (Brinkhurst & Jamieson, 1971). While most of these families contain only a few fissiparous species, some families, such as the Naididae (Oligochaeta) and Aeolosomatidae (Aphanoneura), seem to be entirely comprised of species capable of fission.

Reproduction by fission must involve two processes:

1. Physical separation of an individual into two (or more) pieces, and
2. Reconstitution of a whole individual from each piece (or at least two pieces).

In annelids, two forms of fission are recognized based on the relative order of these two processes (Lasserre, 1975). In architomy (or fragmentation), a worm physically splits into two pieces first, followed by replacement of the structures missing from each piece (i.e. new anterior or posterior ends). In paratomy (or budding), the new anterior and posterior ends of the two zooids develop, at least partially, prior to the physical separation of the two daughter worms.

While the process of forming a new anterior or posterior end during a round of fission is sometimes loosely referred to as 'regeneration', the two processes are biologically very different, and I would argue the terminology should reflect this. Regeneration is triggered when some external force, presumably not predictable to the worm, breaks a worm into pieces. In contrast, fission (either architomy or paratomy) is initiated by the worm itself, and is thus at some level predictable. It is therefore possible, perhaps even likely, that the worm prepares itself physiologically or developmentally for the event.

Nevertheless, in annelids (as well as other groups) a clear relationship exists between fission and regeneration, suggesting that fission relies on regenerative processes. First, there are extensive developmental similarities between fission and regeneration (Galloway, 1899; Berrill, 1952), to the extent that several authors have stated that the two processes are basically one and the same (Dehorne, 1916; Herlant-Meewis, 1953). Second, for a particular species of annelid, the number of segments produced anteriorly during fission is usually the same as the number that can be produced during anterior regeneration. Typically, only the cephalized segments (those that together make up the cephalic region or 'head') are formed anteriorly in either process (Dehorne, 1916). Third, a phylogenetic correlation between the two phenomena is apparent, though a rigorous test of this correlation using the comparative method (e.g. Harvey & Pagel, 1991) is still needed. Annelids that can reproduce by fission typically also have extensive anterior and posterior regenerative abilities, while many groups which have no fissiparous representatives have poor regenerative abilities (e.g. hirudineans, branchiobdellidans) (Hyman, 1940; Berrill, 1952; Needham, 1952; Lasserre, 1975; Schroeder & Hermans, 1975). The distribution of regeneration among taxa is much broader than that of fission, however, and many groups capable of anterior and posterior regeneration do not reproduce by fission.

Given this evidence, a likely scenario for the evolution of fission is that it evolves by recruitment of these regenerative abilities into a new role – that of agametic reproduction. The ancestor in which fission evolved would probably have been capable of both anterior and posterior regeneration, as these represent two separate capabilities which would both be required. Anterior and posterior regeneration may make use of similar developmental programs for producing generic segments, but each must also encode information unique to one end of the animal. For example, anterior regeneration must recreate specifically anterior structures such as the prostomium, cephalic ganglion, mouth, pharynx and salivary glands, while posterior regeneration must recreate specifically posterior structures such as the pygidium (including the anus and any accessory respiratory structures) and the growth zone. A number of annelids can regenerate posteriorly but not anteriorly (Hyman, 1940) further supporting the claim that these are truly separate capabilities.

Thus far, all fissiparous platyhelminths, nemerteans, and annelids whose regenerative abilities have been investigated fully are capable of regenerating both anteriorly and posteriorly (Berrill, 1952; Brondsted, 1969). However, a curious exception to the association between fission and regenerative ability was suggested by the experiments of Martínez (1993) on the annelid *Paranais litoralis* (Müller, 1784) (Oligochaeta: Naididae). In the course of experiments on senescence, Martínez found that *P. litoralis* individuals reproducing by paratomic fission readily regenerated posteriorly when up to 10 posterior segments were removed from the anterior zooid. However, when the anterior 9–18 segments were removed, worms never showed any signs of anterior regeneration. While suggestive, these data are not sufficient to conclude that *P. litoralis* cannot regenerate anteriorly. Most annelids that can regenerate a new anterior end cannot do so if the cut is made too far posteriorly. The region in which anterior regeneration is possible, termed the cephalic territory ('territoire cephalique' *sensu* Herlant-Meewis, 1953), is typically restricted to some anterior fraction of the worm. Thus, two possible explanations could account for the lack of anterior regeneration in Martínez's studies:

1. *P. litoralis* can regenerate anteriorly but its cephalic territory does not extend posteriorly past the eighth segment, or
2. *P. litoralis* cannot regenerate anteriorly.

The regeneration studies presented here test between these two possibilities.

Materials and methods

Field collections and laboratory cultures

Paranais litoralis individuals were originally collected in the summer of 1996 from Flax Pond, a tidal mudflat on the north shore of Long Island, New York, U.S.A. [the same locality studied by Martínez (1993)]. Sexually immature specimens were identified to species using Brinkhurst (1986). Worms were maintained in cultures consisting of sediment from the field site (frozen at −80 °C at least 24 h and thawed prior to use) and 0.75X artificial seawater (~24 ppt) (General Purpose seawater: Strathmann, 1987). Cultures were kept at room temperature in Erlenmeyer flasks with

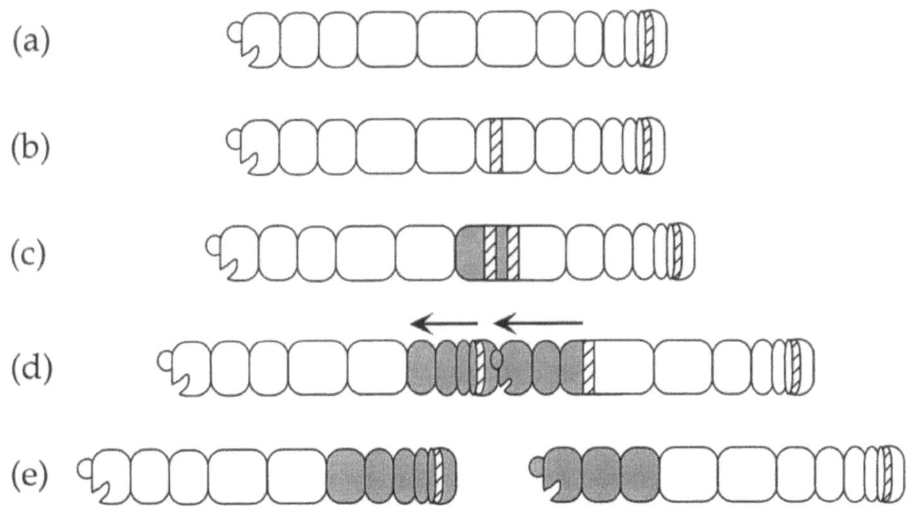

Figure 1. Schematic representation of paratomic fission in naidids. Hatching represents fission or growth zones, and shaded areas represent newly formed tissue. An adult worm (a) develops a fission zone in a midbody segment (b). The fission zone splits into two zones (c), each producing new segments anteriorly (d). The anterior zone produces new posterior segments of the anterior worm, and the posterior zone produces new anterior segments of the posterior worm. The two zooids then detach (e). Worms add segments from subterminal growth zones throughout life. In *P. litoralis*, the fission zone typically forms in the 18th segment. During rapid paratomy, a third zooid may begin developing between the anterior and posterior zooids, before the latter two have separated (see Figure 2).

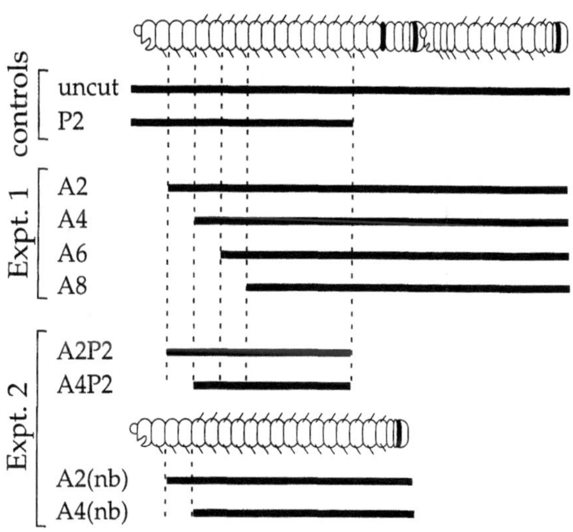

Figure 2. Regeneration treatments. Horizontal bars represent the body region retained in a particular treatment, and hatched vertical lines mark levels of amputation. See text for details.

constant aeration. New cultures were started with fresh sediment every few weeks. Under these conditions, worms reproduce constantly by paratomy (Figure 1).

Regeneration experiments

Experiment 1 was carried out to determine whether worms undergoing paratomy can regenerate anteriorly from various amputation levels. This experiment involved 72 worms from a single healthy culture, all actively reproducing by paratomy and composed of two to three zooids each. Worms were divided into six treatments (n=12 for each treatment) (Figure 2): a control in which no segments were amputated (Uncut Control); a second control in which the two most posterior segments of the anterior zooid were removed (P2 Control), necessary for verifying that worms were capable of regeneration; and four decapitation treatments, in which the two, four, six or eight most anterior segments of a worm were removed (A2, A4, A6, A8). In *P. litoralis*, the four most anterior segments are morphologically distinct and together form the cephalic region of a worm. Therefore, the A2 and A4 treatments removed part of or the entire cephalic region, while the A6 and A8 treatments removed the entire cephalic region plus several trunk segments. Chaetae were used as landmarks in determining the cut sites for all amputations, with cuts made inbetween consecutive chaetal bundles. Because of this method of locating segments, and the fact that in *P. litoralis* the cerebral ganglion (which is technically part of segment I) is shifted posteriorly (typically above the ventral chaetae of segment III), most worms in the A2 and A2(nb) (see below) treatments probably retained their original cerebral ganglia.

Because some evidence suggests that the presence of fission zones can retard or inhibit regeneration (in

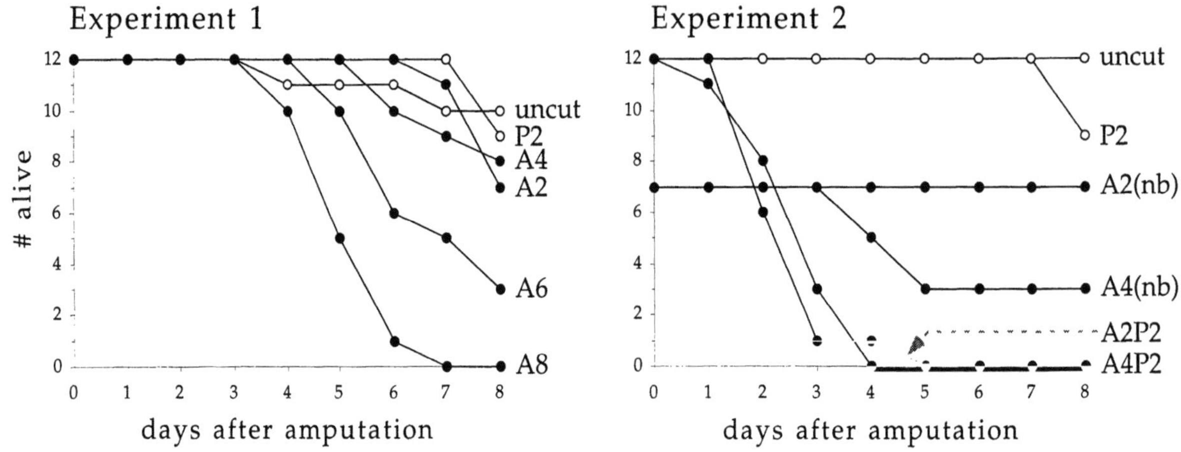

Figure 3. Survival during regeneration experiments. Open circles represent control treatments (Uncut or P2) and filled circles represent decapitation treatments.

annelids: Galloway, 1899; in platyhelminths: Van Cleave, 1929), I initiated a second experiment in which fission zones were either experimentally removed or naturally absent. For Experiment 2, 62 worms from a second culture were divided into six treatments (Figure 2): two control treatments as in Experiment 1 (Uncut Control, P2 Control; n=12 for each); two double amputation treatments (A2P2, A4P2; n=12 for each), in which the two or four most anterior segments were removed, as were the two most posterior segments of the anterior worm (removing the fission zone and all developing posterior zooids); and two decapitation treatments using newly detached zooids (newborns, or 'nb') that had not formed visible fission zones yet (A2(nb), A4(nb); n=7 for each), in which the two or four most anterior segments of the worm were removed. Limited numbers of newborn zooids from the culture prohibited larger sample sizes for these last two treatments.

All worms, including controls, were anesthetized with a carbon dioxide solution (store-bought carbonated water), and worms from the amputation treatments were cut with a scalpel. Anesthesia and amputation never lasted more than 10 min. (I have used these procedures to induce anterior regeneration in several other species of naidids and tubificids.) Worms were then placed singly in individual culture wells with 2 ml 0.75X artificial seawater, where they recovered within a few seconds. Two hours after the start of the experiment, each worm was examined to verify that it had survived the procedure, and any posterior zooids that had detached within that time were noted and removed. Worms were inspected daily for signs of regeneration, and any posterior zooids that had detached were noted and removed. Every two days, worms were transferred to new culture dishes with fresh water. At the end of eight days, all surviving worms were fixed in formaldehyde for subsequent detailed observations.

Results

All worms survived the initial anesthesia and amputation procedure. Survival during the course of each experiment is shown in Figure 3. Nearly all worms (67/72) from the Uncut Controls and the anterior amputation treatments A2, A4, A6 and A8 produced offspring by paratomy during the experiment, while no worms from the other treatments [P2, A2P2, A4P2, A2(nb), A4(nb)] produced any offspring. Decapitation significantly increased the probability of zooid release within the first two hours of the experiment (G-test of independence, $p<0.005$); in Experiment 1, 18/48 decapitated worms released zooids within the first two hours, while 0/12 Uncut Controls released zooids in that time. Although zooid release was initially accelerated by decapitation, by the end of the eighth day, decapitated worms and Uncut Control worms of Experiment 1 had released approximately the same number of zooids, an average of 1.2 and 1.5 zooids respectively (but note that not all worms survived to the end of the experiment).

No decapitated *P. litoralis* exhibited full anterior regeneration, regardless of the amputation level or of the presence or absence of fission zones. Results of Experiments 1 and 2 were qualitatively similar and

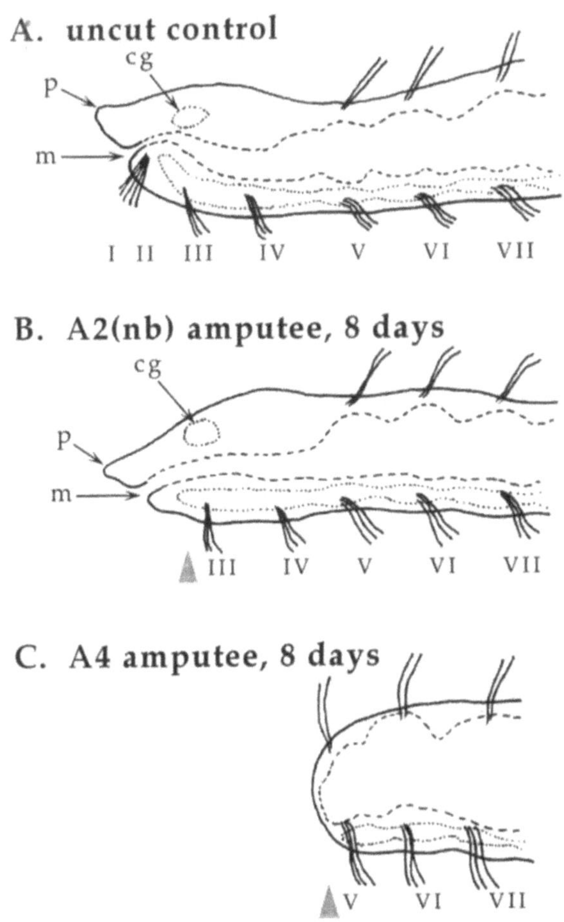

Figure 4. Anterior ends of uncut control and anteriorly amputated worms, on day 8 of the experiment. (A) uncut control; (B) newborn zooid from which the two most anterior segments were removed [A2(nb)]; (C) paratomically reproducing adult from which the four most anterior segments were removed (A4). Roman numerals denote segments (note that segment I does not bear chaetae). Arrowheads mark amputation sites. Stippled lines outline nerve ganglia and gut. Legend: cg – cerebral ganglion; m – mouth; p – prostomium.

will be discussed together. When the two most anterior segments were removed, five of seven newborn zooids [A2(nb)] failed to initiate any kind of regeneration, but two of seven of these newborn zooids and all paratomically reproducing adults (A2) regenerated incompletely, forming a hypomeric cephalic region. These worms regenerated a new prostomium and mouth (within 4–5 days) but failed to regenerate the missing chaetae-bearing segment (segment II) (Figure 4A, B). When the entire cephalic region was removed (A4), when the cephalic region plus several trunk segments were removed (A6 & A8), or when worms were amputated both anteriorly and posteriorly (A2P2 & A4P2), no regeneration of any kind occurred. Most of these worms wound healed normally but never formed a visible blastema or replaced any amputated structures (Figure 4C). Decapitated worms were not physiologically or developmentally dormant; during both experiments, worms continued to grow, forming new segments from the subterminal growth zone (see Figure 1). In addition, nearly all decapitated worms in Experiment 1 produced progeny by paratomy, which requires the development of new anterior segments for the posterior zooid. This demonstrates clearly that worms are physiologically capable of producing a new anterior end, but not at an anterior wound surface. Decaudalized worms (P2) from both experiments exhibited normal posterior regeneration, producing up to seven posterior chaetae-bearing segments by the end of the experiment (day 8).

Discussion

Paranais litoralis appears incapable of full anterior regeneration, regardless of the amputation level. When the four, six or eight most anterior segments are removed, worms wound heal normally but do not form a blastema or replace any missing strucutures. When the two most anterior segments are removed, most worms regenerate a hypomeric head (missing one or more segments), a phenomenon noted in several other oligochaetes (e.g. Krecker, 1910; Gates, 1927). At most, these worms regenerate only a prostomium and mouth. Interestingly, during embryogenesis these two structures arise from presegmental tissue and are not part of a true segment (Shankland & Savage, 1997). The suggestion that regeneration of asegmental and segmental structures might be uncoupled is an intriguing possibility that warrants further study. The presence of limited anterior regeneration and the presence of the original cerebral ganglion is also correlated (both occurred only in worms which had two anterior segments removed), but additional studies are required to determine whether the cerebral ganglion is causally linked to prostomium and mouth regeneration.

While it is well established that most, if not all, of the ~110 species in the family Naididae reproduce by fission (either paratomy or architomy) (Brinkhurst & Jamieson, 1971), regeneration has been studied in only a few of these species. *P. litoralis* is the first naidid reported to be incapable of full anterior regeneration; the other naidid species that have been investigated all exhibit normal anterior regeneration (Table 1), typically within two to four days after amputation. Although a

Table 1. Anterior regeneration in the Naididae. Asterisks (*) denote species names used in current taxonomy. Indented species are probable synonyms of the species (with asterisks) above them (Brinkhurst & Jamieson, 1971). Subfamily and tribe designations are based on Nemec & Brinkhurst (1987)

Species	Anterior regeneration	Reference
Stylarinae		
* *Stylaria lacustris*	normal	Harper (1904); Ortmann (1921) in Herlant-Meewis (1946)
Stylaria fossularis	normal	Chu & Pai (1944)
Nais proboscidea	normal	Rievel (1896); Abel (1902)
Naidinae: Chaetogastrini		
* *Paranais litoralis*	hypomeric/absent	Martínez (1993); this study
Naidinae: Derini		
* *Allonais paraguayensis*		
Nais paraguayensis	normal	Hyman (1938); O'Brien (1946)
* *Dero digitata*	normal	C. D. Drewes, pers. comm.
Dero limosa	normal	Hyman (1916)
Naidinae: Naidini		
* *Nais elinguis*	normal	Consoli (1923) in Berrill (1952)
Naidinae: Pristinini		
* *Pristina longiseta*	normal	Van Cleave (1937)

robust phylogeny for the Naididae is not yet available, and though additional phylogenetic sampling is certainly desirable, some understanding of the evolution of anterior regeneration in naidids is already possible. The taxa sampled to date include six (out of ~21) genera, and fortuitously represent the five major groups of naidids identified by Nemec & Brinkhurst (1987):

1. *Stylaria*, (Stylarinae);
2. *Paranais* (Naidinae: Chaetogastrini);
3. *Allonais*, *Dero* (Naidinae: Derini);
4. *Nais* (Naidinae: Naidini); and
5. *Pristina* (Naidinae: Pristinini).

The data currently available (Table 1) support the hypothesis that the ability to regenerate anteriorly is ancestral for naidid oligochaetes. Consequently, the lack of full anterior regeneration in *P. litoralis* would represent a loss of this ability. In order to determine how old this loss is, anterior regeneration studies must be performed on other populations of *P. litoralis*, other species of *Paranais*, and possibly other closely related genera.

Although countless examples of loss of regeneration must exist in the animal kingdom, only a few have been clearly identified (Scadding, 1977; Vollrath, 1990; Wagner & Misof, 1992), for this requires a careful compilation of comparative data analyzed within a phylogenetic framework. In annelids, the ability to regenerate posteriorly appears to be ancestral (Hyman, 1940; Berrill, 1952) and has been lost only a few times, for example in hirudineans, branchiobdellidans and the polychaete *Polyophthalamus* (Hyman, 1940). Anterior regeneration may have been lost multiple times in the annelids, but identifying these cases will require additional taxon sampling and detailed comparative analyses since anterior regeneration capabilities vary widely within this group.

The loss of full anterior regeneration in *P. litoralis* presents several puzzles. First, why has *P. litoralis* lost such a beneficial ability? And second, given the strong evidence for a close association between regenerative and fission processes (see Introduction), what can account for the decoupling of these two abilities in *P. litoralis*?

Loss of regenerative abilities

While some thought and speculation has been devoted to the conditions under which regeneration should evolve (Morgan, 1901; Needham, 1952; Barr, 1964;

Goss, 1969; Reichman, 1984), there has been much less discussion of why such a seemingly beneficial trait should ever be lost. What changes in an organism's morphology, ecology, or life history might be implicated in the loss of regenerative abilities? Several possibilities have been suggested in the literature. First, regenerative capabilities could be lost as a side effect of other changes. For example, modifications in a structure's size, complexity, degree of innervation, or speed of wound healing might generate constraints (physical, developmental or physiological) that prohibit successful regeneration (Needham, 1952; Goss, 1969; Elder, 1979; Singer et al., 1987). Second, intermediate stages of regeneration might become detrimental, making the suppression of regeneration selectively advantageous (Vollrath, 1990). Third, if a structure becomes too vital or too trivial to an organisms' survival, selection will no longer actively maintain the ability to regenerate that structure (Goss, 1969; Kirkwood, 1981; Reichman, 1984).

None of the hypotheses mentioned above seem useful for explaining the loss of anterior regeneration in *P. litoralis*. The anterior end of *P. litoralis* does not appear significantly different from that of other naidids (which are capable of regenerating anteriorly), in either morphology or functional importance. Additionally, this region of *P. litoralis* does not appear to be too trivial or too essential to an individual worm's survival. The anterior end is required for feeding and is thus essential for an individual zooid's long term survival. However, *P. litoralis* can survive without food for at least two weeks (A. E. Bely, unpublished data), giving it plenty of time to regenerate, since most naidids for which we have data complete anterior regeneration within two to four days.

Two hypotheses which might explain the loss of anterior regeneration in *P. litoralis* are:

1. Anterior amputation frequencies in the field are low for *P. litoralis*, and
2. The life history of *P. litoralis* makes reproduction possible even in the absence of anterior regeneration.

With respect to the first of these, only very limited data are available so far, but it is consistent with *P. litoralis* having low natural frequencies of amputation. A study of regeneration frequencies in oligochaetes performed in a Belgian tidal mudflat found *P. litoralis* present, but recovered no anteriorly or posteriorly regenerating individuals (Seys et al., 1999). Similarly, of the several hundred *P. litoralis* individuals collected from the field site of the present study (Flax Pond), no regenerating individuals have ever been found (pers. obs.). While aquatic oligochaetes are important prey items for many predators (Bouguenec & Giani, 1989), susceptibility to partial predation events (e.g. Drewes & Zoran, 1989) may be quite variable between species, depending strongly on the ecology and behavior of both prey and predator. Therefore, it may be that *P. litoralis* rarely experiences partial predation, even if it is frequently ingested whole. To properly evaluate the hypothesis that *P. litoralis* has lost its capacity for anterior regeneration because it is rarely amputated, we need much more information on absolute and relative frequencies of amputation (by biotic or abiotic factors) in *P. litoralis* as well as other naidids.

The second hypothesis proposed is based on the reasoning that selection for regeneration will decrease if the importance of regeneration to an individual's reproductive output decreases (Kirkwood, 1981). While *P. litoralis* is presumably capable of sexual reproduction (since sexually mature individuals can be found), its primary mode of reproduction is paratomic fission. This fissiparous lifestyle makes it possible for a decapitated individual to reproduce even if it never regenerates: when an anterior zooid is decapitated, posterior zooids continue to develop and detach (Martínez, 1993; this study). For example, in the regeneration experiments described here, 47 out of 48 decapitated *P. litoralis* produced at least one offspring within 8 days. By comparison, in strictly gametically reproducing species, anterior regeneration following decapitation would seem essential to future reproduction. As in several other fissiparous species (Hyman, 1938; Herlant-Meewis, 1953), decapitation in *P. litoralis* actually accelerates posterior zooid detachment. While this phenomenon has historically been attributed to the release of posterior head suppression normally imposed by the anterior cephalic region, the accelerated detachment could also be an adaptive feature. While the anterior zooid is regenerating (or not regenerating, as in *P. litoralis*), feeding ceases completely. The sooner a zooid detaches (provided its mouth is sufficiently developed), the sooner it will be able to obtain food again.

In summary, the loss of anterior regeneration abilities in *P. litoralis* may result from a combination of factors: a low risk of anterior amputation in nature, and a fissiparous lifestyle, in which reproduction is possible even in decapitated individuals. Together, these two factors may have relaxed selection for maintenance of anterior regeneration to the point that this feature could be lost without any significant fitness consequences. There is no reason to believe that regenerative ability is actually disadvantageous in *P. litoralis;* anterior regeneration would surely still be beneficial to an anteriorly amputated individual. However, the fitness advantage this feature imparts may be so minimal in this species that anterior regeneration could be an essentially neutral trait that has been lost through drift (Hartl & Clark, 1989). The maintenance of anterior regeneration in other naidid species (Table 1), all of which also reproduce by fission, may be due to ecological differences in the form of higher frequencies of anterior amputation, or to 'phylogenetic inertia' (the trait is retained simply because it was present in the ancestor). Testing between these two possibilities will require a comparison of anterior amputation frequencies between several naidid species and *P. litoralis*.

Uncoupling of fission and regenerative abilities

The second, perhaps more intriguing puzzle presented by *P. litoralis* is the lack of correspondence between fission and regenerative abilities. This species is the first fissiparous organism shown to be incapable of regenerating a body region that is routinely formed during fission. In *P. litoralis*, the four anterior segments are produced during each round of paratomic fission (Figure 1), yet this study demonstrates that these same four segments cannot be produced by regeneration. At most, only a prostomium and mouth are regenerated anteriorly.

The evolutionary scenario we are left with, then, is that regenerative abilities may have been recruited into a role during fissiparous reproduction in the ancestral naidid (see 'Introduction'), but then subsequently lost in a lineage that retained fission. This suggests that in *P. litoralis* either:

1. Paratomic fission and anterior regeneration use essentially the same developmental program, but this program can only be initiated by intrinsic proximal cues (extrinsic cues, i.e. anterior amputation, can no longer initiate this program); or
2. The developmental programs of paratomic fission and anterior regeneration are significantly different in *P. litoralis* (and possibly other naidids), such that one process may be modified or lost without affecting the other.

Ongoing molecular developmental studies may help to distinguish between these two possibilities.

Acknowledgements

I thank Charles Drewes and Daniel Martínez for useful discussions, and Kathryn Coates, Leo Shapiro, Günter Wagner, Gregory Wray and members of his lab, and one anonymous reviewer for useful comments on the manuscript. This is contribution number 1018 from the Program in Ecology and Evolution, State University of New York at Stony Brook. This work was supported by a Doctoral Dissertation Improvement Grant from the National Science Foundation (DEB-9623565), and a Grant-in-Aid of Research from Sigma Xi, The Scientific Research Society.

References

Abel, M., 1902. Beiträge zur Kenntnis der Regenerationsvorgänge bei den limicolen Oligochäten. Z. wiss. Zool. 1: 1–74.

Adiyodi, K. G. & R. G. Adiyodi, 1993. Reproductive Biology of Invertebrates, Vol. 6. Asexual Propagation and Reproductive Strategies. John Wiley and Sons, West Sussex, 410 pp.

Barr, H. J., 1964. Regeneration and natural selection. Am. Nat. 98: 183–186.

Berrill, N. J., 1952. Regeneration and budding in worms. Biol. Rev. 27: 401–438.

Bouguenec, V. & N. Giani, 1989. Les oligochètes aquatique en tant que proies des invertébrés et des vertébrés: une revue. Acta Oecol. 10: 177–196.

Brinkhurst, R. O., 1986. Guide to the freshwater aquatic microdrile oligochaetes of North America. Can. Spec. Publ. Fish. aquat. Scis 84: 1–259.

Brinkhurst, R. O. & B. G. M. Jamieson, 1971. Aquatic Oligochaeta of the World. Oliver and Boyd, Edinburgh, 860 pp.

Brondsted, H. V., 1969. Planarian Regeneration. Pergamon Press, Braunschweig, 276 pp.

Brusca, R. C. & G. J. Brusca, 1990. Invertebrates. Sinauer, Sunderland (MA), 922 pp.

Christensen, B., 1984. Asexual propagation and reproductive strategies in aquatic Oligochaeta. Hydrobiologia 115: 91–95.

Chu, J. & S. Pai, 1944. The relations between natural fission and regeneration in *Stylaria fossularis* (Annelida). Physiol. Zool. 17: 159–166.

Consoli, L., 1923. La rigenerazione in rapporto con la strobilazione negli Oligocheti limicoli. Bull. Ist. zool. Palermo 1: 23–47.

Dehorne, L., 1916. Les naïdimorphes et leur reproduction asexuée. Arch. Zool. exp. gén. 56: 25–157.

Drewes, C. D. & M. J. Zoran, 1989. Neurobehavioral specializations for respiratory movements and rapid escape from predators in posterior segments of the tubificid *Branchiura sowerbyi*. In J. L. Kaster (ed.), Aquatic Oligochaete Biology, Hydrobiologia/Dev. Hydrobiol. 180: 65–71.

Elder, D., 1979. Why is regenerative capacity restricted in higher organisms? J. theor. Biol. 81: 563–568.

Galloway, T. W., 1899. Observations on non-sexual reproduction in *Dero vaga*. Bull. Mus. comp. Zool. 35: 115–140.

Gates, G. E., 1927. Regeneration in a tropical earthworm *Perionyx excavatus* E. Perr. Biol. Bull. 53: 351–364.

Giese, A. C. & J. S. Pearse, 1975. Reproduction of Marine Invertebrates. Academic Press, New York, vol. I–IX.

Goss, R. J., 1969. Principles of Regeneration. Academic Press, New York, 287 pp.

Harper, E. H., 1904. Notes on regulation in *Stylaria lacustris*. Biol. Bull. 6: 173–190.

Hartl, D. L. & A. G. Clark, 1989. Principles of Population Genetics, 2nd edn. Sinauer Associates, Sunderland (MA), 682 pp.

Harvey, P. H. & M. D. Pagel, 1991. The Comparative Method in Evolutionary Biology. Oxford University Press, New York, 239 pp.

Herlant-Meewis, H., 1946. Contribution a l'étude de la régénération chez les oligochètes: reconstitution du germen chez *Lumbricillus lineatus* (Enchytraeides). Arch. Biol. 57: 197–306.

Herlant-Meewis, H., 1953. Contribution a l'étude de la régénération chez les oligochètes aeolosomatidae. Ann. Soc. r. zool. Belg. 84: 117–161.

Hyman, L. H., 1916. An analysis of the process of regeneration in certain microdrilous oligochaetes. J. exp. Zool. 20: 99–163.

Hyman, L. H., 1938. The fragmentation of *Nais paraguayensis*. Physiol. Zool. 11: 126–143.

Hyman, L. H., 1940. Aspects of regeneration in annelids. Am. Nat. 74: 513–527.

Kirkwood, T. B. L., 1981. Repair and its evolution: survival versus reproduction. In C. R. Townsend & P. Calow (eds), Physiological Ecology: an Evolutionary Approach to Resource Use. Sinauer Associates, Sunderland (MA): 165–189.

Krecker, F. H., 1910. Some phenomena of regeneration in *Limnodrilus* and related forms. Z. wiss. Zool. 95: 383–450.

Lasserre, P., 1975. Clitellata. In A. C. Giese & J. S. Pearse (eds), Reproduction of Marine Invertebrates, Vol 3. Annelids and Echiurans. Academic Press, New York: 215–275.

Martínez, D. E., 1993. On Senescence in Asexual Metazoans. Ph.D. Dissertation, State University of New York at Stony Brook, 181 pp.

Morgan, T. H., 1901. Regeneration. Macmillan & Co., London, 316 pp.

Needham, A. E., 1952. Regeneration and Wound-Healing. Barnes & Noble, New York, 152 pp.

Nemec, A. F. L. & R. O. Brinkhurst, 1987. A comparison of methodological approaches to the subfamilial classification of the Naididae (Oligochaeta). Can. J. Zool. 65: 691–707.

O'Brien, J. P., 1946. Studies on the cellular basis of regeneration in *Nais paraguayensis*, and the effects of X-rays thereon. Growth 10: 25–44.

Ortmann, K., 1921. Ein Beitrag zug Kenntnis der äußeren postoperativen Regenerationsprozesse bei *Stylaria lacustris* L. (*Nais proboscidea* Müll.). Lotos Prague 69: 245–253.

Reichman, O. J., 1984. Evolution of regeneration capabilities. Am. Nat. 123: 752–763.

Rievel, H., 1896. Die Regeneration des Vorderdarmes und Enddarmes bei einigen Anneliden. Z. wiss. Zool. 2: 289–341.

Scadding, S. R., 1977. Phylogenetic distribution of limb regeneration capacity in adult amphibia. J. exp. Zool. 202: 57–68.

Schroeder, P. C. & C. O. Hermans, 1975. Annelida: Polychaeta. In A. C. Giese & J. S. Pearse (eds.), Reproduction of Marine Invertebrates, Vol 3. Annelids and Echiurans. Academic Press, New York: 1–213.

Seys, J., M. Vincx & P. Meire, 1999. Spatial distribution of oligochaetes (Clitellata) in the tidal freshwater and brackish parts of the Schelde estuary (Belgium). Hydrobiologia 406: 119–132.

Shankland, M. & R. M. Savage, 1997. Annelids, the segmented worms. In S. F. Gilbert & A. M. Raunio (eds), Embryology: Constructing the Organism. Sinauer, Sunderland (MA): 219–235.

Singer, M., E. C. Weckesser, J. Geraudie, C. E. Maier & J. Singer, 1987. Open finger tip healing and replacement after distal amputation in Rhesus monkey with comparison to limb regeneration in lower vertebrates. Anat. Embryol. (Berl.) 177: 29–36.

Strathmann, M. F., 1987. Reproduction and Development of Marine Invertebrates of the Northern Pacific Coast. University of Washington Press, Seattle, 670 pp.

Van Cleave, C. D., 1929. An experimental study of fission and reconstitution in *Stenostomum*. Physiol. Zool. 2: 18–58.

Van Cleave, C. D., 1937. A study of the process of fission in the naid *Pristina longiseta*. Physiol. Zool. 10: 299–314.

Vollrath, F., 1990. Leg regeneration in web spiders and its implications for orb weaver phylogeny. Bull. br. arachnol. Soc. 8: 177–184.

Wagner, G. P. & B. Y. Misof, 1992. Evolutionary modification of regenerative capability in vertebrates: a comparative study on teleost pectoral fin regeneration. J. exp. Zool. 261: 62–78.

Hydrobiologia **406:** 253–261, 1999.
B. M. Healy, T. B. Reynoldson & K. A. Coates (eds), Aquatic Oligochaetes.

Autotomy reflex in a freshwater oligochaete, *Lumbriculus variegatus* (Clitellata: Lumbriculidae)

Nalena M. Lesiuk[1] & Charles D. Drewes[2,*]
[1]*Neuroscience Program, Iowa State University, Ames, IA 50011, U.S.A.*
[2]*Department of Zoology and Genetics, Iowa State University, Ames, IA 50011, U.S.A.*
phone: (515)294-8061; E-mail: cdrewes@iastate.edu
*(*author for correspondence)*

Key words: autotomy, fragmentation, reflexes, nicotine, lumbriculid, blackworms

Abstract

A novel apparatus was developed that induced segmental autotomy in the freshwater oligochaete, *Lumbriculus variegatus*. The apparatus delivered a quantifiable amount of focal compression to the dorsal body surface at a selected site along the worm. This resulted in a rapid and stereotyped autotomy sequence, beginning with formation of a lateral fissure in the body just anterior to the compression site. Formation of the fissure usually occurred 100–200 ms after the onset of compression. Autotomy readily occurred in the absence of significant longitudinal tension at the autotomy site and in the absence of direct laceration of the body wall. Autotomy culminated in a complete, transverse separation and sealing of anterior and posterior body fragments with no apparent blood loss from either end. There was a direct relationship between the amount of compression and the probability of autotomy in both midbody and tail regions. However, there was a consistently greater probability of autotomy in tail *versus* midbody regions. Autotomy did not occur if the duration of compression was less than 77 ms. Autotomy responses were suppressed in dose-dependent manner by a 15 min treatment with nicotine prior to compression. In instances where compression just failed to induce autotomy there was no evidence of disruption of impulse conduction in giant nerve fibers. Rapid and clear-cut autotomy, in combination with this worm's significant capacity for regeneration of lost segments, are adaptively significant strategies for surviving predatory attack.

Introduction

Autotomy is a well known behavior among a variety of animal groups such as arthropods (Foelix, 1996), echinoderms (Emson & Wilkie, 1980) and vertebrates (Goss, 1969). Generally, autotomy responses are triggered by direct damage or encounter with threatening stimuli. The result is a stereotyped effector response leading to rapid separation and discarding of a body part (such as an appendage) from the main body. An autotomized body part may serve as a possible sacrifice or a source of distraction to a predator. The latter is especially true in cases where appendages generate substantial movements after autotomy (Arnold, 1990).

Invertebrates without appendages, such as annelid worms, are known to autotomize body segments in response to injury from mechanical or chemical stimuli (Stephenson, 1930; Cameron, 1932; Rasmussen, 1953). In some earthworm species, only anterior fragments are viable and provide substrates for subsequent regeneration of missing segments (Cameron, 1932). However, in a few cases (e.g. the freshwater oligochaete, *Lumbriculus variegatus* Grube, 1884, and marine polychaete, *Pygospio elegans* Claparède, 1863) autotomy results in small body fragments that survive and readily regenerate missing head and/or tail segments, thus forming new individuals (Stephenson, 1930; Rasmussen, 1953). Despite propensities for segmental autotomy in at least a few species, there has been no previous analysis, in any annelid, of the mechanisms and behavioral correlates of segmental autotomy.

Here, we tested the hypothesis that segmental autotomy in a freshwater lumbriculid, *L. variegatus*, is a stereotyped and reflexive motor response which may be consistently induced by adequate stimulation in the form of sudden body compression. Our hypothesis derives from a previous report which noted that these worms are prone to segmental autotomy. This occurred inadvertently while worms were handled and manipulated in preparation for electrophysiological studies of segmental regeneration (Drewes & Fourtner, 1990). Our hypothesis is predicated on the assumption that mechanical compression of the body is a relevant stimulus because living worms may be subject to such stimuli during predatory attack (e.g. pinching or biting) in their littoral environment. After developing and testing an apparatus that delivers precisely controlled amounts of body compression to individual worms, we used frame-by-frame video analysis to examine the timing of autotomy in relation to the amount and duration of body compression. We also examined electrophysiological correlates of ventral nerve cord function in instances in which stimuli just failed to elicit autotomy. Finally, we demonstrated pharmacological blockage of the autotomy response by exposure to nicotine, a cholinergic agonist.

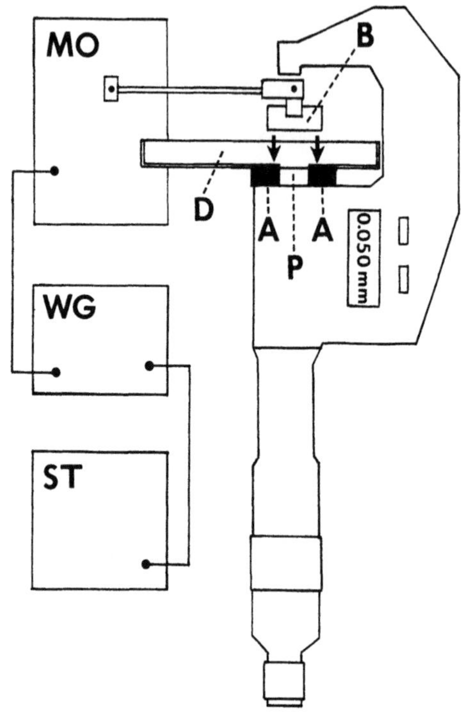

Figure 1. Diagram of apparatus for inducing autotomy. Abbreviations: ST (stimulator); WG (wave function generator); MO (motor); B (blade); P (micrometer post); A (circular brass annulus); D (dish).

Materials and methods

Animals

Lumbriculus variegatus (blackworms), ranging in size from newly hatched to sexually mature adult, were used. Worms were obtained from a variety of sources depending on size; four different size categories were selected.

Mid-sized worms, used for most experiments, were obtained from asexually reproducing laboratory cultures. These were raised in aged (dechlorinated) tap water using brown paper towel as a substrate. Worms were fed twice weekly with sinking fish food pellets. Cultures were aerated and water temperature was held at 22–23 °C. Body diameter (at mid-body) and overall length were about 520 μm and 3.5 cm, respectively. These values represent the approximate upper limit in body size attained before spontaneous fragmentation (asexual reproduction) occurred in the worms in our laboratory cultures. Mid-sized worms were pre-screened for uniformity in segment size and pigmentation. Worms with abrupt discontinuities in segment size and pigmentation were not used; typically such discontinuities are indicative of very recent tail segment regeneration.

Larger, sub-adult worms were obtained from a local pet shop which receives weekly shipments of field-collected worms originating near Fresno, California. Body diameter (at mid-body) and length of these worms were about 900 μm and 5 cm, respectively.

Sexually mature (adult) worms were obtained in May from a marsh in Gull Point State Park (West Lake Okoboji, Iowa). Body diameter (at mid-body) and lengths of these worms were 1150 μm and 8 cm, respectively.

Newly hatched worms were obtained from cocoons collected from the same site as adult worms. Transparent cocoons, about 5 mm in length and containing 4–12 orange eggs, were found attached to submerged leaf litter. After approximately one week of development in the laboratory (22–23° C), worms emerged from cocoons. Segment diameter (at mid-body) and worm length were about 170 μm and 0.5 cm, respectively.

Apparatus for inducing autotomy

We designed a novel apparatus for delivering a brief mechanical stimulus that reliably induced segmental autotomy in individual worms (Figure 1). The stimulus consisted of localized compression caused by rapid application of focal pressure, in a transverse orientation, to the dorsal surface of the body. Compression was caused by direct pressure to the dorsal body wall with the smooth, flattened edge of an aluminum 'blade' (blade thickness = 0.27 mm). The blade was attached to a pintle which permitted a pendulum-like movement of the blade, as shown in Figure 1. The holder for the pintle was attached by a short rod to the axial post of a pen driver motor (originally from a strip chart recorder).

Rectangular pulses from a Tektronix FG 501 wave function generator provided direct electrical input to the pen motor. The output of the function generator was, in turn, controlled by the square-pulse output from a Grass SD9 stimulator (Figure 1). When the single-pulse switch of the stimulator was pressed, a stimulus pulse (20 ms duration) was routed to the gating input of the function generator. This resulted in production of exactly one cycle of rectangular wave output from the function generator. The cycle duration of this wave was controlled by the frequency control on the function generator. A low frequency value (= long cycle time) resulted in a compression stimulus with long duration. In most experiments, the cycle duration was adjusted to 5 s. In some experiments, cycle duration was varied between 65 ms and 5 s. Exact measures of compression durations were made by placing a piezoelectric disk beneath the blade. Transient outputs from the disk indicated the timing of blade contact and release, as measured on an oscilloscope.

The output pulse from the function generator resulted in axial rotation of the pen motor shaft and 0.6 cm downward excursion of the blade. Blade movement was limited by its contact with the flat surface of a brass annulus (outside diameter = 2.4 cm; inside diameter = 0.7 cm). As the blade contacted the annulus, the pintle to which the blade was attached rotated slightly. This compensated for the slight curvilinear motion of the blade and assured that blade contact with the annulus was flush with each compression stimulus. The center space of the brass annulus was occupied by the optically flat end of the movable post from an electronic digital micrometer (Fred V. Fowler Co., Inc.).

Prior to delivery of the compression stimulus, the micrometer post was set exactly level with the surface of the annulus and the zero button on the micrometer was pressed. Then, the micrometer post was adjusted to the desired distance below the level of the annulus, according to the digital readout on the micrometer (readout to nearest μm). This distance was equal to

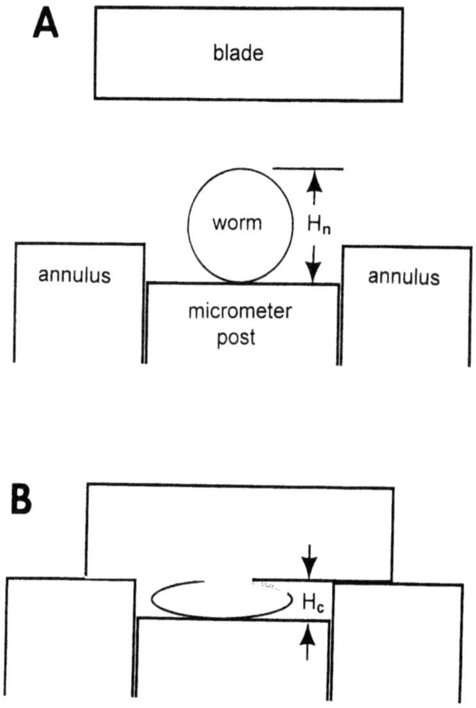

$$100 - (H_c / H_n \times 100) = \%\ \text{compression}$$

Figure 2. Quantitation of compression stimulus. (A) The normal cross-sectional height of the body (Hn) is shown before transverse compression with the narrow blade. (B) The compression height of the body (Hc) is shown with the blade resting on the annulus. The formula for % compression is given below.

the desired height of the body, under the blade, at the time of compression. For example, if the desired compression level was 90% for a worm, whose segment diameter was 1000 μm, then the micrometer post was correspondingly set to 100 μm below the level of the annulus. Micrometer settings were recalculated with each compression stimulus according to the worm diameter and desired compression value (Figure 2). Once calculations and appropriate micrometer settings were made, a worm was transferred along with a small amount of water to the autotomy platform. A fine glass hook was used to position the worm so that segments from a specified location rested across the micrometer post and the body axis was perpendicular to the blade above.

Then, the single-pulse switch on the stimulator was manually pressed, resulting in immediate downward movement of the blade. Blade velocity at the time of contact with the worm was 0.66 m/s, as measured by electronic sensors. Blade velocity remained constant, regardless of changes in the duration of the compression stimulus.

Autotomy responses to the compression stimulus were recorded on video tape using a JVC Model TK 1070U color video camera and Model HRS7300U video cassette recorder. The camera was mounted on an Olympus Model SZ CTV dissecting microscope. Camera frame rate was 30 frames/s. Camera shutter-speed was adjusted to 1/500 s and the autotomy platform was illuminated with a fiber optic illuminator. This combination of equipment and specifications insured relatively clear, freeze-frame images of worms immediately before and after contact with the blade. Video images were electronically captured and then photographed with a SONY UP1200A video printer.

Nicotine preparation

Nicotine solutions were made with double distilled water the day of use. Worms were exposed via immersion in a 20-ml aliquot of nicotine solution in a glass container (one worm/container). Based on preliminary range-finding experiments, the following nicotine concentrations and exposure time were selected for final experiments: 0.002–0.250 mM nicotine and 15 min exposure. After nicotine exposure, worms were immediately rinsed with double distilled water and placed on the autotomy chamber.

Electrophysiology

Worms that did not autotomize in response to compression stimuli were electrophysiologically tested to determine whether the velocity of impulse conduction in giant nerve fibers was affected by compression. All-or-none action potentials from the medial and lateral giant nerve fibers in the ventral nerve cord were recorded non-invasively, as previously described (Drewes & Fourtner, 1990; Rogge & Drewes 1993). The worm was placed on a printed circuit board recording grid. Giant fiber conduction velocity was then measured immediately before compression, over a 1 cm distance at a midbody position. Medial and lateral giant fiber spikes were evoked by light tactile stimulation to the worms' head or tail, respectively, using a hand-held probe. Amplified signals were displayed and measured using a digital storage oscilloscope. The same protocols were used when measuring conduction velocity immediately after the compression, making sure that the compression site was positioned between the two pairs of recording electrodes.

In non-autotomized worms that were pre-treated with nicotine, measurements of the conduction velocity were made 4 h after the compression stimulus, a time that was sufficient for recovery from nicotine effects on escape reflex pathways and function. During recovery, worms were placed in aged tap water. Conduction velocity values after recovery were then compared to those from a corresponding location before nicotine treatment and compression.

Results

Videotape analysis provided direct support for our hypothesis that autotomy is a reflexive response which is consistently induced by brief compression of the body. This response consists of a stereotyped sequence of mechanical and behavioral events. Initially, full excursion of the blade (from fully up to fully down position) occurred in less than one video frame (<33 ms). Once the blade was in the fully down position, body segments under the blade (approximately 1–2 segments in mid-sized worms) were distended laterally by the pressure but these segments were not lacerated or ruptured. Segments immediately adjacent to the compression point, on both sides of the blade, bulged slightly (Figure 3A). Then, during the next several frames, one or several lateral bending movements occurred anterior to the compression point. Similar movements, though slower and weaker, also occurred in segments posterior to the compression.

After this, one of two results occurred: autotomy success or failure. That is, autotomy appeared to be an all-or-none process in which segments quickly and cleanly separated from one another near the point of compression or remained fully connected and intact. There were never instances in which autotomy was incomplete or segments remained partially attached to one another.

In cases of successful autotomy, a transverse fissure formed along one side of the body wall. Invariably, this fissure occurred just anterior to the compression point, at a site approximately midway between the anterior and posterior segmental groove of one segment. Then, during the next few frames, the fissure became circumferential, anterior and posterior portions separated, and autotomy was completed (Figure 3A). Importantly, at the instant of separation, there was little or no pulling back or recoil of the free anterior portion of the worm away from the blade (Figure 3A, frame f7). This indicates that there was little or no longitudinal tension exerted by the anterior end upon the compression point. Given the experimental conditions, establishment of such tension would be

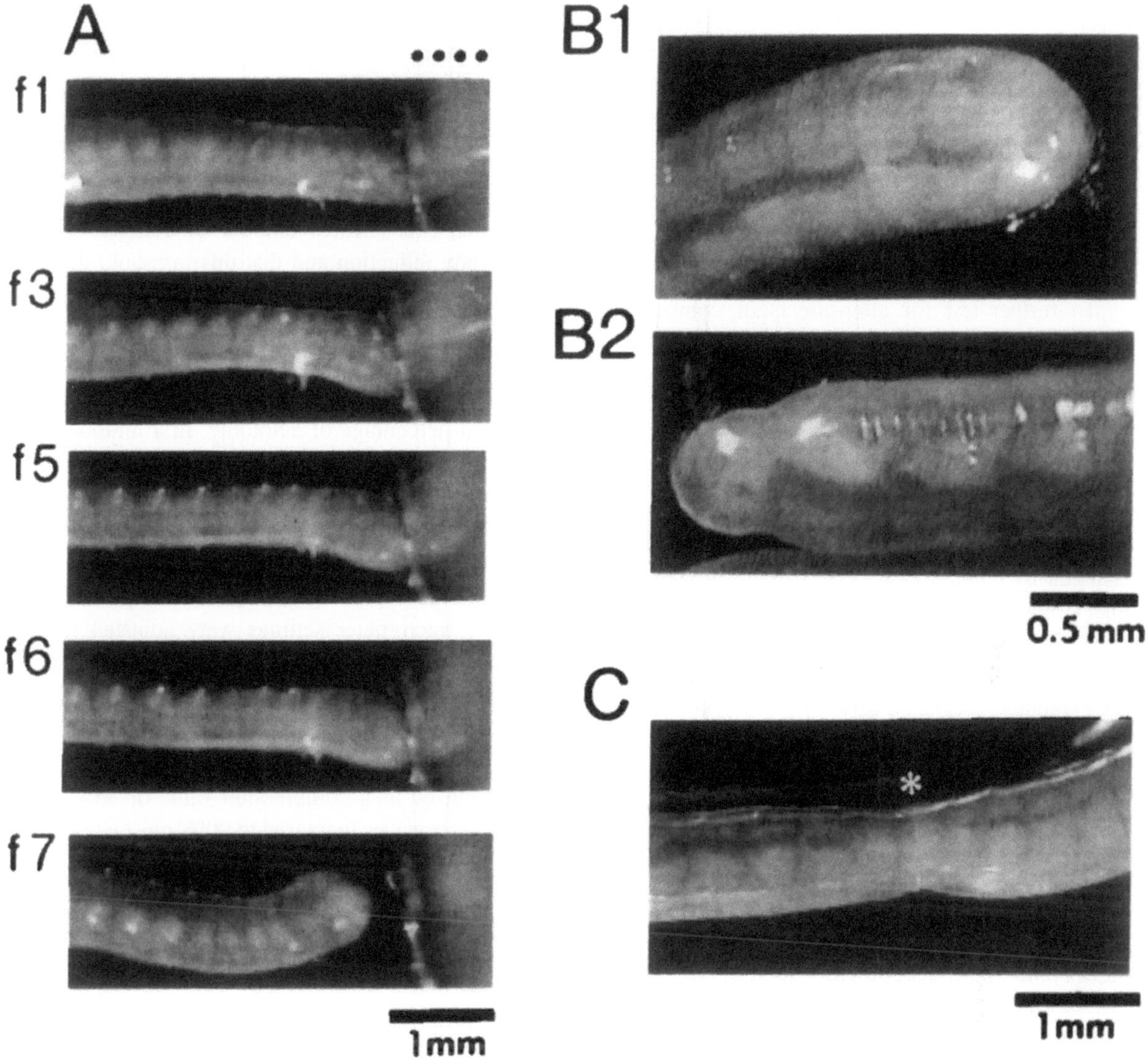

Figure 3. Freeze-frame video images of posterior segments during and after body compression. In A, panel f1 shows the first frame in which the blade is lowered and the body is compressed. The location of the blade is shown by the dots. Panels f3, f5, and f6 show the weak lateral bending movements that precede autotomy. Separation of the anterior portion end of the worm is shown in f7. Panels B1 and B2 show the rounded and sealed ends of the anterior body fragment and posterior body fragment, respectively, 10 min after autotomy. Panel C shows the appearance of body segments immediately after 85% compression (delivered at asterisk), which failed to trigger autotomy.

necessarily limited and unexpected due to the worms' inability to obtain traction against the smooth metal surfaces on the post and annulus. After autotomy was completed, but before the termination of compression, the posterior fragment remained briefly trapped under the blade until the blade lifted. Video images at the time of autotomy showed no indication of blood leakage from anterior or posterior ends of autotomized segments.

Immediately after autotomy, a rapid circumferential closure of the body wall occurred, resulting from a crimping of the terminal segment of each fragment. By 10 min after autotomy, the exposed terminal surfaces on both anterior and posterior fragments appeared rounded and smooth (Figure 3B). However, the first one or two segments of the posterior fragment were pale and thinner in diameter than the rest of the fragment. The unusual appearance of these segments was apparently related to their brief entrapment under the blade, as previously described. Pulsations in the dorsal blood vessel of these pale segments were re-established approximately 20 min after autotomy.

In worms that failed to autotomize in response to compression, there was no apparent laceration or rupture of the body wall after the blade lifted, although segments under the blade were clearly affected by the compression. These segments were pale and thinner in diameter than neighboring segments (Figure 3C). Gradually, over the next few hours, segment diameter appeared normal and dorsal blood vessel pulsations were re-established throughout these segments. There was no sign of blood vessel rupture or internal hemorrhage after compression.

To further test for after-effects of compression in non-autotomized worms, the functional integrity of the nervous system was assessed by non-invasive neurophysiological recording of giant nerve fiber spikes. Sub-adult worms were electrophysiologically tested before and immediately after a compression stimulus (85%). The conduction velocities of medial and lateral giant nerve fibers were not significantly changed by compression. The average medial giant fiber conduction velocity before compression was 11.4 ± 0.9 SD m/s; immediately after compression, it was 11.2 ± 0.7 SD m/s. Similarly, the average lateral giant fiber conduction velocity was 7.3 ± 0.5 SD m/s before compression and 7.3 ± 0.5 SD m/s after. Thus, compression stimuli caused no significant alteration in the conduction properties of nerve fibers.

Next, quantitation of the timing of the autotomy response was done using frame-by-frame analysis of video images. Time taken to autotomize was determined by counting the number of elapsed frames between the first frame in which the blade was down and the first frame in which a lateral fissure of the body wall was evident (= onset of autotomy). Figure 4 shows a compilation the elapsed times from compression to onset of autotomy in 50 mid-sized worms. The median and mode of this distribution were both 133 ms (four video frames). Autotomy onset occurred at this time in twelve of 50 worms. Nine worms required more than 200 ms to autotomize, thus slightly skewing the distribution to the right. The shortest time for autotomy onset was 67 ms (two video frames); four of 50 worms autotomized at this time.

These results raise questions about the minimal stimulus duration required for autotomy. To test this, the duration of compression was varied from 65–540 ms using 95% compression. No worms autotomized when durations were set to 65 ms (n = 20) and 77 (n = 20) ms. Two of 20 worms autotomized when the duration was 91 ms. The percent autotomy increased as the duration increased, until 100% (20 of 20) autotomized when the duration was 540 ms (approximately 16 frames). These results demonstrate that there is a lower limit in the duration of the stimulus compression required for autotomy; this limit was approximately 65–77 ms.

Effect of amount and location of body compression

We next tested the possibility that there is a threshold stimulus (i.e. critical amount of compression) for autotomy induction and that this threshold differs in relation to the locus of compression. Figure 5 shows the relationship between changes in percent autotomy and amount of compression in mid-sized worms. This graph shows a direct effect of the amount of compression on percentage of autotomy. In a midbody location where body diameter is approximately 520 μm, about 30% of worms autotomized when the compression was 90%. However, all worms autotomized with 90–95% compression.

For studying compression effects in posterior segments, micrometer settings were adjusted to compensate for the relatively smaller segment diameter (approximately 430 μm). As with midbody compression, autotomy increased as the compression in the tail region increased. However, some worms (36%) autotomized at a compression value of 85%. When compression was increased to 90%, most worms autotomized (71%). Further increase in compression to 95% resulted in 100% autotomy. Thus, autotomy responses were more easily induced in tail than midbody regions.

Nicotine effects

Since chemical synaptic transmission in annelid neuromuscular junctions, and in at least some central synapses, is cholinergic (Gerschenfeld, 1973), we examined the possibility that the cholinergic agonist, nicotine, may somehow disrupt autotomy effector responses. Nicotine effects on autotomy were tested in mid-sized worms using a 15 min treatment prior to compression. A value of 95% compression was selected for all treated worms since this concentration normally results in 100% autotomy. There was a dose-dependent reduction in autotomy in relation to increased nicotine concentration (Figure 6). This trend was clearly evident for both midbody and tail regions. When nicotine concentration was higher than 0.01 mM, no worms autotomized in response to midbody compression. Likewise, autotomy was suppressed by nicotine in the tail region, where only 5% autotomized when exposed to 0.01 mM and none with 0.1 mM nicotine.

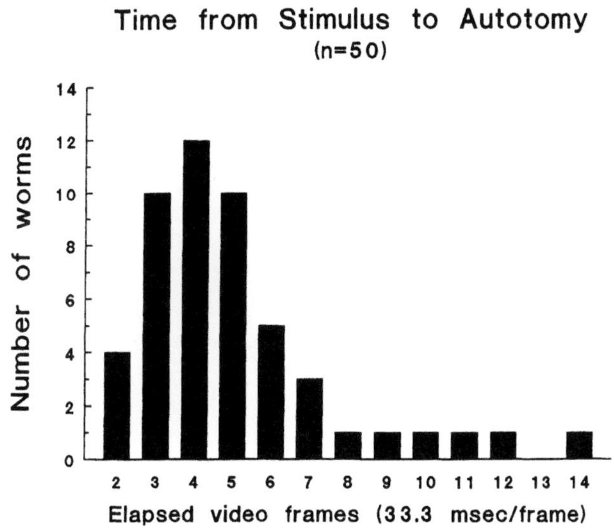

Figure 4. Histogram showing elapsed time from delivery of compression stimulus to onset of autotomy in 50 worms. Each bar indicates the number of worms whose autotomy onsets correspond to the number of video frames. The shortest time for autotomy onset was two video frames. Median and mode of this distribution were four video frames, corresponding to an elapsed time of 133 ms.

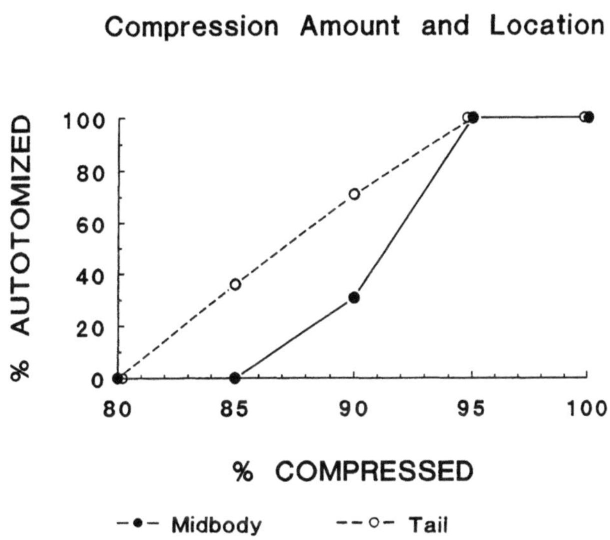

Figure 5. Graph showing relationship between compression amount and percent autotomy for two locations along the worm. Each point is a mean for 50 worms.

The question then arose as to whether nervous system function in non-autotomized, nicotine-treated worms was affected by 95% compression. A group of worms was selected for non-invasive electrophysiological testing before and after compression. The average medial giant fiber conduction velocity, before nicotine treatment and compression, was 8.3 ± 0.6 SD m/s. After recovery from nicotine (4 h), velocity was 8.1 ± 0.8 SD m/s. Similarly, the lateral giant fiber conduction velocity was 5.5 ± 0.8 SD m/s before treatment and compression, and 5.2 ± 0.6 SD m/s after recovery. Neither of these changes was statistically significant. Thus, there was no sign of functional damage to giant nerve fibers in worms following nicotine treatment and 95% compression.

Size effect

Additional experiments were done to compare tendencies for autotomy in newly hatched, mid-sized, subadult, and adult worms in response to a 90% compression stimulus. Autotomy occurred in 57% (n = 7) of newly hatched, 71% (n = 50) of mid-sized, 80% (n = 20) of subadult and 80% (n = 10) of adult worms.

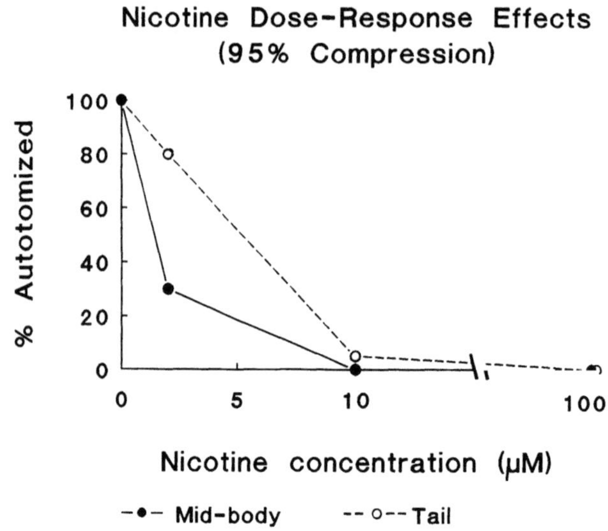

Figure 6. Dose-response effects of nicotine on autotomy. Each point indicates the percentage of worms (20 worms/ point) that autotomized at a particular concentration and body location.

Videotape analysis revealed that the timing and appearance of the autotomy responses in all of these stages were similar to those shown in Figure 3.

Discussion

We have developed an apparatus for reliably inducing and quantifying segmental autotomy in *L. variegatus*. Our results indicate that autotomy is a rapid response that may be reflexively triggered by sudden application of mechanical stimulation. Particularly effective stimulus parameters included a combination of focal application of transverse body compression at a level of at least 85% (see Figure 2) and a duration of at least 65 ms. These stimuli were severe enough to cause some overt trauma to segments but not severe enough to directly lacerate the body wall or disrupt functional integrity of the ventral nerve cord. With adequate stimulation, autotomy was initiated by the worm itself as a stereotyped, all-or-none response in which the body is cleanly and quickly separated into two fragments.

Since any segment in mid-body or tail regions appeared capable of initiating autotomy in all body sizes, it seems reasonable to hypothesize that sensory and effector elements involved in autotomy are metamerically arranged and present at all stages of postembryonic growth. These results raise questions regarding the reflex mechanisms of autotomy. Are there mechanosensory neurons that specifically detect compression or other traumatizing stimuli? What muscles or other effector structures are involved in quickly separating and sealing off the ends of body fragments? The rapid formation of a lateral fissure and the clean separation of anterior and posterior fragments (Figure 3) suggests that each segment may contain a pre-formed, circumferential autotomy site, perhaps comparable to the autotomy planes seen in the limbs of several arthropod groups (Bliss, 1960; Needham,1965; Foelix, 1996). Investigations are currently underway to examine effector mechanisms involved in autotomy, as well as possible similarities between stimulus-induced autotomy and spontaneous self-fragmentation that occurs in asexually reproducing cultures in this species.

The capability of initiating rapid autotomy in response to mechanical trauma likely provides numerous survival advantages in relation to predatory pressures that undoubtedly exist in the worm's littoral habitat. First, rapid autotomy could enhance the probability of escape and survival of one or both worm fragments. Since the site of autotomy is just anterior to the site of body compression, anterior segments would be more likely to be spared if the predatory attack involved a bite or pinch at mid-body or posterior locations. Second, the rapid sealing of the body wall could minimize the potential chemo-attractiveness of autotomized segments to predators. Rapid sealing could also reduce blood loss, osmotic stress and necrosis in segments adjacent to the autotomy plane. These factors, in combination with the consistent formation of a circumferential autotomy plane, would help to normalize the subsequent processes of blastema formation and segment regeneration.

The pharmacological effects of nicotine on responses to compression (Figure 6) raise the possibility that some controlling step(s) in the autotomy reflex pathway may utilize nicotinic cholinergic receptors, but the specific site and mechanism of nicotine action are unknown. We are currently using nicotine as a pharmacological tool (paralytic agent) for routinely blocking the autotomy response. Circumvention of autotomy by nicotine treatment is especially useful because it allows surgical invasion and manipulation of the central nervous system, thus introducing significant opportunity for studying processes of neural repair and regeneration.

References

Arnold, E. N., 1990. The throwaway tail. New Scientist 125: 42–45.

Bliss, D. E., 1960. Autotomy and regeneration. In T. H. Waterman (ed.), Physiology of Crustacea, Academic Press, New York: 561–589.

Cameron, G. R., 1932. Inflammation in earthworms. J. Pathol. Bacteriol. 35: 933–973.

Drewes, C. D. & C. R. Fourtner, 1990. Morphallaxis in an aquatic oligochaete, *Lumbriculus variegatus*: reorganization of escape reflexes in regenerating body fragments. Dev. Biol. 138: 94–103.

Emson, R. H. & I. C. Wilkie, 1980. Fission and autotomy in echinoderms. Oceanogr. mar. Biol. Ann. Rev. 18: 155–250.

Foelix, R. F., 1996. Biology of Spiders. Oxford University Press, New York: 229–230.

Gerschenfeld, H. M., 1973. Chemical sysnaptic transmission in invertebrate central nervous systems and neuromuscular junctions. Physiol. Rev. 53:1–119.

Goss, R. J.,1969. Principles of Regeneration. Academic Press, New York: 214–217.

Needham, A. E., 1965. Regeneration in the Arthopoda and its endocrine control. In V. Kiortsis & H. A. L. Trampusch (eds), Regeneration in Animals and Related Problems. North-Holland Publishing Company, Amsterdam: 284–287.

Rasmussen, E., 1953. Asexual reproduction in *Pygospio elegans* Claparède (Polychaeta sedentaria). Nature 171: 1161–1162.

Rogge, R. W. & C. D. Drewes, 1993. Assessing sublethal neurotoxicity effects in the freshwater oligochaete, *Lumbriculus variegatus*. Aquat. Toxicol. 26: 73–90.

Stephenson, J., 1930. The Oligochaeta. Clarendon Press, Oxford: 537–538: 639–640.

Hydrobiologia **406:** 263–269, 1999.
B. M. Healy, T. B. Reynoldson & K. A. Coates (eds), Aquatic Oligochaetes.

Helical swimming and body reversal behaviors in *Lumbriculus variegatus* (Annelida: Clitellata: Lumbriculidae)

Charles D. Drewes
Department of Zoology and Genetics, Iowa State University, Ames, IA 50011, U.S.A.
Phone: (515)294-8061; E-mail: cdrewes@iastate.edu

Key words: swimming, locomotion, reflexes, biomechanics, Lumbriculidae, blackworms, Clitellata, oligochaetes, avoidance behavior

Abstract

Two unusual locomotor behaviors (body reversal and helical swimming) are described and related to postembryonic body size in the freshwater oligochaete, *Lumbriculus variegatus* (Annelida: Clitellata: Lumbriculidae). Both behaviors occur as responses to tactile stimulation when worms are on smooth substrates that offer little or no traction or protection. Body reversal, evoked by touch to anterior segments, involves a stereotyped sequence of bending movements that effectively reverse head and tail positions in about 0.4 s in newly hatched worms and 0.6 s in juvenile and adult worms. Though little net shift in the body center occurs, reversal may optimize body positioning in preparation for swimming away from predatory threat. In contrast to reversal, swimming is evoked by touch to posterior segments and consists of a rapid, rhythmic sequence of helical body waves (frequency $\approx$ 9–11 Hz). Waves alternate between clockwise and counterclockwise helical orientations, with posterior passage of each wave providing forward thrust. Swim velocity and wave velocity increase with body size. Though total distance and duration of each swim episode is short ($\approx$ 1–2 body lengths in < 2 s), swimming may be an important means of predator avoidance in the littoral environment of these worms.

Introduction

Swimming is a relatively uncommon form of locomotion among aquatic oligochaetes (Annelida: Clitellata). It is apparently non-existent in the Family Tubificidae, but relatively common in the Naididae. Sperber (1948), for example, reviewed the systematics and natural history of approximately 100 naidid species, noting more than a dozen that were capable of swimming movements. Although she did not analyze or quantify swim movements, she noted two apparently different patterns of swimming, describing these as 'spiral' movements in some species and 'lateral' (or horizontal) movements in others.

Recently, Drewes & Fourtner (1993) provided the first detailed analysis of swimming behavior in any aquatic oligochaete. They showed that the so-called 'spiral' pattern of movements in the naidid, *Dero digitata* Oken, 1815, actually consisted of a rhythmic series of helical body waves. Curiously, successive waves alternated between left-handed and right-handed helical bending of the body, with posterior passage of each wave along the body providing forward thrust. Swimming in *Dero* occurred spontaneously or in response to posterior touch. However, it only occurred when worms were unconfined, on smooth substrates, or in 'open-water' conditions. Capabilities for touch-evoked swimming were also studied in relation to asexual fission, the normal mode of reproduction in *D. digitata*. Touch-evoked swimming responses occurred in both anterior and posterior zooids immediately after the zooids separated by transverse fission.

Here, I describe helical swimming in *Lumbriculus variegatus* Grube, 1884 (Family Lumbriculidae), the only representative of the Order Lumbriculida in which such behavior has been analyzed. In addition, I document a novel reflex response in which worms rapidly reverse head and tail ends in response to anterior tactile stimulation. Finally, I examine the dynamics of these unusual behaviors in relation to increases in body size that occur during postembryonic growth from hatching to sexually maturity (cf., Drewes & Brinkhurst, 1990).

Materials and methods

Lumbriculus variegatus were obtained in May from a marsh in Gull Point State Park (West Lake Okoboji, Iowa, U.S.A). *Lumbriculus* cocoons, each containing 4–12 orange embryos, were found attached to submerged, decaying leaves in shallow water at the edge of the marsh. Cocoons were maintained in the laboratory (21–22 °C) for about one week until worms emerged. Maximal length of these hatched worms (at rest) was about 5–6 mm. Juvenile and sexually mature adult worms were collected within the same habitat as cocoons. Resting lengths of these worms ranged from 30–40 mm and 70–85 mm (juvenile and adult, respectively).

Worms were placed in individual shallow containers filled with spring water (21–22 ° C). Containers for juvenile and adult worms were flat-bottomed, white pans (23×23 cm), each containing about 1500 ml water (≈ 3 cm water depth). Containers for newly hatched worms were 5.5 cm diameter Petri dishes containing 20 ml of spring water (≈ 1 cm water depth).

Touch stimuli were delivered with the tip of a handheld probe. The probe tip consisted of a thin rubber band loop (0.5 thickness for newly hatched worms, or 1.0 mm thickness for juvenile and adult worms). The loop was attached to the end of a wooden applicator stick (2 mm diam) and extended approximately 15 mm beyond the end of the stick. To evoke swimming or reversal responses, the probe was held just above the worm at a 30° angle to its long axis. Then, the head or tail was gently touched with the rubber loop. The flexibility of the loop ensured that worms were not injured when tactile pressure was applied. Touch stimuli were delivered at 15 min intervals to minimize habituation and optimize consistency in evoked responses.

Behavioral responses were recorded on videotape using a color video camera attached to a tripod. The camera lens was directly above the worm and shutter speed was 1/500 s. Jog and shuttle controls on the video cassette recorder allowed single-frame analysis of locomotor movements. Photographs of video images were obtained with a Sony UP1200A video printer.

The following parameters of swimming were examined:

(a) *number of helical body waves* produced per swim episode;
(b) *wave frequency,* expressed in Hz;
(c) *swim distance*, expressed as a percentage of resting body length and measured by determining the net translocation of the worm's head relative to a fixed point in its container; and
(d) *wave velocity*, determined by measuring rearward progress of the helical wave along the worm's body axis during two consecutive video frames.

Parameters related to reversal behavior included:

(a) estimated *angular change* in the worm's longitudinal axis caused by reversal, the maximum possible being 180°;
(b) net positional *change in the worm's body center* caused by reversal, expressed as a percentage of total body length, but without vectorial designation; and
(c) estimated *time from onset to completion of reversal,* as measured by the number of elapsed video frames (0.033 s/frame).

Results

Helical swimming responses

Spontaneous swimming, in the absence of experimenter-applied stimulation, was never seen. However, light tactile stimulation with the stimulus loop anywhere within the posterior one-half of the worm (i.e. between the tail tip and the geometric middle of the body) consistently evoked swimming responses in all sizes of worms. Tactile stimulation slightly anterior to the mid-body sometimes evoked swimming and/or body reversal, as described later. Thus, the sensory field for swimming behavior is slightly greater than one-half the worm's body length.

Swimming movements in all worms consisted of a series of rapid and rhythmic waves of helical body bending. Each wave began at the anterior end and rapidly progressed posteriorly along the body, thus propelling the worm forward (Figure 1). Analysis of freeze-frame video images showed that each successive wave alternated from clockwise (right-handed) to counterclockwise (left-handed) helical orientation. Each episode of swimming was always brief, often lasting less than 1 s. Occasionally, episodes were longer, but none lasted more than 3 s.

As shown in Table 1, the longest swim episodes were seen in juvenile worms in which a mean of

Table 1. Parameters related to swimming in worms of different sizes. Significance of differences between means were determined by *t*-tests

Parameter	(A) Newly hatched mean	± SD	(*n*)	(B) Juvenile mean	± SD	(*n*)	(C) Adult mean	± SD	(*n*)	Significance ($p <$) A–B	B–C	A–C
Number of waves per swim episode	5.9	± 2.4	(29)	14.0	± 6.2	(16)	6.4	± 3.4	(24)	0.001	0.001	ns
Wave frequency (Hz)	11.1	± 1.5	(29)	9.9	± 1.1	(16)	9.5	± 1.5	(21)	0.05	ns	0.001
Swim distance (% of body length)	89	± 45	(17)	128	± 43	(12)	80	± 43	(23)	0.05	0.01	ns
Swim velocity (mm/s)	11	± 2	(15)	47	± 9	(14)	74	± 12	(10)	0.001	0.001	0.001
Wave velocity (mm/s)	40	± 11	(21)	154	± 28	(16)	201	± 48	(16)	0.001	0.01	0.001
Ratio of swim velocity to wave velocity	0.30	± 0.11	(14)	0.31	± 0.08	(14)	0.38	± 0.11	(9)	ns	ns	ns

14 helical waves occurred per episode, or more than twice the number in newly hatched or adult worms. All of these values reflect the most vigorous swimming responses produced by individual worms in each size class. However, some responses to stimulation in every worm, regardless of size, were less vigorous and consisted of as few as 2–3 helical waves.

Despite wide variations in the number of swim waves per episode, the frequency of swim waves was remarkably invariant for all swim episodes in each size class (Table 1). The highest frequency (11.1 Hz) occurred in newly hatched worms. Progressively lower frequencies occurred in juvenile and adult worms. Although frequency in adults was only about 15% less than in hatched worms, this difference was significant ($p < 0.001$). Generally, wave frequency data shown in Table 1 indicate that the inherent frequency of outputs from the central pattern generator for swimming is relatively constant throughout postembryonic growth.

Comparisons of relative swim distance revealed that juvenile worms swam further than either newly hatched or adult worms, and these differences were correspondingly reflected in the number of waves per swim episode. Nevertheless, even juvenile worms rarely swam a distance of more than two body lengths.

A comparison of swim velocity in different worm sizes is shown in Table 1. Velocity significantly increased in relation to body size – there was nearly a seven-fold difference in velocity between the smallest and largest worms. The higher velocity in larger worms may result from greater propulsive forces due to any or all of the following factors:

(a) larger surface area of the body incorporated into the helical wave (i.e. larger stroke size);
(b) greater distance along the body over which each wave acts (i.e. longer stroke length); and
(c) faster wave velocity along the body (Table 1).

The velocity that helical waves move along the body of a worm provides an indirect measure of the rate that the neural excitation controlling swimming propagates along the ventral nerve cord. As shown in Table 1, mean wave velocity in newly hatched worms was 40 mm/s, or about 100 times slower than the conduction velocity of giant nerve fibers in these worms (Drewes & Brinkhurst, 1990). Wave velocity in adult worms was about 200 mm/s, or about 50 times slower than giant nerve fiber velocity in these worms. These differences make it highly unlikely that the worm's giant nerve fibers mediate the wave of central neural excitation that controls swimming.

Calculation of the ratio of swim velocity to wave velocity provides an estimate of the 'efficiency' with which posteriorly directed helical waves translate into forward progress of the body through water. Ratios that approach 1.0 indicate high translational efficiency and low slippage while those approaching 0.0 indicate

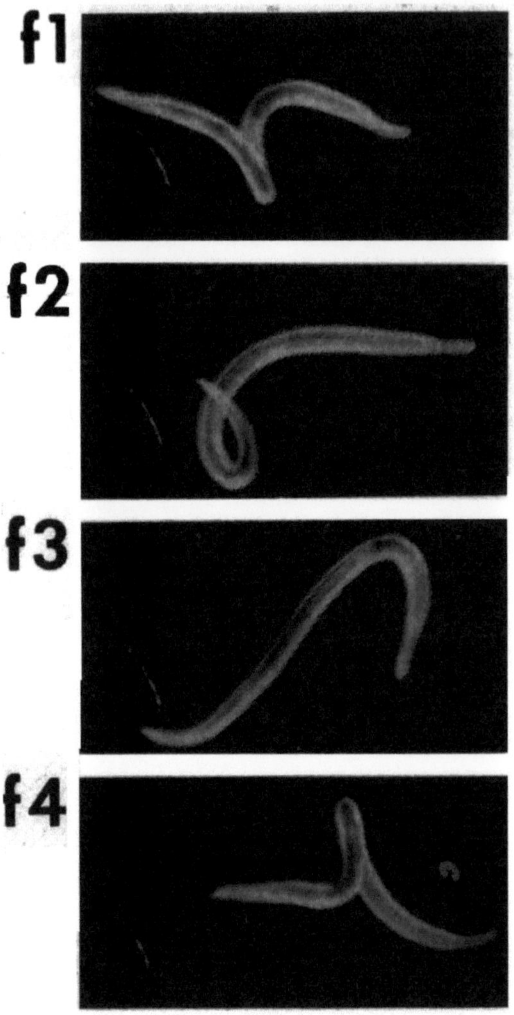

Figure 1. Four consecutive freeze-frame images of body movements during one complete cycle of swimming in a small worm. Frame 1 shows a helical wave (clockwise orientation) at a mid-body position. In frame 2, the same wave has progressed to near the tail end. In frame 3, a second wave begins near the head end. Frame 4 shows this second wave (counter-clockwise orientation) at a mid-body position. Thus, one complete cycle occurs in three video frames, a wave frequency = 10 waves/s.

low efficiency and high slippage. The ratios shown in Table 1 were similar in all sizes, ranging from 0.30–0.38. These values are similar to those derived for helical swimming in *D. digitata* (Drewes & Fourtner, 1993). Comparable values occur in other invertebrates (leeches, certain nematodes, polychaetes and insect larvae), which all utilize two-dimensional, sinusoidal undulatory movements (Taylor, 1952; Gray & Lissmann, 1964; Clark & Hermans, 1976).

Determination of a worm's swim velocity and body length allows calculation of a Reynolds number (R_e), a unit-less measure of the relative influences of viscous and pressure drag during locomotion in fluid or air (Nachtigall, 1983; Vogel, 1994). The R_e for newly hatched worms was 50, a value indicating that both viscous drag and pressure drag influence the dynamics and forward progress of swimming. This R_e is comparable to that of swimming copepods and very small flying insects (Strickler, 1975; Vogel, 1994). It is nearly identical to that in *D. digitata* (Drewes & Fourtner, 1993), a worm whose body size and swim velocity are similar to newly hatched *Lumbriculus*. In contrast, the R_e for swimming in adult *Lumbriculus* was 5000, a value indicative of less viscous drag and relatively stronger influences of pressure drag and inertial forces during its swimming. These values are comparable to those of medium and large flying insects (Nachtigall, 1983).

Body reversal responses

Light tactile stimulation with the stimulus loop anywhere within the anterior-most 30% of the worm's body consistently evoked body reversal responses in all sizes of worms. Tactile stimulation in locations somewhat posterior to this evoked either swimming or reversal, or occasionally both. Thus, the sensory field for reversal behavior is the spatial complement to the field for swimming, although there is an area ($\approx$ 10% of the worm's body length) in which the two fields overlap and either or both responses may be evoked.

Reversal behavior in newly hatched worms involved three main components of movement (Figure 2). The first component was rapid, end-to-end body shortening that occurred within the first few video frames following delivery of the touch stimulus. The second component was a rapid, J-shaped bending of the posterior end of the body; this bending was concurrent with body shortening. The third component was an anteriorly-directed, wave-like shift in the locus of body bending. As this wave of bending progressed to the mid-body, the worm formed an 'alpha-shaped' loop (see Figure 2, upper frames 4–6). With further progress of the loop, the head end of the worm swung to an open and more straightened position. Passage of this wave of bending was relatively gradual, requiring about 0.3 s (9 video frames) for completion. The net result of reversal was a change of approximately 150° in the longitudinal axis of the body but only a slight translocation in the center point of the worm's body (Table 2).

Juvenile and adult worms also exhibited body reversal in response to anterior touch, although the sequence of reversal movements required significantly

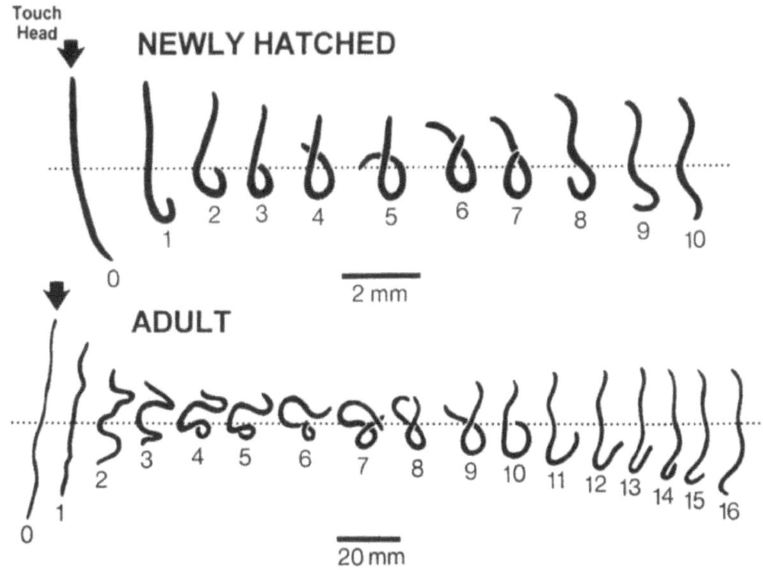

Figure 2. Freeze-frame images (silhouettes) of body reversal responses to head touch in newly hatched and adult worms (upper and lower sequences, respectively). Note the wave of body bending that begins in posterior segments and progresses anteriorly as a moving loop, as shown in frames 3–7 (upper sequence) and 6–10 (lower sequence). Generation of this moving loop is a key event for reversal of the body axis.

more time and was more complex and variable in appearance than in hatched worms (Table 2; Figure 2). Three phases of movement were evident. The first phase was rapid, end-to-end body shortening that occurred within one video frame after the touch stimulus. The second phase involved simultaneous and oppositional body bending in head, mid-body, and tail regions, thus resulting in an 'omega-shaped' appearance of the body. Acquisition of this shape often required several video frames of time. The third phase of movement involved a wave-like, anterior progression of the tail bending initiated during the preceding phase. As this bending progressed anteriorly, it was preceded by straightening (unbending) of mid-body and head regions. Thus, the body formed a moving loop, as shown in Figure 2 (see frames 6–9 in bottom sequence). Gradually, over the next few frames, anterior progression of the wave opened the loop, and reversal was complete.

Execution of the entire reversal sequence required about 0.6 s (18 video frames) in both juvenile and adult worms. The net result of reversal was a change of 140–150° in the longitudinal axis, with only a slight translocation in the center point of the worm's body (Table 2).

Discussion

Adaptive significance of swimming and reversal

Swimming is a rather uncommon means of locomotion in aquatic oligochaetes; it does not occur in tubificids, which often occupy tubes or tunnels in muddy sediments (Stephenson, 1930). However, it is seen in a few naidid species that occupy shallow habitats containing emergent vegetation and vegetable debris (Stephenson, 1930; Sperber, 1948). One possible significance of swimming in such habitats is that it may allow worms to be transiently planktonic. This may provide a faster and more direct means, compared to the alternative of crawling, for escaping adverse conditions and locating more suitable habitat.

Although the mechanics and frequency of helical swimming movements in *Lumbriculus* are similar to those previously described in *D. digitata* (Drewes & Fourtner, 1993), swim behaviors in these two species differ with respect to several important features. First, swimming in *Dero* is often spontaneous, whereas it is always stimulus-evoked in *Lumbriculus*. Second, individual swim episodes in *Dero* were much longer in duration and resulted in much greater swim distances than in *Lumbriculus,* in which swim episodes were brief (usually less than 2 s) and swim distances short (about 1–2 body lengths; Table 1). Thus, in contrast to *Dero*, it seems highly unlikely that the adaptive significance of swimming in *Lumbriculus* is

Table 2. Parameters related to body reversal in worms of different sizes. Significance of differences between means were determined by *t*-tests

	(A) Newly hatched			(B) Juvenile			(C) Adult			Significance ($p <$)		
Parameter	mean	± SD	(n)	mean	± SD	(n)	mean	± SD	(n)	A–B	B–C	A–C
Angular change in body axis (°)	153	± 23	(25)	149	± 23	(15)	139	± 29	(24)	ns	ns	0.05
Change in center of body (% of body length)	16	± 13	(25)	13	± 8	(15)	12	± 10	(24)	ns	ns	ns
Time to reverse (s)	0.37	± 0.07	(17)	0.59	± 0.19	(12)	0.59	± 0.18	(24)	0.001	ns	0.001

related to achieving either sustained vertical movements in the water column (i.e. planktonic existence) or long-distance, horizontal migration.

A more likely explanation is that swimming and reversal responses, separately or in combination, are important behavioral strategies for eluding predators throughout the worm's postembryonic life. Predatory threat to the worm's head results in body reversal so that the worm is well positioned to execute swimming in a direction *away* from the predator. Swimming responses enable rapid, short-distance retreat from a threatening stimulus in instances when worms are on a smooth substrate without the benefit of either protection or traction.

In the laboratory, swimming often occurs when worms are massed together, in the open, around a food source. Then, disturbance by repeated tactile stimulation evokes a relatively rapid and stereotyped sequence of events, which begins with worms executing body reversal responses while still *en masse*; this effectively re-positions the worm's heads away from the food source. The sequence culminates as many worms simultaneously swim away from the food, resulting in rapid dis-aggregation of the assembled mass (Figure 3). The result is an abrupt and dramatic burst of radiating locomotor movements. It seems reasonable to hypothesize that such movements, collectively, may startle or confuse a predator while, at the same time, enhance the probability of the survival of each worm. In summary, helical swimming and body reversal behaviors appear well suited for an oligochaete, such as *Lumbriculus variegatus*, which opportunistically feeds and freely moves about within freshwater, littoral habitats.

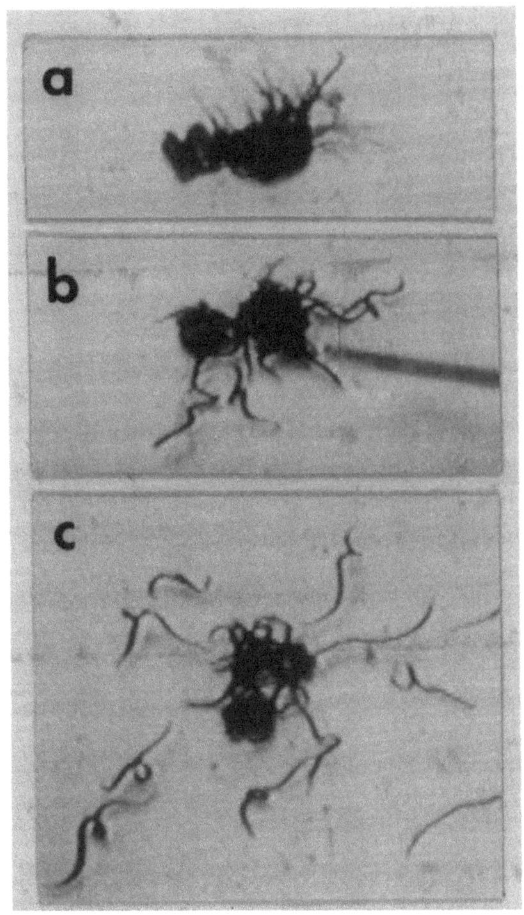

Figure 3. Video images of disaggregation in feeding worms. (a) Before stimulation worms were tightly clustered around a food pellet and their tails were extended outward. (b) After tactile stimulation with an applicator stick, a few worms reversed head and tail positions and began to swim away. (c) With additional stimulation, more worms began to swim, creating an abrupt, radiating burst of swimming movements.

References

Clark, R. B. & C. O. Hermans, 1976. Kinetics of swimming in some smooth-bodied polychaetes. Proc. zool. Soc. Lond. 178: 147–159.

Drewes, C. D. & R. O. Brinkhurst, 1990. Giant nerve fibers and rapid escape reflexes in newly hatched aquatic oligochaetes, *Lumbriculus variegatus* (Family Lumbriculidae). Invert. Reprod. Devel. 17: 91–95.

Drewes, C. D. & C. R. Fourtner, 1993. Helical swimming in a freshwater oligochaete. Biol. Bull. 185: 1–9.

Gray, J. & H. W. Lissmann, 1964. The locomotion of nematodes. J. Exp. Biol. 41: 135–154.

Nachtigall, W., 1983. The biophysics of locomotion in water. In W. Hoppe, W. Lohmann, H. Markl & H. Ziegler (eds), Biophysics. Springer-Verlag, New York: 587–600.

Sperber, C., 1948. A taxonomical study of the Naididae. Zool. Bidr. Upps. 28: 3–296.

Stephenson, J., 1930. The Oligochaeta. Clarendon Press, Oxford: 625–626.

Strickler, J. R., 1975. Swimming of planktonic *Cyclops* species (Copepoda, Crustacea): pattern, movements and their control. In T. Y.-T. Wu, C. J. Brokaw & C. Brennan (eds), Symposium on Swimming and Flying in Nature, vol. 2. Plenum, New York: 599–613.

Taylor, G., 1952. Analysis of the swimming of long and narrow animals. Proc. r. Soc., Lond. (A)214: 158–183.

Vogel, S., 1994. Life in Moving Fluids: The Physical Biology of Flow. Princeton University Press, Princeton, New Jersey, pp. 67-73, 243–244.

Hydrobiologia **406**: 271–280, 1999.
B. M. Healy, T. B. Reynoldson & K. A. Coates (eds), Aquatic Oligochaetes.

Bioaccumulation test with Tubificid Sludgeworms in artificial media – development of a standardisable method

Philipp Egeler[1], J. Römbke[1], M. Meller[1], Th. Knacker[1] & R. Nagel[2]
[1]ECT Oekotoxikologie GmbH, Böttgerstr. 2-14, D-65439 Flörsheim, Germany
[2]Technische Universität Dresden, Institut f. Hydrobiologie, D-01062 Dresden, Germany

Key words: tubificidae, bioaccumulation, artificial sediment, risk assessment

Abstract

Lipophilic chemicals entering aquatic ecosystems often tend to adsorb to sediments. Although often not bioavailable to pelagic organisms, these compounds can be subject to bioaccumulation and thus pose a potential threat for sediment-dwelling organisms. The assessment of bioaccumulation is currently based on the extrapolation of fish bioconcentration data. Thus, measuring the bioaccumulation in benthic organisms gives more detailed information on the hazard of chemicals for aquatic ecosystems. The freshwater tubificids *Tubifex tubifex* and *Limnodrilus hoffmeisteri* were chosen as test organisms. Breeding of the oligochaetes as well as bioaccumulation studies were carried out in a system consisting of artificial sediment and reconstituted water. ^{14}C-lindane (γ-hexachlorocyclohexane, γ-HCH) and ^{14}C-hexachlorobenzene (HCB) served as model substances. Additionally, ^{14}C-3,4-dichloroaniline (3,4-DCA) was tested. Uptake and elimination of lindane and HCB were examined. Bioaccumulation factors (BAF) were determined from the ratio of concentration in oligochaete tissue to concentration in sediment. The tubificids accumulated ^{14}C-lindane, ^{14}C-HCB, and ^{14}C-3,4-DCA by factors of 4.7 (mean value; $n = 4$), 6.6 (mean value; $n = 4$) and 13.2, respectively, based on wet weight and radioactive concentrations. No major metabolites were detected in worms, sediment and water. Elimination kinetics were examined for ^{14}C-lindane and ^{14}C-HCB. When transferred to clean sediment, the worms eliminated both test substances rapidly and nearly completely. The tubificid bioaccumulation factors are contrasted with corresponding fish bioconcentration factors from literature. The presented data indicate once more that the assessment of bioaccumulation based exclusively on lipophilicity and on the extrapolation of fish bioconcentration data to other organisms or environmental compartments is not appropriate.

Introduction

Lipophilic substances entering aquatic ecosystems often have a strong affinity for suspended particulate matter and subsequently become associated with sediments. Thus, endobenthic animals are usually more exposed to such chemicals than fish, and there is the possibility of bioaccumulation in these sediment dwelling organisms (e.g. Oliver, 1987; Reynoldson, 1987; Klerks & Bartholomew, 1991). This means that the current bioaccumulation model 'fish' (e.g. OECD, 1996) might not be adequate for other aquatic animals, especially for sediment organisms. The measurement of bioaccumulation in benthic animals may therefore improve the assessment of ecotoxicological risks of sediment associated chemicals for aquatic ecosystems (e.g. Ingersoll et al., 1994).

The main goals of the presented work were:

1. Development, performance and standardisation of a method for measuring bioaccumulation of sediment-associated chemicals in benthic organisms in a sediment water system.

2. Elaboration of an artificial sediment for the use in both culture of the test animals and in the tests, with special respect to the standardisation of the method.

3. Selection of suitable test species and standardisation of laboratory culture.

4. Examination of the influence of hydrophobicity or lipophilicity on bioaccumulation from sediment by testing of chemicals with different lipophilicity.
5. Discussion of the resulting bioaccumulation of the test substances from the sediment with respect to risk assessment procedures currently used in Europe.

Definitions

Before going into methodological details, some important definitions connected with bioaccumulation in accordance with ASTM (1995), U.S. EPA (1995) and OECD (1996) are given. Bioaccumulation, which is defined as the increase of a chemical in an organism relative to the concentration in the ambient medium or the food, consists of several processes and arises from the sum of all possible exposure routes. These processes are on one hand the bioconcentration, which results exclusively from uptake from the ambient medium via the body surface. In fish, this may be the gills or the integument. Biomagnification on the other hand is defined as the increase of body concentration resulting from contaminated food, usually along a predator-prey series. The loss of a chemical from the organism by active or passive processes is called elimination. When an equilibrium is reached between uptake and elimination processes, the concentration in the organism remains constant and one talks of the steady state. During this steady state the main criteria associated with bioaccumulation can be estimated:

1. The bioconcentration factor (BCF), which is the quotient of the chemical's concentration in the organism, for example a fish, and its concentration in the ambient water at steady state, and
2. the bioaccumulation factor (BAF), which is defined as the chemical's concentration in the organism, for instance a benthic oligochaete, divided by its concentration in the medium, for example the sediment, which serves both as medium and food; the BAF takes into account all uptake pathways.

The ratio of the chemical's concentration in the medium and its concentration in the organism at any given time during the exposure phase when no steady state is reached is termed accumulation factor (AF).

Sediment and water

The artificial sediment used in this study (Egeler et al., 1997) was to consist of the major components of field sediments, including sand, clay silt, and an organic fraction (e.g. Walsh et al., 1992; Suedel & Rodgers, 1994). The composition is mainly based on the OECD artificial soil used for earthworm toxicity tests (OECD, 1984), and is meant as a compromise between natural conditions, standardisability and practicability. The sediment contents (referring to dry weight) are listed in Table 1. The pH of the sediment was adjusted to approximately 7 using $CaCO_3$ according to OECD (1984). The sediment was moistened with demineralised water to a percentage of about 46% of the dry weight, which gave a homogenous concistency and minimised water separation after settling.

Reconstituted water as described by OECD (1992) was chosen as the overlying medium, again with respect to standardisation. The water contained CaCl (294 mg l^{-1}), $MgSO_4$ (123.25 mg l^{-1}), $NaHCO_3$ (64.75 mg l^{-1}) and KCl (5.75 mg l^{-1}). Its pH was about 8, its total hardness between 1.8 and 2.7 mmol l^{-1}, and the oxygen content ranged between 80 and 100% saturation.

Test organisms

The freshwater oligochaetes *Tubifex tubifex* Müller and *Limnodrilus hoffmeisteri* Claparède (Tubificidae, Oligochaeta) were chosen as test organisms. Both are endobenthic species, which means they live in the sediment and feed on it (e.g. Wachs, 1967), therefore exhibiting maximum contact with the substrate. All exposure routes can contribute to bioaccumulation in these organisms (Reynoldson, 1987). Due to their reputation of being relatively tolerant of stress caused by organic pollutants (e.g. Chapman & Brinkhurst, 1984; Giesy & Hoke, 1989) they were considered suitable for bioaccumulation testing. Additionally these cosmopolitan aquatic oligochaetes (e.g. Brinkhurst, 1980) are known to be cultured easily in the laboratory (e.g. Kennedy, 1966, Kosiorek, 1974, Wiederholm et al., 1987; Reynoldson et al., 1991; Lotufo & Fleeger, 1996).

For this study, the worms were bred in the artificial sediment at 20 ± 2 °C in single species cultures. The worms were fed a suspension of finely ground TetraMin flakes. The water body was gently aerated and changed once a week. The artificial sediment was

Table 1. Composition of the artificial sediment

Sediment fraction	Constituent	Percentage of sediment dry weight	Characteristics
Sand	Quartz sand	76.0%	Grain size: more than 80% of the particles 0.05–0.125 mm
Clay	Kaolinite clay	22.0%	Kaolinite content $\geq$ 30%
Organic fraction/Silt	Sphagnum peat	2.0%	Sphagnum peat, in accordance with the OECD Guideline 207 (OECD, 1984)
	$CaCO_3$	$\leq$0.1%	$CaCO_3$, pulverized, chemically pure
	Aqua demin. added	$\approx$ 46.0%	conductivity $\leq 10\ \mu S\ cm^{-1}$

renewed at least every three months to avoid excessive build-up of ammonia in the stock cultures. Under these conditions, both species showed good reproduction and a synchronised culture was established with a generation cycle of approximately 10 weeks (c.f. Egeler et al., 1997). The test conditions were nearly identical to the culture conditions, but the worms (exposed and control animals) were not fed and the water was not exchanged during the exposure period (up to 12 d). In some of the experiments, the water body was not aerated to minimise evaporation of test substances.

Model chemicals

The model substances were chosen according to their wide distribution and different lipophilicity. The lipophilicity is represented by the logarithm of the octanol-water-partition coefficient, the log K_{ow}, which is for some chemical classes in good correlation with adsorptivity and bioaccumulation. The insecticide lindane (γ-hexachlorocyclohexane, γ-HCH) was chosen as a chemical of medium lipophilicity. It has a log K_{ow} of about 3.6 (Rippen, 1991). The second test substance, hexachlorobenzene, which was formerly used as a fungicide, shows a significantly higher lipophilicity than lindane (log K_{ow} 5.7 according to Rippen, 1991).

Besides these two model substances 3,4-Dichloroaniline (3,4-DCA), which belongs to the weakly lipophilic environmental chemicals, has to be mentioned. Its log K_{ow} is 2.7 according to BUA (1994). This substance was not originally selected for this study, but since there was indication of a relatively strong affinity to sediments in spite of its weak lipophilicity (BUA, 1994; Schmitz, 1997), an accessory short experiment (only uptake) was performed with *T. tubifex*.

All tests were done with ^{14}C-radiolabelled test substances. The tracers were applied to the sediment by coating the quartz sand fraction. The radioactivity in the samples was measured using Liquid Scintillation Counting. The test substance concentrations in the whole sediment were more than 10 times below the LOECs for sublethal parameters derived from previously performed short term toxicity tests (Meller et al., 1998).

Test apparatus and procedures

To prevent loss of volatile test substances, a closed test system was used, which is shown in Figure 1. In this closed chamber, the test vessels were set up. For each sampling date there were four replicate vessels (100 ml-glass tubes) containing water, spiked sediment (appr. 30 g wet weight) and 10 oligochaetes. Background activities were measured in four additional replicates containing clean instead of spiked sediment. The system was supplied with moistened air using a pressure pump. A suction pump passed the exhaust air through a series of ice cooled washing bottles. The first bottle was empty and retained evaporated water. The following bottles contained toluene

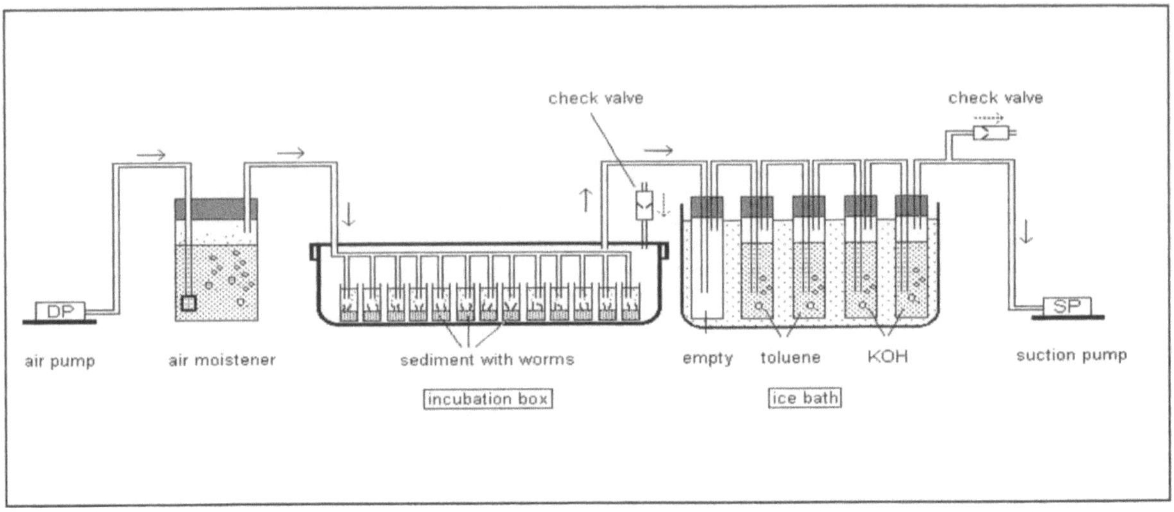

Figure 1. Test apparatus, schematic setup (from Egeler et al., 1997).

in order to trap evaporating test substance, and in the last traps, there was a solution of potassium hydroxide to catch radioactive carbon dioxide, which might result from degradation of the test substance. With this system, recoveries of the test substances of more than 95% were obtained.

At each of at least six sampling dates, four replicates were analysed for radioactivity in worms, sediment and water according to methods described by Kukkonen & Landrum (1994) and Egeler et al. (1997). Based on log K_{ow} and pre-tests, the exposure period was chosen to be 6 d for lindane and 3,4-DCA, and 12 d for HCB. For several practical reasons the exposure in our study did not exceed 12 d: The worms were not fed during the study, in order to avoid influence of added food on chemical adsorption and on build-up of ammonia; thus, exposure time (and subsequent elimination time) was limited. Additionally, the development of a test method always underlies compromises with respect to cost-benefit considerations, and a longer experimental period would require much more expenses including amount of test substance, replicates, animals, etc..

When steady state was reached, the worms of the remaining replicates were transferred to test vessels containing uncontaminated sediment and water to eliminate accumulated test substance. Again, radioactivity in the compartments was measured over time. Accumulation factors (AF) were calculated from the ratio of radioactive concentrations in worm and sediment samples of the uptake phase. For the elimination phase, the radioactive concentration relative to the steady state concentration in the worms was used to follow the kinetics. Nonlinear regression was applied to fit a curve to the data. A detailed description of sampling and analysis is given in Egeler et al. (1997).

So far, kinetics and bioaccumulation factors (BAF) were based on total radioactivity, and metabolites were not taken into account. To determine the bioaccumulation factor of the mother compound, samples of oligochaetes, sediment and water, taken at steady state, were analysed by High Performance Liquid Chromatography (HPLC) for their percentage of parent compound.

Results and discussion

Using the following Figures as examples, the experimental procedure to determine uptake and elimination kinetics can be described. In Figure 2, the uptake and elimination kinetics of lindane in *Tubifex tubifex* are shown. Figure 3 demonstrates uptake and elimination of hexachlorobenzene in *Limnodrilus hoffmeisteri*. As reported elsewhere (Egeler et al., 1997), the different species did not exhibit significantly different kinetics.

The Figures show the time course of test substance accumulation (part a) and elimination in the worms during the experiments (part b). The data points in the plots represent the mean AF (uptake) and the mean percentage of steady state activity (elimination phase) in the oligochaetes derived from the mentioned four replicates of each sampling date. The error bars stand for the corresponding standard deviation. A rapid increase of radioactivity in the worms occurred during

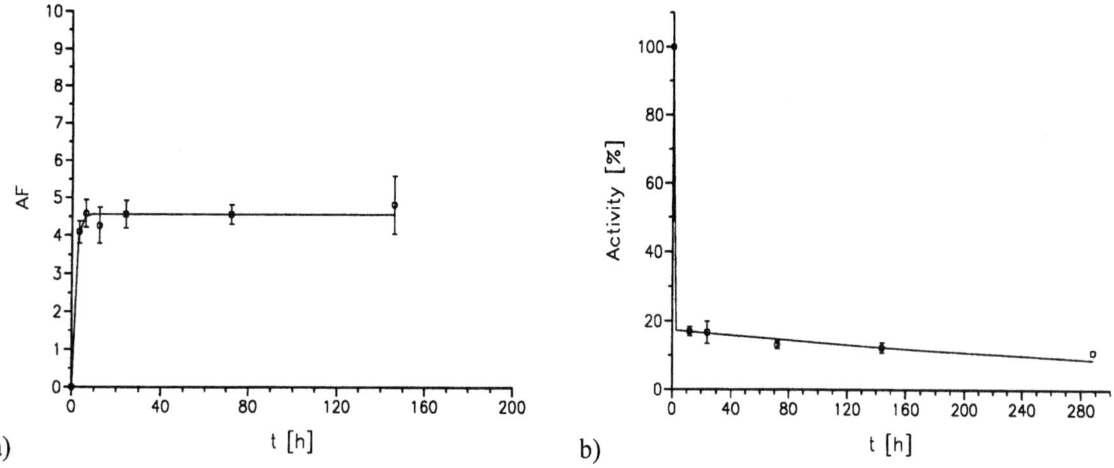

Figure 2. ^{14}C-lindane kinetics in *T. tubifex* during uptake (a) and elimination phase (b); AF = accumulation factor; activity = ^{14}C-activity in% of steady state activity in worms; $n = 4$; error bars = sd.

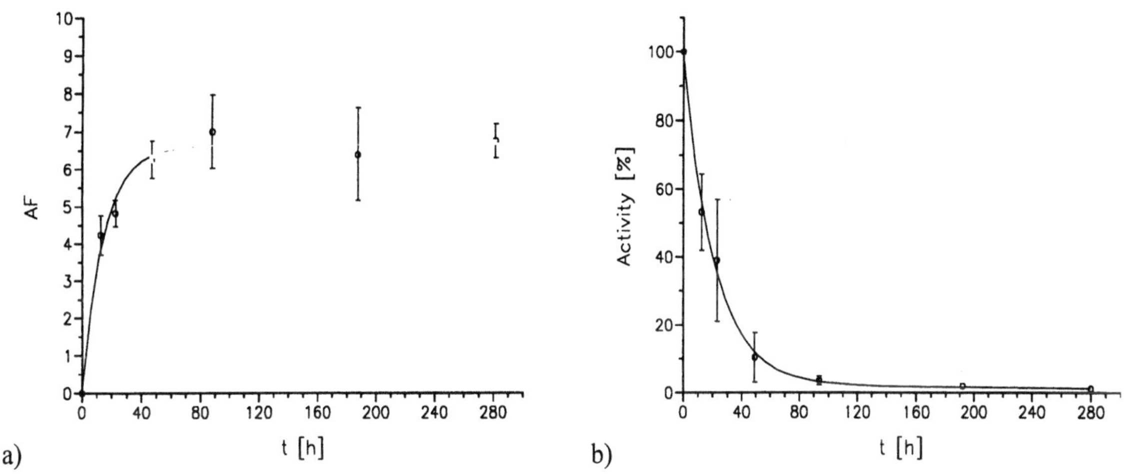

Figure 3. ^{14}C-HCB kinetics in *L. hoffmeisteri* during uptake (a) and elimination phase (b); AF = accumulation factor; activity = ^{14}C-activity in% of steady state activity in worms; $n = 4$; error bars = sd.

the initial hours of the uptake phase. This increase appeared much faster for lindane (Figure 2 a) than for hexachlorobenzene (Figure 3 a). After a relatively short period of uptake (15–24 h for lindane, 109–152 h for HCB), the plateau or steady state was reached, and the bioaccumulation factor was calculated dividing the concentration in the oligochaetes by the concentration in sediment. The curves were fitted to these data using a 1-compartment model and nonlinear regression analysis. The uptake rate constants were determined for the lindane and HCB experiments. They are mentioned in Egeler et al. (1997).

There is evidence from some studies (e.g. Reynoldson et al., 1989; Füll, 1996), that during exposure periods exceeding the time needed to establish a (first) plateau the accumulation factor (AF) of certain substances in tubificids and earthworms may increase once more. This phenomenon, however, was accompanied – at least in the mentioned studies – by a strong increase of variation in individual AF. Füll (1996) found metabolism of the ^{14}C-labelled test substance (lindane) to be the reason for the second increase of AF in *Lumbricus rubellus*. Since in our study only negligible amounts of lindane metabolites were found, the steady state for the parent compound is assumed to have been established within the exposure period. Reynoldson et al. (1989) examined the uptake of ^{14}C-Trichlorophenol in *T. tubifex* from a natural sediment and observed a second increase of activity connected with rising variability in the oligochaetes after 336 h. They suggested the more variable biological uptake processes and growth of the worms to be responsible for the large variability. For this high variation, the resulting BAF should be looked at carefully with respect to the uncertain equilibrium. HCB, however, was found to reach 90% of steady state after approximately 139 hours (Reynoldson et al., 1989), which is quite similar to our findings.

Having reached the plateau, the remaining worms were transferred to vessels containing clean water and sediment for the elimination phase (Figures 2 b and 3 b). They rapidly lost the test substance, so that after 12 h for lindane, the worms contained only 19% of the steady state body concentration. After this rapid decrease, the radioactivity in the worms declined very slowly. Thus, the elimination appears clearly biphasic, showing a rapid and a slower phase.

This biphasic elimination may be explained with the organism model consisting of two body compartments. After having entered the organism, the substance is distributed to these compartments. When the organism, in this case the worm, is transferred to uncontaminated sediment, the chemical is lost from these compartments with different velocities (e.g. Spacie & Hamelink, 1982).

In Table 2, the initial loss of the substance (from 80 to 90% of the steady state concentration) is then assigned to compartment A, for example the body wall, with a high elimination rate constant. Since gut clearance was not measured, and the worms including gut contents were processed and analysed for radioactivity, the question, if the initial, rapid lindane elimination could be mainly due to eliminated chemical adsorbed to gut contents, is assessed by the following considerations. The measured lindane concentration in the whole oligochaete body was more than four times the concentration in the sediment. According to Oliver (1987), the gut contents contained nearly the same concentration of test substance as the ambient sediment. The average portion of gut contents was measured by Oliver (1987) to be 15% of worm dry weight. From this value, a portion of 2.9% of worm wet weight can be calculated. Thus, assuming the gut contents were the compartment in which the lindane was accumulated, the concentration in gut contents (C_g) must have been higher than the contaminant concentration in the ambient sediment (C_s) by a factor of more than 100. This factor was calculated using the following equation:

$$C_g = C_{worm}{}^*(\%\ of\ worm\ wet\ weight/\%\ of\ gut\ content\ of\ worm\ wet\ weight),$$

$$C_g = C_{worm}{}^*(100/2.9)$$

e.g.:

$$C_g = 40\,000 dpm\ g^{-1*}(100/2.9) = 1\,379\,310\, dpm\ g^{-1} = C_s{}^*137.9$$

Although selective adsorption of lindane to certain sediment fractions might occur (e.g. Wirth, 1985), the selective ingestion of exclusively these fractions by the tubificids can virtually be ruled out according to microscopic examination of tubificid gut content in our lab. Findings on substrate selection reported by McMurtry et al. (1983) gave no indication of such a highly selective ingestion by the worms. If lindane was not concentrated in the sediment during gut passage by some extremely effective physiological mechanism, then elimination of lindane contaminated gut material is not the main source for the rapid elimination from compartment A, and the major portion of accumulated lindane was associated with worm tissue, not with gut content.

The second compartment B, from which the substance is lost very slowly – indicated by a very low elimination rate constant – could be the lipid fraction of the worms, in which a part of the substance has been stored. According to the low lipid content of tubificid oligochaetes (Whitten & Goodnight, 1966; Oliver, 1987) this compartment appears much smaller than compartment A. Although the elimination of hexachlorobenzene appears much slower and smoother than the lindane elimination, the application of a 2-compartment regression model showed a better fit than a 1-compartment model. With the coefficient of determination, $r^2 > 0.99$ the 2-compartment elimination model describes the elimination of both lindane and HCB very well.

1- as well as 2-compartment models were tested for description of both uptake and elimination kinetics. For the lindane and HCB uptake phase, a 2-compartment model could not be fitted to the data by the calculating software. The 1-compartment uptake model, however, showed very good fit ($r^2 = 0.78$–0.99 for lindane, and 0.99–1.00 for HCB), so this model was assumed to describe the uptake sufficiently.

Table 3 shows the bioaccumulation factors of each of the experiments relative to the sediment concentration based on total radioactivity as well as the BAF of the parent compound. Lindane was accumulated by both species by a factor of about five, while hexachlorobenzene reached a bioaccumulation factor of about six. Since the HPLC analysis revealed no major metabolites for lindane and hexachlorobenzene, the bioaccumulation factors corrected for parent com-

Table 2. Elimination characteristics of ^{14}C-lindane and ^{14}C-HCB in *Tubifex tubifex* and *Limnodrilus hoffmeisteri*

Elimination	A [%]	k_a [h^{-1}]	B [%]	k_b [h^{-1}]	residue [%]	r^2
Lindane						
T. tubifex	84	0.352	16	0.001	11	0.999
L. hoffmeisteri	90	0.195	10	0.002	7	1.000
HCB						
T. tubifex	87	0.037	13	0.004	4	0.998
L. hoffmeisteri	97	0.047	3	0.002	1	0.997

A = 'fast' compartment [% of steady state concentration]; k_a = elimination rate constant for compartment A; B 'slow' compartment [% of steady state concentration]; k_b = elimination rate constant for compartment B; residue = concentration in worms after 12 d elimination [% of steady state concentration]; r^2 = coefficient of determination.

Table 3. Uptake of ^{14}C-lindane, ^{14}C-HCB and ^{14}C-3,4-DCA in *Tubifex tubifex* and *Limnodrilus hoffmeisteri*

	BAF	BAF_{pc}
Lindane		
T. tubifex	4.7 (0.14)	4.37 (0.13)
L. hoffmeisteri	4.6 (0.71)	4.52 (0.63)
HCB		
T. tubifex	6.2 (1.98)	6.16 (1.97)
L. hoffmeisteri	7.0 (0.42)	6.92 (0.49)
3,4-DCA		
T. tubifex	13.2 (2.22)[1]	–

[1]Accumulation factor (AF) after 146.5 h, no steady state. BAF = bioaccumulation factor for ^{14}C-activity (mean values; sd in brackets); BAF_{pc} = BAF corrected for parent compound; (mean values; sd in brackets).

pound – termed BAF_{pc} – do not diverge substantially from the ones based on total radioactivity.

Oliver (1987) reported lindane and HCB uptake by tubificids in spiked field sediments, and gave worm and sediment concentrations referring to dry weight. Corrected for worm and sediment wet weight (assuming 87% and 80% water content, respectively), these values were 4.2 after 4 d, 6.3 after 11 d, and 4.9 after 79 d for lindane, and 2.6 after 4 d, 4.5 after 11 d and 2.0 after 79 d for HCB. Reynoldson et al. (1989) measured HCB uptake in *T. tubifex* from a spiked field sediment and found a similar BAF of approximately 2, calculated for worm and sediment wet weight (assuming 87% and 75% water content, respectively). The lindane and HCB BAF of our study are in the same order of magnitude as the values by Oliver (1987).

The initially mentioned ^{14}C-3,4-DCA was accumulated by *Tubifex tubifex* by a factor of 13 relative to the sediment after 146 h. In this case, no determination of mother compound or metabolites was performed, which might have indicated possible reasons for the relatively high accumulation.

Aspects of risk assessment

Compared to threshold values indicating moderate to high bioaccumulation (Franke et al., 1994), the tubificid BAF are very low. The following considerations are meant to discuss these low values in the context of risk assessment. Table 4 shows the test substances and their bio*accumulation* factors for tubificid oligochaetes (*T. tubifex* and *L. hoffmeisteri*) in contrast to the bio*concentration* factors for a test fish recommended by OECD (1996) for bio*concentration* testing, the zebrafish *Brachydanio rerio* (Hamilton-Buchanan). This comparison of tubificid BAF with fish BCF appears quite inappropriate, but nevertheless fish BCF often serve as the only measured endpoints when assessing bioaccumulation. For example, the only guideline, which is accepted by the European Commission (EU, 1996) to measure and evaluate bioaccumulation in aquatic systems in general, is still the fish bioconcentration guideline No. 305 "Bioconcentration: Flow-through fish test" (OECD, 1996). By putting these parameters (BCF and BAF) together in one Table, this unsuitable comparison demonstrates that the extrapolation of bioaccumulation from fish BCF to benthos BAF is not possible.

Table 4. Accumulation and body concentrations of 3,4-DCA, lindane and HCB in tubificids and fish

Substance		Tubificids (*T. tubifex* and *L. hoffmeisteri*)			Fish (*B. rerio*)		
	log K_{ow}	C_s [μg kg^{-1}]	BAF (mean)	C_a [μg kg^{-1}]	C_w [μg l^{-1}]	BCF	C_a [μg kg^{-1}]
3,4-DCA	2.69[1]	26 (1)	13.2 (2.2)*	342 (50)	0.802	30[3]	24
Lindane	3.63[2]	29 (5)	4.4 (0.4)	127 (35)	1.237	798[4]	987
HCB	5.72[2]	206 (53)	6.5 (1.3)	1,360 (518)	0.205	1,761[4]	361

[1] BUA (1994); [2] Rippen (1991); [3] Kalsch et al. (1991); [4] Nagel (1988); * AF (146.5 h, *T. tubifex*); C_s = sediment concentration; C_a = concentration in animal; C_w = concentration in water; sd in brackets.

Apparently, the values of sediment based worm-BAF are notably lower than the corresponding water based fish-BCF. Looking at the substances and their lipophilicity, represented by log K_{ow}, it is obvious that the accumulation in fish – as expected from literature data on fish bioconcentration (e.g. Neely et al., 1974; Nagel, 1988) – is positively correlated with increasing lipophilicity. The tubificid bioaccumulation factors do *not* show this clear relationship, especially in the case of 3,4-DCA, which was accumulated higher than expected from log K_{ow}.

This situation can be explained by the medium taken as a basis for the bioaccumulation or bioconcentration, the sediment or the water. Chemicals which adsorb to particles or sediments because of their lipophilicity or any other characteristic, usually show very low concentrations in the water body. Their high affinity to organic or biological material causes high body concentrations in fish. Thus, the resulting BCF is also high. In the sediment, the same characteristics produce relatively low BAFs, since the denominator of the quotient, the sediment concentration, is high. This means, that even when relatively high body concentrations in benthic animals are reached, the bioaccumulation factor remains low.

The conclusion, that bioaccumulation by a factor smaller than 30, which is currently considered a threshold value separating low from moderate bioaccumulation (e.g. Franke et al., 1994), equals low risk, may therefore not be adequate for sediment organisms, as demonstrated by the data in Table 4. Multiplying the concentration in the ambient medium by the bioaccumulation or bioconcentration factor produces the body concentration in the animal. Based on substance concentrations in sediment and water from this study and the available bioaccumulation and bioconcentration factors for tubificids and fish, the resulting body concentrations were calculated.

It is obvious that those substances with high affinity to sediments and low partitioning into the overlying water (HCB: 0.1%, and 3,4-DCA: 3.1% of sediment concentration) – in spite of their low oligochaete bioaccumulation factors – result in remarkably high body concentrations in the tubificids. For those chemicals, the tubificid body concentrations are by several factors higher than in fish, when water concentrations and the high bioconcentration factors are multiplied. Lindane shows opposite behaviour, which is due to its relatively high partitioning from the sediment into the overlying water(4.3% of sediment concentration). Again, this phenomenon is obviously *not* simply correlated with lipophilicity, as shown in the case of 3,4-DCA, which was accumulated in the worms to an unexpectedly high extent considering its low lipophilicity. More investigations are necessary to understand the mechanisms of this finding.

Summary

A sediment-water-system for standardised measuring of bioaccumulation of chemicals in endobenthic organisms was established. The method appeared practicable and gave reproducible results. The presented test system is meant as a tool to be used in risk assessment of chemicals with high affinity to sediments and/or with evidence of high potential for bioaccumulation. Sediment based tubificid bioaccumulation factors appear notably lower than the corresponding fish bioconcentration factors. Especially the previous considerations concerning body concentrations show that lipophilicity as a measure for estimation of bioaccumulation from sediment is not sufficient. This demonstrates again, that the extrapolation of fish BCF to other organisms or other environmental compartments (e.g. sediments) during environmental risk assessment (e.g. EU, 1996) is not generally possible. The tests with artificial sediment probably cannot replace studies with natural sediments, but they can provide indication of a "relative bioaccumulation potential" of sediment-associating chemicals.

Further research is needed concerning the relation between the characteristics of sediment-associating chemicals and their bioaccumulation, the improvement of extrapolation of bioaccumulation data between different taxa, maybe using directly comparable test systems. Since only few of such directly comparable data exist, the expansion of the database using the presented method would contribute to a better understanding of bioaccumulation in aquatic systems.

Acknowledgements

This study was performed at ECT Oekotoxikologie GmbH, Flörsheim, Gemany in co-operation with Johannes Gutenberg-University Mainz, Germany. It was sponsored by the Federal Environmental Agency (Umweltbundesamt, UBA) Berlin, Germany. Once again the authors would like to thank A. Anlauf, R. J. Aston, R. O. Brinkhurst, P. M. Chapman, C. Erséus, W. Goedkoop, C. H. Högger, K. H. Hoffmann, P. L. Klerks, J. Kukkonen, T. B. Reynoldson, P. Rodriguez, T. Timm and A.-M. Wiederholm for their generous help and advice concerning identification, keeping and breeding of the tubificids as well as for stimulating suggestions on several aspects of the method. Moreover, two anonymous reviewers are thanked for their inspiring comments on the manuscript.

References

ASTM, 1995. Standard guide for determination of the bioaccumulation of sediment-associated contaminants by benthic invertebrates. American Society for Testing and Materials, E 1688–95.

Brinkhurst, R. O., 1980. Taxonomy, pollution and the sludge worm. Marine Pollution Bulletin 11: 248–251.

BUA, 1994. 2,4-Dichloranilin, 2,5-Dichloranilin und 3,4-Dichloranilin. BUA-Stoffbericht 140. Hrsg.: Beratergremium für umweltrelevante Altstoffe (BUA) der Gesellschaft Deutscher Chemiker. Hirzel, Wissenschaftliche Verlagsgesellschaft, Stuttgart.

Chapman, P. M. & R. O. Brinkhurst, 1984. Lethal and sublethal tolerances of aquatic oligochaetes with reference to their use as a biotic index of pollution. Hydrobiologia 115: 139–144.

Egeler, Ph., J. Römbke, M. Meller, Th. Knacker, C. Franke, G. Studinger & R. Nagel, 1997. Bioaccumulation of lindane and hexachlorobenzene by tubificid sludgeworms (Oligochaeta) under standardised laboratory conditions. Chemosphere 35: 835–852.

EU, 1996. Technical Guidance Documents in support of the Commission Directive 93/67/EEC on Risk Assessment for New Notified Substances, and the Commission Regulation (EC) 1488/94 on Risk Assessment for Existing Substances.

Franke, C., G. Studinger, G. Berger, S. Böhling, U. Bruckmann, D. Cohors-Fresenborg & U. Jöhnke, 1994. The assessment of bioaccumulation. Chemosphere 29: 1501–1514.

Füll, C., 1996. Bioakkumulation und Metabolismus von γ-1,2,3,4,5,6-Hexachlorcyclohexan (Lindan) und 2, 4-(2, 4-Dichlorphenoxy)-Propionsäure (Dichlorprop) beim Regenwurm *Lumbricus rubellus* (Oligochaeta, Lumbricidae). PhD-Thesis, Johannes Gutenberg-University Mainz.

Giesy, J. P. & R. A. Hoke, 1989. Freshwater sediment toxicity bioassassment: Rationale for species selection and test design. J. Great Lakes Res. 15: 539–569.

Ingersoll, C. G., G. T. Ankley, D. A. Benoit, E. L. Brunson, G. A. Burton, F. J. Dwyer, R. A. Hoke, P. F. Landrum, T. J. Norberg-King & P. V. Winger, 1995. Toxicity and bioaccumulation of sediment-associated contaminants using freshwater invertebrates: a review of methods and applications. Envir. Toxicol. Chem. 14: 1885–1894.

Kalsch, W., R. Nagel & K. Urich, 1991. Uptake, elimination, and bioconcentration of ten anilines in zebrafish (*Brachydanio rerio*). Chemosphere 22: 351–363.

Kennedy, C. R., 1966. The life history of *Limnodrilus hoffmeisteri* Clap. (Oligochaeta, Tubificidae) and its adaptive significance. Oikos 17: 158–168.

Klerks, P. L. & P. R. Bartholomew, 1991. Cadmium accumulation and detoxification in a Cd-resistant population of the oligochaete *Limnodrilus hoffmeisteri*. Aquatic Toxicology: 19: 97–112.

Kosiorek, D., 1974. Development cycle of *Tubifex tubifex* Müll. in experimental culture. Pol. Arch. Hydrobiol. 21: 411–422.

Kukkonen, J. & P. F. Landrum, 1994. Toxicokinetics of sediment-associated Pyrene to *Lumbriculus variegatus* (Oligochaeta). Envir. Toxicol. Chem. 13: 1457–1468.

Lotufo, G. R. & J. W. Fleeger, 1996. Toxicity of sediment-associated Pyrene and Phenanthrene to *Limnodrilus hoffmeisteri* (Oligochaeta: Tubificidae). Envir. Toxicol. Chem. 15: 1508–1516.

McMurtry, M. J., D. J. Rapport & K. E. Chua, 1983. Substrate selection by tubificid oligochaetes. Can. J. Fish. aquat. Sci. 40: 1639–1646.

Meller, M., Ph. Egeler., J. Römbke, H.-J. Schallnass, R. Nagel & B. Streit, 1998. Short-term toxicity of lindane, hexachlorobenzene and copper sulfate to tubificid sludgeworms (Oligochaeta) in artificial media. Ecotoxicol. Environ. Saf. 39: 10–20.

Nagel, R., 1988. Umweltchemikalien und Fische – Beiträge zu einer Bewertung. Habilitation Thesis, Johannes Gutenberg-Universität Mainz.

Neely, W. B., D. R. Branson & G. E. Blau, 1974. Partition Coefficient to measure Bioconcentration Potential of Organic Chemicals in Fish. Envir. Sci. Technol. 8: 1113–1115.

OECD, 1984. Earthworm, acute toxicity test. Guideline No. 207. Paris.

OECD, 1992. Fish, cute toxicity test. Guideline No. 203. Paris.

OECD, 1996. Bioconcentration: Flow-through fish test. Guideline No. 305. Paris.

Oliver, B. G., 1987. Biouptake of chlorinated hydrocarbons from laboratory-spiked and field sediments by oligochaete worms. Envir. Sci. Technol. 21: 785–790.

Reynoldson, T. B., 1987. Interactions between sediment contaminants and benthic organisms. Hydrobiologia 146: 53–66.

Reynoldson T. B., D. Van de Valk, L. Stervoski & S. Thompson, 1989. Uptake and depuration of ^{14}C-labelled organic contaminants by oligochaeta: development of methods. Pers. communication (1996), unpublished data.

Reynoldson T. B., S. P. Thompson & J. L. Bamsey, 1991. A sediment bioassay using the tubificid oligochaete worm *Tubifex tubifex*. Envir. Toxicol. Chem. 10: 1061–1072.

Rippen, G., 1991. Handbuch der Umweltchemikalien: Physikalisch-chemische und ökotoxikologische Daten ausgewählter chemischer Stoffe. Ecomed Verlagsgesellschaft, Landsberg.

Schmitz, A., 1997. Bioakkumulation und Verteilung von 3,4-Dichloranilin und alpha-Endosulfan in aquatischen Laborsystemen – ein Vergleich zwischen Einzelspezies – und Mikrokosmos-Experimenten. Dissertation, Johannes Gutenberg-Universität Mainz.

Spacie, A. & J. L. Hamelink, 1982. Alternative models for describing the bioconcentration of organics in fish. Envir. Toxicol. Chem. 1: 309–320.

Suedel, B. C. & J. H. Rodgers, jr., 1994. Development of formulated reference sediments for freshwater and estuarine sediment testing. Envir. Toxicol. Chem. 13: 1163–1175.

Timm, T., 1974. On the life cycles of the aquatic oligochaeta in aquaria. Inst. Zool. Bot. Acad. Sci. Est. SSR Hydrobiol. Investigations 6: 97–118.

U.S. EPA, 1995. Great Lakes Water Quality Initiative Technical Support Document for the procedure to determine bioaccumulation factors. EPA 820-8-95-005.

Wachs, B., 1967. Die Oligochaeten-Fauna der Fließgewässer unter besonderer Berücksichtigung der Beziehung zwischen der Tubificiden-Besiedlung und dem Substrat. Arch. Hydrobiol. 63: 310–386.

Walsh, G. E, D. E. Weber, L. K. Esry, M. T. Nguyen, J. Noles & B. Albrecht, 1992. Synthetic substrata for propagation and testing of soil and sediment organisms. Pedobiologia 36: 1–10.

Whitten, B. K & C. L. Goodnight, 1966. The comparative chemical composition of two aquatic oligochaetes. Comp. Biochem. Physiol. 17: 1205–1207.

Wiederholm T., A. M. Wiederholm & G. Milbrink (1987). Bulk sediment bioassays with five species of fresh-water oligochaetes. Wat. Air Soil Pollut. 36: 131–154.

Wirth, H., 1985. Zur Sorption von ^{14}C-markiertem γ-Hexachlorcyclohexan (Lindan) im ng/L-Bereich an geogenen Adsorbern. GKSS-Forschungszentrum Geesthacht GmbH, Report No. GKSS 85/E/30.

Hydrobiologia **406:** 281–290, 1999.
B. M. Healy, T. B. Reynoldson & K. A. Coates (eds), Aquatic Oligochaetes.

Lumbriculids, branchiobdellidans and leeches: an overview of recent progress in phylogenetic research on clitellates

Ralph O. Brinkhurst
205 Cameron Court, Hermitage, TN, 37076-1353, U.S.A.

Key words: evolution, Branchiobdellida, Hirudinea, Clitellata, Lumbriculidae

Abstract

The debates about the relationship among clitellate taxa at research conferences on aquatic oligochaetes have provided two alternative hypotheses. While both involve cladistic analyses, the differences focus on questions of homology. A single unique synapomorphy supports a taxon consisting of Branchiobdellida plus Lumbriculidae. Muscle ultrastructure provides a potential autapomophy for Lumbriculidae within the clade, while Branchiobdellida has many autapomophies. Some proposed synapomorphies for a Branchiobdellida plus Hirudinea grouping (jaws, oral sucker) are not based on homologous structures. Others (loss of chaetae, median male pore) are not unique, or are based on potentially convergent characters. Initial molecular analyses support a Branchiobdellida - Hirudinea grouping. A preliminary manually drawn cladogram based on re-evaluated characters supports Lumbriculidae, Branchiobdellida, *Acanthobdella* and Euhirudinea as a monophyletic taxon. Observations on sperm ultrastructure support this hypothesis. If substantiated, a very considerable change in nomenclature will result. However, before any nomenclatural changes are proposed, a resolution of the apparently paraphyletic Oligochaeta is necessary.

Introduction

The relative ranks assigned to oligochaetes, leeches, branchiobdellidans and polychaetes have been debated at intervals for a century. Despite a decade of close scrutiny, evolutionary relationships among the oligochaetes,[1] leeches and branchiobdellidans is still a contentious issue. For that reason, and because this is an informal attempt to summarize conflicting ideas, I have chosen mostly to adopt adjectival rather than taxonomic names for the groups involved.

Sister relationships between lumbriculids and branchiobdellidans, and between branchiobdellidans and leeches have been claimed, both on the basis of cladistic analyses. Resolution of this issue depends on the definition of homology that is applied in defining the character states to be used, as will be discussed later.

Cladistic analyses of the branchiobdellidans (Gelder & Brinkhurst, 1990) and lumbriculids (Brinkhurst, 1989) reveal a unique synapomorphy, the semi-prosoporous condition of the male reproductive apparatus (Figure 1). Long-standing concepts of the relationship between leeches and branchiobdellidans are based on obvious shared characters such as loss of chaetae, fixed segment number, possession of suckers, median male pores and jaws. Cladistic analyses employing most of these (excluding jaws, for example, which are obviously convergent – see below) support the relationship of leeches and branchiobdellidans (Siddall & Burreson, 1995, 1996). Previously, Brinkhurst & Gelder (1989) had reviewed some of the anatomical data on which these characters are based. They concluded that several of the supposed synapomorphies of branchiobdellidans and leeches might not withstand critical evaluation, but that they were likely convergences related to the very specialized habitus of these derived taxa. Siddall & Burreson (1995, 1996) suggest that those who are critical of their hypothesis are rejecting it on authoritarian and/or phenetic rather than phylogenetic evidence.

The alternative hypotheses are reiterated here, with some new interpretations, and the opinions of those colleagues who responded to a circular letter between the 6th and 7th Symposia. The intent is to report at least some of the debates that have taken place at our

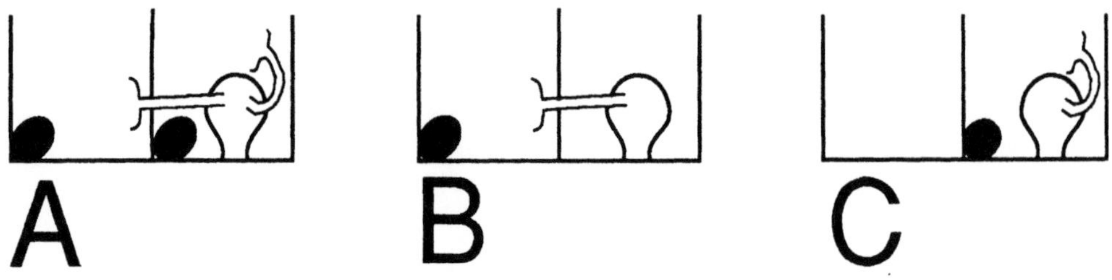

Figure 1. (A) semi-prosoporous male ducts of Lumbriculidae and Branchiobdellida (B) supposed plesioporous male ducts of *Acanthobdella* (C) supposed prosoporous male ducts of Euhirudinea.. (B) and (C) are based on assumptions regarding anatomy and ontogeny. Gonads and male ducts are paired, atria are paired in most Lumbriculidae, median in the other taxa.

meetings, and to focus future discussion for the sake of those less exposed to the details. This is more of an essay than a scientific report of discovery, and so this seems appropriate, especially as this may well be the author's final contribution to the issue.

Basis for re-evaluation

When I first became interested in revising the concepts of relationship among the clitellate taxa, it was apparent that several steps needed to be taken. These were:

1. Attempt to assess the hypothesis that the monotypic Baikal endemic lumbriculid genus *Agriodrilus* was ancestral to leeches, as proposed by Michaelsen (1928).
2. Translate the original long description of *Acanthobdella*, obtain new material and confirm the significant anatomical details to clarify the phylogenetic position of the taxon.
3. Perform a phylogenetic analysis on Branchiobdellida in order to evaluate phenetic classifications and polarize character states.
4. Perform a phylogenetic analysis on Lumbriculidae for the same reasons.
5. Perform a phylogenetic analysis on Euhirudinea for the same reasons, and solicit input from leech taxonomists.
6. Examine existing characters to confirm homology and search new fields of evidence for additional character states.

Completed work and its consequences

Predatory lumbriculids and reduced coelom

Material of *Agriodrilus* is not available. A new genus of predatory lumbriculids (*Phagodrilus*), was described from Oregon, U.S.A. (McKey-Fender & Fender, 1988) and compared with *Agriodrilus*. The obliteration of the coelom in anterior segments of both genera is due to the development of a modified muscular pharynx (Brinkhurst & McKey-Fender, 1991). This is an obvious convergence of *Phagodrilus* with *Agriodrilus* as both the pharyngeal structure and the male reproductive systems are completely different in the two taxa. The obliteration of part of the coelom in *Agriodrilus* was said to indicate a relationship to the leeches (Michaelsen, 1928). Near obliteration of the coelom is a synapomorphy for all euhirudineans, but not *Acanthobdella* or branchiobdellidans. As their common ancestor would therefore have a capacious coelom, the predatory lumbriculids are irrelevent to the debate. The oligochaetes *Chaetogaster* and *Haplotaxis*, both of which have modified pharyngeal structures for this same purpose, retain a complete coelom (Brinkhurst & Gelder, 1989; Brinkhurst & McKey-Fender, 1991).

Acanthobdella

The Livanow (1906) monograph was obtained and translated. A study based on fresh material resulted in a detailed reinvestigation of *Acanthobdella* (Purschke et al., 1993). There are 31 pairs of ganglia although only 29 apparent segments. The first chaetae are on the first annulus, and so the usual pre-chaetigerous prostomium and peristomium (at least) are missing. This narrows the difference in segment number between *Acanthobdella* and leeches, and would seem to strengthen the relationship between them. However, the male ducts may well be plesioporous, (Figure 1) as the funnels appear to lie in XI with the pores in XII, although the position of the testes is still unclear. If this originated from a semi-prosoporous ancestor, loss of one pair of male ducts and testes must be presumed. Ontogenetic work might elucidate this point.

Many synapomorphies of *Acanthobdella* and Euhirudinea were identified (characters 12–31 of cladogram A, Purschke et al., 1993). Some of these will be discussed below. This also applies to the proposed synapomorphies of Branchiobdellida and Hirudinea (characters 12–16, cladogram B).

Branchiobdellidan phylogeny

A phylogenetic analysis was performed on genera of Branchiobdellida (Gelder & Brinkhurst, 1990), using traditional characters. Three groups of genera, which broadly represent the three most numerous families, could be identified, but monophyletic subgroups could not be identified by synapomorphies. Ancestral character states for the group were determined, and can be used in subsequent analyses of higher taxa. The semi-prosoporous condition of the male reproductive system is plesiomorphic.

Lumbriculids

A phylogenetic analysis of Lumbriculidae resulted in a new hypothesis about their ancestry (Brinkhurst, 1989). Phenetic classifications had assumed that the completely prosoporous genus *Lamprodrilus* represented at least living descendants of the ancestor of lumbriculids, if not ancestral forms directly (Michaelsen, 1928). The restriction of the majority of species to Lake Baikal created some suspicion about that proposal, in addition to the more recent awareness that it was unwise to look to living taxa to be direct ancestors. A new hypothesis derived from the phylogenetic analysis is that the commonest form of male duct is the plesiomorphic condition within the group. I now term this the semi-prosoporous condition (Figure 1). Semi-prosoporous male systems consist of two pairs of testes and two pairs of funnels and vasa deferentia, but only a single pair of atria in the second testicular segment (i.e. GI and GII are present with atria in GII). This condition is a now a synapomorphy for Branchiobdellida and Lumbriculidae.

Leeches

Lack of consistency among generic definitions by Sawyer (1986) prevented S. R. Gelder and I from completing a cladistic analysis of Euhirudinea. We both attended the 2nd International Conference of Leech Scientists in Ottawa, Canada, in 1988, and solicited attendance of leech taxonomists at our subsequent conferences. The characteristics of oligochaetes as discussed by Sawyer (1986) were critically reviewed in an unpublished contribution. Subsequently, Siddall & Burreson (1995, 1996) have provided phylogenetic analyses of Hirudinea.

Characters and character states

In examining the traditional characters and their states used to relate oligochaetes, leeches and branchiobdellidans, Brinkhurst & Gelder (1989) determined that several were not homologous and that others did not provide unique synapomorphies. The characters concerned and those used by Purschke et al. (1993: numbered in parentheses) are:

(A) jaws,
(B) anterior sucker,
(C) reduced coelom,
(D) posterior sucker (14),
(E) constant segment number (12),
(F) median male pore (16),
(G) loss of chaetae,
(H) position of gonopores ,
(I) gonoducts and gonads. Purschke et al. (1993) also used:
(J) chaetae restricted to anterior segments (13), and
(K) typical muscle cells (15) as synapomorphies for Branchiobdellida - Hirudinea.

Homologies

As the major controversy in this area involves the question of homologies, it seems necessary to reiterate and present some new interpretations and information in relation to each character involved. The determination of homology for each of the characters involved is as follows.

(A) Jaws. Paired jaws are a synapomorphy for Branchiobdellida. Jaws represent an apomorphic condition for Hirudiniformes, not for all Hirudinea. When present in some hirudinean taxa, the plesiomorphic character state appears to be tri-radiate jaws, with a subsequent apomorphic condition of dual jaws. Therefore, jaws are irrelevant to the discussion of a branchiobdellidan - leech relationship.

(B) Anterior sucker. The anterior attachment organs in leeches and branchiobdellidans are clearly not homologous. In leeches, there is an oral sucker in most taxa (in the cladogram below it will be shown as an autapomorphy for leeches in order to emphasize

its absence in the other relevant taxa). In branchiobdellidans, there is a duo-gland adhesive organ on the ventral surface of the ventral peristomial lip. The mouth is apparently sometimes used to assist in locomotion (Brinkhurst & Gelder, 1969), but in view of the presence of the specific anterior attachment organ, it seems erroneus to assume that both leeches and branchiobdellidans share a homologous anterior sucker. The anatomy of the anterior attachment structure was described by Weigl (1994), who confirmed the duo-gland nature of the structure. She indicated that, while muscles are associated with the anterior structure "the anterior attachment site was not muscularized heavily and did not assume a disc- or sucker-form..." The structure seems to fail both the structural and positional tests of homology. Recent attempts to include function as a test of homology seem exceedingly dangerous, as this is the predominant characteristic of analogous structures. Weigl concludes "The dissimilarity between branchiobdellids and true leeches in gross morphology of both anterior and posterior attachment sites supports the interpretation that they represent convergent, rather than homologous, structures."

Acanthobdella has lost both prostomium and peristomium, and perhaps some additional anterior segments, because the first chaetiger contains the terminal mouth. The chaetae are said to aid in locomotion and attachment, but detailed study is badly needed. It seems eminently possible, if we resort to unsubstantiated speculation, that the chaetae may be used for feeding, perhaps locomotion, but probably not attachment in this instance. There is no oral sucker in the anatomical sense comparable to those found in leeches.

(C) Reduced coelom. There is no reduced coelom in branchiobdellidans, apart from that due to the pharyngeal muscles. The very rare condition of reduced coelom in predatory lumbriculids is related to the adaptation of the pharyngeal musculature, which is convergent in the two taxa involved. "The coelom of *Acanthobdella* is well developed, and consists of segmentally arranged coelomic cavities lined by a peritoneum......The volume of the coelomic cavities is comparatively small because of the voluminous musculature, the connective tissue, and the large gland cells" (Purschke et al., 1993).

(D) Posterior sucker. The nature of the posterior attachment structure in branchiobdellidans, leeches and *Acanthocephala* has caused the greatest disagreement between proponents of the competing hypotheses. On the one hand, it seems quite simple to state that all three taxa have a posterior sucker. The argument, if I cite it correctly, is that it has the same structure, is in the same terminal location, and serves the same function in all three taxa. I will now attempt to set out those facts that appear to this author, at least, to challenge that assumption without necessarily completely negating it.

In branchiobdellidans, the posterior attachment organ operates as a duo-gland adhesive pad in the same way as the anterior attachment organ. Worms can be seen to be attached even when the disc is only partially in contact with the substratum (S. R. Gelder, pers. com.), a situation that would cause a muscle-operated sucker to fail. The structure is composed of a single segment, based on the presence of a single pair of ventral nerve cord ganglia. Weigl (1994) stated that "In the posterior adhesive organ...paired groups of viscid gland cell bodies were located in body segments 9 and 10...... Extensive circular and longitudinal muscles were associated with the attachment disk." Repeated dislodging of the worms revealed a latent period during which the worms of one species would not reattach, even though the central region of the disc was protruded and retracted. Later we find "In branchiobdellids, the possibility that muscles are used to assist in release from the substratum should be considered. ....at the posterior attachment disc, branchiobdellids may be able to supplement chemical adhesion with suction and / or use of muscles to assist in release.....[or] release from the attachment surface might be chemical, with the muscles utilized to lift the surface of the attachment disc ....perhaps to avoid premature reattachment". While this is, of necessity, speculation, the conclusion reached by Weigl is that this structure is "convergent rather than homologous" (see (B) *Anterior sucker*, above). Sawyer (1986) emphasized the muscular nature of the organ, even though research on the duo-adhesive gland hypothesis antedated his publication.

The posterior position of the attachment disc ensures that it is terminal, as clitellates generate new segments posteriorly, and this specialization clearly inhibits further segment development. The actual position in terms of segment number of branchiobdellidans differs from that of leeches (segment XV *versus* XXXII or XXXIV depending on interpretation). The function is clearly similar to that of leeches, but the relevance of this criterion is challenged herein.

The extent to which a duo-gland system is involved in leech posterior suckers is, I believe, unknown in detail, and this area badly needs new research. Appy

& Cone (1982) performed a histological study of two leech species in relation to the host tissue, but the methods are different from those used by Weigl (1994) and the results are not fully comparable. They found a PAS-positive acellular layer between the host tissue and the caudal sucker. They reported the effects of Mallory's and Alcian Blue stains on preserved tissue. Two types of tubular ducts emptied into this acellular layer from the subepidermal secretory cells. The disc had two types of pores associated with it. "At least two secretory products are present" according to Appy & Cone, but these form two separate layers of the acellular layer. They go on to state that "such secretions are thought to be an acid mucus and may function in adhesion perhaps by repelling water.... An oily caudal sucker secretion ... may augment suctorial pressure or even allow an initial adhesion of the caudal sucker. Absence of host epithelium at the attachment site also suggests a lytic function ...." It is not known if the various authors cited were aware of the possibility of a duo-gland adhesion system or not, but at this point we have many reasons for considering the 'suckers' in these taxa to be a necessary convergent adaptation to a specialized life style. This does not exclude the possibility that the leech style of sucker originated in the simpler branchiobdellidan style of attachment structure, but the question of segment number would then confound a cladistic solution to the issue.

Acanthobdella has a sucker comprised of four segments which Purschke et al. (1993), suggest is a reduction from the condition in leeches, which have one of seven segments. This structural difference suggests caution in accepting the hypothesis of homology.

Differences between authors as to their interpretations of all of these structures is reasonable, but detailed work on leech suckers might help resolve the issue.

(E) Constant segment number. As clitellates have a posterior growing point, this characteristic is simply a consequence of modification of the posterior end of the worm as an adaptation to locomotion. The situation is further challenged by the fact that the number of segments differs markedly (see (D) *Posterior sucker*, above). The same argument may be applied to loss of pygidium, cited by Purschke et al. (1993).

(F) Median male pore. This character is found in both branchiobdellidans and leeches, but also sporadically among several subfamilies and genera of tubificids, naidids, and lumbriculids (at least). It is not therefore a unique synapomorphy.

(G) Loss of chaetae. This character is found in branchiobdellidans and Hirudinea but as it occurs in other clitellate taxa (some species of Enchytraeidae, Lumbriculidae, Narapidae) it does not provide a unique synapomorphy.

The chaetae in *Acanthobdella*, present in five segments, are always cited as attachment organs, but should also be considered as possible feeding aids, lacerating skin of the fish to provide access to soft tissue and fluids (see (B). *Anterior sucker*).

(H) Position of gonopores. This character varies widely in clitellates, especially in Lumbriculidae and it also varies within Hirudinea. The whole sequence of gonads, associated male and female ducts, and spermathecae can be seen to be shifted forward relative to a common basic position in X–XIII in many oligochaetes. Brinkhurst (1991) argued for the recognition of 'shift in gonad position' as an objective character, as opposed to scoring the positions of every single element of the whole set when it was obvious that two taxa differed by this single characteristic. Recalculation of *Acanthobdella* segmentation suggests that the position of their pores is quite close to that of some leeches.

(I) Gonoducts and gonads.

1. Branchiobdellida and lumbriculids share the presence of GI–GIII with semi-prosoporous male ducts. This is a unique synapomorphy.
2. Branchiobdellida has fused median male pores that arise in a manner that differs from the median fusion observed in several lumbriculids (Brinkhurst & Gelder, 1989). The lack of a second pair of testes makes it impossible to evaluate fusion in *Acanthobdella* and Euhirudinea in respect to this character.
3. In *Acanthobdella*, the testes may be in GI, the ducts may be plesiopore. In Hirudinea we assume the testes, which appear to be posterior to the ovaries, are actually located in GII, but this hypothesis has not been verified by ontogenetic study to my knowledge. Either condition could be developed independently from that pertaining to branchiobdellidans (Figure 1).
4. The position of ovaries in respect to female pores differs between *Acanthobdella* and Euhirudinea on one hand, and the oligochaetes and branchiobdellidans on the other. Fusion of female pores of both *Acanthobdella* and leeches is unknown in oligochaetes and branchiobdellidans.

(J) Character revisions. Some updating of the cladograms presented by Purschke et al. (1993) is re-

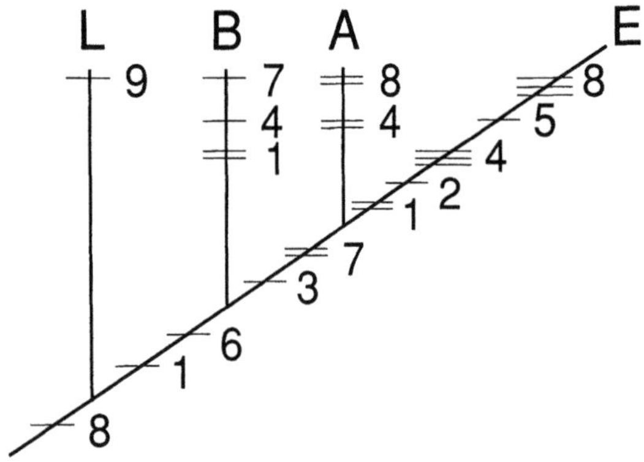

Figure 2. Phylogenetic analysis of Lumbriculidae, Branchiobdellida, *Acanthobdella*, and Euhirudinea based on character states discussed in the text. L: Lumbriculidae B: Branchiobdellida A: *Acanthobdella* E: Euhirudinea. One bar indicates an apomorphy (state 1), two bars state 2, three bars state 3. States 2 and 3 for character 8 are actually reversions from the semi-prosoporous condition of the male ducts, and are assumptions (see Figure 1).

quired. Apomorphies for the clade Branchiobdellida - Hirudinea used by Purschke et al. (1993, Figure 8B) included 'chaetae restricted to anterior segments (Figure 8B # 13)'. This is illustrated as an apomorphy for branchiobdellids and leeches but as a subsequent loss of chaetae in Euhirudinea (Figure 8B # 47), but was not mentioned for Branchiobdellida. It is an autapomorphy of *Acanthobdella*. Another character, typical structure of muscle cells (Figure 8A # 21), needs updating, as noted below.

Revised characters and their states presented as a cladogram

I will now indicate how a monophyletic assemblage can be supported by a hand-drawn cladogram. There are implied assumptions, such as loss of male reproductive elements in leeches and *Acanthobdella*, and an ancestor of branchiobdellidans with some chaetae, as there are in any such analysis.

1. Chaetae present (0), reduced to 5 segments (1), absent (2)
2. Coelom present (0), reduced (2)
3. Female pore paired (0), median (1)
4. Posterior attachment organ absent (0), 1 segment, duo-gland (1), 4 segment sucker (2), 7 segment sucker (3) (unordered multistate)
5. Oral sucker absent (0), present (2)
6. Male pore paired (0), median (1)
7. Segment number inconstant (0), 15 segments (1) more than 30 (2) (unordered multistate)
8. Male ducts paired plesiopore (0), paired semi-prosopore (1), ? single plesiopore (2), ? single prosopore (3) [states 2 and 3 assumed, not proven]
9. Lumbriculid style muscles absent (0), present (1)

Other autapomorphies have been ignored for this purpose.

A matrix of character states for the taxa appears as follows, with an hypothetical ancestor with all states plesiomorphic.

Lumbriculids	00000 0011
Branchiobdellidans	20010 1210
Acanthobdellids	10120 1120
Leeches	21131 1130

The resulting cladogram is shown in Figure 2. The intent is to show that the relationships can be resolved when the character states are used with respect to the reappraisal of homology. This indicates a monophyletic prosopore group with independent reversion to plesioporous and prosoporous male ducts in the Hirudinea, independent development of posterior attachment organs, and a remote ancestor to branchiobdellidans that had acanthobdellid chaetae, a coelom, and semi-prosoporous male ducts. As the posterior attachment organ arises independently in each group, differences in segment number between branchiobdellidans and leeches is no longer an issue. These are assumptions, and there may be other unconscious assumptions that remain to be revealed by others.

New approaches

New characteristics include sperm and muscle ultrastructure and molecular approaches. This author recognizes these data, but is unable to evaluate such evidence as critically as traditional anatomy.

1. Lumbriculid muscle ultrastructure so far observed is unique and provides a needed potential apomorphy for Lumbiculidae as a sister group to Branchiobdellida (de Eguileor et al., 1987). Branchiobdellidan muscles arrangements are similar to those of oligochaetes, and not leeches. A phylogeny based on muscle structure alone would suggest a separation of branchiobdellidans from a common clitellate stock before the separation of oligochaetes and leeches (Valvassori et al., 1994). Characters in common between leeches and branchiobdellidans are shared with *Haplotaxis*, Sipunculida, Priapulida and other annelids. This corrects statements made by Purschke et al. (1993) based on earlier literature.

2. The initial study of sperm ultrastructure of a single lumbriculid confirmed a growing understanding that the family is defined by many apomorphic character states *vis a vis* other microdriles, rather than representing an ancient stock as previously supposed (Ferraguti & Jamieson, 1987). Some characteristics of the sperm of this lumbriculid are unique, others suggest a relationship to megadriles, the manchette geometry is similar to that of Hirudinea. Branchiobdellidan sperm are typically clitellate (Ferraguti & Gelder, 1991). The electron dense, central sheath in the flagellum was thought to be shared only by branchiobdellidans and leeches, but is now known from oligochaetes also. Each of four branchiobdellidans initially studied (S. R. Gelder, pers. com.) have a unique assortment of hirudinean characters, but the phylogenetic implications of this are unclear once other taxa are involved in the comparison. Some oligochaete characters are present in branchiobdellidan sperm. The end result is that, while oligochaete and leech sperm can be defined by apomorphic character states, the branchiobdellidan sperm exhibits a mixture of characters of both other taxa as well as unique characters. Marco Ferraguti presented a cladogram at the symposium which, if memory serves, supported the position of branchiobdellidans between lumbriculids and leeches.

3. Molecular data suggest a sister relationship between branchiobdellidans and leeches within a monophyletic Clitellata based on analysis of only one or two representative taxa (Moon et al., 1996; Kim et al., 1996). Taxa studied in the first of these contributions were not included in the second data base. As in all such studies at the early stage of development of a new technique, intra-taxon variations have to be ignored as each is represented by only one or two species. Siddall & Burreson (1998) included one branchiobdellidan and *Acanthobdella* in a phylogenetic evaluation using mitochondrial Cytochrome c Oxidase subunit 1, with two polychaetes and four oligochaetes used as outgroup taxa. While claiming that these data are "free of the subjective evaluation of morphological characters for the first time" the authors "remain skeptical about a few" of the "striking departures from conventional systematic groupings". Their results lead to the interpretation that *Acanthobdella* is the sister taxon to a branchiobdellidans plus leech clade. The lumbriculid used is a somewhat atypical species, being parthenogenetic, and one of the few very widely distributed species in the family and one of only two to reach the southern hemisphere. Despite this, it was placed immediately below the leech-branchiobdellidan clade. It was shown associated with a tubificid, and the pair appears between the megadriles and the inclusive leech group. Molecular data requires confirmation, as different analyses can produce conflicting results. Other studies are ongoing and are awaited with interest. Conversations with colleagues familiar with these methods suggest that the results can be affected by the lengths of the sequences involved, the choice of the DNA fragments to amplify, and the complexity of the computer matching programs employed. Only time will tell if these initial results are substantiated, but as of now, molecular work supports a version of the traditional relationship among the four taxa.

Input from colleagues

The following information is derived from a circular letter sent to colleagues after the symposium held in Sweden.

Christer Erséus agreed that Branchiobdellida-Lumbriculidae is defined by a unique apomorphy, as is Branchiobdellida, but Lumbriculidae was bereft of an apomorphy. The muscle cells of lumbriculids may provide this, as presented above.

Tarmo Timm stated that leech male ducts are regarded as self-evidently prosoporous. That is true if the testes are derived from the segment bearing the male pore, as supposed. This is one basis for the traditional link between leeches and lumbriculids that predates phylogentic analysis by parsimony methods. Otherwise he sees no synapomorphic characters that unite branchiobdellidans and leeches, but suggests both are likely derived from a common ancestor with the lumbriculids.

Enrique Martinez-Ansemil provided extracts from his doctoral dissertation in which he accepts a relationship between lumbriculids and branchiobdellidans.

Discussion

I have attempted to review anatomical characters as carefully as possible. This is done not just to make any a priori judgement, but to make sure that characters entered into a data matrix are as demonstrably homologous as possible. All cladistic analyses of oligochaetes have revealed a great deal of homoplasy, which is why the taxonomy of the group remains debatable. The results of these analyses depend on the identification of truly homologous character states, the choice of a great many optional procedures, and even the assumption of parsimony itself. The analyses are not infallible. The questions raised about various character states that could be homoplasious are meant as a caution and a spur to further research, not as a "resort to speculative arguments or baseless adaptationist scenarios" (Siddall & Burreson, 1996). While interpretations of anatomical characters may differ, they are not simply 'assumptions' (Siddall & Burreson, 1998) if we continually re-evaluate them and use analytical results to stimulate more detailed research, and if homology statements are supported by sound logical and consistent argument. There are still conscious assumptions, such as the traditional view that the posterior testes of leeches are derived from an ancestor with oligochaete-like testes. My purpose here is not to refute the possibility that there may be a monophyletic assemblage consisting of lumbriculids, branchiobdellidans, leeches and *Acanthobdella,* but to examine the existing evidence critically in order to know how to phrase the terms applied to characters and their states. This resulted in a cladistic - style statement of the character states that is in accord with the facts as presented. This will be an ongoing process if we continue to research, not if we simply argue about homology. We also need to ensure that we are working with compatible definitions of homology, which, for this author, should not place too much reliance on agreement of function of the structures concerned.

In an attempt to convey earlier phylogenetic hypotheses in a cladistic manner, Brinkhurst (1994) included a single taxon 'Hirudinea' in an analysis of clitellate families. This was recognized as a very preliminary 'sketch' of a phylogeny. As 11 of 23 character states for leeches were coded 'unknown', there was a reasonable basis for an a posteriori suggestion that the relationship to branchiobdellids and lumbriculids that resulted might not be meaningful.

There is a problem in coding character states for a whole family. The plesiomorphic character state for a family (such as chaetae present) may conceal individuals within the family (such as *Achaeta,* Enchytraeidae) which share a homoplasious apomorphy with another entire taxon (Branchiobdellida). Ignorance of this frequent evidence of homoplasy may lead to undue confidence in an analytical result.

Purschke et al. (1993) proposed two alternative phylogenetic dendrograms. One (cladogram A) placed Branchiobdellida and Oligochaeta as a joint sister group to the Hirudinea, which included *Acanthobdella* and Euhirudinea as sister groups. There were no synapomorphic character states to support Branchiobdellida-Oligochaeta. Neither are there synapomorphies that support Oligochaeta in the alternative, cladogram B. In cladogram B, Branchiobdellida and Hirudinea are purported sister groups. The supposed synapomorphies used to define that monophyletic assemblage (characters 12–16 of Purschke et al., 1993) are now challenged, as discussed above. If we had substituted Lumbriculidae for Oligochaeta in the analysis, accepting the paraphyletic state of Oligochaeta, we may have made more progress. The cladogram presented here, based on limited but carefully evaluated character states, with some conscious assumptions, supports a monophyletic assemblage for the four taxa concerned. It is presented here to illustrate the arguments in a cladistic context, and to emphasize the underlying assumptions needed in an assumption of monophyly.

Siddall & Burreson (1995, 1996) discuss relationships among the Hirudinea. In the first of these studies, *Acanthobdella* is used as a sister group and included in the character matrix. The Branchiobdellida were not included in the analysis, but their relationship to Hirudinea was considered. The authors claim that continued debate about the position of the Branchiobdellida is due to "a persistent lack of formal recognition of the paraphyletic relationship of the oligochaetes and a propensity for formulating *ad hoc* hypotheses of convergence rather than assuming that observed characters are homologues". Further, "The tendency to hypothesize rampant convergence again stems from accepting independent origins for taxa of equal rank. A posterior attachment organ, fused genital pores, and a constant number of body somites are all evident in *Acanthobdella peledina*, and in branchiobdellids. Even with the benefit of a well-supported cladogram clearly showing these characters as synapomorphies for the three groups, Purschke et al. (1993) continue in the tradition of hypothesizing convergent acquisition of each of these characters for the branchiobdellids and leeches ....". In the later paper (1996), the branchiobdellids appear on the cladogram but no data matrix is presented, and so their position on homology of 'oral suckers' is not established, though it has been suggested as homologous in an unpublished document. When only Oligochaeta, Branchiobdellida and Hirudinea are included, Branchiobdellida is bound to appear as a sister group to Hirudinea. The authors stated that "arguments proposing the inclusion of branchiobdellids within the oligochaetes and exclusion of leeches are purely phenetic and have no phylogenetic support." The suggestion that previous authors do not work in a phylogenetic context is easily refuted by reference to the groundwork on branchiobdellids and lumbriculids. Many authors have proposed schemes in which Hirudinea and/or Branchiobdellida are contained within Oligochaeta (Timm, 1981; Brinkhurst, 1984,1992; Brinkhurst & Nemec, 1987; Jamieson et al., 1987; Erséus, 1987; for example). What these authors have not done is to convert these ideas into a classification. The retention of older terminology and ranks (such as Class Hirudinoidea for a putative sister taxon to Order Lumbriculida, Brinkhurst & Nemec, 1987) is more an indication of a preliminary attempt to determine relationships than of ignorance of the nomenclatural consequences (Jamieson, 1988), although this should have been explicitly stated as it was by Brinkhurst & Gelder (1989). Until we are ready to resolve a number of issues, the terminology would become cumbersome if changed prematurely. For example, if we abandon the taxon Oligochaeta in favour of Clitellata, we have no way to refer to non-hirudinean, non-branchiobdellidan clitellates until new evidence proposes an alternate systematics. Continued use of the term for convenience should not be confused with a closed mind. The debate is likely to continue without progress until research solves the problem of apparent *versus* real homology.

Conclusion

We are all in agreement that there is a single taxon Clitellata. We should also be able to agree that there is conflicting evidence about the relationship of Branchiobdellida. At our last meeting I tried to indicate that I am quite willing to entertain the idea that lumbriculids and branchiobdellidans are a single taxon because of their synapomorphy. There are, as yet, no unequivocal synapomorphies relating branchiobdellidans and leeches, although initial molecular work does support the relationship. Homoplasies, such as loss of chaetae and presence of median male pores, are consistent with the hypothesis. Constant segment number and lack of pygidium are not independent characters, being the inevitable consequence of locomotion with an attachment organ that removes the growing point at the posterior end of the body. Other supposed synapomorphies are certainly not based on homologous characters. This is not to deny that there are relationships among branchiobdellidans, *Acanthobdella*, leeches and Lumbriculidae. A preliminary cladogram illustrates how such a monophyletic group could be defined given certain assumptions, particularly about the evolution of male reproductive systems.

It is difficult to see why there is a need to force a conclusion to a puzzle we are very little nearer solving than were our predecessors who, despite their lack of access to modern methods, could outline the same systematic alternatives that we face today.

Notes

[1]The use of adjectival rather than taxonomic terms in places reflects the unresolved state of the nomenclature and taxonomy. Currently, Oligochaeta includes all non-leech, non-branchiobdellidan Clitellata, Hirudinea includes true leeches (Euhirudinea) and *Acanthobdella*. Euhirudinea and Branchiobdellida are certainly monophyletic.

Acknowledgements

I am indebted to those collegues who took time to respond to my circular letter and to requests for information; C. Watson and L. M. Wood kindly reviewed the manuscript.

References

Appy, R. G. & D. K. Cone, 1982. Attachment of *Myzobdella lugubris* (Hirudinea: Pisicolidae) to Logperch, *Percina caprodes*, and Brown Bullhead, *Ictalurus nebulosus*. Trans. Am. microsc. Soc. 101: 135–141.

Brinkhurst, R. O., 1984. The position of the Haplotaxidae in the evolution of oligochaete annelids. Hydrobiologia 115: 25–36.

Brinkhurst, R. O., 1989. A phylogenetic analysis of the Lumbriculidae (Annelida, Oligochaeta). Can. J. Zool. 67: 2731–2739.

Brinkhurst, R. O., 1991. Ancestors (Oligochaeta). Mitt. hamb. zool. Mus. Inst. 88: 97–110.

Brinkhurst, R. O., 1992. Evolutionary relationships within the Clitellata. Soil Biol. Biochem. 24: 1201–1205.

Brinkhurst, R. O., 1994. Evolutionary relationships within Clitellata: an update. Megadrilogica 5: 109–112.

Brinkhurst, R. O. & S. Gelder, 1989. Did the lumbriculids provide the ancestors of the branchiobdellidans, acanthobdellidans and leeches? Hydrobiologia 180: 7–15.

Brinkhurst, R. O. & D. McKey-Fender, 1991. The anatomy of the pharynx of two predatory aquatic oligochaetes. Can. J. Zool. 69: 669–675.

Brinkhurst, R. O. & A. F. L. Nemec, 1987. A comparison of phenetic and phylogenetic methods applied to the systematics of Oligochaeta. Hydrobiologia 155: 65–74.

de Eguileor, M., G. Lanzavecchia, R. Valvassori & P. Lanzavecchia, 1987. Unusual mode of lumbriculids' helical muscles: comparison with body wall muscles in other microdriles. Hydrobiologia 155: 135–144.

Erséus, C., 1987. Phylogenetic analysis of the aquatic Oligochaeta under the principle of parsimony. Hydrobiologia 155: 75–89.

Ferraguti, M. & S. R. Gelder, 1991. The comparative ultrastructure of spermatozoa from five branchiobdellidans (Annelida: Clitellata). Can. J. Zool. 69: 1945–1956.

Ferraguti, M. & B. G. M. Jamieson, 1987. Spermiogenesis in *Bythonomus lemani* and the phylogernetic position of the Lumbriculidae (Oligochaeta, Annelida). Hydrobiologia 155: 123–134.

Gelder, S. & R. O. Brinkhurst, 1990. An assessment of the phylogeny of the Branchiobdellida (Annelida: Clitellata), using PAUP. Can. J. Zool. 68: 1318–1326.

Jamieson, B. G. M., C. Erséus, & M. Ferraguti, 1987. Parsimony analysis of the phylogeny of some Oligochaeta (Annelida) using spermatozoal ultrastructure. Cladistics 3: 145–155.

Jamieson, B. G. M., 1988. On the phylogeny and higher classification of the Oligochaeta. Cladistics 4: 367–410.

Kim, C. B., S. Y. Moon, S. R. Gelder & W. Kim, 1996. Phylogenetic relationships of annelids, molluscs, and arthropods evidenced from molecules and morphology. J. Mol. Evol. 43: 207–215.

Livanov, N., 1906. *Acanthobdella peledina* Grube, 1851. Zool. Jb. Anat. Ontog. Tiere 22: 637–866.

McKey-Fender, D. & W. M. Fender, 1988. *Phagodrilus* gen. nov. (Lumbriculidae): Systematics and ecology of a predaceous oligochaete from Western North America. Can. J. Zool. 66: 2304–2311.

Michaelsen, W., 1928. Clitellata-Gürtelwürmer. 3. Klasse der Vermes polymera (Annelida). Kükenthal and Krumbach. Handb. Zool. Berlin 2: 1–118.

Moon, S. Y., C. B. Kim, S. R. Gelder & W. Kim, 1996. Phylogentic positions of the aberrant branchiobdellidans and aphanoneurans within the Annelida as derived from 18S ribosomal RNA gene sequences. Hydrobiologia 324: 229–236.

Purschke, G., W. Westheide, D. Rhode & R. O. Brinkhurst, 1993. Morphological reinvestigation and phylogenetic relationship of *Acanthobdella peledina* (Annelida, Clitellata). Zoomorph. 113: 91–101.

Sawyer, R. T., 1986. Leech biology and behaviour. Clarendon Press, Oxford. 1065 pp.

Siddall, M. E., & E. Burreson, 1995. Phylogeny of the Euhirudinea: Independent evolution of blood feeding by leeches? Can. J. Zool. 73: 1048–1064.

Siddall, M. E. & E. Burreson, 1996. Leeches (Oligochaeta?: Euhirudinea), their phylogeny and the evolution of life-history strategies. Hydrobiologia 334: 277–285.

Siddall, M. E. & E. Burreson, 1998. Phylogeny of leeches (Hirudinea) based on mitochondrial Cytochrome c Oxidase subunit 1. Molec. Phylog. Evol. 9:156–162.

Timm, T., 1981. On the origin and evolution of aquatic Oligochaeta. Eesti NSV Tead. Akad. Toimet. Biol. 30: 174–181.

Valvassori, R., M. Eguileor, G. Lanzavecchia & S. Gelder, 1994. Comparative body wall musculature and muscle fibre ultrastructure in branchiobdellidans (Annelida:Clitellata), and their phylogenetic significance. Hydrobiologia 278: 189–199.

Weigl, A. M., 1994. Ultrastructure of the adhesive organs in branchiobdellids (Annelida: Clitellata). Trans. am. microsc. Soc. 113: 276–301.

MIX
Papier aus verantwortungsvollen Quellen
Paper from responsible sources
FSC® C105338

If you have any concerns about our products,
you can contact us on
ProductSafety@springernature.com

In case Publisher is established outside the EU,
the EU authorized representative is:
Springer Nature Customer Service Center GmbH
Europaplatz 3, 69115 Heidelberg, Germany

Printed by Libri Plureos GmbH
in Hamburg, Germany